Quantum-Mechanical Signal Processing and Spectral Analysis

About the author

Dževad Belkić is a theoretical physicist. He is Professor of Mathematical Radiation Physics at Karolinska Institute in Stockholm, Sweden. His current research activities are in atomic collision physics, radiation physics, radiobiology, magnetic resonance physics and mathematical physics. In atomic collision physics, he has worked on many problems including major challenges such as the theory of charge exchange and ionization at high non-relativistic energies. *Inter alia* he used distorted wave methods, paying special attention to treatments with correct boundary conditions for scattering particles which interact through Coulomb potentials.

In radiation physics, Professor Belkić has worked on the passage of fast electrons and multiply charged ions through tissue as needed in radiation therapy in medicine. Here he has employed both deterministic methods through the Boltzmann equation and stochastic simulations via Monte Carlo computations. In radiobiology, he has worked on mathematical modelling for cell survival, and has focused on mechanistic modelling by including the main pathways for survival of cells under irradiation during radiotherapy.

In magnetic resonance physics, Professor Belkić has worked on nuclear magnetic resonance in medicine where he focused on high-resolution parametric signal processors which go beyond the conventional shape estimations of spectra. In mathematical physics, he has worked on many problems including the derivation of analytical expressions for scattering integrals or bound-free form factors, for rational response functions in signal processing, for coupling parameters in the nearest neighbour approximation which is one of the most frequently used methods in physics and chemistry, etc.

He has published more than 160 scientific publications which have received over 2000 citations. His book *Principles of Quantum Scattering Theory* has been published in 2004 by the Institute of Physics Publishing. He has received a number of international awards including the triple Nobel grantee status for research grants in atomic collision theory from the Royal Swedish Academy of Sciences as approved by the Nobel Committee for Physics.

Series in Atomic and Molecular Physics

Quantum-Mechanical Signal Processing and Spectral Analysis

Dževad Belkić

Karolinska Institute, Stockholm, Sweden

Institute of Physics Publishing
Bristol and Philadelphia

© IOP Publishing Ltd 2005

All rights reserved. No part of this publication may be reproduced, stored in a retrieval system or transmitted in any form or by any means, electronic, mechanical, photocopying, recording or otherwise, without the prior permission of the publisher. Multiple copying is permitted in accordance with the terms of licences issued by the Copyright Licensing Agency under the terms of its agreement with Universities UK (UUK).

British Library Cataloguing-in-Publication Data

A catalogue record for this book is available from the British Library.

ISBN 0 7503 1019 7

Library of Congress Cataloging-in-Publication Data are available

Commissioning Editor: Tom Spicer
Editorial Assistant: Leah Fielding
Production Editor: Simon Laurenson
Production Control: Sarah Plenty
Cover Design: Victoria Le Billon
Marketing: Louise Higham, Kerry Hollins and Ben Thomas

Published by Institute of Physics Publishing, wholly owned by The Institute of Physics, London

Institute of Physics Publishing, Dirac House, Temple Back, Bristol BS1 6BE, UK

US Office: Institute of Physics Publishing, The Public Ledger Building, Suite 929, 150 South Independence Mall West, Philadelphia, PA 19106, USA

Typeset in LaTeX 2_ε by Text 2 Text Limited, Torquay, Devon
Printed in the UK by MPG Books Ltd, Bodmin, Cornwall

Contents

	Preface	ix
	Acknowledgments	xii
	Acronyms used in the text	xiii
1	Introduction	1
2	Auto-correlation functions	20
3	The time-independent Schrödinger equation	23
4	The time-dependent Schrödinger equation	27
5	Equivalence: auto-correlation functions and time signals	31
6	Orthonormality and completeness of expansion functions	38
7	Difference equations and the harmonic inversion	44
8	The Schrödinger eigenproblem in global spectral analysis	56
9	Dimensionality reduction in the frequency domain	60
10	Dimensionality reduction in the time domain	67
11	The basic features of the Padé approximant (PA)	73
12	Gaussian quadratures and the Padé approximant	81
13	Padé–Schur approximant (PSA) with no spurious roots	86
14	Spectral analysis and systems of inhomogeneous linear equations	102
15	Exact iterative solution of a system of linear equations	104
16	Relaxation methods and sequence accelerations	120
17	Quantification: harmonic inversion in the time domain	122
18	Enhanced convergence of sequences by nonlinear transforms	129
19	The Shanks transform as the exact filter for harmonic inversion	133
20	The base-transient concept in signal processing	150

21	**Extraction of the exact number of transients from time signals**	**152**
	21.1 Explicit filtering of one transient $c_n = d_1 u_1^n$	152
	21.2 Explicit filtering of two transients $c_n = d_1 u_1^n + d_2 u_2^n$	154
	21.3 Explicit filtering of three transients $c_n = d_1 u_1^n + d_2 u_2^n + d_3 u_3^n$	156
	21.4 Explicit filtering of K transients $c_n = d_1 u_1^n + d_2 u_2^n + d_3 u_3^n + \cdots + d_K u_K^n$	158
22	**The Lanczos recursive algorithm for state vectors $|\psi_n\rangle$**	**160**
23	**The Lanczos orthogonal polynomials $P_n(u)$ and $Q_n(u)$**	**168**
24	**Recursions for derivatives of the Lanczos polynomials**	**174**
25	**The secular equation and the characteristic polynomial**	**176**
26	**Power series representations for two Lanczos polynomials**	**180**
27	**The Wronskian for the Lanczos polynomials**	**184**
28	**Finding accurate zeros of high-degree polynomials**	**186**
29	**Recursions for sums involving Lanczos polynomials**	**189**
30	**The Lanczos finite-dimensional linear vector space $\mathcal{L}_M$**	**194**
31	**Completeness proof for the Lanczos polynomials**	**197**
32	**Duality: the states $|\psi_n\rangle$ and polynomials $Q_n(u)$**	**203**
33	**An analytical solution for the overlap determinant $\det \mathbf{S}_n$**	**205**
34	**The explicit Lanczos algorithm**	**207**
35	**The explicit Lanczos wave packet propagation**	**210**
36	**The Padé–Lanczos approximant (PLA)**	**213**
37	**Inversion of the Schrödinger matrix $u\mathbf{1} - \mathbf{U}$ by the PLA**	**218**
38	**The exact spectrum via the Green function series**	**222**
39	**Uniqueness of the amplitudes $\{d_k\}$**	**227**
40	**Partial fractions in the PLA only from the denominator polynomial**	**230**
41	**The Lanczos continued fractions (LCF)**	**232**
42	**Equations for eigenvalues u_k via continued fractions**	**235**
43	**The exact analytical expression for the general Padé spectrum**	**237**
44	**The exact analytical result for any continued fraction coefficients**	**243**
	44.1 Calculation of coefficients a_1 and a_2	244
	44.2 Calculation of coefficient a_3	245
	44.3 Calculation of coefficient a_4	246
	44.4 Calculation of coefficient a_5	247
	44.5 Calculation of coefficient a_6	249

	44.6 Calculation of coefficient a_7	251
45	Stieltjes' formula for any continued fraction coefficients	258
46	The exact analytical expressions for any Lanczos couplings $\{\alpha_n, \beta_n\}$	261
47	The Lanczos continued fractions and contracted continued fractions	264
48	Exact retrieval of time signals from the Padé polynomial coefficients	266
49	Auto-correlation functions at large times	273
50	The power moment problem in celestial mechanics	278
51	Mass positivity and the power moment problem	286
52	The power moment problem and Gaussian numerical quadratures	290
53	The power moment problem of the generalized Stieltjes type	292
54	Matrix elements $(u^m | Q_n(u))$ and the Stieltjes integral	294
55	The modified moment problem	298
56	The mixed moment problem	304
57	Proof for zero-valued amplitudes of spurious resonances	310
58	Mapping from monomials u^n to polynomials $Q_n(u)$	317
59	Mapping of state vectors: Schrödinger $\longleftrightarrow$ Lanczos	319
60	The Gram–Schmidt *versus* the Lanczos orthogonalization	323
61	Analytical inversion of the Schrödinger overlap matrix	328
62	A variational principle for a quadratic Hankel form	334
63	Eigenvalues without eigenproblems or characteristic equations	337
64	Tridiagonal inhomogeneous systems of linear equations	341
65	Delayed time series	350
66	Delayed Green function	356
67	The quotient-difference (QD) recursive algorithm	358
68	The product-difference (PD) recursive algorithm	362
69	Delayed Lanczos continued fractions	368
70	Delayed Padé–Lanczos approximant	375
71	Delayed Padé approximant convergent outside the unit circle	379
72	Delayed Padé approximant convergent inside the unit circle	384
73	Illustrations	390
74	The uncertainty principle and resolution improvement	398

75	Prediction/extrapolation for resolution improvement	400
76	Main advantages of the Padé-based spectral analysis	402
77	Conclusions	411
A	Linear mappings among vector spaces	416
B	Non-classical polynomials in the expansion methods	423
C	Classical polynomials in the expansion methods	436
	References	439
	Index	454

Preface

This book is on quantum-mechanical signal processing based upon the Padé approximant (PA). We here link the PA and the Lanczos algorithm to design the Padé–Lanczos approximant (PLA). The PLA is operationalized with the recursive algorithm called the fast Padé transform (FPT) for both parametric and non-parametric estimations of spectra. The FPT for any given power series is defined by the unique quotient of two polynomials. This processor provides a meaningful result even when the original expansion diverges. It can significantly accelerate slowly converging sequences/series. As opposed to a single polynomial, e.g. the fast Fourier transform (FFT), the FPT can analytically continue general functions outside their definition domains. Moreover, we show that the FPT is an efficient solver of generalized eigenproblems, e.g. the quantum-mechanical evolution/relaxation matrix $\mathbf{U}$ comprised of auto-correlation functions. These generic functions can be either computed theoretically or measured experimentally. Such a concept, put forward as a computational tool, surpasses its initial purpose. Indeed auto-correlation functions represent a veritable alternative formulation of quantum mechanics. This is not just because all the major observables, e.g. complete energy spectra, local density of states, quantal rate constants, etc, are expressible through the auto-correlation functions. It is also because these and other observables could be given completely in terms of some appropriate, relatively small informational parts that can be singled out and analysed separately from the unwanted/redundant remainder of the full data set of auto-correlation functions. The required dimensionality reduction of original large problems treated by the FPT can be achieved by, e.g. windowing using the band-limited decimation. Alternatively, as done in this book, the Lanczos tridiagonalization can be employed yielding sparse Jacobi matrices in terms of the Lanczos coupling parameters $\{\alpha_n, \beta_n\}$ that have very important physical interpretations. The FPT is naturally ingrained in the Schrödinger picture of quantum mechanics and in the total time-independent Green function for the studied system. This yields a versatile framework for a unified treatment of spectroscopy and collisions within signal processing and quantum mechanics. In the quantum-mechanical method of nearest neighbours or tight bindings, we use time signal points $\{c_n\}$ as the only input data to derive the exact analytical expressions for the FPT, the continued fractions, the Lanczos polynomials

$\{P_n(\omega), Q_n(\omega)\}$, the couplings $\{\alpha_n, \beta_n\}$, the Lanczos state vectors ψ_n, the total wavefunction $\Upsilon(\omega)$ at any frequency ω and the 'Hamiltonian' operator $\hat{\Omega}$.

Within signal processing we also analyse ordinary difference equations, base-quotient concept, iterative relaxations, nonlinear accelerations, rational approximations, regularizations of spurious roots and the methods of moments. The analysis is performed within the context of the so-called harmonic inversion problem, or equivalently, the spectral decomposition or quantification problem. This is an inverse/reconstruction problem which is aimed at retrieving uniquely the whole information from the input time signal $\{c_n\}$, such that the output data contain the number K, position $\text{Re}(\omega_k)$, height $|d_k|$, width $\text{Im}(\omega_k)$ and phase $\text{Arg}(d_k)$ of every complex harmonic including all possible degeneracies/multiplicities of normal mode frequencies $\{\omega_k\}$ due to overlapping resonances[1]. The FPT has the unique virtues of unequivocally providing all these parameters for the non-degenerate and degenerate spectra with the least computational effort by attaining the required accuracy via robust and stable signal processing. This could be fully accomplished in only two simple steps: (1) solving a system of linear equations to extract the expansion/prediction coefficients of the characteristic/secular polynomial $Q_K(\omega)$ directly from $\{c_n\}$ and (2) rooting $Q_K(\omega)$ to obtain all the sought complex frequencies $\{\omega_k\}$ ($1 \leq k \leq K$). Only step (1) is needed in the non-parametric FPT, which is given by the polynomial quotient $P_L(\omega)/Q_K(\omega)$. This yields the shape spectrum in the FPT at any ω. In the quantification problem, the parametric FPT is required by carrying out step (2) which gives all the K complex frequencies $\{\omega_k\}$. The corresponding complex amplitudes $\{d_k\}$ are obtained from the two separate explicit expressions for the non-degenerate and degenerate spectrum as the residues of $P_L(\omega)/Q_K(\omega)$ taken at one single frequency $\omega = \omega_k$. This produces the K isolated Lorentzian resonances in a non-degenerate spectrum if the roots $\{\omega_k\}$ of $Q_K(\omega)$ are all distinct (no multiplicities). When some of the roots $\{\omega_k\}$ of $Q_K(\omega)$ coincide, i.e. have multiplicities, a degenerate non-Lorentzian spectrum $P_L(\omega)/Q_K(\omega)$ is obtained in the FPT which describes all isolated and completely or partially overlapping resonances. By contrast, other parametric methods, e.g. linear predictor (LP) or the Hankel–Lanczos singular value decomposition (HLSVD), are much more computationally demanding as they obtain the set $\{d_k\}$ by solving another system of linear equations using *all* the elements $\{\omega_k\}$ ($1 \leq k \leq K$), where errors in some $\{\omega_k\}$ might cause severe inaccuracies in d_k. Moreover, to arrive at d_k the method of filter diagonalization (FD) performs additional computations for the full state vectors $\{\Upsilon_k\}$ from the generalized eigenproblem of the evolution/relaxation matrix, i.e. the Hankel data matrix. Relative to the linear FFT, the advantages of the nonlinear FPT are in the significantly increased resolution for the same signal length or in the same resolution for shorter signal lengths. Furthermore, as opposed to other nonlinear

---

[1] Hereafter the symbols $\text{Re}(z)$ and $\text{Im}(z)$, respectively, denote the real and imaginary parts of a complex number z.

methods that typically undergo wild oscillations before they eventually stabilize, the FPT exhibits a strikingly stable convergence with the increased signal length. This is the present finding via an illustrative spectral analysis of a time signal, which has been experimentally measured via Magnetic Resonance Spectroscopy (MRS) at the magnetic field strength of 7T encoded from the brain of a healthy volunteer.

We also analyse unphysical (spurious or extraneous) resonances in the parametric estimation of spectra. Spurious peaks stem from noise corruption of the time signal and from the so-called overdetermination problem. If the signal of length N happens to have less than $N/2$ peaks, the problem becomes algebraically overdetermined, since there are more equations than unknowns. This leads to singular values associated with false peaks that represent spurious resonances. In the PA, extraneous roots could appear in both the numerator and denominator polynomial. Spurious roots in the denominator polynomial of the PA are undesirable, since they lead to unphysical spikes in the Padé spectrum. Spurious roots in the numerator polynomial of the PA are also unwelcome, since they fill in the valleys with unphysical anti-resonances and this destroys the phase minimum as well as the uniqueness feature of the PA. We examine this important problem by using the so-called constrained root reflection, which is an analytical procedure for regularizing spurious roots. First, we separate unequivocally the genuine from spurious resonances in the Padé power spectrum, which is itself the Padé–Chebyshev approximant (PCA). Second, the unstable PCA containing diverging and converging exponentials is properly stabilized. This is done by a special root reflection, which reverses the sign of diverging exponentials, so that they are relocated on the side of genuine resonances. Such a procedure is accomplished under the constraint that the parameter and the shape spectra of the PA are the same. The ensuing constrained root reflection has its physical justification in the preservation of the total energy of the signal via the Parseval identity. The resulting method is called the Padé–Schur approximant (PSA) which, as a stable estimator, possesses only converging exponentials. The PSA describes the Padé power spectrum as the unique ratio of two Schur polynomials whose computed roots are all adequately regularized by being placed on the side of physical resonances. Thus, rather than trying to eliminate or reduce the noise content from the measured data, as has often been attempted previously with a potential risk of losing weak genuine spectral features, the PSA processes noise, as measured, together with the physical signal.

Dževad Belkić
Professor of Mathematical Radiation Physics
Karolinska Institute
Stockholm, Sweden
November 2004

Acknowledgments

Thanks are due to Professor Ivan Mančev for thoroughly checking the presented material and to Professor Karen Lynne Belkić for language editing. The author thanks Tom Spicer and Simon Laurenson from the Institute of Physics Publishing (IOPP) and Victor Riecansky from Cambridge International Science Publishing (CISP) for fruitful cooperations. The work on this book was supported by The Royal Swedish Academy of Sciences, The Swedish National Science Foundation (Vetenskapsrådet) and Stockholm's County Council for Research, Development & Education (FoUU).

Acronyms used in the text

ABC:	Absorbing Boundary Conditions
AR:	Auto-Regressive
ARMA:	Auto-Regressive Moving Average
a.u.:	arbitrary units
bl:	band limited
bld:	band limited decimated
CF:	Continued Fractions
CCF:	Contracted Continued Fractions
CPU:	Central Processing Unit
CSI:	Chemical Shift Imaging
DFT:	Discrete Fourier Transform
DLP:	Decimated Linear Predictor
DOS:	Density of States
DPA:	Decimated Padé Approximant
DSD:	Decimated Signal Diagonalization
DVR:	Discrete Variable Representation
EHIP:	Extended Harmonic Inversion Problem
ESPRIT:	Estimation of Signal Parameters via Rotation Invariance
FD:	Filter Diagonalization
FFT:	Fast Fourier Transform
FID:	Free Induction Decay
FPT:	Fast Padé Transform
FWHM:	Full Width at the Half Maximum
GHIP:	Generalized Harmonic Inversion Problem
GSO:	Gram–Schmidt Orthogonalization
HIP:	Harmonic Inversion Problem
HLSVD:	Hankel–Lanczos Singular Value Decomposition
ICR:	Ion Cyclotron Resonance
IDFT:	Inverse Discrete Fourier Transform
IDOS:	Integrated density of states
LCF:	Lanczos Continued Fractions
LCModel:	Linear Combination of Model *in vitro* Spectra

lhs:	left-hand side
LP:	Linear Predictor
LPC:	Linear Predictor Coding
MA:	Moving Average
MR:	Magnetic Resonance
MRI:	Magnetic Resonance Imaging
MRS:	Magnetic Resonance Spectroscopy
MRSI:	Magnetic Resonance Spectroscopic Imaging
MUSIC:	Multiple Signal Classification
NMR:	Nuclear Magnetic Resonance
ODE:	Ordinary Differential Equations
OΔE:	Ordinary Difference Equations
OPA:	Operator Padé Approximant
PA:	Padé Approximant
PD:	Product-Difference
PCA:	Padé–Chebyshev Approximant
PLA:	Padé–Lanczos Approximant
ppm:	parts per million
PVL:	Padé via Lanczos
PSA:	Padé–Schur Approximant
RCM:	Rotated Coordinate Method
RF:	Radio Frequency
rhs:	right-hand side
rms:	root mean square
ROPEM:	Recursive Orthogonal Polynomial Expansion Method
RRGM:	Recursive Residue Generation Method
SPECT:	Single Photon Emission Computerized Tomography
SVD:	Singular Value Decomposition
TE:	Echo Time
QD:	Quotient-Difference

Chapter 1

Introduction

Among sciences and engineering marking the beginning of the new millennium, the versatile concept of *signals* and *images* ranks high along with a few outstanding leading strategies. This is because signals are indeed universal responses of objects to perturbations ranging from galactic interstellar phenomena, helio-seizmology to the central nervous system or to the genetic code, etc. Such responses can vary from, e.g. intercellular communications all the way down to subatomic particle levels and beyond. Besides having the unique potential to successfully intertwine and indeed unify some of the major interdisciplinary research fields in basic and applied sciences with the necessary bridging to engineering and technology, signals are also directly amenable to innumerable practical applications including the issue of the health of human beings. For an illustration, it suffices to mention only a few of the major fields where signals play a key role, e.g. molecular beam magnetic resonance [1], nuclear magnetic resonance (NMR) [2]–[6], ion cyclotron resonance (ICR) mass spectroscopy [7]–[9], magnetic resonance spectroscopy (MRS) [10]–[29], magnetic resonance imaging (MRI) [30]–[32], magnetic resonance spectroscopic imaging (MRSI) [33]–[39] which is also called chemical shift imaging (CSI), single photon emission computerized tomography (SPECT) [40]–[42], etc.

These unique tools for magnetic resonance (MR) physics are based upon inverse reconstruction problems that are deeply rooted in quantum-mechanical scattering and spectroscopy. Moreover, they represent a sequence of indispensable non-invasive diagnostic techniques of paramount importance in modern medical practice. This was made possible by a remarkable cooperation of mathematics, physics, medicine, computing science and technologies. Development of software was followed by hardware implementations and it is this coupling which represents the key driving force of signal and image processing. When it comes to imaging of, e.g. a patient, severe constraints must be put onto the applied magnetic field strengths that cause perturbations of the examined organ of the human body. These constraints might limit the resolution of the obtained spectra or images, and it is precisely here where recent vigorous advances in the mathematical software become critical to further progress. However, all the commercial software built into, e.g. spectrometers is presently based upon the fast Fourier transform (FFT) [43]–[55], despite the availability of a myriad of high-resolution signal and image processing methods. The reason for this is that most of the existing parametric methods in signal processing are inherently *unstable* in the sense of Schur [56]. This means that some of the complex eigenvalues or roots of characteristic polynomials are unphysical, i.e. spurious within many parametric methods such as the linear predictor (LP) [57]–[64], the Padé approximant (PA) [65]–[92], the Lanczos algorithm [93]–[104], the multiple signal classification (MUSIC) [105], the estimation of signal parameters via rotation invariance technique (ESPRIT) [106], the Padé–Laplace transform [107], the filter diagonalization (FD) [108]–[112], the decimated signal diagonalization (DSD) [113]–[116], the fast Padé transform (FPT) [16]–[29], etc. In signal processing, these extraneous eigenvalues of, e.g. the evolution **U**-matrix lead to exponentially increasing harmonics of the corresponding free induction decay (FID) or time signal c_n with the increase of time. These diverging harmonics are also encountered in the decimated Padé approximant (DPA) and decimated linear predictor (DLP) [115] through the emergence of unstable, i.e. non-Schur polynomials with some of their roots located on the 'wrong side' of the unit circle with exploding exponentials as the time increases infinitely [56].

The main difficulty with spurious roots is not so much in that they are regularly encountered in practice, but rather it is the implementation of a theoretically devised procedure which, by construction, would guarantee that the spectral density (power or magnitude spectrum) is invariant under the selected regularization. Recently, the FD has been applied to a variety of problems [108]–[112], but it was found that this method also generates spurious frequencies with Im(ω_k) > 0 leading to exponentially diverging natural or fundamental harmonics in the time signal. However, both the FD and the DSD have a technique for handling spurious eigenenergies [108]–[116]. This procedure consists of comparing the eigenvalues of the powers of the matrix $\mathbf{U}^{(s)}$ (comprised of the matrix elements of the sth power of the evolution operator, $\hat{U}^s$) for two values of the integer s, e.g. $s = 1$ and $s = 2$. From these two separate diagonalizations only those eigenvalues are retained that are within a prescribed level of accuracy and the other fluctuating solutions are rejected. Similarly, the recursive Lanczos algorithm [93]–[104] for solving large eigenproblems produces extraneous eigenvalues, but they are also managed by an adequate procedure [99]. In the Padé–Laplace transform [107] and DPA [115] spurious roots are recognized by comparing the results between the main diagonal and several successive paradiagonals of the Padé table. Within a prescribed threshold of accuracy, some of the roots are virtually unaltered when passing from the main diagonal to several neighbouring paradiagonals. These stable roots are classified as physical and the corresponding peaks are retained in the final spectrum. However, the same comparison also finds some peaks whose parameters change considerably beyond the prescribed accuracy level. These resonances are considered as unphysical, i.e.

spurious and they are discarded from the spectrum. Diagonalization of the matrix $\mathbf{U}^{(s)}$ in the DSD for an integer s is mathematically equivalent to the DPA for the main diagonal ($s = 1$) and the paradiagonals ($s > 1$). Therefore, the techniques for recognizing spurious resonances in the DSD or FD and DPA are the same as the procedure used previously in the Padé–Laplace transform [107].

As an alternative to the mentioned numerical techniques for dealing with extraneous resonances, we presently elaborate an *analytical* tool for unambiguous identifications of all existing spurious roots. To this end, within the generic PA, we use the so-called constrained root reflection [18] for regularization of *all* the encountered spurious roots. In particular, the procedure [18] automatically assures that *both* the numerator and denominator polynomials of the PA are stable or Schur polynomials [56]. This generalizes the related previous work from [117] where only the Padé denominator polynomial was regularized by the constrained root reflection. In [18], a variant of the root reflection has been analysed by subjecting all the spurious roots of the two Padé polynomials P_L/Q_K to complex conjugation, but with the important constraint that magnitude $|P_L/Q_K|$ and, likewise, power spectra $|P_L/Q_K|^2$ remain strictly unaltered. This is the constrained root reflection which regularizes all the encountered extraneous poles by relocating them to the site of the genuine poles. In other words, this method deliberately *avoids eliminating* any of the encountered spurious roots. This is because, in practice, some of the spurious roots might well be associated with the genuine, physical transients that were supposed to decay, but nevertheless remained in the signal for the given duration of the acquisition time of the FID. The presence of these 'spurious' resonances may influence the spectral information in the signal and, as such, they should not be eliminated from the analysis. The constrained root reflection as a power-spectrum-preserving regularization of spurious resonances should be contrasted to the usual root reflection, which replaces $\mathrm{Im}(\omega_k)$ by $-|\mathrm{Im}(\omega_k)|$ for $\mathrm{Im}(\omega_k) > 0$, but without a simultaneous requirement that the resulting magnitude and/or power spectra remain the same as if this replacement were absent.

Originally, signal and image processing developed autonomously, and remained for a long time mainly within the realm of applied sciences and engineering. Yet, here, very similar or nearly the same methods have been used as in the basic sciences. For example, in electric circuit theory [118]–[120], one of the most frequently used methods is the so-called rational response of the examined system to an external perturbation. This response is, in fact, the PA in the frequency domain. Precisely the same type of PA can also be found in various fields that use signal processing (speech patterns, system theory, optimizations, heart rate variability, etc) or in mathematical statistics under different names such as the auto-regressive moving average (ARMA) [121]–[129]. Reconstructing spectra and images from the received signals is a general inverse problem with its intrinsic time development, which is ideally suited for description by the Schrödinger picture of quantum mechanics. Therefore, it could be beneficial for signal and image processing to intertwine with a quantum-mechanical description of stationary and time-independent phenomena. It has been only within a few recent years [108]–[116] that the situation improved through using quantum mechanics and quantum resonant scattering theory in signal processing, by relying upon the concept of auto-correlation functions as the amplitude probabilities for survival of the Schrödinger or Krylov states. In such an approach, the Schrödinger ordinary or generalized eigenvalue equations and the spectral problem of the resolvent or the Green operator play the pivotal roles. From this framework many more opportunities might open up with the advantage of furthering signal and image processing themselves, but also of providing new tools for studying generic spectra irrespective of their experimental or theoretical origins. The spin-offs from this cross fertilization of the invoked research fields is anticipated to have a twofold benefit, e.g. emergence of new high-resolution signal and image processors for applied sciences as well as engineering and, in turn, the usage of these very same methods in basic sciences for analysing spectra of large physical, chemical or biological systems. Naturally, the literature on signal and image processing in applied mathematics and engineering [118]–[123] is abundant with specially designed methods for robust performance in industry and technology under rigorous requirements for accuracy, stability and reliability. To achieve such strict goals the most advanced mathematical methods have been used with the possibility for export to applied and basic sciences for versatile applications. For this to happen in a manner which is systematic rather than sporadic, more cross disciplinary interactions are needed with the mutual benefits that would *inter alia* reduce unnecessary duplication of the results in non-overlapping fields. In addition to other existing efforts along these lines, the two recent topical issues on spectral analysis [130, 131] have already paved a new road for cross disciplinary exchange of ideas. The present book is also on spectral analysis with the main goal of attempting to convey to the reader some of the common aspects, strategies and mathematical methods in seemingly disjoint research fields.

Notwithstanding the importance of the practical implications in arriving at highly resolved spectra and images, the key merit of introducing quantum mechanics into signal processing is in providing the fundamental framework of a complete theory of physics. This is crucial due to the possibility of directly relating arbitrary signals to the *dynamics* of the examined system and its time evolution described by the first principles of physics. For overly complex systems the dynamics might be either unknown or unmanageable for direct theoretical treatments. In such cases, whenever experimental data are available as, e.g. counts per channel, time signals and the like, quantum mechanics can peer into the dynamics of the system and extract the full spectral information. This is possible due to, e.g. the established equivalence between time signals that lead to Lorentzian spectra and auto-correlation functions [108]–[115]. This equivalence replaces the usual nonlinear fitting problem of experimental data by the standard quantum-mechanical search for eigenspectra of the studied system. The obtained results reconstruct the unknown dynamics and interactions in the system which has undergone certain transitions under the influence of a perturbation, before generating the recorded time signal. This is at once recognized as the well-studied inverse scattering problem in quantum mechanics, where one is given certain experimentally measured data (e.g. phase shift, etc) and the task is to retrieve the interaction potentials that are the key to every physics problem. It is this unfolded dynamics which provides the sought information about the studied system, e.g. the proton spin density $\rho(x, y)$ yielding the emission data in MRI, MRS, MRSI, etc. This accomplishes the task of locating the accurate spatial positions (x, y) of protons. It is a plot of the reconstructed observable $\rho(x, y)$ which directly gives an image of the object whose signals were acquired via, e.g. MRI or MRSI. Slices of such images can be spectroscopically analysed, as done within MRSI, to gain invaluable quantitative information about peak parameters, such as abundance of various metabolites, their relaxation times, etc. For example, relaxation times of lipids, whose peaks are superimposed on top of 'noisy' background, exhibit very different patterns in tumorous and healthy tissue, pointing at the important diagnostic role of MRS and MRSI in medicine [16]–[29], [33]–[39].

There is tremendous interest in medicine for MR physics whose multitude of retrieval or reconstruction techniques has nowadays attained the status of an essential part of diagnostic radiology. The ultimate goal of MRI is to generate two-dimensional images of pre-assigned sections of the examined human organ. The two distinct aspects of MRI are advantageous relative to other competitive methods: (i) an arbitrary orientation/position of the imaged area and (ii) a large contrast among soft tissues. In addition to imaging static anatomy, many current clinical tasks use MRI for imaging blood vessels without contrast agents, cardiac imaging, dynamic imaging of the musculoskeletal system as well as for measuring tissue temperature and recording diffusion in tissue. The common denominator in all the MR encoding techniques is radio frequency (RF) excitation of the sample and a reliance upon the elementary magnetic fields that originate from the nuclear magnetic moments. The Larmor precession of these magnetizations around the external constant magnetic field, combined with the additional magnetic field gradients and RF pulses, generates a small current in the receiver coil which surrounds the relevant part of the examined human body. This is the pathway through which the sample responds to the external perturbation, after which de-excitation of the sample takes place, leading to an echo with a composite electromagnetic time signal. Spectral and image analysis of this complex valued signal can reveal the types and the spatial two-dimensional positions (x, y) of the nuclei that are imaged within the selected sample.

The current key problems of MRI are long imaging times and insufficient spatial resolution. The reasons for such an unsatisfactory situation is that advances in MRI have thus far been tied to upgrading hardware, whereas the commercially implemented signal and image processing software is limited exclusively to the FFT. However, the FFT is known to be a low-resolution estimator unless exceedingly long time signals are recorded, which is in clinical practice precluded by high levels of noise and, more importantly, by intolerably long exposure of patients to external magnetic fields. This situation could be significantly improved by simply complementing the FFT from commercially available instrumentation with a separate implementation of several new high-resolution signal and image processors imported from the sciences and engineering. However, many of the mathematically and technically different processors lead to spurious, i.e. unphysical resonances. Such unphysical peaks are often not easy to detect and regularize. These methods should be subjected to convergent validation, after which only the features of a spectrum/image that are faithfully reproduced by several processors could eventually be considered as candidates for reliable estimates. Other characteristics that fluctuate from one theory to another should be considered with caution. The cross verification of various methods should be done after signal encoding and, therefore, no changes are needed to the conventional scanning protocol. Hence, processing of the recorded signals could be done independently of scanners and this would lead to a greater flexibility in applying different spectral analysers. This could help, e.g. non-invasive diagnostics in medicine within the realm of MRS, MRI and MRSI, as clinicians would have more options than relying solely upon FFT images from commercial scanners. Needless to say, signal processing methods that have no recipe for regularization of spurious resonances are particularly undesirable in medical diagnostics. For example, in MRS the lactate and some lipid resonances usually have small intensities relative to the concentration of other metabolites and are, in fact, embedded in a large background. Spurious resonances could easily interlace with lipid peaks and cause considerable confusion in the identification pattern for diagnostic purposes.

The drawback of the mentioned convergent validation of different parametric estimators is that there is no guarantee that resonances reproduced by several methods are indeed the genuine ones. This is due to the intrinsic *instability* of all the existing parameter estimators. Therefore, alternative avenues should be explored, as done, e.g. in the present book, where a reliable procedure is designed to ensure stability of the PA by having to deal only with stable or Schur polynomials in the Padé numerator and denominator. This is the Padé–Schur approximant (PSA), with a rational polynomial ansatz which can be generated in any of the selected procedures including the Lanczos algorithm as elaborated in this book. A number of favourable capabilities (stability, robustness, efficiency, etc) of the PSA may prove to be well-suited for the task of adequately supplementing the FFT software in commercial scanners. The PSA might be conceived as an interface to new experiments in scattering and spectroscopy through signal/image processing with the possibility of reaching a substantially higher resolving power than the one currently available. Such an achievement could be exploited in the future to build new spectrometers that are based on, e.g. the PSA which also provides *en route* the conventional FFT spectra. To optimally focus on the region of interest, various segmentations of spectra and images could be done *a posteriori* along with resampling the signal at different sampling rates, but with preservation of the indigenous information in the selected region, etc.

Such a multi-faceted analysis is possible at the different equivalent acquisition times, but without having to repeat the experimental scanning itself. This high-resolution processing is aimed at extracting more information from measurements than is actually feasible by the FFT for a given fixed signal length. At first, this goal might seem paradoxical in view of the fact that the FFT is a linear transform which, therefore, must hold the same information as recorded in the measured time signal. However, this paradox is only apparent. Indeed the FFT does fully preserve the measured information, but a part of this contingent might be hidden in many Fourier spectra and images. This is due to a low resolution power of the FFT. Therefore the main task of any new high-resolution processor is to convert this obscure part of the FFT with its inaccessible information into the corresponding transparent information with potentially improved results.

One of the most reliable algorithms capable of providing the recommended, cross validated spectral information for the purpose of assisting, e.g. medical diagnostics is the FPT. This is the generic name for the PA applied to signal processing irrespective of which concrete computational algorithm is used to obtain the expansion coefficients of the Padé rational polynomial. In fact, the FPT is unique among all the existing signal processors, since it performs a self-contained cross validation. This stems from the fact that the FPT, as a system function designed to respond optimally to an external excitation, has two complementary variants. They are denoted by FPT^+ and FPT^- to point at their expansion variables $z^{\pm 1}$ and to the two entirely different convergent regions located inside ($|z| < 1$) and outside ($|z| > 1$) the unit circle, respectively. Both FPT^+ and FPT^- rely upon the same input spectrum given by the truncated and scaled Green function $\sum_{n=0}^{N-1} c_n z^{-n}$ where $z = \exp(i\omega\tau)$ with $\text{Im}(\omega_k) > 0$, $\tau > 0$ and the set $\{c_n\}(0 \leq n \leq N-1)$ of the total length N represents the collection of all the encoded signal points. Here, the variant FPT^- is the standard diagonal PA given as the rational polynomial $A_K^-(z^{-1})/Q_K^-(z^{-1})$ in which both the input series development $\sum_n c_n z^{-n}$ and the PA have the same expansion variable z^{-1}. On the other hand, the FPT^+ formally coincides with the so-called causal z-transform given by the diagonal PA via a rational polynomial in variable z, i.e. $A_K^+(z)/B_K^+(z)$ for the input sum $\sum_n c_n z^{-n}$, which is itself an expansion in powers of z^{-1}. The fundamental frequencies $\{\omega_k^{FPT\pm}\}$, counted with their proper multiplicities (if any), are computed respectively via the $FPT^{\pm}$ by rooting the characteristic equations $B_K^{\pm}(z^{\pm 1}) = 0$ in the complex z-plane[1]. Both variants $FPT^{\pm}$ must give the identical spectral parameters that can jointly be denoted by $\{\omega_k^{FPT}, d_k^{FPT}\}$ and this occurrence stems from the uniqueness of the FPT for the same input sum $\sum_n c_n z^{-n}$. Then, by necessity, the only common location in the complex frequency plane, where the eigenvalues $\{\omega_k^{FPT\pm}\}$ could possibly be very near each other $\omega_k^{FPT+} \approx \omega_k^{FPT-}$, is a segment located infinitesimally close to the unit circle, i.e. $\omega_k^{FPT\pm} \approx \omega_k^{FPT} \mp \varepsilon_k^{FPT}$. Here, the quantity $\varepsilon_k^{FPT} = \varepsilon_{k,R}^{FPT} + i\varepsilon_{k,I}^{FPT}$ is

---
[1] Hereafter, in order to avoid double superscripts, whenever the acronyms $FPT^{\pm}$ are used in the superscript of a given quantity, say X, we shall write $X^{FPT\pm}$ instead of $X^{FPT^{\pm}}$.

an 'infinitesimally small complex number' which is defined by the corresponding pair of infinitesimally small real numbers $\varepsilon_{k,R}^{FPT} > 0$ and $\varepsilon_{k,I}^{FPT} > 0$. Thus, in principle, by selecting the accuracy thresholds $\varepsilon_{k,R} > 0$ and $\varepsilon_{k,I} > 0$ with $\varepsilon_k = \varepsilon_{k,R} + i\varepsilon_{k,I}$, the optimal frequencies $\omega_k \mp \varepsilon_k$ could be extracted with fidelity from those computed values $\{\omega_k^{FPT\pm}\}$ that are infinitesimally different from each other $\omega_k^{FPT\pm} \approx \omega_k^{FPT} \mp \varepsilon_k^{FPT}$ provided that $\varepsilon_{k,R}^{FPT} \leq \varepsilon_{k,R}$ and $\varepsilon_{k,I}^{FPT} \leq \varepsilon_{k,I}$. Moreover, in practice, we do not even need to guess the error estimates $\{\varepsilon_{k,R}, \varepsilon_{k,I}\}$. Instead, the sought genuine frequencies $\{\omega_k\} \approx \{\omega_k^{FPT}\}$ will automatically be identified as those values $\{\omega_k^{FPT+}\}$ that are equal to $\{\omega_k^{FPT-}\}$ through a pre-assigned number of decimal places (say, two). Likewise, those frequencies $\{\omega_k^{FPT\pm}\}$ that do not pass this stringent accuracy threshold are classified as spurious. Of course, this assumes that most of the computed frequencies and amplitudes $\{\omega_k^{FPT\pm}, d_k^{FPT\pm}\}$ in both variants have converged as a function of, e.g. the truncated signal length $N/M (M > 1)$. In this inherently controlled and self-contained manner, the FPT is able to simultaneously provide the genuine and identify the spurious resonances.

The power of this novel type of cross validation is in bypassing altogether comparisons among different methods that could yield approximately the same results, which nevertheless might turn out to be wrong. By contrast, the FPT$^+$ and FPT$^-$ share a common conceptual design of a system response function via rational polynomials $A_K^+(z)/B_K^+(z)$ and $A_K^-(z^{-1})/B_K^-(z^{-1})$ within the same kind of a mathematical modelling, namely the FPT. Yet, the two variants FPT$^+$ and FPT$^-$ are computationally so complementary to each other that they could rightly be considered as distinct strategies. This is because the FPT$^-$ is an accelerator of a given slowly convergent series, whereas the FPT$^+$ induces/forces convergence into an initially divergent series by means of analytical continuation [28, 72, 75]. The FPT methodology accomplishes these two diametrically opposite tasks through its two complementary wings, the FPT$^+$ and FPT$^-$ that converge inside ($|z| < 1$) and outside ($|z| > 1$) the unit circle, for the same input $\sum_n c_n z^{-n}$ which is itself divergent and convergent for $|z| < 1$ and $|z| > 1$, respectively. The optimal performances of the FPT$^-$ and FPT$^+$ are achieved for $N = 2K$ and $N > 2K$, respectively, where K is the degree of the numerator and denominator Padé polynomials from the quotients $A_K^\pm(z^{\pm 1})/B_K^\pm(z^{\pm 1})$. The cases $K = N/2$ and $N > 2K$ correspond respectively to an algebraically determined and over-determined system of linear equations for obtaining the expansion coefficients of the polynomials $A_K^\pm(z^{\pm 1})$ and $B_K^\pm(z^{\pm 1})$. These different systems of linear equations encountered in the FPT$^+$ and FPT$^-$ employ the same input information, i.e. truncated and scaled Green function $\sum_{n=0}^{N-1} c_n z^{-n}$.

Notice that for the purpose of illustration, the above highlights focused on explaining the Padé self-consistent cross validation mainly for complex frequencies $\{\omega_k^{FPT\pm}\}$. However, the same type of stringent accuracy thresholds are used also for cross validation of the associated complex amplitudes $\{d_k^{FPT\pm}\}$ that are computed from their analytical expressions in the FPT$^\pm$. As the final results, we obtain the spectral parameters $\{\omega_k^{\rm FPT}, d_k^{\rm FPT}\}$ that are the intersection of the two computed sets $\{\omega_k^{\rm FPT+}, d_k^{\rm FPT+}\}$ and $\{\omega_k^{\rm FPT-}, d_k^{\rm FPT-}\}$. Once the sought spectral pairs $\{\omega_k^{\rm FPT}, d_k^{\rm FPT}\}$ are faithfully extracted from the encoded data $\{c_n\}$, the Padé complex mode shape spectra (called hereafter the Padé parametric shape spectra) given by the Heaviside partial fractions [132]–[141] could be at once constructed. These latter shape spectra determined from the estimated pairs of parameters $\{\omega_k^{\rm FPT\pm}, d_k^{\rm FPT\pm}\}$ must, up to random noise differences, coincide with the corresponding complex mode shape spectra (called hereafter the Padé non-parametric shape spectra) computed using only the quotients $A_K^\pm(z^{\pm 1})/B_K^\pm(z^{\pm 1})$ which do not rely at all upon the complex frequencies and amplitudes $\{\omega_k^{\rm FPT\pm}, d_k^{\rm FPT\pm}\}$. Any difference between the Padé parametric and non-parametric shape spectra detected above the signal-to-noise ratio (determined as, e.g. the rms—the root mean square of the complex mode FFT spectrum) would indicate that the estimated parameters $\{\omega_k^{\rm FPT\pm}, d_k^{\rm FPT\pm}\}$ need to be refined via an appropriate optimization. We accomplish this optimization via an extremely powerful, iterative, self-correcting and stable algorithm based on the Maclaurin expansion for the spectral parameters, by minimizing the squared difference between the Padé parametric and non-parametric shape spectra within the same variant, e.g. the FPT$^+$ [142]. Of course, we proceed likewise with this kind of optimization while using the FPT$^-$. The initial values of such two separate and independent iterative optimizations in FPT$^\pm$ are the spectral pairs $\{\omega_k^{\rm FPT\pm}, d_k^{\rm FPT\pm}\}$ that stem from the application of the explained parametric versions of the FPT$^\pm$. The iterations are stopped when the difference between the Padé parametric and non-parametric shape spectra is indistinguishable from the random noise, or equivalently, the computed rms. The output spectral parameters, denoted again for simplicity by the same labels as the initial values $\{\omega_k^{\rm FPT\pm}, d_k^{\rm FPT\pm}\}$ are optimal in the mentioned least-square sense. This constitutes an internal validation of the two sets of estimates $\{\omega_k^{\rm FPT+}, d_k^{\rm FPT+}\}$ and $\{\omega_k^{\rm FPT-}, d_k^{\rm FPT-}\}$ in an entirely independent manner without any direct reference of one set of spectral parameters from the FPT$^+$ to the other from the FPT$^-$. It is only after accomplishing such intrinsic validations of parameters $\{\omega_k^{\rm FPT+}, d_k^{\rm FPT+}\}$ and $\{\omega_k^{\rm FPT-}, d_k^{\rm FPT-}\}$ that we proceed towards the critical cross validation by identifying the intersection between these latter two sets with the emergence of the final estimates $\{\omega_k^{\rm FPT}, d_k^{\rm FPT}\}$ for an accurate and maximally reliably quantification of the unknown exact physical resonances $\{\omega_k, d_k\}$.

Overall, there is a strong need for concepts and methods from quantum mechanics and quantum scattering theory [143]–[157] in many basic sciences other than physics, as well as in applied sciences, including life sciences and beyond. This is not just because quantum mechanics is a complete physics concept, which is among the best established strategies in all sciences, thoroughly validated in innumerable comparisons with high-precision experimental measurements that provide the most stringent tests, it is also because the versatile mathematical tools of this theory are well-suited for direct implementations in other fields. Even when similar tools are used elsewhere, as indeed they are not necessarily invented in quantum mechanics, but rather in pure and/or applied mathematics, a rich research experience gained in quantum-mechanical complicated applications and sophisticated computations could be highly beneficial to research beyond physics. A partial account on this point is given in the present book, which illuminates the quantum-mechanical tight binding model or, equivalently the nearest neighbour approximation [96, 97], using spectral analysis based upon the Lanczos recursive algorithm [93]–[104]. Working along these latter lines as well as intertwining the Lanczos and Padé methods in the setting of the tight bindings, we have recently [17, 18] derived the practical and general exact analytical formulae for the following key quantities: (i) the expansion coefficients $\{a_n\}$ of the continued fractions (CF), (ii) the Lanczos coupling parameters $\{\alpha_n, \beta_n\}$, (iii) the Lanczos polynomials $\{Q_n(u), P_n(u)\}$, (iv) the fast Padé transform $(c_0/\beta_1)P_n(u)/Q_n(u)$, (v) the Lanczos state vectors ψ_n, (vi) the total wavefunction $\Upsilon(u)$ of the studied general system at any complex variable u and (vii) the 'Hamiltonian' operator $\hat{\Omega}$. All the closed formulae (i)–(vii) rely only upon the given input data of experimentally measured time signals $\{c_n\}$ or theoretically generated auto-correlation functions $\{C_n\}$ of arbitrary length N. Moreover, a multi-dimensional extension of the FPT has also been formulated and implemented to single, double and triple quadratures [157, 158] with an unprecedented accuracy outperforming spectacularly the FFT by ten decimal places. In [24], the bivariate FPT has been devised and successfully applied to two-dimensional magnetic resonance spectroscopy (2D-MRS) for cross correlation plots using biomedical time signals $\{c_{n_1,n_2}\}$ encoded from patients. The first application of the FPT to phantom and volunteer data from MRI reported in [34] also shows great promise especially for reducing the truncation artefacts known as the Gibbs oscillations.

The spectral analysis or frequency analysis problem has a long history which can be traced back to the harmonic inversion problem of Prony [57]. This problem is a decomposition of a given function into its components comprised of complex harmonics whose parameters are unknown. It is stated as follows: given a set of input data, via, e.g. time signal points $\{c_n\}$ or auto-correlation functions $\{C_n\}$, one is required to find the number of the constituent resonances, their complex positions and the corresponding residues. This is recognized as one of the main topics in the quantum-mechanical resonant scattering theory and spectroscopy that are among the chief physics methods for studying the structure of matter on different levels and scales [143]–[157]. This observation alone could justify the effort in attempting to tie together signal processing with quantum-mechanical resonance scattering and spectroscopy [108]–[116].

An apparently different subject, the power moment problem from celestial mechanics is set up to search for the mass distribution/concentration on different locations, i.e. spatial positions in a given constellation or a system or a body for the known total mass [159, 160]. However, this is mathematically equivalent to the harmonic inversion in processing time signals and, in particular, to mass spectroscopy in ICR in analytical chemistry or to MRS in medical diagnostics of human tissue. The method of the power moment is not frequently studied in quantum mechanics [161]. By contrast, this method is abundantly used in mathematical probability theory and in statistics [162]–[175], whose results could give a qualitatively new insight into ordinary or generalized quantum-mechanical eigenvalue problems as well as into the harmonic inversion in signal processing and the like. For example, the characteristic polynomials, i.e. eigenpolynomials that originate from the given secular equation, as the building blocks of a number of spectral analysers, are known to be exponentially ill-conditioned when constructed using the power moments. The situation can be dramatically improved when the power moments are weighted with one [162]–[170] or two [171]–[175] orthogonal polynomials that lead to the modified and mixed moments, respectively. This is analysed in chapters 53–56. In the case of the mixed moments, one of the two polynomials is given in advance and, as such, could be advantageously selected from the family of the known, classical orthogonal polynomials (Legendre, Laguerre, Lagrange, Hermite, Chebyshev, Gegenbauer, Neumann, etc) [176]–[180]. The mixed moments enable an extremely accurate and stable construction of the needed characteristic polynomials of very high orders. This numerically robust method, especially in the formulation of Wheeler [174], is capable of producing the complete set of the coupling parameters $\{\alpha_n, \beta_n\}$ that are encountered in the three-term contiguous recursions for polynomials[2]. This capability yields high precision which is achieved systematically within machine accuracy, as has been demonstrated by Gautschi [175] for moments defined at finite integration limits.

It has been stated in *Numerical Recipes* [176] that the power moment is notoriously ill-conditioned and, as such, should not be used for computations. Indeed, the computations of Gautschi [175] based on the power moments shows that the coupling constants $\{\alpha_n, \beta_n\}$ can lose all their significant digits for the values of n that are of a relatively low order, e.g. 10–15. However, Gordon [170] also used the power moments in a computation of the constants $\{\alpha_n, \beta_n\}$ via the product-difference (PD) recursive algorithm and arrived at machine accurate results. Therefore, it is not the use of power moments which should be abandoned, but rather the way in which they are incorporated into the subsequent generation of the parameters $\{\alpha_n, \beta_n\}$. The unprecedentedly powerful PD algorithm of Gordon [170] was published in 1968, but this seminal work passed unnoticed in the mentioned papers of Wheeler [174] from 1974 and Gautschi [175] from 1978. The PD algorithm is similar to the more familiar quotient-difference (QD) algorithm from 1957 attributed to Rutishauser [169], although it was known to Frobenius in 1879 [66], as pointed out previously by Wynn [73]. The PD algorithm performs the division only once at the end of the recursion. However,

---

[2] The same constants $\{\alpha_n, \beta_n\}$ are also encountered in the Lanczos algorithm [93]–[104] for the state vectors $\{\psi_n\}$.

the QD algorithm uses divisions in each iteration and in a finite-precision arithmethic this could lead to more round-off errors than in the PD algorithm.

The Lanczos polynomials that emerge from the algorithms of [170] and [174] are expected to preserve orthonormality as well as the *linear independence* throughout the computation for high orders. Once the coupling constants are generated, and the polynomials constructed, a *complete* basis set of the expansion functions with conserved orthonormalization becomes automatically available. No numerical quadrature is required for the generation of the key ingredient of the method, namely the coupling parameters $\{\alpha_n, \beta_n\}$. The methods of Gragg [171] and Wheeler [174] for the modified and mixed moments, respectively, have their roots in the original Chebyshev [162] algorithm. The mentioned demonstrations by Gordon [170] and Gautschi [175] of the performances of the PD and the Wheeler algorithms, respectively, should be contrasted to one of the most frequently employed methods in quantum chemistry and in solid state physics, i.e. the Lanczos recursive algorithm [93]–[104], which generates the state vectors that are observed to suffer from severe numerical instabilities in striving to maintain orthogonalization and linear independence. An accurate and stable extraction of all the required coupling parameters $\{\alpha_n, \beta_n\}$ and the associated expansion basis set functions from the mixed moment problem are of direct use in versatile applications, e.g. solving multi-channel Schrödinger eigenvalue problems in quantum mechanics, scattering problems, signal processing, etc. For example, the constants $\{\alpha_n, \beta_n\}$ can instantaneously provide the resolvent or the Green functions, Lanczos continued fractions (LCF) and the Padé–Lanczos approximant (PLA) [19, 85, 97, 120]. These three mathematically equivalent methods can yield the complete spectral information (eigenenergies, eigenfunctions, residues) by rooting the characteristic polynomials. This is the parametric aspect of the mentioned estimators. Alternatively, they could provide shape spectra without rooting the characteristic polynomials and this role is reminiscent of transforms in the spirit of the Fourier transform. Moreover, we shall demonstrate in the present book that the coupling parameters $\{\alpha_n, \beta_n\}$ are also sufficient for a complete construction of auto-correlation functions $\{C_n\}$ without the knowledge of either u_k or d_k [19]. This is in sharp contrast with the current practice, which necessitates the entire collection of the eigenvalues and residues $\{u_k, d_k\}_{k=1}^{K}$ for computations of auto-correlation functions [181].

Root searching is a nonlinear procedure which represents a bottleneck of all spectral methods based upon polynomials. Therefore, a direct rooting of high-order polynomials should be avoided whenever possible. Fortunately, this is feasible, since finding all the roots of any polynomial is completely equivalent to diagonalization of the corresponding Hessenberg matrix [176]–[179]. This latter matrix is extraordinarily sparse, since it has mainly zeros throughout its structure, except in one of the diagonals with all the elements equal to unity and in the first row containing the expansion coefficients of the characteristic polynomial. Diagonalization of such a matrix is computationally manageable even for a very large dimension and, therefore, root search of polynomials can be completely bypassed, i.e. replaced by an equivalent diagonalization of the associated Hessenberg matrix, as has recently be done in signal processing [115]. All told, a versatile and universal role of non-classical orthogonal polynomials becomes exceedingly attractive when established with the help of the stable, accurate and recursive algorithms of Gordon [170] and Wheeler [174] for moment problems. The unprecedented success of these two algorithms within the moment problem are indeed invaluable and their full exploration in physics and chemistry as well as in signal processing and beyond is yet to come. Alternative and/or complementary to the numerical algorithms from [170] and [174], the analytical/closed and simple (i.e. non-determinantal) formulae have been reported in [17] for the Lanczos coupling constants $\{\alpha_n, \beta_n\}$ in the general case of an arbitrary value of the non-negative integer n. This implies that, e.g. the resulting shape spectra are also available from an analytical formula in the form of a finite-order continued fraction which is defined solely in terms of the constants $\{\alpha_n, \beta_n\}$.

Recall here that continued fractions are among the most stable and most robust computational tools from numerical analysis as well as from theory and practice of approximations of functions [182]–[192]. This statement is readily substantiated by merely a cursory inspection of any computer main frame and/or the leading mathematical software libraries for computations of nearly every elementary function (exponentials [187], trigonometric functions [188], etc) and special functions (Bessel functions, gamma-function, probability/error function [189, 190], Gauss hypergemetric functions [183], Coulomb regular/irregular functions [192], etc) [176]–[179]. For example, a computation of the exponential $\exp(-x)$ for e.g. $x = 20$ should, at first sight, appear as trivial through the corresponding Maclaurin expansion which converges for every x. Nevertheless, the results obtained in this manner are entirely unsatisfactory, as opposed to the corresponding continued fractions which work perfectly well for any x including $x = 20$ from this particular example [193].

One of the goals of the present book is to carry out spectral analysis from the viewpoint of the power, modified and mixed moment problems. This is particularly motivated by the accuracy, stability and efficiency of the algorithms of Gordon [170] and Wheeler [174] for the power and mixed moment problem. On top of this, we add the present analytical formula for the coupling parameters $\{\alpha_n, \beta_n\}$. The advantages of these three methods are expected to lead to significant improvements in harmonic inversion, as well as in obtaining complete spectral information for any given Hamiltonian or other dynamical operators in chemistry, physics and beyond. This is highly relevant to applications across MR physics and especially in MRS. For example, the so-called quantification problem in MRS, i.e. determining concentrations, relaxation times and chemical shifts of metabolites in a given tissue of the human body is entirely equivalent to the moment problem, and yet this line of thought remains virtually unused in applications within MR physics. This could be achieved through a number of accurate, stable and simple algorithms analysed in the present book.

Among the most commonly used nonlinear *fittings* for processing *in vivo* FIDs recorded within MRS is the so-called LCModel [11]. This acronym stands for the linear combination of model *in vitro* spectra from individual metabolite solutions. In the preprocessing stage, the LCModel usually doubles the number of data points in the time and frequency domains by using the zero-filling in both *in vitro* and *in vivo* data. If a spectrum in the frequency domain is approximated by a sum of Gaussians or Lorentzians, then the corresponding form of the FID in the time domain will be given by a linear combination of damped exponentials. In such a case, any fittings in the time and frequency domains are necessarily nonlinear. This should be contrasted to linear mappings, such as FFT, in which the fundamental Fourier frequencies in the exponentials are fixed prior to processing. Nonlinear fits are iterative and some initial values are needed for all the freely fitted parameters of a chosen mathematical model. Since these values are unknown prior to fitting, they must be either guessed or produced by an auxiliary mathematical model or acquired via a separate measurement of FIDs in MRS on, e.g. some related phantoms. The latter alternative is adopted in the LCModel where the initial values for a nonlinear fit are taken from data banks or libraries of those *in vitro* measured spectra that correspond to the examined *in vivo* spectrum. Here, it is assumed that the measuring instrument, magnetic field and the encoding protocol/sequence are the same for acquiring both the *in vitro* and *in vivo* spectra. The LCModel does not assume *a priori* that the measured *in vitro* and *in vivo* spectra are of the Lorentzian or Gaussian type or of any other pre-assigned form. However, an overall exponential factor of the model for the complex spectrum is dependent upon two unknown phase corrections that are treated as the fitting parameters of the LCModel. This implies that the LCModel is nonlinear even if everything else were dependent upon the unknown parameters in a linear way. Prior to using the LCModel in each laboratory, one must measure a set of the *in vitro* spectra on a selected phantom, which should match as closely as possible the human or animal organ whose *in vivo* spectrum will be subsequently subjected to signal processing. Then a mathematical model is set up as a linear combination of these measured *in vitro* spectra from metabolite solutions. The unknown constants in such a sum are the metabolite concentrations, phases, line broadening shifts, line shapes and baseline polynomial coefficients. The latter coefficients are used to construct a cubic B-spline piece-wise polynomial for description of the background contribution to the given spectrum. Resonances are superimposed on top of this rolling baseline of the spectrum. The above sum of the *in vitro* model spectra is fitted to the given *in vivo* spectrum by using the nonlinear Levenberg–Marquardt [194] iterative algorithm with the Newton–Gauss type of a constrained least-square minimization. The final results of this fitting are the *in vivo* spectral parameters that were the mentioned unknown constants of the LCModel.

The common limitation of all nonlinear fits is *non-uniqueness*. This is to say that, within the same prescribed uncertainty, a measured spectral structure could be fitted by, e.g. two, three or four shape lines (Gaussians and/or Lorentzian, etc), without a sound criterion determining which of the fits should be retained in the end of the analysis. Such a lack of a criterion of established validity could readily yield spurious metabolites (over-fitting) or cause some genuine metabolites to pass undetected (under-fitting). Either outcome could lead to difficulties that would inevitably limit diagnostics. The capabilities of the LCModel are in the absence of assumptions about any pre-assigned line shape of the studied spectrum and also in the usage of the Levenberg–Marquardt nonlinear optimization for determination of the peak parameters. One of the limitations of the LCModel is that it necessitates the prior information on model *in vitro* spectra from data bases acquired by a subsidiary measurement on the related phantom. If the existing data banks do not contain the needed model spectra, supplementary encoding is necessary for the purpose of initiating the LCModel. Rather than carrying out these additional measurements on phantoms, Monte Carlo simulations have occasionally been employed [12, 13] for creating data banks that would provide the Levenberg–Marquardt algorithm with alternative sources of the initial spectra. In the LCModel the number of metabolites N_M is fixed in advance and it is usually set to be equal to 15 or so. It could be better to consider this number N_M as one of the unknown parameters of the problem. A suitable pre-assigned number N_{M_0} could play the role of the initial value, in which case the final number N_M would come out from the list of the optimized parameters from the Levenberg–Marquardt algorithm. Even with this potential refinement which has not been implemented in the literature thus far, the ensuing number N_M of estimated metabolites could at best be only a rough approximation, which is controlled solely by the appearance of the FFT shape spectrum, rather than the genuine information hidden in the measured time signal. By contrast, this true information about the exact number of resonances or metabolites is extractable directly from the raw FID by one of the variants of the Shanks transform [72] which belongs to the Padé methodology [19]. Practice has shown that removing, e.g. only one metabolite from an *in vitro* spectra leads to poor fits in a part of the *in vivo* spectrum where the missing model resonance normally occurs [11]. An obscure interpretation for this latter situation has been put forward stating that 'there should be no reason for such a removal, since generally a metabolite which has a good possibility of being detected should be included in the basis set' [11]. This interpretation itself is at variance with a proclaimed objectivity of the LCModel [11]. In reality, what one hopes to obtain from an *in vivo* spectrum by using the LCModel is directly proportional to what one puts by hand into the basis set through *in vitro* model spectra as has recently been emphasized [26]. Hence subjectivity of the LCModel. Even in normal tissue there is a noticeable degree of variation of metabolite concentrations that could roughly be estimated from the product of the peak hight and peak width. In tumorous tissue some of the metabolite resonances might be unidentifiable from a spectrum, since they could be buried in noise or found at some other shifted positions or detected with heights and widths having quite different values from those in normal tissues. This is especially true for the two lactate resonances that are usually difficult to see due to a large background in spectra of the brain gray matter measured at a long echo time (TE), e.g. TE = 136 ms or TE = 272 ms as analysed in [14]. In such a case, no spectral fitting techniques would be adequate, including the LCModel. Lactate and lipid resonances are clustered together, but the relaxation times of the latter are shorter than those of the former. Hence, using larger values of TE, as alluded to above, appears to be better, since it would allow the lipid resonances to decay and this should facilitate the identification of the lactate peaks. On the other hand, a larger TE would enter the tail of the FID envelope where the the signal is weak and, as such, comparable with the intensities of random noise. No fitting of FFT spectra of this type should be expected to give any reliable results. Therefore, an alternative to fitting of low-resolution FFT spectra is sought. Such an alternative must be capable of identifying even the weakest resonances and, e.g. the Padé-type methods are well suited for this purpose [27]–[29]. As already mentioned, in a normal spectrum of the healthy brain gray matter, the lactate and lipid resonances are practically unrecognizable from the background. However, lactate elevations are discovered in a number of brain disorders, e.g. ischemia, progressive leukodystrophies, severe inflammation, malignancies, etc. Lipid enhancements are also often related to tumours, multiple sclerosis plaques, etc. Precise theoretical estimators of all the peak parameters of other resonances are also needed in helping diagnostics in the case of data recorded at short values of TE. For example, at TE $\sim$ 20 ms, the detection of a mere decrease of the height of the NAA (Nitrogen–Acetyl–Aspartate) metabolite, which is the strongest resonance after the giant water peak, could be diagnostically important. Such an indicator from cerebral proton MRS spectra could be advantageously available before any pathological alteration has been observed on the corresponding anatomical scans from MRI.

The field of MRS needs a more vigorous access to methods that are alternative to fitting devices and these can be found within a number of parametric estimators of spectra [108]–[116]. One such estimator is the PSA [18] which can be used both as a parametric and/or spectral processor. The parametric version of the PSA could be employed in, e.g. MRS which aims at quantification of spectra, yielding the precise quantitative characteristics of the principal resonances via determination of their positions, widths, heights, phase, relaxation times, etc. The non-parametric version of the PSA is pertinent to, e.g. MRI where one is not interested in evaluating the numerical values of the reconstructed features, but rather attention is focused onto the anatomy of the imaged organ of the human body. There is also a newly emerging hybrid discipline MRSI. Here, one is interested in topological structures of the imaged subject in a chosen segment as well as in quantitative evaluations of the purely spectroscopic type of the selected slices. Both of these important diagnostic tasks in current medical practice could be successfully accomplished by the application of, e.g. the mentioned two versions of the PSA. The DPA [9, 115] is also a parametric and a non-parametric method, but it cannot cover the whole Nyquist range in one run, since it relies upon windowing, as do the FD, DSD or DLP processors. The common price paid by FD, DSD, DPA and DLP methods for their achieved dimensionality reduction of large data matrices associated with long signals is limitation to only local spectral analysis in one run. The PSA uses no windowing or decimation and, therefore, both local and global spectral analyses are possible in a single application of the method without patching together the separate frequency intervals. These four methods (FD, DSD, DLP and DPA) refer to windowed, i.e. local spectra in a selected frequency interval $[\omega_{min}, \omega_{max}]$ which is a part of the the full Nyquist range $[-\pi/\tau, +\pi/\tau]$. Every type of windowing in either the time or frequency domain ignores the possible effects of frequencies outside the chosen local range $[\omega_{min}, \omega_{max}]$ onto the informational content within this interval of interest. Such effects exist especially for *in vivo* signals encoded by e.g. proton MRS from human brain where detection of dilute metabolites is required in the presence of a giant water resonance which is about 10^4 times larger than other peaks in the corresponding frequency spectrum. To cover the entire Nyquist range, the windowing methods perform the spectral analysis window-by-window, such that all the local spectra are afterwards glued together in the consecutive order to yield the complete spectrum. This window patching often leads to some unwanted edge effects requiring special care to avoid artefacts, e.g. aliasing phenomena, the Gibbs oscillations and the like [64]. Such occurrences are enhanced for overly narrow frequency intervals imposed by exceedingly long time signals and it is, therefore, desirable to have a generic method, such as PSA, which does not need to resort to windowing.

The experimental resolution power of many instruments such as spectrometers is limited in part by a particular theory, which is the FFT. This method should be complemented by other more powerful high-resolution parametric estimators. The latter processors are able to extract, directly from the measurements of $\{c_n\}$, without any post-processing fits, the main spectral features, i.e. information about resonances. For example, in the parametric version of the PSA, the final outcome is a set of tabular numerical values of the peak parameters of each resonance found in the analysis using only the raw signal points $\{c_n\}$. This provides an alternative to fitting procedures that are customarily in use within MRS. When dealing with conventional spectrometers, one is usually unaccustomed to such tables of spectral results, since the first thing which is viewed directly on the screen is the magnitude or power spectrum. In such a case, if the peak parameters are required, one would fit each of the FFT peaks to a pre-assigned line shape comprised of, e.g. one or more Lorentzian and/or Gaussians[3]. As emphasized earlier, such a standard post-processing analysis requires prior information, which is the knowledge of the initial values of the parameters for each individual resonance before undertaking the task of the peak parameter search. By contrast, the PSA does not necessitate any prior knowledge, since the fundamental frequencies *as well as their total number*, and the corresponding amplitudes, are extracted directly from the raw signal. Of course, the peak

---

[3] One of the alternatives to this peak-by-peak fit would be adjustments of a linear combination of a premeasured sequence of spectra to a studied spectrum as done, e.g. in the LCModel [11].

parameters obtained in such a computation can be used to construct spectra in various modes, e.g. magnitude, power, absorption, etc. This change of strategy via 'extracting first all the relevant peak parameters from the signal and then constructing any desired spectrum', constitutes a fundamental advantage of the class of parametric methods [108]–[116], relative to the standard Fourier-based spectroscopy and the accompanying nonlinear fittings [11, 194]. Using various initial guesses while post-processing FFT spectra through a nonlinear fitting, could produce significantly different final outcomes that might deviate from the expected exact results. Furthermore, often very complicated and congested spectra with tightly packed peaks and overlapping resonances are encountered in MRS, hampering reliable fittings and this reduces fidelity in the whole processing via adjustable parameters. Such fittings become particularly difficult for degenerate or quasi-degenerate spectra, where several resonances can emerge with nearly equal frequencies, as is actually the case with a number of molecules[4]. Moreover, the literature witnessed no attempts thus far by the LCModel aimed at quantification of 2D-MRS cross-correlation plots which, by contrast, have recently been successfully quantified within the Padé method [24]. These listed drawbacks of the LCModel are clearly absent from the PSA which invokes no adjustable quantities whatsoever and, therefore, requires no initial values for the sought peak parameters. Of course, this is also the case with other parametric, non-iterative estimators, e.g. the FD [108]–[112], DSD, DPA or DLP [115], but the present emphasis is on the Padé methodology due to its unique reliability in handling the spurious roots entirely by analytical means, i.e. not through some numerical procedures. This is operationalized by one of the Padé algorithms, the already mentioned PSA, which can describe both Lorentzian (non-degenerate) and non-Lorentzian (degenerate) spectra. Experimentally measured spectral profiles could behave like the Gauss or the Voigt distribution functions [195]–[202]. It should be pointed out that it is possible to extend the PSA and encompass the Gauss and/or Voigt shape functions [142].

In the present illustrations, the Padé approximant is used for an accurate quantification of proton 1D-MRS spectra of the occipital gray matter of the human brain. The corresponding FID have been measured at the magnetic field strengths of 7T [13]. We obtain maximally accurate peak parameter values by relying only upon the measured raw FID with no other prior information and without zero-filling, preprocessing or post-processing the data [16]–[29]. The baseline of the FFT-MRS spectrum is considerable [13]. Nevertheless, we do not resort to any separate modelling for the background to account for this baseline, but nevertheless the complete spectral information is unfolded from the experimentally measured FID. *Before* drawing a graph to construct the shape of the investigated spectrum, all the metabolites are unequivocally identified with their corresponding peak parameters for every individual resonance. The total number of resonances is also one of the outcomes of the analysis within the

---

[4] An example of a molecule with a quasi-degenerate spectrum is CO_2.

Padé approximant. This number is determined exactly by a precise determinantal (Hankel) condition from the system of linear equations encountered in the theoretical development.

Chapter 2

Auto-correlation functions

In the present book, we focus on a unified theoretical treatment of collisions and spectroscopy using the concept of the auto-correlation function $C(t) = (\Phi(0)|\Phi(t))$ with a natural link to signal processing. From the onset, this strategy is rooted in basic quantum mechanics [143, 144] and quantum-scattering theory [145]–[157], since the state vector $|\Phi(t))$ is the solution of the time-dependent Schrödinger equation[1] $\hat{\Omega}|\Phi(t)) = i(\partial/\partial t)|\Phi(t))$. In quantum mechanics of genuine bound states and a pure continuum, the dynamics of a considered physical system is described by a Hamiltonian $\hat{H}$, which as a Hermitean operator $\hat{H}^\dagger = \hat{H}$ coincides with $\hat{\Omega}$ in the above Schrödinger equation for $|\Phi(t))$. The eigenvalues of a Hermitean operator are real. However, to include resonances via complex energy eigenvalues, as the most important part of scattering, one can generalize the notion of a 'Hamiltonian' and extend it to encompass a non-Hermitean dynamic operator $\hat{\Omega}$ in the same Schrödinger formalism. Hereafter, non-Hermiticity of $\hat{\Omega}$ might imply that the scalar product in the vector space $\mathcal{H}$ is defined as the symmetric inner product $(\zeta|\xi) = (\xi|\zeta)$. No conjugation via the star superscript is placed onto either of the two state vectors or 'orbitals' $(\zeta|$ or $|\xi)$ that both belong to $\mathcal{H}$. To symbolically indicate this special feature of the symmetry of the scalar product, soft round brackets $|\cdots)$ and $(\cdots|$ are used in place of the usual Dirac 'bra-ket' notation $\langle\cdots|$ and $|\cdots\rangle$ with $\langle\zeta|\xi\rangle = \langle\xi^*|\zeta^*\rangle$ and $\langle\zeta^*|\xi\rangle = (\zeta|\xi)$. An abstract formulation of quantum mechanics and its deterministic postulate imply that, if the wave packet $|\Phi(0))$ of *any generic system* under study is well prepared/controlled at

---
[1] As Pauli [203] remarked, since relativity and quantum mechanics should be among the hypotheses of every adequate physics theory, it is plausible to redefine their respective basic quantities c and $\hbar = h/(2\pi)$ as unity, where c is the speed of light in vacuum and h is the Planck constant. In the ensuing so-called natural units ($c = 1 = \hbar$) all other physical quantities will have the dimensions of a power of length. If such reduced dimensions are multiplied by the requisite powers of c and $\hbar$, the true dimensions of the given quantity would follow. For example, in the natural units, the energy E has the dimension cm^{-1}. However, in reality, the energy divided by $c\hbar$ is a quantity with the dimension cm^{-1}. Throughout this book, unless otherwise stated, the natural units (also called the atomic units) $c = 1 = \hbar$ will be employed. This is convenient especially for the present analysis, since we can interchangeably use the frequency (ω) and energy (E) as two synonyms, $E = \hbar\omega = \omega$.

the initial time $t = 0$, and if its further development is propagated by the given dynamics/interactions, then the state $|\Phi(t)\rangle$ will be known exactly at any later instant t [157]. The state $|\Phi(t)\rangle$ is an element of the abstract vector space $\mathcal{H}$. Since $|\Phi(t)\rangle$ is used to derive $C(t)$, it follows that this latter quantity also represents an abstract concept. This in itself means that the auto-correlation functions $C(t)$ are independent of the origin from which they are generated and, therefore, could be computed theoretically or measured experimentally as, e.g. time signals $c(t)$. Moreover, the mathematical equivalences $C(t) = c(t)$ and $C_n = c_n$ exist for both the continuous (t) and discrete ($t = t_n \equiv n\Delta t$) variables[2], if a given time signal is a sum of damped (attenuated) complex exponentials that yield a spectrum as a linear combination of pure Lorentzians [108]–[115]. In either case, the auto-correlation functions $C(t)$ or the time signals $c(t)$ physically represent the instantaneous *survival probability amplitude* of the corresponding time-dependent state or wave packet $|\Phi(t)\rangle$ of the examined system. This is important for at least two reasons: (1) experimental raw signals $c(t)$ can be used directly, without necessarily relying upon the theory, to deduce *by computations* the basic observables for scattering and spectroscopy, such as cross sections, rate coefficients or the like, and (2) measured time signals $c(t)$ that are also identifiable as counts per channel, can be directly and dynamically intertwined with the theory on a deeper fundamental level to yield more valuable spectral information.

It should be emphasized that the concept of auto-correlation functions first emerged as a computational tool, but soon afterwards surpassed its initial purpose. This is because the auto-correlation functions represent a veritable alternative formulation of quantum mechanics and quantum-mechanical resonant scattering theory. Many of the major observables, e.g. complete energy spectra, local density of states, quantal rate constants and other related quantities are expressible through auto-correlation functions or their suitable transforms. Other important observables could be given completely in terms of some appropriate, relatively small informational parts that might be singled out and analysed separately from the unwanted/redundant remainder of the full data set of auto-correlation functions. In order to theoretically generate auto-correlation functions $C(t)$, that otherwise play a critical role in analysis of spectra of physical systems, all the parametric methods require that the peak parameters $\{\omega_k, d_k\}$ are computed first as pairs of complex numbers. These are the fundamental frequencies $\{\omega_k\}$ and the associated residues $\{d_k\}$ that constitute the natural harmonics whose linear combination of K terms represents the building block of every individual time signal point c_n from the set $\{c_n\}$. Here, the elements of the triple $\{\text{Re}(\omega_k), \text{Im}(\omega_k), |d_k|\}$ are the position, width and height of the kth peak/resonance. One can use either the dynamic matrix $\boldsymbol{\Omega}$ or the related evolution/relaxation matrix $\mathbf{U} = \exp(-i\boldsymbol{\Omega}\tau)$ to generate the corresponding Jacobi matrix $\mathbf{J}$ whose diagonalization [176]–[179] can give the required spectral

---

[2] Here, $C_n \equiv C(n\tau)$ and $c_n \equiv c(n\tau)$, where $\tau = \Delta t$ is the sampling time, i.e. the segment between any two equidistantly sampled points on the real time axis t.

set $\{\omega_k, d_k\}$. However, this is not the only way as, in fact, the non-parametric version of the FPT is *also* able to generate any auto-correlation function C_n [18]. Such a version of the FPT does not compute the pair $\{\omega_k, d_k\}$ at all. Instead, it merely generates the general expansion coefficients $\{a_n\}$ in the continued fractions (CF) of any finite order. Nevertheless, it has been shown [18] that this is fully sufficient for the exact determination of the auto-correlation functions $\{C_n\}$ *without ever rooting* the characteristic equation $Q_K(\omega) = 0$, which otherwise gives the eigenvalues $\{\omega_k\}$ of $\hat{\Omega}$. Our procedure represents an important advantage relative to the current practice in many parametric estimators, e.g. the FD [108]–[112], the DSD, DLP or DPA [113]–[115], the Lanczos algorithm [181] where the eigenvalues and residues $\{\omega_k, d_k\}$ are determined first before constructing the auto-correlation functions $\{C_n\}$.

Chapter 3

The time-independent Schrödinger equation

Irrespective of whether one is concerned with spectroscopy or collisions, the whole physics of any given system is ingrained in one single quantity, which is the full Green function $(\zeta|\hat{G}(\omega)|\xi)$ in a given representation $\{(\zeta|, |\xi)\}$ with $\hat{G}(\omega)$ being a resolvent known as the Green operator [145]–[157]

$$\hat{G}(\omega) = \frac{1}{(\omega + i\eta)\hat{1} - \hat{\Omega}}. \qquad (3.1)$$

Further, $\hat{\Omega} \in \mathcal{H}$ is a dynamical operator which governs the development of the studied physical system and $\mathcal{H}$ is the underlying vector space of operators and state vectors. In quantum mechanics, operator $\hat{\Omega}$ is the standard Hamiltonian $\hat{H}$ and in classical physics it is a Lagrangean or Liouvillian, etc. Operator $\hat{\Omega}$ need not be Hermitean and, thus, we shall consider $\hat{\Omega}^{\dagger} \neq \hat{\Omega}$ throughout, so that generally the eigenvalues $\{\omega_k\}_{k=1}^{K}$ are complex numbers. The total number K of frequencies $\{\omega_k\}$ is any finite or infinite positive integer. The FPT determines this number exactly from the uniqueness conditions of the Padé polynomial quotient for the Green function, which is a power series with the given time signal points as the expansion coefficients. This is in sharp contrast to guessing K as customarily done in many other parametric estimators, e.g. LP, HLSVD, etc. Even the usual Hamiltonians are often converted to complex operators through the concept of non-Hermitean optical-type absorbing potentials that could mimic absorption of the incoming particle flux by the target. Customarily, a complex 'Hamiltonian' is constructed by introducing a purely imaginary absorbing potential $i\mathrm{W}$ according to the substitution $\hat{\Omega} \longrightarrow \hat{\Omega} + i\mathrm{W}\hat{1}$. In the interaction region, the value W is selected to be negligibly small and it is taken as being positive in the asymptotic region of scattering. The net effect of these so-called absorbing boundary conditions (ABC) [204]–[206] is to remove, i.e. to absorb the outgoing wave packet corresponding to direct scattering states lying in the asymptotic domain. The potential $i\mathrm{W}$ is otherwise artificial and rather arbitrary. What matters is that iW does not change the physics at all. Rather, a flexible choice of W could help numerical computations. The introduction of complex 'Hamiltonians' is essential for treating resonances that are inherently different from genuine (physical) bound states. True square-integrable bound-state eigenvectors are associated with discrete negative eigenenergies and Hermitean Hamiltonians. Resonant states are also square-integrable, but they are linked to complex energies with positive real parts that belong to a continuum. Continuum states of Hermitean Hamiltonians are not normalizable and, therefore, do not represent physical states, so that they cannot describe particles. The concept of the ABC comes to rescue the situation via the introduction of complex 'Hamiltonians' with normalizable scattering wavefunctions as proper physical states. Note that complex 'Hamiltonians' could also follow from the Dyson device which is based upon the so-called adiabatic theorem from formal scattering theory [145]–[157]. This amounts to a modification consisting of including the damping factor $i\eta$ from $\hat{G}(\omega)$ directly into the total interaction potential $\hat{V} \longrightarrow \hat{V}e^{i\eta}$ which is a part of $\hat{\Omega}$. Alternatively, one could use the well-known rotated coordinate method (RCM) [207]–[210], which is also called the complex scaling method, to construct a complex 'Hamiltonian' $\hat{\Omega}$, which is non-Hermitean[1]. The crucial practical advantage of these circumstances that render operator $\hat{\Omega}$ complex is that its spectrum does not need to include explicitly continuum states that are known as difficult to handle in computations. Such states could be approximately represented by a pseudo-continuum, which is a collection of pure discrete states at complex energies encompassing both bound states and resonances. A pseudo-continuum or discretized continuum can be built from complex energies as a surrogate for a true continuum. For example, the behaviour of the Green function $\hat{G}(\omega)$ in the close vicinity of cuts or branch point singularities, can be described remarkably well by a sequence of poles of the FPT for $\hat{G}(\omega)$. This is essential, since no theory can pretend to describe scattering phenomena unless cuts and branch point singularities are treated in a satisfactory manner. An approximate description of a genuine continuum can also be achieved by the RCM in which a mapping from the real radial coordinate r to its complex counterpart $re^{i\theta}$ with a real scaling parameter θ leads to a non-Hermitean Hamiltonian $\hat{H}^{\dagger}(\theta) \neq \hat{H}(\theta)$. The full widths at half maxima of all the spectral peaks, that are embedded in a continuum to represent localized positive energy wave packets, determine the inverse lifetimes of every resonance in the spectrum of $\hat{\Omega}$. The infinitesimal number $\eta > 0$ secures regularity of $\hat{G}(\omega)$ for those values of ω that belong to the set of the eigenfrequencies $\{\omega_k\}$ of $\hat{\Omega}$. Once the calculation has been completed, the limit $\eta \rightarrow 0^+$ or $\eta \rightarrow 0^-$ should be taken depending upon whether the outgoing or incoming boundary conditions are imposed. The superscripts $\pm$ indicate that η should tend to zero through positive/negative numbers, respectively. In principle, we could omit η hereafter

---

[1] It has been often been stated that the RCM was introduced in 1971 by Baslev and Combes [210]. However, in reality, the RCM has been published at least three times before 1971, e.g. in 1960 by Zel'dovich [207], in 1961 by Dykhne and Chaplik [208] and in 1964 by Lovelace [209].

The time-independent Schrödinger equation

with the understanding that ω will stand for $\omega + i\eta$. However, this is unnecessary and we shall keep considering ω as being purely real, since $\hat{\Omega}$ will be conceived as a complex 'Hamiltonian'. In such a case, the eigenfrequencies ω_k of $\hat{\Omega}$ are complex valued, so that the resolvent $1/(\omega\hat{1} - \hat{\Omega}) = \sum_{k=1}^{K} \hat{\pi}_k/(\omega - \omega_k)$ is not singular, provided $\text{Im}(\omega_k) \neq 0$, with $\hat{\pi}_k$ being the projection operator $\hat{\pi}_k = |\Upsilon_k)(\Upsilon_k|$ where $|\Upsilon_k)$ is the complete eigenfunction of the studied system obeying the stationary Schrödinger equation $\hat{\Omega}|\Upsilon_k) = \omega_k|\Upsilon_k)$. Clearly, if $\text{Im}(\omega_k) = 0$, the Dyson damping $i\eta$ should be reintroduced.

A method which can provide an adequate spectral representation of the total Green operator or resolvent $\hat{G}(\omega)$ would be one of the key inputs into an invaluable practical quantum-mechanical theory for scattering ($E \geq 0$) *and* spectroscopy ($E < 0$). If the operator $\hat{G}(\omega)$ is available then all the observables for a collisional and a spectroscopic phenomenon could be obtained from the general Green function $\mathcal{G}_{if}(\omega) = (\Phi_{0f}|\hat{G}(\omega)|\Phi_{0i})$. The diagonal elements are obtained for $|\Phi_{0i}) = |\Phi_{0f}) \equiv |\Phi_0)$ as $\mathcal{G}_{00}(\omega) \equiv \mathcal{G}(\omega)$

$$\mathcal{G}(\omega) \equiv (\Phi_0|\hat{G}(\omega)|\Phi_0) \quad (3.2)$$

where $|\Phi_0)$ is the initial state of the system. It is possible to show that even the state vector $|\Phi_0)$ can be reconstructed from the known $\{C_n\}$ or $\{c_n\}$ [142]. Hereafter, we shall use the symmetric definition of the scalar product

$$(\zeta|\xi) = (\xi|\zeta) = \int dV \zeta\xi \quad (3.3)$$

without conjugation of either of the two functions $(\zeta|$ and $|\xi)$. Here, dV is the differential of the volume element. The so-called local density of states $\rho_{00}(\omega) \equiv \rho(\omega)$ can be computed from the residues of $\mathcal{G}(\omega)$ at its singularities

$$\rho(\omega) = -\frac{1}{\pi} \lim_{\eta \to 0^+} \text{Im}\{\mathcal{G}(\omega)\}. \quad (3.4)$$

An entirely generic system is presently subjected to our analysis by the general stationary and time-dependent methods of quantum mechanics. In the stationary methods, the complete set of eigenstates $\{|\Upsilon_k)\}$ of the Schrödinger operator $\hat{\Omega}$ will be available by obtaining the pair of the solutions $\{\omega_k, |\Upsilon_k)\}$ of the time-independent eigenvalue problem $\hat{\Omega}|\Upsilon_k) = \omega_k|\Upsilon_k)$ or by solving a more general eigenequation $f(\hat{\Omega})|\Upsilon_k) = f(\omega_k)|\Upsilon_k)$

$$\hat{\Omega}|\Upsilon_k) = \omega_k|\Upsilon_k) \qquad f(\hat{\Omega})|\Upsilon_k) = f(\omega_k)|\Upsilon_k) \quad (3.5)$$

where $1 \leq k \leq K$ and $f(\hat{\Omega})$ is any operator analytic function. *The main postulate of quantum mechanics is that the whole information of a general system under study is contained in the total wavefunctions* $\{|\Upsilon_k)\}$. This is known as completeness of the quantum-mechanical description of phenomena in nature. Such a circumstance is coherent with (3.5), since $\hat{\Omega}$ is assumed to carry the whole information of the investigated object. This is formally expressed through the local closure relation for the exact orthonormalized basis $\{|\Upsilon_k)\}$

$$\sum_{k=1}^{K} \hat{\pi}_k = \hat{1} \qquad \hat{\pi}_k = |\Upsilon_k)(\Upsilon_k| \qquad (\Upsilon_{k'}|\Upsilon_k) = c_0 \delta_{k,k'} \qquad (3.6)$$

where $\hat{\pi}_k$ is the projection operator and $\delta_{k,k'}$ is the Kronecker δ-symbol

$$\delta_{k,k'} = \delta_{kk'} = \begin{cases} 1 & k' = k \\ 0 & k' \neq k. \end{cases} \qquad (3.7)$$

Notice that the local completeness is used in (3.6) by limiting the summation to the K terms only. Nevertheless, as already pointed out, the non-negative integer K need not be finite in the subsequent analysis. The norm $\|\Upsilon_k\|$, as the number whose square is defined by the scalar product of $|\Upsilon_k)$ with $(\Upsilon_k|$, is set to be equal to $c_0 \neq 0$

$$\|\Upsilon_k\|^2 \equiv (\Upsilon_k|\Upsilon_k) = c_0 \qquad (k = 1, 2, 3, \ldots). \qquad (3.8)$$

This norm is not necessarily positive definite and it could be even complex for a complex c_0. The sum over k in (3.6) should, in principle, include integration over the continuum part of the spectrum of $\hat{\Omega}$. This is presently omitted, since resonances are taken into account through the spectrum of the non-Hermitian dynamic operator $\hat{\Omega}$. Inserting the representation (3.6) of the unit operator $\hat{1}$ into $\hat{G}(\omega) = \hat{G}(\omega)\hat{1}$ leads at once to the following spectral representation of the Green resolvent $\hat{G}(\omega)$ and the Green function $\mathcal{G}(\omega)$ from (3.2)

$$\hat{G}(\omega) = \sum_{k=1}^{K} \frac{\hat{\pi}_k}{\omega - \omega_k} \qquad \mathcal{G}(\omega) = \sum_{k=1}^{K} \frac{d_k}{\omega - \omega_k}. \qquad (3.9)$$

Here, d_k are the complex amplitudes representing the residues associated with the eigenfrequencies $\{\omega_k\}$ that are the poles of the Green function $\mathcal{G}(\omega)$

$$d_k = (\Phi_0|\Upsilon_k)^2 \qquad (3.10)$$

where (3.3) is employed with the property $(\Phi_0|\Upsilon_k) = (\Upsilon_k|\Phi_0)$. Let us temporarily assume that we are given the eigensolutions $\{\omega_k, |\Upsilon_k)\}$, but *not* the 'pseudo-Hamiltonian' $\hat{\Omega}$ itself. Then, $\hat{\Omega}$ is retrieved from its implicit definition

$$\hat{\Omega} = \sum_{k=1}^{K} \omega_k \hat{\pi}_k \qquad f(\hat{\Omega}) = \sum_{k=1}^{K} f(\omega_k) \hat{\pi}_k \qquad (3.11)$$

where $f(\hat{\Omega})$ is any operator analytic function. In the particular case with $f(\hat{\Omega}) = (\omega\hat{1} - \hat{\Omega})^{-1} = \hat{G}(\omega)$, the result (3.9) follows again from (3.11). According to the Cauchy theorem, the local spectral representation of the Green operator $\hat{G}(\omega)$ is fully determined by the set $\{\omega_k, \hat{\pi}_k\}$ of its singularities $\{\omega_k\}$ and the corresponding operator residues $\{\hat{\pi}_k\}$.

Chapter 4

The time-dependent Schrödinger equation

A formalism which is equivalent to the preceding stationary treatment is provided by the time-dependent Schrödinger equation [143]–[157]

$$i\frac{\partial}{\partial t}|\Phi(t)\rangle = \hat{\Omega}|\Phi(t)\rangle. \tag{4.1}$$

In the Schrödinger picture of quantum mechanics, operators are stationary and wavefunctions are time-dependent. The Schrödinger states $|\Upsilon\rangle$ and $|\Phi\rangle$ are interrelated by the usual one-sided Fourier integral with the semi-infinite limits [145, 157]

$$|\Upsilon(u)\rangle = \frac{1}{2\pi}\int_0^\infty dt\, e^{i\omega t}|\Phi(t)\rangle \qquad |\Upsilon_k\rangle = \frac{1}{2\pi}\int_0^\infty dt\, e^{i\omega_k t}|\Phi(t)\rangle \tag{4.2}$$

where $|\Upsilon_k\rangle \equiv |\Upsilon(u_k)\rangle$. Here $u = \exp(-i\omega\tau)$ and $u_k = \exp(-i\omega_k\tau)$ are present through $\exp(i\omega t)$ and $\exp(i\omega_k t)$ written as $\exp(i\omega t) = u^{-t/\tau}$ and $\exp(i\omega_k t) = u_k^{-t/\tau}$, respectively. For the stationary ansatz $\hat{\Omega}$ we know that (4.1) possesses the solution of the type

$$|\Phi(t)\rangle = \hat{U}(t)|\Phi_0\rangle \qquad |\Phi_0\rangle \equiv |\Phi(0)\rangle \tag{4.3}$$

where $\hat{U}(t)$ is the dynamical evolution/relaxation linear operator of the system

$$\hat{U}(t) = e^{-i\hat{\Omega}t}. \tag{4.4}$$

In (4.3), $|\Phi_0\rangle$ represents the initial unnormalized state of the studied system at the time $t = 0$

$$\|\Phi_0\|^2 \equiv (\Phi_0|\Phi_0) = C_0 = c_0 \neq 0 \tag{4.5}$$

where the constant c_0 from (3.8) is not necessarily unity, but rather it could be a complex number. These preliminaries imply that, if $\hat{\Omega}$ and $|\Phi_0\rangle$ are known, then the determinism of quantum mechanics prescribes the exact knowledge of the state $|\Phi(t)\rangle$ of the system at any later instant $t > 0$. The spectral representation of the operator $\hat{U}(t)$ follows from using (3.11) where we set $f(\hat{\Omega}) = e^{-i\hat{\Omega}t}$

$$\hat{U}(t) = \sum_{k=1}^{K} e^{-i\omega_k t} \hat{\pi}_k \qquad \text{Im}(\omega_k) < 0. \tag{4.6}$$

The state $|\Phi(t)\rangle$ at the instant t is obtained by propagating the initial well-prepared wave packet $|\Phi_0\rangle$ from $t = 0$ to t via $\hat{U}(t)$. Given $|\Phi_0\rangle$ at $t = 0$, there will be a non-zero probability amplitude to find the system in the state $|\Phi(t)\rangle$ at the later time $t > 0$, if the two wave packets have a non-vanishing overlap. This overlap is found quantum-mechanically by projecting $|\Phi(t)\rangle$ onto $\langle\Phi_0|$ by means of their scalar product

$$C(t) = \langle\Phi_0|\Phi(t)\rangle = \langle\Phi_0|\hat{U}(t)|\Phi_0\rangle \tag{4.7}$$

where (4.3) is used. The quantity $C(t)$ is called the auto-correlation function, since it measures the degree of correlations between the states $|\Phi(t = 0)\rangle$ and $|\Phi(t \neq 0)\rangle$ under the influence of the dynamical operator $\hat{\Omega}$. It is the presence of the operator $\hat{\Omega}$ which makes $|\Phi(t)\rangle$ differ from $|\Phi_0\rangle$, as is obvious from (4.3). By switching the dynamics off ($\hat{\Omega} = \hat{0}$), the system would be allowed to remain indefinitely in the initial state, so that $C(t) = \langle\Phi_0|\Phi_0\rangle = c_0 \neq 0$ as in (4.5) at any time t. The initial state $|\Phi_0\rangle$ is assumed to be a non-zero state vector $|\Phi_0\rangle \neq |0\rangle$, as is customary in quantum mechanics. Propagating the initial 'zero vector' $|\Phi_0\rangle = |0\rangle$ from $t = 0$ and onward would inevitably lead to the trivial Schrödinger state $|\Phi(t)\rangle = |0\rangle$ at any later time $t \neq 0$ and, therefore, this possibility is excluded as uninteresting. At two times t' and t where $t' < t$, the state vector $|\Phi(t')\rangle$ can be viewed as a delayed 'copy' of $|\Phi(t)\rangle$. Thus the overlap between the states $|\Phi(t)\rangle$ and $|\Phi(t')\rangle$ for $t' < t$, as per the inner product $\langle\Phi(t')|\Phi(t)\rangle$, represents a measure of correlation between the state and its delayed copy. One such example follows from taking the time t' all the way back to the initial moment $t' = 0$. This yields the auto-correlation function $C(t) = \langle\Phi(0)|\Phi(t)\rangle$.

At large times t, the auto-correlation function $C(t)$ could be numerically unreliable due to instabilities that stem from considerable oscillations of the overlap $\langle\Phi_0|\Phi(t)\rangle$ as t increases. This could cause a heavy corruption of the quantity $C(t)$ with computational noise, e.g. round-off errors, ill-conditioning, etc. In general, the Green function $\mathcal{G}(\omega)$ in (3.2) exhibits singularities (poles, cuts, etc) because of the presence of the resolvent operator $\hat{G}(\omega)$ from (3.1). It is often claimed in the literature that $C(t)$ is free from such singularities. This is untrue. The reason being that the severe oscillations of $C(t)$ for large t act as disguised singularities entirely similar to those encountered more transparently in $\mathcal{G}(\omega)$ [211]. This is obvious from the fact that both functions $C(t)$ and $\mathcal{G}(\omega)$ are built from the same 'Hamiltonian' $\hat{\Omega}$ which is a generator of an infinitesimal unitary transformation described by the evolution operator $\hat{U}(t)$ [157]. Moreover, the limits $t \longrightarrow \pm\infty$ in $C(t)$ are strictly equivalent to $\eta \longrightarrow 0^{\mp}$ in $\hat{G}(\omega)$ in accordance with the so-called Abel limit from the formal scattering theory [142]. As a matter of fact, if one does not encounter instabilities in $C(t)$ in producing a spectrum, this could only mean that the asymptotic region $t \longrightarrow \pm\infty$ has not been approximately reached, with the consequence that some of the longer-lived transients did not have sufficient time to decay [211].

The term 'transient' usually refers to a time developing phenomenon which dies out after a sufficiently long time lag has elapsed ($t \longrightarrow \infty$) [72]. Such are the envelopes of, e.g. experimentally encoded time signals as a linear combination of damped exponentials $\{\exp(-i\omega_k t)\}$ with constant amplitudes $\{d_k\}$ such that all complex frequencies $\{\omega_k\}$ must have the negative imaginary parts $\mathrm{Im}(\omega_k) < 0$. Strictly speaking, these are *stable* transients. A more general meaning of the term transient in sequence-to-sequence transformations has been encountered in quantum-mechanical signal processing [17, 18]. There, the term *unstable* or *secular* transients has been used following [72, 212] whenever referring to, e.g. a time signal $c(t)$ in which one or more fundamental frequencies ω_k are exponentially diverging ($\mathrm{Im}(\omega_k) > 0$). Any finite or infinite sum of harmonics $\{\exp(-i\omega_k t)\}$ would diverge if at least one of the complex fundamental frequencies is secular $\mathrm{Im}(\omega_k) > 0$. In such a case, this whole mentioned sum of exponentials $\{\exp(-i\omega_k t)\}$, i.e. the signal $c(t)$ would cease to have any direct physical meaning, since $|c(t)| \longrightarrow \infty$ as $t \longrightarrow \infty$. This situation will remain incurable as long as one keeps adding up directly the partial sums of the time signal, as a linear combination of exponentials $\{\exp(-i\omega_k t)\}$. However, such a divergent set of partial sums could still be computed with a finite result $|c(t)| < \infty$ by means of certain sequence-to-sequence nonlinear transformations that are capable of converting divergent into convergent sequences through the concept of the anti-limit [72]. One such transformation is the PA [65]–[92], which can also accelerate convergence of slowly converging multi-dimensional sequences and/or series [158], [213]–[216].

The above-introduced two concepts, the stationary and the time dependent representations, are interrelated by means of the exact one-sided Fourier integrals (*not* to be confused with the Fourier transforms)

$$\mathcal{F}(\omega) = \frac{1}{2\pi} \int_0^\infty dt\, e^{i\omega t} C(t) \qquad C(t) = \int_0^\infty d\omega\, e^{-i\omega t} \mathcal{F}(\omega). \qquad (4.8)$$

Both (3.1) and (4.4) contain the same operator $\hat{\Omega}$ which is the source of the complete information about the system. This feature, together with the unitarity and linearity of the standard Fourier operator, guarantees that the information is preserved when passing from the time to the frequency domain. Inserting (4.6) into (4.7) yields the result

$$C(t) = \sum_{k=1}^{K} d_k e^{-i\omega_k t} \qquad (4.9)$$

where the residues $\{d_k\}$ are given by (3.10). By definition, the quantities $\{d_k\}$ measure the extent of the squared projection of the state $|\Upsilon_k\rangle$ onto $|\Phi_0\rangle$. Thus the amplitudes $\{d_k\}$ are the weights that carry information about the strength of the contributions of individual normal/natural mode frequencies $\{\omega_k\}$ to $C(t)$ in (4.9). The magnitudes $\{|d_k|\}$ are the intensities of the non-orthogonal 'harmonics' $\{\exp(-i\omega_k t)\}$ featuring as the principal components that constitute the auto-correlation function $C(t)$. Moreover, $\phi_k = \text{Arg}(d_k)$ is the phase of $C(t)$. Substituting $C(t)$ from (4.9) into (4.8) and carrying out the time integral exactly gives a *theoretically generated* complex spectrum $\mathcal{F}(\omega)$ as

$$\mathcal{F}(\omega) = -i \sum_{k=1}^{K} \frac{d_k}{\omega - \omega_k} \qquad (4.10)$$

provided that $\text{Im}(\omega_k) < 0$. The result (4.10) for $i\mathcal{F}(\omega)$ agrees fully with the Green function $\mathcal{G}(\omega)$ from (3.9) as it should. The real quantities, e.g. the magnitude $|\mathcal{F}(\omega)|$, power $|\mathcal{F}(\omega)|^2$, absorption and dispersion spectra can all be obtained directly from $\mathcal{F}(\omega)$. Under ideal conditions of absence of noise and with no initial time delay, the absorption and dispersion spectra are given by $\text{Re}\{\mathcal{F}(\omega)\}$ and $\text{Im}\{\mathcal{F}(\omega)\}$, respectively. Of course, in practice, these conditions are not fulfilled and the absorption and dispersion spectra might differ from $\text{Re}\{\mathcal{F}(\omega)\}$ and $\text{Im}\{\mathcal{F}(\omega)\}$, respectively. In such cases certain alternative definitions of the absorption and dispersion spectra could be used [9, 115]. The resonance parameters from (4.10) are the position, width, height and phase of the kth peak given by $\text{Re}(\omega_k)$, $\text{Im}(\omega_k)$, $|d_k|$, $\text{Arg}(\omega_k)$, respectively. The auto-correlation function $C(t)$ from (4.7) is identified with an instantaneous transition probability amplitude for the passage of the system from $|\Phi(t = 0)\rangle$ to $|\Phi(t \neq 0)\rangle$. Then the survival probability amplitude for the state $|\Phi(t)\rangle$ can be obtained in the limit of $C(t)$ for large times $|t| \longrightarrow \infty$ [17]. Such a time limit is crucial for any collision problem in order to secure that the full scattering states are reduced to the appropriate *free* wave packets [157, 211]. In spectroscopy, the total acquisition time or epoch (T) of time-dependent observables should be sufficiently long to facilitate decays of all transient states, so that the physically relevant transitions could be unambiguously detected.

Chapter 5

Equivalence: auto-correlation functions and time signals

In practice, one equidistantly discretizes (digitizes) the continuous (analog) time variable t as $t = t_n = n\Delta t \equiv n\tau$ ($n = 0, 1, 2, \ldots, N-1$), where the non-negative integer n counts the time. The quantity τ is the time increment (the time lag) or the sampling time, which is also called the dwell time $\tau = T/N$ in ICR mass spectroscopy [7, 9]. As in chapter 2 we shall write

$$C_n \equiv C(n\tau) \qquad |\Phi_n) \equiv |\Phi(n\tau)) \tag{5.1}$$

for the discrete counterparts of $C(t)$ and $|\Phi(t))$, respectively. In practice, T is finite which implies that the time Fourier integrals from (4.8) should have the upper limits equal to T with $T < \infty$. In such a case, the Fourier integral $F(\omega)$ of the auto-correlation function $C(t)$ is introduced as

$$F(\omega) = \frac{1}{T} \int_0^T dt\, C(t)\, e^{i\omega t}. \tag{5.2}$$

This expression can be discretized if the integral in (5.2) is represented by its Riemann sum F_k evaluated at the Fourier grid $\omega = \tilde{\omega}_k$

$$F_k = \frac{1}{N} \sum_{n=0}^{N-1} C_n e^{in\tilde{\omega}_k \tau} = \frac{1}{N} \sum_{n=0}^{N-1} C_n e^{2i\pi nk/N} \qquad \tilde{\omega}_k \equiv \frac{2\pi k}{T} = \frac{2\pi k}{N\tau} \tag{5.3}$$

where ($0 \le k \le N-1$) and $F_k \equiv F(\tilde{\omega}_k)$. This is the complex form of the discrete Fourier *transform* (DFT) [64], which is defined only at the Fourier frequencies $\omega = \tilde{\omega}_k$. If we multiply (5.3) by $(1/N)\exp(-im\tilde{\omega}_k \tau)$ and sum the result over k from 0 to $N-1$, we shall retrieve all the elements from the original set $\{C_n\}$ through the inverse discrete Fourier transform (IDFT)

$$C_n = \sum_{k=0}^{N-1} F_k e^{-in\tilde{\omega}_k \tau} = \sum_{k=0}^{N-1} F_k e^{-2i\pi nk/N} \qquad (0 \le n \le N-1) \tag{5.4}$$

where the orthogonalization property of the harmonic basis set functions $\{|\exp(-i\tilde{\omega}_k\tau))\}_{k=0}^{N-1}$ is employed [64]

$$\frac{1}{N}\sum_{k=0}^{N-1} e^{2i\pi(n-m)k/N} = \delta_{n,m} \qquad (5.5)$$

where $\delta_{n,m}$ is the Kronecker δ-symbol from (3.7). The Fourier frequencies $\{\tilde{\omega}_k\}$ from (5.3) are purely real, so that the exponentials in (5.4) are unattenuated. Unlike (4.9), where the elements ω_k from the set $\{\omega_k\}$ are the unknown peak parameters, the frequencies $\tilde{\omega}_k$ from (5.3) are fixed in advance and this latter feature leads to linearity of both the DFT and the IDFT. The spectrum F_k^{bl}, which is equal to F_k within a fixed frequency interval/band and zero elsewhere, is called the band-limited (bl) Fourier transform [64, 113]. In (5.3) and (5.4) one encounters three sequences $\{C_n, F_k, |\exp(-i\tilde{\omega}_k\tau))\}_{n,k=0}^{N-1}$, each of which is of length N. A direct computation by means of the DFT would require N^2 multiplications that are drastically reduced to only $N\log_2 N$ multiplications in the fast Fourier transform (FFT) [176].

Here, it is pertinent to briefly recall several important efforts aimed at accomplishing the 'fast' $N^2 \longrightarrow N\log_2 N$ reduction. Certainly, significant computational costs required by N^2 multiplications were the main reason for the lack of widespread usage of the DFT up to 1965. Nevertheless, there was another reason which delayed computerized applications of the DFT, and that was a lack of awareness of certain of the most critical achievements in the Fourier analysis from the past. For example, in 1942 Danielson and Lanczos [46] showed in a transparent way, as recapitulated more recently in [53, 54, 176], that the N^2 computational complexity of the DFT can be drastically reduced to only $N\log_2 N$ multiplications, thus yielding the FFT, provided that the signal length N is a composite number, $N = 2^m (m = 0, 1, 2, 3, \ldots)$. Danielson and Lanczos [46] refer to the papers by Runge [44] from 1903 for the original source of their algorithm. In the period 1939–1963 there have been several other revivals of the $N^2 \longrightarrow N\log_2 N$ reduction with the ensuing transformation of the DFT to the FFT (Stumpff [45], Good [47], Thomas [48], etc). Despite this vigorous development, which originally started by Gauss in 1805 [43], and published in his collected works, it was not until 1965 that the FFT became widely known through its reinvention by Cooley and Tukey [49] who were apparently unaware of any of the previous related works from [43]–[48]. However, soon after publication of the paper by Cooley and Tukey [49], an important comment was made in 1966 by Rudnick [50] who pointed out that the $N^2 \longrightarrow N\log_2 N$ computational saving in the FFT had been achieved already in 1942 by Danielson and Lanczos [46]. Following Rudnick [50], in 1967 Cooley *et al* [51] attempted to make a chronological overview of the introductions of the FFT prior to Cooley and Tukey [49]. They stated that [49] was more general in that it considered other alternatives to $N = 2^m$ (e.g. N being a prime number), while still preserving the crucial $N\log_2 N$ expediency. However, subsequently, it has been categorically recommended, in e.g. *Numerical Recipes* [176], that one should use the FFT *only* with the signal length N being a power of two, in which case the algorithm from the work of Cooley and Tukey [49] coincides with that of Danielson and Lanczos [46]. It has been further pointed out in [176] that when N is not a power of two one should do either zero padding to the nearest power of two or use the LP method to predict the missing time signal points and thus achieve the needed length $N = 2^m$. Although Cooley and Tukey [49] were not the first to invent the FFT, they were the first in the modern computer era to succeed in reviving an unprecedented interest in the FFT, which after 1965 indeed revolutionized many sciences and technologies yielding profound benefits, especially to MR physics with most prominent applications in chemistry, biology and medicine. This comes as no surprise, since a computation by means of the DFT (N^2) for, e.g. $N = 10^6$ which might take even two weeks could be done within 30 seconds of the CPU (Central Processing Unit) through the use of the FFT ($N\log_2 N$) [176].

Due to the exponential nature of the evolution operator (4.4), construction of its discrete counterpart $\hat{U}(t) = \hat{U}(t_n) = \hat{U}(n\tau)$ at the time $t = t_n \equiv n\tau$ is done simply through raising the ansatz $\hat{U}(\tau)$ to the nth power

$$\hat{U}(n\tau) = \hat{U}^n(\tau) \qquad \hat{U}(\tau) = e^{-i\hat{\Omega}\tau} \implies |\Phi_n\rangle = \hat{U}^n(\tau)|\Phi_0\rangle. \qquad (5.6)$$

The set $\{|\Phi_n\rangle\}$ represents the Schrödinger basis which could be used for diagonalization of the evolution/relaxation matrix **U**. In linear programming and engineering literature [217]–[220], the set $\{|\Phi_n\rangle\}$ is called the Krylov basis [221]. The same name is also used in quantum chemistry [108]–[116]. In quantum-mechanical signal processing, the term Schrödinger basis is more transparent, since it points directly at the quantum-mechanical origin of the state functions $\{|\Phi_n\rangle\}$ that *do stem from the Schrödinger equation* (4.1). Nevertheless, to avoid potential confusion across interdisciplinary fields, one should always bear in mind that the Schrödinger and Krylov basis are two different nomenclatures for the same set $\{|\Phi_n\rangle\}$ and, for this reason, we shall interchangeably use both terminologies. The discrete/digital counterpart of (4.6) can be written as

$$\hat{U}^n(\tau) = \sum_{k=1}^{K} e^{-in\omega_k \tau} \hat{\pi}_k \qquad \text{Im}(\omega_k) < 0. \qquad (5.7)$$

For any operator analytic function $f(\hat{U})$, the eigenproblems follow as in (3.5)

$$\hat{U}(\tau)|\Upsilon_k\rangle = u_k|\Upsilon_k\rangle \qquad u_k = e^{-i\omega_k \tau} \qquad f(\hat{U})|\Upsilon_k\rangle = f(u_k)|\Upsilon_k\rangle. \qquad (5.8)$$

Use of (4.7), (4.9) and (5.6) gives the discrete auto-correlation function

$$C_n = \langle\Phi_0|\Phi_n\rangle = \langle\Phi_0|\hat{U}^n(\tau)|\Phi_0\rangle \qquad C_n = \sum_{k=1}^{K} d_k u_k^n. \qquad (5.9)$$

The same result $C_n = \sum_{k=1}^{K} d_k u_k^n$ is obtained by inserting (5.7) for $\hat{U}^n(\tau)$ into $C_n = (\Phi_0|\hat{U}^n(\tau)|\Phi_0)$. The decomposition (5.9) of the auto-correlation functions $\{C_n\}$ into the sum of K products, i.e. transients $\{d_k u_k^n\}$ of a geometric progression type constitutes the harmonic inversion. The set of at least $2K$ points $\{C_n\}$ is given in advance, whereas the unknown pair of $2K$ parameters $\{u_k, d_k\}_{k=1}^{K}$ represent the solution of the problem. The representation (5.9) of each element C_n of the given set $\{C_n\}$ with the task of finding the unknown pairs $\{\omega_k, d_k\}$ and the order K is hereafter called the 'harmonic inversion problem' (HIP) [108, 109]. A large fraction of experimentally encoded time signals $c(t)$ in many fields yields spectra that are well approximated by sums of damped exponentials

$$c(t) = \sum_{k=1}^{K} d_k e^{-i\omega_k t}. \tag{5.10}$$

In other words, such signals are just like the auto-correlation functions (4.9). This means that in the time domain, each signal point $c_n \equiv c(n\tau)$ from the experimentally recorded sequence $\{c_n\}$ is built from a linear combination of discrete attenuated exponentials

$$c_n = \sum_{k=1}^{K} d_k e^{-in\omega_k \tau} = \sum_{k=1}^{K} d_k u_k^n \tag{5.11}$$

precisely as in (5.9), provided that the amplitudes $\{d_k\}$ are identified via the prescription $d_k = (\Phi_0|\Upsilon_k)^2$ from (3.10). Hence, such experimentally measured signals are mathematically equivalent to the auto-correlation functions in both the digital and analog representations [108, 109]

$$c_n = C_n \qquad c(t) = C(t). \tag{5.12}$$

Throughout this book we shall most frequently use the symbols $c_n \equiv c(n\tau)$ when referring to both the time signals *and* to the auto-correlation functions $C_n \equiv C(n\tau)$. The set $\{\exp(-i\omega_k t)\}$ is exceedingly non-orthogonal and *this property causes numerical difficulties in all recipes for nonlinear fittings of experimental time signals* $\{c_n\}$ to the form (5.11) as discussed in [77]. If instead of the matrix element $(\Phi_0|\hat{U}^s(\tau)|\Phi_0)$, we consider a more general scalar product involving two different Schrödinger states $(\Phi_m|\hat{U}^s(\tau)|\Phi_n)$, then (3.3) and (5.12) would immediately lead to

$$U_{n,m}^{(s)} \equiv (\Phi_m|\hat{U}^s(\tau)|\Phi_n) = C_{n+m+s} = c_{n+m+s} \qquad S_{n,m} = c_{n+m}. \tag{5.13}$$

These signal points $\{c_{n+m+s}\}$ can be used to set up the following matrix

$$\mathbf{H}_n(c_s) = \mathbf{U}_n^{(s)} = \begin{pmatrix} c_s & c_{s+1} & c_{s+2} & \cdots & c_{s+n-1} \\ c_{s+1} & c_{s+2} & c_{s+3} & \cdots & c_{s+n} \\ c_{s+2} & c_{s+3} & c_{s+4} & \cdots & c_{s+n+1} \\ \vdots & \vdots & \vdots & \ddots & \vdots \\ c_{s+n-1} & c_{s+n} & c_{s+n+1} & \cdots & c_{s+2n-2} \end{pmatrix} \tag{5.14}$$

Equivalence: auto-correlation functions and time signals   35

and its determinant

$$\mathbf{H}_n(c_s) = \det \mathbf{H}_n(c_s) = \begin{vmatrix} c_s & c_{s+1} & c_{s+2} & \cdots & c_{s+n-1} \\ c_{s+1} & c_{s+2} & c_{s+3} & \cdots & c_{s+n} \\ c_{s+2} & c_{s+3} & c_{s+4} & \cdots & c_{s+n+1} \\ \vdots & \vdots & \vdots & \ddots & \vdots \\ c_{s+n-1} & c_{s+n} & c_{s+n+1} & \cdots & c_{s+2n-2} \end{vmatrix} \quad (5.15)$$

where $\mathbf{U}_n^{(s)} = \{U_{i,j}^{(s)}\}_{i,j=0}^{n-1}$. The 'argument' c_s in the small parenthesis in $\mathbf{H}_n(c_s)$ denotes the leading element from the first row and first column in the matrix (5.14). The matrix $\mathbf{H}_n(c_s)$ is called the Hankel matrix [222]. In signal processing, the experimentally measured signal points are the input data to a theoretical analysis and, as such, matrix (5.14) is named the data matrix. The determinant $\mathbf{H}_n(c_s)$ from (5.15), which is associated with matrix $\mathbf{H}_n(c_s)$ is called the Hankel determinant [83, 87, 223]. For $s = 0$ and $s = 1$ the Hankel matrices $\mathbf{H}_n(c_0)$ and $\mathbf{H}_n(c_1)$ become the overlap matrix $\mathbf{S}_n = \mathbf{U}_n^{(0)}$ and the evolution/relaxation matrix $\mathbf{U}_n = \mathbf{U}_n^{(1)}$ in the Schrödinger basis set $\{|\Phi_n\rangle\}$

$$\mathbf{S}_n = \mathbf{H}_n(c_0) = \begin{pmatrix} c_0 & c_1 & c_2 & \cdots & c_{n-1} \\ c_1 & c_2 & c_3 & \cdots & c_n \\ c_2 & c_3 & c_4 & \cdots & c_{n+1} \\ \vdots & \vdots & \vdots & \ddots & \vdots \\ c_{n-1} & c_n & c_{n+1} & \cdots & c_{2n-2} \end{pmatrix} \quad S_n = \det \mathbf{S}_n$$

(5.16)

$$\mathbf{U}_n = \mathbf{H}_n(c_1) = \begin{pmatrix} c_1 & c_2 & c_3 & \cdots & c_n \\ c_2 & c_3 & c_4 & \cdots & c_{n+1} \\ c_3 & c_4 & c_5 & \cdots & c_{n+2} \\ \vdots & \vdots & \vdots & \ddots & \vdots \\ c_n & c_{n+1} & c_{n+2} & \cdots & c_{2n-1} \end{pmatrix} \quad U_n = \det \mathbf{U}_n.$$

(5.17)

The results of the first four determinants $\mathbf{H}_n(c_0)$ and $\mathbf{H}_n(c_1)$ with $1 \leq n \leq 4$ are given by

$$\det \mathbf{S}_1 = H_1(c_0) = c_0$$

$$\det \mathbf{S}_2 = H_2(c_0) = \begin{vmatrix} c_0 & c_1 \\ c_1 & c_2 \end{vmatrix} = c_0 c_2 - c_1^2$$

$$\det \mathbf{S}_3 = H_3(c_0) = \begin{vmatrix} c_0 & c_1 & c_2 \\ c_1 & c_2 & c_3 \\ c_2 & c_3 & c_4 \end{vmatrix} = 2c_1 c_2 c_3 + c_0 c_2 c_4 - c_0 c_3^2 - c_1^2 c_4 - c_2^3$$

$$\det \mathbf{S}_4 = H_4(c_0) = \begin{vmatrix} c_0 & c_1 & c_2 & c_3 \\ c_1 & c_2 & c_3 & c_4 \\ c_2 & c_3 & c_4 & c_5 \\ c_3 & c_4 & c_5 & c_6 \end{vmatrix} = 2c_0 c_3 c_4 c_5 + 2c_1 c_2 c_3 c_6 - 2c_1 c_2 c_4 c_5$$

$$+ c_0c_2c_4c_6 - 3c_2c_3^2c_4 + 2c_1c_3c_4^2 + 2c_2^2c_3c_5 - 2c_1c_3^2c_5 - c_0c_2c_5^2$$
$$- c_0c_3^2c_6 - c_0c_4^3 - c_1^2c_4c_6 + c_1^2c_5^2 - c_2^3c_6 + c_2^2c_4^2 + c_3^4 \tag{5.18}$$

$\det \mathbf{U}_1 = \mathrm{H}_1(c_1) = c_1$

$\det \mathbf{U}_2 = \mathrm{H}_2(c_1) = \begin{vmatrix} c_1 & c_2 \\ c_2 & c_3 \end{vmatrix} = c_1c_3 - c_2^2$

$\det \mathbf{U}_3 = \mathrm{H}_3(c_1) = \begin{vmatrix} c_1 & c_2 & c_3 \\ c_2 & c_3 & c_4 \\ c_3 & c_4 & c_5 \end{vmatrix} = 2c_2c_3c_4 + c_1c_3c_5 - c_1c_4^2 - c_2^2c_5 - c_3^3$

$\det \mathbf{U}_4 = \mathrm{H}_4(c_1) = \begin{vmatrix} c_1 & c_2 & c_3 & c_4 \\ c_2 & c_3 & c_4 & c_5 \\ c_3 & c_4 & c_5 & c_6 \\ c_4 & c_5 & c_6 & c_7 \end{vmatrix} = 2c_1c_4c_5c_6 + 2c_2c_3c_4c_7$

$$- 2c_2c_3c_5c_6 + c_1c_3c_5c_7 - 3c_3c_4^2c_5 + 2c_2c_4c_5^2 + 2c_3^2c_4c_6 - 2c_2c_4^2c_6$$
$$- c_1c_3c_6^2 - c_1c_4^2c_7 - c_1c_5^3 - c_2^2c_5c_7 + c_2^2c_6^2 - c_3^2c_7 + c_3^2c_5^2 + c_4^4. \tag{5.19}$$

These formulae and especially (5.18) and (5.19), are unwieldy and the results for the higher-order determinants $\mathrm{H}_n(c_0)$ and $\mathrm{H}_n(c_1)$ become quickly unmanageable. The common feature of (5.18) and (5.19) is that they do not provide any clue as to what the general result for the Hankel determinants $\mathrm{H}_n(c_0)$ and $\mathrm{H}_n(c_1)$ of the arbitrary order n would look like after these determinants have been explicitly calculated. Additionally, no obvious regular pattern is present in the results for any of the particular determinants $\mathrm{H}_n(c_0)$ and $\mathrm{H}_n(c_1)$ when going from, e.g. $\mathrm{H}_2(c_s)$ via $\mathrm{H}_3(c_s)$ to $\mathrm{H}_4(c_s)$ for $s = 0$ and $s = 1$. This is at odds with the evident symmetry in the definition of the Hankel determinants $\mathrm{H}_n(c_0)$ and $\mathrm{H}_n(c_1)$ from (5.16) and (5.17). However, there ought to be a way to restore a matching symmetry also in the results for $\mathrm{H}_n(c_0)$ and $\mathrm{H}_n(c_1)$. An attempt along these lines could lead to a general formula for the result of $\mathrm{H}_n(c_0)$ and $\mathrm{H}_n(c_1)$ for any non-negative integer n. This is indeed possible, as has been shown in [17, 224, 225] by using the theory of orthogonal polynomials. The same theory will also provide the general cofactors of the Hankel determinants $\mathrm{H}_n(c_0) = \det \mathbf{S}_n$ and $\mathrm{H}_n(c_1) = \det \mathbf{U}_n$. These analytical results would then automatically lead to the closed expressions for the inverse Schrödinger matrices $\mathbf{S}^{-1}$ and $\mathbf{U}^{-1}$ [17].

Returning to (5.11) we set the time t to zero, i.e. $n = 0$ and derive the sum rule for the residues

$$\sum_{k=1}^{K} d_k = c_0 \neq 0 \tag{5.20}$$

where the local closure (3.6) is used. Here, the first recorded signal point c_0 is related to the norm of the initial state $|\Phi_0\rangle$ according to (4.5). Along with the Green function $\mathcal{G}(\omega)$ from (3.9), associated with $\hat{\Omega}$, we shall also consider the

Equivalence: auto-correlation functions and time signals   37

resolvent $\hat{R}(u)$, which corresponds to $\hat{U}(\tau)$

$$\hat{R}(u) = [u\hat{1} - \hat{U}(\tau)]^{-1} \qquad u = e^{-i\omega\tau} \tag{5.21}$$

where the frequency ω is real as before. The exact quantum-mechanical spectrum of the operator $\hat{R}(u)$ is defined through the following Green function

$$\mathcal{R}(u) = (\Phi_0|\hat{R}(u)|\Phi_0). \tag{5.22}$$

Then, within the local representation of operators and matrices of rank K, we can use the projection operator $\hat{\pi}_k = |\Upsilon_k)(\Upsilon_k|$ from (3.6) to write down at once the following approximations

$$\hat{R}(u) \approx \sum_{k=1}^{K} \frac{\hat{\pi}_k}{u - u_k} \qquad \mathcal{R}(u) \approx \sum_{k=1}^{K} \frac{d_k}{u - u_k}. \tag{5.23}$$

In the exact spectrum $\mathcal{F}(\omega)$ from (4.8), an infinite set $\{c_n\}_{n=0}^{\infty}$ is needed. However, generally any set of time signals $\{c_n\}_{n=0}^{N-1}$ or auto-correlation functions $\{C_n\}_{n=0}^{N-1}$ is of a finite length $N < \infty$. Thus, the FFT can only produce either a band-limited spectrum $\{F_k^{bl}\}$ or a periodically repetitive spectrum, after exhausting all available c_n. The former case assumes that c_n represents a function of compact support with $c_n = 0$ for $n \geq N$, whereas the latter case supposes that the quantities $\{c_n\}$ are periodic with period N, so that $c_{n+N} = c_n$ for $n \geq N$ [64]. Both assumptions are arbitrary. Parametric methods can go beyond the FFT, since they do not make such assumptions. Instead, they draw inferences about the unknown exact spectrum built from an infinite set $\{c_n\}$ while dealing only with a finite number of available c_n. This leads to a resolution which can be better than the Rayleigh limit $2\pi/(N\tau)$ fixed by the FFT. For instance, the LP method [57]–[64] for stationary signals can generate any number of c_n for $n \geq N$ from the given finite set $\{c_n\}_{n=0}^{N-1}$. This is equally true for the ARMA method, which combines the auto-regressive (AR) estimator and the moving average (MA) model. The ARMA is known as the method of choice in mathematical statistics, speech processing [124]–[128] and in other fields [64]. The LP coincides with the AR, whereas from a theoretical (but not computational) viewpoint the PA is identical to the ARMA or to the Prony method [57]–[64].

Chapter 6

Orthonormality and completeness of expansion functions

Given an infinite-dimensional orthonormalized basis set $\{\xi_k(x)\}_{k=1}^{\infty}$ defined, e.g. on a real segment $x \in [a, b]$ the closure is introduced by the relation

$$\delta(x - y) = \sum_{k=1}^{\infty} w_k \xi_k^*(x) \xi_k(y) \tag{6.1}$$

where the star superscript denotes the complex conjugation. Here, $\delta(x - y)$ is the Dirac δ-function and w_k is a given weight function

$$f(x) = \int_a^b dy f(y) \delta(x - y). \tag{6.2}$$

Moreover, the associated orthonormality condition is

$$\int_a^b dx \, W(x) \xi_{k'}^*(x) \xi_k(x) = \delta_{k,k'} \tag{6.3}$$

where $W(x)$ is a positive function in $[a, b]$ and $\delta_{n,m}$ is the Kronecker δ-symbol given in (3.7). Applying one of the quadrature rule to the lhs of (6.3), one would have

$$\int_a^b dx \, W(x) \xi_m^*(x) \xi_n(x) \approx \sum_{k=1}^{K} w_k \xi_m^*(x_k) \xi_n(x_k) \tag{6.4}$$

where $\{x_k, w(x_k)\} \equiv \{x_k, w_k\}$ are the zeros and the corresponding weights (residues) of a set of classical polynomials that are orthonormal in $[a, b]$ relative to a specific function $W(x)$ as stated in (6.3). Every classical orthogonal polynomial [180] has its own function $W(x)$ and the quadruple $\{a, b; x_k, w(x_k)\}$. Of course, $W(x_k)$ is different from w_k. In fact, according to (6.4) the weight w_k includes the function $W(x_k)$. In order to have $x_k \in [a, b]$ for $1 \leq k \leq K$, the function $W(x)$ must be positive in $[a, b]$. The weights w_k are also positive for classical polynomials. Let $f(x)$ be any function defined on the interval $[a, b]$ such that the following expansion can be set up in terms of some other functions $\{\xi_k(x)\}$

$$f(x) = \sum_{k=1}^{\infty} a_k \xi_k(x) \tag{6.5}$$

where a_k are the constant expansion coefficients. In such a case, whenever (6.5) is satisfied, the set $\{\xi_k(x)\}$ is said to be complete. Multiplication of (6.5) by $W(x)$ and $\xi_{k'}(x)$, followed by integration of the result over the region $x \in [a, b]$ with the help of the orthogonality relation (6.3) yields

$$a_k = \int_a^b dx\, W(x) \xi_k^*(x) f(x). \tag{6.6}$$

The closure implies completeness, as may be verified at once by inserting (6.1) into the Dirac identity (6.2). For a more general case of complex variables, the last five equations should be extended to encompass contour integrals. The usefulness of orthonormality and completeness can be readily appreciated in, e.g. spectral analysis of time signals. In signal processing, one of the key concerns is obtaining an adequate approximation for the energy of the signal in every small time interval. For this purpose, it is very useful to use the concept of an 'approximation in the mean' over, e.g. the time interval ranging from zero to infinity with respect to a positive weight function $W(t)$. Here, we are given a function $c(t)$ from, e.g. the Lebesgue–Stieltjes space $L^2(0, \infty)$ and a collection of orthonormal basis set functions $\{\varphi_n(t)\}$ with elements $\varphi_n(t) \in L^2(0, \infty)$ such that the following approximation is introduced

$$c(t) \approx \sum_{n=0}^{N-1} \gamma_n \varphi_n(t) \tag{6.7}$$

where $\{\gamma_n\}$ are the unknown constant expansion coefficients. We say that the function $c(t)$ belongs to the space $L^m(C)$ $(m = 2, 3, 4, \ldots)$ if the following two conditions are fulfilled: (i) the Lebesgue–Stieltjes integral $\oint_C |c(t)|^m d\sigma(t)$ exists where $d\sigma(t)$ is a distribution function which is also known as the Lebesgue measure, and if (ii) the contour C includes all the singularities of $c(t)$. In the case of a real variable t defined on the segment $[a, b]$ of the real axis $\{C : [a, b]\}$ with the continuously differentiable function $\sigma(t) = W(t)$, the measure $d\sigma(t)$ becomes equal to $W(t)dt$ in which case the Lebesgue–Stieltjes integral reduces to the ordinary Riemann integral $\int_a^b |c(t)|^m W(t) dt$ where $c(t) \in L^m(a, b)$. The problem in the above-mentioned 'approximation in the mean' is to find the minimum value of the integral I defined by

$$I = \int_0^\infty dt\, W(t) \left| c(t) - \sum_{n=0}^{N-1} \gamma_n \varphi_n(t) \right|^2. \tag{6.8}$$

Solving this problem reveals that the minimum of the integral I is possible to attain if γ_n are the Fourier coefficients defined through the usual, i.e. unsymmetric scalar product

$$\gamma_n = \langle \varphi_n(t) | c(t) \rangle \equiv \int_0^\infty dt\, W(t)\varphi_n^*(t)c(t). \tag{6.9}$$

In such a case, the resulting minimum value $I_{\min}(N)$ of the integral I is

$$I_{\min}(N) = \int_0^\infty dt\, W(t)|c(t)|^2 - \sum_{n=0}^{N-1} |\gamma_n|^2 \equiv \|c(t)\|^2 - \sum_{n=0}^{N-1} |\gamma_n|^2 \tag{6.10}$$

where the symbol $\|c(t)\|$ denotes the norm of $c(t)$. Since $\int_0^\infty |c(t)|^2 W(t) dt > 0$ for a positive weight function $W(t)$ within the integration range $0 \leq t \leq \infty$, it follows from (6.10) that

$$\sum_{n=0}^{N-1} |\gamma_n|^2 \leq \|c(t)\|^2 \tag{6.11}$$

and, by the Bessel inequality

$$\sum_{n=0}^{\infty} |\gamma_n|^2 \leq \|c(t)\|^2. \tag{6.12}$$

The convenient simplicity of the obtained expression for $I_{\min}(N)$ in (6.10) points at a clear practical advantage of using an orthogonal basis set $\{\varphi_n(t)\}$ for expansion functions. More importantly, when orthogonal functions $\{\varphi_n(t)\}$ are employed, the best value of an individual γ_k is not dependent upon the number of the expansion coefficients calculated. This crucial feature of orthogonal functions can best be appreciated from the argument which runs as follows. Suppose that instead of the approximation (6.7) we consider the ansatz

$$c(t) \approx \sum_{n=0}^{N+M} \gamma_n \varphi_n(t) \quad (M = 0, 1, 2, \ldots). \tag{6.13}$$

Then from a new calculation of the enlarged set of coefficients $\{\gamma_n\}_{n=0}^{N+M}$ for a fixed value of $M \geq 0$, the subset of the first N parameters $\{\gamma_0, \gamma_1, \ldots, \gamma_{N-1}\}$ will still retain the same values as those obtained previously with $M = -1$ using (6.7). This is the meaning of the adjective *optimal* in conjunction with the coefficients $\{\gamma_0, \gamma_1, \ldots, \gamma_{N-1}\}$ and precisely for this reason orthonormal basis set functions are predominantly in use in the theory of approximations. Yet, orthonormality of basis sets is a matter of convenience, not necessity[1]. By contrast, completeness of a collection of expansion functions in $L^m(0, \infty)$ is a *vital* prerequisite for

---

[1] For example, the Schrödinger, i.e. the Krylov basis set is not orthogonal.

usefulness of an approximation to a given function [193, 226, 227]. There is also an alternative definition of completeness pointing at a more quantitative assessment as follows. A set $\{\varphi_n\}_{n=0}^{N-1}$ with $\varphi_n(t) \in L^m(0, \infty)$ is said to be *complete* in $L^m(0, \infty)$ if for each infinitesimally small positive number $\epsilon > 0$ one can construct a linear combination $\phi_{N-1}(t)$ of N functions $\{\varphi_n(t)\}_{n=0}^{N-1}$ with constant coefficients $\{\gamma_n\}$

$$\phi_{N-1}(t) \equiv \sum_{n=0}^{N-1} \gamma_n \varphi_n(t) \tag{6.14}$$

with the property

$$\int_0^\infty dt\, W(t) |c(t) - \phi_{N-1}(t)|^m < \epsilon. \tag{6.15}$$

This implies that if, e.g. an orthogonal set is complete in $L^2(0, \infty)$, then the Bessel *inequality* (6.12) becomes the Parseval *equality*

$$\sum_{n=0}^\infty |\gamma_n|^2 = \|c(t)\|^2 \tag{6.16}$$

in which case the minimum value $I_{\min}(N)$ from (6.8) is found to approach zero when N tends to infinity

$$\lim_{N \to \infty} I_{\min}(N) = 0. \tag{6.17}$$

If an enumerable set of functions $\{f_n(t)\} \in L^2(a, b)(n = 0, 1, 2, \ldots)$ is complete in $L^2(a, b)$, then it follows that the same set is also closed in the space $L^2(a, b)$. Assume that we are given a function $g(t) \in L^2(a, b)$ and a sequence $\{f_n(t)\} \in L^2(a, b)(n = 0, 1, 2, \ldots)$. Then the set $\{\varphi_n(t)\}$ is said to be *closed* in $L^2(a, b)$ if

$$\int_a^b dt\, g(t) f_n(t) = 0 \tag{6.18}$$

provided that $g(t)$ is zero throughout the interval $[a, b]$ with the exception of at most a set of measure zero. It is important to realize that a set can be complete and closed without necessarily being orthonormal. For example, an infinite set of power functions

$$\{t^n\} \qquad (n = 0, 1, 2, \ldots) \tag{6.19}$$

is complete and, hence, closed in the interval $(0,1)$, but it is not orthogonal. By definition, using (6.18) it follows that the equality

$$\int_0^1 dt\, g(t) t^n = 0 \qquad (n = 0, 1, 2, \ldots) \tag{6.20}$$

is valid only if $g(t) = 0$ for every $t \in [0, 1]$ except for at most a set of measure zero. This infinite set remains closed even without the first $M < \infty$ terms

$$\{t^n\} \qquad (n = M, M+1, M+2, \ldots). \tag{6.21}$$

If the set (6.21) is closed in the interval (0,1), we shall have

$$\int_0^1 dt\, g(t) t^n = 0 \qquad (n = M, M+1, M+2, \ldots) \tag{6.22}$$

where $g(t)$ is zero for each $t \in [0, 1]$, with the exception of at most a set of measure zero. Likewise (6.18) implies that the following relation is also fulfilled

$$\int_0^1 dt\, g_M(t) t^n = 0 \qquad (n = 0, 1, 2, \ldots) \tag{6.23}$$

if and only if $g_M(t) \equiv g(t) t^M = 0$ for every $t \in [0, 1]$, except at most a set of measure zero. Therefore, it follows that $g(t) = 0$ for $t \neq 0$, which shows that the set $\{t^n\}(n = M, M+1, M+2, \ldots)$ with the first M terms discarded is still closed[2] in the interval $(0, 1)$. Notice that the value $g(t)$ at $t = 0$ is not a problem, since $g(0)$ does not contribute to the integral (6.23) [227]. Such a situation is due to the fact that the set (6.19) is not orthogonal, so that the dimension of the Hilbert space associated with $\{t^n\}(n = 0, 1, 2, \ldots)$ is not necessarily reduced by removal of a subset $\{t^n\}_{n=0}^{M-1}$ [227]. This feature of a set remaining to be closed after dropping its first M terms is relevant to signal processing. Namely, using the Cayley–Hamilton theorem [228], we can formally replace the scalar power function t^n by the nth power of the evolution operator applied to the initial state $|\Phi_0\rangle$ leading to the Schrödinger basis set $\{|\Phi_n\rangle\} = \{\hat{U}^n(\tau)|\Phi_0\rangle\}$. Due to, e.g. a time delay from deexcitation to the beginning of encoding the investigated FID, or because of unreliability of a number of the first signal points from ICR experiments, one is often obliged to drop a number of earliest measured data points from the set $\{c_n\}$. Discarding the first M signal points within, e.g. FD or DSD methods can be adequately compensated by introducing a phase factor with the opposite sign in the evolution operator [109, 115].

Here, it is important to highlight a connection between expansion basis sets with the so-called completeness and separability axioms of the Hilbert eigenspace $\mathcal{H} = \{\Psi\}$ in quantum mechanics [157, 193]. This vector space has a positive definite asymmetric scalar product $\langle\Psi|\chi\rangle$ with the properties $\langle\Psi|\chi\rangle = \langle\chi|\Psi\rangle^*$ and $\|\Psi\| = \langle\Psi|\Psi\rangle^{1/2} > 0$ where $\Psi \neq \emptyset$ (the zero state vector is excluded from consideration as physically uninteresting). In addition to the standard features of linearity of the inner product, it is also assumed that $\mathcal{H}$ is separable and that it contains all the limits of its sequences that are convergent in the norm (the so-called strong convergence [157]). More specifically, the latter two statements read as follows.

- *The completeness*: if the space $\mathcal{H}$ contains a Cauchy sequence $\{\chi_k\}(k = 1, 2, \ldots)$ such that $\|\chi_m - \chi_n\| < \epsilon$, whenever $m > n > N_\epsilon$, then there exists a unique element $\Psi \in \mathcal{H}$ such that $\|\Psi - \chi_n\| < \epsilon$ for $n > N_\epsilon$, where ϵ

---

[2] Of course, the 'reduced' set $\{t^n\}(n = M, M+1, M+2, \ldots)$ remains infinite after removal of its M terms where M is any finite integer.

- is an infinitesimally small positive number and $N_\epsilon > 0$ is another number dependent on ϵ.
- *The separability*: the vector space $\mathcal{H} = \{\Psi\}$ contains an enumerable set $\mathcal{B} = \{\chi_k\}$ which is everywhere dense in $\mathcal{H}$, meaning that for every function Ψ there exists at least one element $\chi_k \in \mathcal{B}$ with the property $\|\Psi - \chi_k\| < \epsilon$ [157, 193]. The completeness and the separability axioms ensure the existence of at least one complete orthonormal set $\{\varphi_k\}$ which can be used as a basis for the expansion theorem $\Psi = \sum_{k=1}^{\infty} \alpha_k \varphi_k$ where the series converges strongly and $\alpha_k = \langle \varphi_k | \Psi \rangle$. The orthonormal basis set $\{\varphi_k\}$ can be found by subjecting the non-orthogonal sequence $\{\chi_k\}$ to the Gram–Schmidt or the Lanczos orthogonalization. Even if the expansion of Ψ is truncated by including a finite number N of terms ($N < \infty$), the separability axiom guarantees that every element $\Psi \in \mathcal{H}$ could still be approximated by $\Psi = \sum_{k=1}^{N} \alpha_k \varphi_k$ to any desired accuracy ϵ. Misleading claims have often been advanced according to which the separability axiom is hardly of any practical importance in physics. Quite the contrary, however, it is precisely the practice of computing, which has thus far abundantly demonstrated that the separability axiom is, in fact, the fundamental building block for nearly all the important numerical applications of physics [157, 193].

Chapter 7

Difference equations and the harmonic inversion

In the harmonic inversion problem, we are given a set of digitized auto-correlation functions $\{C_n\}$ or signal points $\{c_n\}$ and the task is to find the unknown triple $\{K, u_k, d_k\}_{k=1}^{K}$ such that (4.9) or (5.11) is valid, where $u_k = \exp(-i\omega_k \tau)$ with τ being the sampling time or the time lag. To see why this is an inverse problem from the viewpoint of differential calculus, consider the following homogeneous ordinary differential equation (ODE) of a general order K given by

$$L_K(\hat{D})c(t) = 0 \qquad \hat{D} = i\frac{d}{dt} \qquad (7.1)$$

with the quantity $L_K(\hat{D})$ being a linear operator

$$L_K(\hat{D}) = b_0 + b_1\hat{D} + \cdots + b_{K-1}\hat{D}^{K-1} + b_K\hat{D}^K \qquad (7.2)$$

where the constant coefficients $\{b_k\}$ and the initial conditions at the time $t = 0$ are assumed to be known. With the ODE from (7.1), one associates the scalar characteristic polynomial $L_K(u)$ which is the rhs of (7.2) such that the nth differential operator $\hat{D}^n$ is formally replaced by the power function u^n according to the prescription of the Heaviside symbolic calculus [132]–[141]

$$L_K(u) = b_0 + b_1 u + \cdots + b_{K-1} u^{K-1} + b_K u^K \qquad u = e^{-i\omega\tau}. \qquad (7.3)$$

The scalar polynomial $L_K(u)$ is associated with the characteristic equation

$$L_K(u_k) = 0 \qquad u_k = e^{-i\omega_k \tau}. \qquad (7.4)$$

The characteristic polynomial of the degree K has precisely K roots. The Cayley–Hamilton theorem [228] states that the operator $\hat{D}$ satisfies its own characteristic equation

$$L_K(\hat{D}) = \hat{0}. \qquad (7.5)$$

Difference equations and the harmonic inversion

If all the roots $\{u_k\}$ of $L_K(u)$ were distinct, the exact *unique* solution of (7.1) would be given by a linear combination of precisely K exponentials

$$c(t) = \sum_{k=1}^{K} d_k e^{-i\omega_k t} \qquad \text{Im}(\omega_k) < 0 \tag{7.6}$$

where d_k are the constant amplitudes independent of t. The solution (7.6) coincides with the continuous representation (5.10) of the time signal $c(t)$ which corresponds to a Lorentzian spectrum. Once the set $\{u_k\} = \{\exp(-i\omega_k \tau)\}$ has been found, the amplitudes $\{d_k\}$ can be determined via the boundary conditions to the ODE from (7.1). This well-known procedure constitutes the so-called *direct problem* in the context of the ODE. On the other hand, the inverse problem relative to the ODE consists of the exchanged roles of the knowns and unknowns. In the *inverse problem*, the ODE itself is unknown and so are the eigenroots $\{u_k\}$ and the accompanying boundary conditions. Here, the *global* solution $c(t)$ is known and the task is to reconstruct the ODE and the boundary conditions. The term 'global' is used to emphasize that we are given the solution $c(t)$ as a whole in a tabular or graphical form and this is the lhs of (7.6), such that the rhs of the same equation is unknown. In this case, the parameters $\{u_k, d_k\}$ and their number K must be determined. This inverse problem is the continuous version of the harmonic inversion. A retrieval of the ODE and the boundary condition is essential for unfolding the hidden dynamics that produced the data $c(t)$ and that govern the evolution of the examined system.

For numerical purposes, one discretizes (7.1) by setting $t = t_n$ with $t_n = n\tau \, (0 \leq n \leq N-1)$ where $\tau = \Delta t$ is the sampling time as before. Then by abbreviating $c(n\tau) \equiv c_n$ as in chapter 5, we set up a discrete counterpart of (7.1) as the ordinary difference equation (OΔE) of the Kth order with the constant coefficients as

$$b_0 c_n + b_1 c_{n+1} + \cdots + b_{K-1} c_{n+K-1} + b_K c_{n+K} = 0. \tag{7.7}$$

Let $\{c_j\}$ be a set whose elements are the power functions $c_n = cu_k^n$ where $c \neq 0$ is a constant and $u_k \neq 0$. Then inserting these functions into (7.7), we have

$$b_0 c u_k^n + b_1 c u_k^{n+1} + \cdots + b_{K-1} c u_k^{n+K-1} + b_K c u_k^{n+K} = 0 \tag{7.8}$$

which after division by $cu_k^n \neq 0$ becomes

$$L_K(u_k) = 0 \qquad L_K(u) \equiv b_0 + b_1 u + \cdots + b_{K-1} u^{K-1} + b_K u^K \tag{7.9}$$

and this is the characteristic equation (7.4). If the K roots of the polynomial $L_K(u)$ are all different, the general solution of the OΔE from (7.7) will be a linear combination of these roots $u_k = \exp(-i\omega_k \tau)$, each of which is raised to the same power n

$$c_n = \sum_{k=1}^{K} d_k e^{-in\omega_k \tau} \tag{7.10}$$

where $\text{Im}(\omega_k) < 0$. Hence, the solution of the OΔE from (7.7) coincides with the discrete representation (5.11) of the time signal c_n associated with a Lorentzian spectrum. The necessary and sufficient conditions for the existence of the unique solution (7.10) of (7.7) for $n \in [0, N-1]$ are [76, 218]

$$H_K(c_n) \neq 0 \tag{7.11}$$
$$H_{K+1}(c_n) = 0 \tag{7.12}$$

where $H_n(c_s)$ is the Hankel determinant of rank n from (5.15). The constants $\{d_k\}_{k=1}^{K}$ from (7.10) can be determined using the first K signal points $\{c_0, c_1, \ldots, c_{K-1}\}$. Once the set $\{u_k\}$ becomes available, the constants $\{d_k\}$ can be obtained by solving the system of linear equations

$$\begin{pmatrix} 1 & 1 & \cdots & 1 \\ u_1 & u_2 & \cdots & u_K \\ u_1^2 & u_2^2 & \cdots & u_K^2 \\ \vdots & \vdots & \ddots & \vdots \\ u_1^{K-1} & u_2^{K-1} & \cdots & u_K^{K-1} \end{pmatrix} \begin{pmatrix} d_1 \\ d_2 \\ d_3 \\ \vdots \\ d_K \end{pmatrix} = \begin{pmatrix} c_0 \\ c_1 \\ c_2 \\ \vdots \\ c_{K-1} \end{pmatrix}. \tag{7.13}$$

The inverse problem regarding the OΔE consists of reconstructing the set of $2K$ parameters $\{u_k, d_k\}$ from (7.10) for a given input sequence $\{c_n\}$ of the length $N \geq 2K$. This inverse problem is the discrete variant of the harmonic inversion. By solving (7.9), one obtains the roots $\{u_k\}$ that upon insertion into (7.10) lead to a system of linear equations (7.13) for the remaining unknowns $\{d_k\}$. The square matrix from (7.13) with the elements $u_k^n (1 \leq k \leq K, 0 \leq n \leq K-1)$ is called the Vandermonde matrix [176]. This matrix is non-singular, since the associated Vandermonde determinant is equal to the product of all the differences $(u_k - u_{k'}) \neq 0$ with $k > k'$

$$V_K \equiv \begin{vmatrix} 1 & 1 & \cdots & 1 \\ u_1 & u_2 & \cdots & u_K \\ u_1^2 & u_2^2 & \cdots & u_K^2 \\ \vdots & \vdots & \ddots & \vdots \\ u_1^{K-1} & u_2^{K-1} & \cdots & u_K^{K-1} \end{vmatrix} = \prod_{k=1}^{K} \prod_{k'=1 (k>k')}^{K} (u_k - u_{k'}) \tag{7.14}$$

where $V_k \neq 0$, since the roots $\{u_k\}$ are unequal $u_{k'} \neq u_k (k' \neq k)$.

The difference equation (7.7) can advantageously be considered as an inverse problem in which the quantities $\{c_n\}$ are known and given by (7.10), or equivalently, by $c_n = \sum_{k=1}^{K} d_k u_k^n$ as per (5.11), where d_k and $u_k = \exp(-in\omega_k \tau)$ are the fixed parameters. In such a case, the unknowns would be the coefficients $\{b_r\} (0 \leq r \leq K)$. To determine these K coefficients we need K linear equations. The leading coefficient b_K can be determined from normalization of the characteristic polynomial. The remaining K coefficients $\{b_r\} (0 \leq r \leq K-1)$ can be obtained by rewriting (7.7) as

$$b_0 c_n + b_1 c_{n+1} + \cdots + b_{K-1} c_{n+K-1} = \tilde{c}_{n+K} \qquad n \in [0, K-1] \tag{7.15}$$

where $\tilde{c}_n = -b_K c_n$. The determinant of the system (7.15) is equal to the Hankel determinant $H_K(c_0)$ which is given in (5.15) for $s = 0$. The inhomogeneous system (7.15) has the unique solution given as the column vector $\mathbf{b} = \{b_r\}$ provided that the determinant $H_K(c_0)$ is not zero

$$H_K(c_0) \neq 0. \tag{7.16}$$

The general element b_r of the set $\{b_r\}$ is given by the Cramer formula (see Appendix A)

$$b_r = \frac{H_{K,r}(c_0)}{H_K(c_0)} \equiv b_r^{(K)} \tag{7.17}$$

$$H_{K,r}(c_n) = \begin{vmatrix} c_n & c_{n+1} & \cdots & c_{n+r-1} & \tilde{c}_{n+K} & c_{n+r+1} & \cdots & c_{n+K-1} \\ c_{n+1} & c_{n+2} & \cdots & c_{n+r} & \tilde{c}_{n+K+1} & c_{n+r+2} & \cdots & c_{n+K} \\ \vdots & \vdots & \vdots & \vdots & \vdots & \vdots & \ddots & \vdots \\ c_{n+K-1} & c_{n+K} & \cdots & c_{n+K-2+r} & \tilde{c}_{n+2K-1} & c_{n+K+r} & \cdots & c_{n+2K-2} \end{vmatrix}.$$

$$\tag{7.18}$$

It is also convenient to introduce an auxiliary set $\{\tilde{b}_r\}$ with the general element $\tilde{b}_r$ as

$$\tilde{b}_r = \frac{H_{K,r}(c_n)}{H_K(c_n)} \equiv \tilde{b}_r^{(K)}. \tag{7.19}$$

Using (5.11) for c_n we shall derive the analytical results for both $H_K(c_n)$ and $H_{K,r}(c_n)$. Then the corresponding analytical results for b_r and $\tilde{b}_r$ will follow from (7.17) and (7.19). This can easily be achieved by a simple calculation of only a few explicit Hankel determinants. From the obtained particular findings the general results can readily be deduced. Before proceeding further, we want to check whether the condition (7.16) for the existence and uniqueness of the solution of the system (7.15) is satisfied. To this end, we first take the 2×2 determinant

$$H_2(c_n) = c_n c_{n+2} - c_{n+1}^2. \tag{7.20}$$

Then we set $K = 2$ in (5.11), i.e. $c_n = d_1 u_1^n + d_2 u_2^n$ which after insertion into (7.20) yields

$$H_2(c_n) = d_1 d_2 u_1^n u_2^n (u_2 - u_1)^2. \tag{7.21}$$

Similarly, the 3×3 determinant reads

$$H_3(c_n) = c_n(c_{n+2}c_{n+4} - c_{n+3}^2) - c_{n+1}(c_{n+1}c_{n+4} - c_{n+2}c_{n+3})$$
$$+ c_{n+2}(c_{n+1}c_{n+3} - c_{n+2}^2). \tag{7.22}$$

The rank $K = 3$ in (5.11) gives $c_n = d_1 u_1^n + d_2 u_2^n + d_3 u_3^n$ which is substituted in (7.22) to produce

$$H_3(c_n) = d_1 d_2 d_3 u_1^n u_2^n u_3^n (u_2 - u_1)^2 (u_3 - u_1)^2 (u_3 - u_2)^2. \quad (7.23)$$

Likewise, the 4×4 determinant is given by

$$\begin{aligned} H_4(c_n) =\ & 2 c_n c_{n+3} c_{n+4} c_{n+5} + 2 c_{n+1} c_{n+2} c_{n+3} c_{n+6} - 2 c_{n+1} c_{n+2} c_{n+4} c_{n+5} \\ & + c_n c_{n+2} c_{n+4} c_{n+6} - 3 c_{n+2} c_{n+3}^2 c_{n+4} \\ & + 2 c_{n+1} c_{n+3} c_4^2 + 2 c_{n+2}^2 c_{n+3} c_{n+5} \\ & - 2 c_{n+1} c_{n+3}^2 c_{n+5} - c_n c_{n+2} c_{n+5}^2 - c_n c_{n+3}^2 c_{n+6} - c_n c_{n+4}^3 \\ & - c_{n+1}^2 c_{n+4} c_{n+6} + c_{n+1}^2 c_{n+5}^2 - c_{n+2}^3 c_{n+6} + c_{n+2}^2 c_{n+4}^2 + c_{n+3}^4. \end{aligned} \quad (7.24)$$

The case $K = 4$ from (5.11) simplifies as $c_n = d_1 u_1^n + d_2 u_2^n + d_3 u_3^n + d_4 u_4^n$ which together with (7.24) generates

$$\begin{aligned} H_4(c_n) =\ & d_1 d_2 d_3 d_4 u_1^n u_2^n u_3^n u_4^n \\ & \times (u_2 - u_1)^2 (u_3 - u_1)^2 (u_4 - u_1)^2 (u_3 - u_2)^2 (u_4 - u_2)^2 (u_4 - u_3)^2. \end{aligned} \quad (7.25)$$

The clear pattern appearing in (7.21), (7.23) and (7.25) implies that the result of the general $K \times K$ Hankel determinant $H_K(c_n)$ has the following analytical expression

$$\left. \begin{aligned} H_K(c_n) &= V_K^2 \{\bar{d}\, \bar{u}^n\} \neq 0 \\ \bar{d} = \prod_{k=1}^{K} d_k \neq 0 \quad &\bar{u} = \prod_{k=1}^{K} u_k \neq 0 \\ u_k \neq 0 \quad u_k \neq 1 \quad &d_k \neq 0 \quad (1 \leq k \leq K) \end{aligned} \right\} \quad (7.26)$$

where $V_K \neq 0$ is the Vandermond determinant (7.14). Therefore, the result (7.26) shows that indeed $H_K(c_n) \neq 0$, as it was set to prove (QED). In (7.26), the term $\bar{d}\, \bar{u}^n (0 \leq n \leq N-1)$ is seen as *a single* transient with the effective harmonic parameters $\bar{d}$ and $\bar{u}$ where $\bar{d}\, \bar{u}^n$ [229].

Next we consider the first few particular cases of the determinant (7.18). For instance, for $K = 2$, it follows

$$H_{2,1}(c_n) = c_{n+1} c_{n+3} - c_{n+2}^2 \quad (7.27)$$

$$H_{2,1}(c_n) = d_1^n d_2^n u_1 u_2 (u_2 - u_1)^2 \quad (7.28)$$

$$\therefore \quad \tilde{b}_2^{(2)} = u_1 u_2 = b_2^{(2)} \quad (7.29)$$

$$H_{2,2}(c_n) = c_{n+1} c_{n+2} - c_n c_{n+3} \quad (7.30)$$

$$H_{2,2}(c_n) = - d_1 d_2 u_1^n u_2^n (u_1 + u_2)(u_2 - u_1)^2 \quad (7.31)$$

$$\therefore \quad \tilde{b}_1^{(2)} = -(u_1 + u_2) = b_1^{(2)}. \quad (7.32)$$

Difference equations and the harmonic inversion

In the same fashion, for $K = 3$, we have

$$H_{3,1}(c_n) = c_{n+1}(c_{n+4}^2 - c_{n+3}c_{n+5})$$
$$- c_{n+3}(c_{n+2}c_{n+4} - c_{n+3}^2) - c_{n+2}(c_{n+3}c_{n+4} - c_{n+2}c_{n+5}) \quad (7.33)$$

$$H_{3,1}(c_n) = -d_1 d_2 d_3 u_1^n u_2^n u_3^n u_1 u_2 u_3 (u_2 - u_1)^2 (u_3 - u_1)^2 (u_3 - u_2)^2 \quad (7.34)$$

$$\therefore \quad \tilde{b}_3^{(3)} = -u_1 u_2 u_3 = b_3^{(3)} \quad (7.35)$$

$$H_{3,2}(c_n) = c_{n+3}(c_{n+1}c_{n+4} - c_{n+2}c_{n+3}) - c_n(c_{n+4}^2 - c_{n+3}c_{n+5})$$
$$- c_{n+2}(c_{n+1}c_{n+5} - c_{n+2}c_{n+4}) \quad (7.36)$$

$$H_{3,2}(c_n) = d_1 d_2 d_3 u_1^n u_2^n u_3^n (u_1 u_2 + u_1 u_3 + u_2 u_3)$$
$$\times (u_2 - u_1)^2 (u_3 - u_1)^2 (u_3 - u_2)^2 \quad (7.37)$$

$$\therefore \quad \tilde{b}_2^{(3)} = u_1 u_2 + u_1 u_3 + u_2 u_3 = b_2^{(3)} \quad (7.38)$$

$$H_{3,3}(c_n) = c_{n+1}(c_{n+1}c_{n+5} - c_{n+2}c_{n+4}) - c_n(c_{n+2}c_{n+5} - c_{n+3}c_{n+4})$$
$$- c_{n+3}(c_{n+1}c_{n+3} - c_{n+2}^2) \quad (7.39)$$

$$H_{3,3}(c_n) = -d_1 d_2 d_3 u_1^n u_2^n u_3^n (u_1 + u_2 + u_3)(u_2 - u_1)^2 (u_3 - u_1)^2 (u_3 - u_2)^2 \quad (7.40)$$

$$\therefore \quad \tilde{b}_1^{(3)} = -(u_1 + u_2 + u_3) = b_1^{(3)}. \quad (7.41)$$

Therefore, inspecting (7.28), (7.31), (7.34), (7.37) and (7.40), we can easily infer the following general result

$$H_{K,r}(c_n) = H_K(c_n) \left\{ (-1)^r \sum_{k_1=1}^{K-r-1} \sum_{k_2=k_1+1}^{K-r-2} \cdots \sum_{k_r=k_{r-1}+1}^{K} u_{k_1} u_{k_2} \cdots u_{k_r} \right\} \quad (7.42)$$

where (7.26) is used. We see that the only n-dependence of (7.26) and (7.42) is through $\bar{u}^n$ and this implies

$$H_{K+m}(c_n) = 0 \iff d_{K+m} = 0 \quad m \geq 1. \quad (7.43)$$

Finally, dividing the rhs of (7.42) by $H_K(c_n)$ according to (7.17) and (7.19), we obtain the following analytical results for $b_r \equiv b_r^{(K)}$ which is equal to $\tilde{b}_r \equiv \tilde{b}_r^{(K)}$

$$b_r = (-1)^r \underbrace{\sum_{k_1=1}^{K-r+1} \sum_{k_2=k_1+1}^{K-r+2} \cdots \sum_{k_r=k_{r-1}+1}^{K}}_{r \text{ summations}} u_{k_1} u_{k_2} \cdots u_{k_r} = \tilde{b}_r. \quad (7.44)$$

This result depends only upon the roots $\{u_k\}(1 \leq k \leq K)$. All the roots $\{u_k\}$ from (7.44) stem from the elements $c_n = \sum_{k=1}^{K} d_k u_k^n$ of the Hankel determinants $H_K(c_n)$ and $H_{K,r}(c_n)$. Inserting the signal points $\{c_n\}$ from (5.11) into these two latter determinants and using the definition (7.17) for b_r, we show that the general solution of the inhomogeneous system (7.15) of linear equations has the unique solution given by (7.44). It follows from (7.44) that the result for b_r is the sum of all the possible products of distinct roots $\{u_k\}(u_k \neq u_{k'}, k' \neq k)$ with an overall multiplication factor which is ± 1, depending on whether r is even/odd, respectively. The accomplished derivation shows that (7.44) does not contain any amplitudes $\{d_k\}$. As seen from (7.26) and (7.42), both $H_K(c_n)$ and $H_{K,r}(c_n)$ contain the same product of such amplitudes $(d_1 d_2 \cdots d_K)$ that in the end cancel out when the determinantal quotient (7.17) is taken to arrive at b_r.

The result (7.44) can be checked by an entirely different derivation. To this end, we recall that the system of linear equations (7.15) is derived from the assumption that the coefficient $\{b_r\}$ are unknown, whereas the quantities $\{c_n\}$ are known and given as the signal points taken from (5.11). Alternatively, the same system of equations (7.15) can be viewed as a single equation, which is the characteristic equation (7.4), where this time the coefficients $\{b_r\}$ are known, whereas the roots $\{u_k\}$ are unknown. Once these roots become available, we can rewrite (7.4) equivalently in terms of $\{u_k\}$ only

$$L_K(u) = \frac{1}{b_K} \prod_{k=1}^{K} (u - u_k) \equiv \sum_{r=0}^{K} \bar{b}_r u^r = 0. \qquad (7.45)$$

We can explicitly multiply all the differences $u - u_k$ for $1 \leq k \leq K$ on the rhs of (7.45) to collect the ensuing coefficients, say $\{\bar{b}_r\}$, of the power function u^r. Finally, this procedure will reveal that

$$\bar{b}_r = (-1)^r \underbrace{\sum_{k_1=1}^{K-r+1} \sum_{k_2=k_1+1}^{K-r+2} \cdots \sum_{k_r=k_{r-1}+1}^{K}}_{r \text{ summations}} u_{k_1} u_{k_2} \cdots u_{k_r} = b_r. \qquad (7.46)$$

The equality $\bar{b}_r = b_r$ is expected from the uniqueness theorem of polynomial expansions. This theorem states that if the same given polynomial $L_K(u)$ has two expansions (7.9) and (7.45) in powers of its variable u, then the corresponding general coefficients b_r and $\bar{b}_r$ must necessarily be identical to each other, as in (7.46). This completes the mentioned independent check (QED). Note that the sum on the rhs of (7.44) can also be computed through an efficient recursive algorithm which will be given in chapter 17.

In signal processing, one deals with both continuous $c(t)$ and discrete c_n time signals. One often measures directly a continuous function $c(t)$ as customarily done, e.g. in electroencephalograms or electrocardiograms [129]. Subsequently these analog data are digitized by converting them to the corresponding discrete counterpart $\{c_n\}$. The discrete time signal $\{c_n\}(0 \leq n \leq N-1)$ is sampled at the rate τ such that the signal length N and the epoch T are related through $\tau = T/N$. The corresponding spectra for many experimentally recorded time signals can often be well approximated by sums of pure Lorentzians or Gaussians or Voigtians in the frequency domain [195]–[199]. Any two functions of the conjugate

Difference equations and the harmonic inversion 51

variables $\{t, \omega\}$ are linearly interrelated by the Fourier integral. Therefore, a linear combination of Lorentzians in the frequency domain leads to a sum of damped exponentials in the time representation, irrespective of whether we are dealing with continuous or discrete functions. In such a case, the functions $c(t)$ and c_n that are given by (5.10) and (5.11) coincide with (7.6) and (7.10) for the ODE and OΔE, respectively. In the harmonic inversion, the set $\{c_n\}(0 \leq n \leq N-1)$ is known in advance and the unknown quantities are $\{u_k, d_k\}$ and the order K. Therefore, the harmonic inversion is mathematically equivalent to the inverse OΔE problem.

The purpose of the above-outlined parallel being drawn between the harmonic inversion problem and the inverse problems in differential calculus is both practical and fundamental. Practical, because the equivalence between the two different problems is beneficial, especially when combining the methods for their solutions from the two areas. Fundamental, because the representations (5.10) and (5.11) for the analog and digital signals $c(t)$ and c_n could be traced back to the formulae (7.6) and (7.10) from the dynamics governed by the ODE and OΔE, respectively. This opens up the possibility to give the processing of experimentally recorded data the status of a deductive theory, as an alternative to customary fitting as largely practised in, e.g. MRS [11]. For example, fitting measured data $\{c_n\}$ to a sum of damped exponentials (7.10) inevitably yields non-unique sets of parameters $\{u_k, d_k\}$. However, the spectral analysis originating from the inverse OΔE provides the *uniquely* determined fundamental frequencies and residues $\{u_k, d_k\}$ as well as their number K. For example, to determine unequivocally the order K, one should compute the set of the Hankel determinants $\{H_m(c_n)\}$ for $0 \leq n \leq N - 2m + 2$ for increasing values of the positive integer $m (m = 1, 2, \ldots)$. The computation is stopped when m reaches a value, say m', at which both conditions (7.11) and (7.12) are simultaneously satisfied. Then this value of m' *will be* the sought order K, i.e. $K = m'$ which is thus found without any ambiguity. Of course, in practice, the determinant in (7.12) is never exactly zero due to either insufficient accuracy in the input signal points $\{c_n\}$ or because of the round-off errors in computations of Hankel determinants. Therefore, a suitable practical criterion is desirable to avoid determining a false order K. Suppose that the elements of the set $\{c_n\}$ are real and that they are known within the error margin $\sigma_n > 0$ such that the expected true value c_n should lie in the interval $[c_n - \sigma_n, c_n + \sigma_n]$. Let us assume that a computation shows that $H_m(c_n) \neq 0$ for a fixed m and that $H_{m+1}(c_n)$ is bound according to the inequality [76]

$$|H_{m+1}(c_n)| \leq \sigma_{n+2m}|H_m(c_n)|. \tag{7.47}$$

To interpret this relationship, we use the Cauchy method to expand the Hankel determinant $H_{m+1}(c_n)$ from (5.15) by the last row containing the elements $\{c_{n+m}, c_{n+m+1}, \ldots, c_{n+2m}\}$. This will express $H_{m+1}(c_n)$ as a sum of m terms representing the cofactors that are multiplied by the corresponding signal point from the row $\{c_{n+m}, c_{n+m+1}, \ldots, c_{n+2m}\}$. Obviously, one such term in this linear combination is exactly $c_{n+2m}H_m(c_n)$. Then, the inequality (7.47) tells us that the errors in computing $H_{m+1}(c_n)$ and $c_{n+2m}H_m(c_n)$ are comparable to each other. Since $c_{n+2m}H_m(c_n)$ is simply one of the constituents of $H_{m+1}(c_n)$ in the performed Cauchy expansion, it follows that, whenever the inequality (7.47) is satisfied, we must consider $H_{m+1}(c_n)$ as zero, i.e. $H_{m+1}(c_n) \approx 0$. This method has been implemented in [76] in a representation of experimental data from biological problems by sums of exponential functions of time. In practice, for long signals, direct computations of a sequence of high-order determinants aimed at finding the order K as well as the pair $\{u_k, d_k\}$ may prove to be a difficult numerical problem. An alternative and indirect way of accurate and efficient computations of the Hankel determinant is provided by the use of the Lanczos coupling constants $\{\beta_n^2\}$, as has been analysed in [17]. There it was seen that, e.g. the general Hankel determinant $H_K(c_0)$ is given by $H_K(c_0) = c_0^K \prod_{m=1}^{K-1} \beta_m^{2K-2m}$. The whole set of the needed parameters $\{\beta_n\}$ can be obtained from, e.g. the recursive numerical algorithms of Gordon [170] or Wheeler [174] via efficient, robust and accurate computations. Moreover, in [17] an explicit and exact analytical formula is given for $\{\alpha_n, \beta_n^2\}$ in a general case of any non-negative value of the integer n. This also leads to an analytical result for the Hankel determinant $H_K(c_0)$ in a practical form for any rank K.

Further, it is possible to show that no explicit computation of the determinants from (7.11) and (7.12) is needed at all [17]. Moreover, the corresponding alternative method [17] is much more efficient than the above-described procedure from [76] where the explicit values of a number of possibly high-order determinants are required for a verification of the condition (7.47). The method from [17] is based upon the observation that both (7.11) and (7.12) can be written together as

$$\frac{H_{K+1}(c_n)}{H_K(c_n)} = 0. \qquad (7.48)$$

Here, the denominator is assumed to be non-zero as in (7.11). Rather than computing the determinants $H_K(c_n)$ and $H_{K+1}(c_n)$ separately, as required in (7.47), we can obtain their ratio $H_{K+1}(c_n)/H_K(c_n)$ in an easy and recursive manner without ever evaluating explicitly either $H_K(c_n)$ or $H_{K+1}(c_n)$. To achieve this goal, we shall use $H_K(c_n) = V_K^2\{\bar{d}\,\bar{u}^n\} \neq 0$ from (7.26) as well as the following readily derivable formula

$$H_K(\Delta^2 c_n) = W_K^2 H_K(c_n) \neq 0 \qquad (7.49)$$

$$W_K = \tilde{Q}_K(1) = \prod_{k=1}^{K}(u_k - 1) \neq 0 \qquad (7.50)$$

where $\Delta^2 c_n = \Delta c_{n+1} - \Delta c_n$ and $\Delta c_n = c_{n+1} - c_n$. The effective transient parameters $\bar{d}$ and $\bar{u}$ are defined in (7.26). Using (7.49), we can write (7.48) as

Difference equations and the harmonic inversion 53

$W_K^2 H_{K+1}(c_n)/H_K(\Delta^2 c_n)$, so that

$$0 = \frac{H_{K+1}(c_n)}{H_K(c_n)} = W_K^2 \frac{H_{K+1}(c_n)}{H_K(\Delta^2 c_n)} \equiv W_K^2 e_K(c_n). \tag{7.51}$$

Here, $e_K(c_n)$ is the Shanks transform for the geometric sequence of time signal points $\{c_n\}$ from (5.11). The general expression for $e_m(c_n)$ is given by the ratio of two Hankel determinants $H_{m+1}(c_n)$ and $H_m(\Delta^2 c_n)$ [71, 72]

$$e_m(c_n) \equiv \frac{H_{m+1}(c_n)}{H_m(\Delta^2 c_n)} = \frac{\begin{vmatrix} c_n & c_{n+1} & \cdots & c_{n+m} \\ c_{n+1} & c_{n+2} & \cdots & c_{n+m+1} \\ \vdots & \vdots & \ddots & \vdots \\ c_{n+m} & c_{n+m+1} & \cdots & c_{n+2m} \end{vmatrix}}{\begin{vmatrix} \Delta^2 c_n & \Delta^2 c_{n+1} & \cdots & \Delta^2 c_{n+m-1} \\ \Delta^2 c_{n+1} & \Delta^2 c_{n+2} & \cdots & \Delta^2 c_{n+m} \\ \vdots & \vdots & \ddots & \vdots \\ \Delta^2 c_{n+m-1} & \Delta^2 c_{n+m} & \cdots & \Delta^2 c_{n+2m-2} \end{vmatrix}}. \tag{7.52}$$

Dividing (7.51) by $W_K^2 \neq 0$, it follows $H_{K+1}(c_n)/H_K(\Delta^2 c_n) \equiv e_K(c_n) = 0$ and, hence, for each n from the interval $0 \leq n \leq N - 1$, we have

equations (7.11), (7.12) & (7.48) are equivalent to: $e_K(c_n) = 0$. (7.53)

In the considered direct problem, the conditions (7.11) and (7.12) guarantee uniqueness of the solution (7.10) for the difference equation (7.7). In the harmonic inversion, the same conditions (7.11) and (7.12) guarantee that the given time signal c_n associated with a Lorentzian spectrum will have exactly K attenuated complex exponentials $\{\exp(-i\omega_k \tau)\}$ with constant amplitudes $\{d_k\}$. We show above that (7.11) and (7.12) are also the two conditions for the existence of the Shanks transform $e_K(c_n)$ which must be equal to zero for the sequence $\{c_n\} = \{\sum_{k=1}^K d_k \exp(-in\omega_k \tau)\}$ [142]. The transform $e_K(c_n)$, which is yet another form of the PA, will be analysed in detail in chapter 18. In practice, the transform (7.52) is not computed directly from its definition invoking the determinantal quotient, but rather via a very efficient recursion called the Wynn ε-algorithm [73] which will also be elaborated in chapter 18. Briefly, Wynn [73] realized that the Shanks transform [71, 72] from (7.52) would be of a limited practical value unless an alternative procedure is found to alleviate explicit computations of the Hankel determinants whose rank can be large. One of such alternatives is the Wynn epsilon algorithm [73] which computes an auxiliary quantity $\varepsilon_m(c_n) \equiv \varepsilon_n^{(m)}$ via the following recursion connecting the four adjacent elements of the Padé table

$$\varepsilon_n^{(m+1)} = \varepsilon_{n+1}^{(m-1)} + \frac{1}{\varepsilon_{n+1}^{(m)} - \varepsilon_n^{(m)}} \tag{7.54}$$

$$\varepsilon_n^{(m)} \equiv \varepsilon_m(c_n) \qquad \varepsilon_n^{(-1)} = 0 \qquad \varepsilon_n^{(0)} = c_n. \tag{7.55}$$

The ε-algorithm (7.54) can be programmed to work only with one dimensional arrays as all the intermediate results can be safely overwritten [89]. Wynn [73] has shown that the even numbered elements of the ε-table give the Shanks transform according to the relationship

$$\varepsilon_n^{(2m)} = e_m(c_n). \tag{7.56}$$

Furthermore, the odd numbered elements $\varepsilon_n^{(2m+1)}$ of the same ε-table are given by the following formula

$$\varepsilon_n^{(2m+1)} = \frac{1}{e_m(\Delta c_n)}. \tag{7.57}$$

The expression (7.51) for determining the order K can now take the form

$$\varepsilon_n^{(2K)} = 0. \tag{7.58}$$

The requirement (7.51) or (7.58) determines the rank K uniquely. Therefore, given the set $\{c_n\}_{n=0}^{N-1}$ of N signal points, we compute recursively the ε-vectors $\{\varepsilon_n^{(2m)}\} = \{e_m(c_n)\} (n, m = 0, 1, 2, \ldots)$. Then the value $m = K$ for which we detect that $\varepsilon_n^{(2K)} = 0$, as required by (7.58), will yield exactly the rank K. Such an obtained rank K shall coincide with the exact number of harmonics/transients/resonances (that represent metabolites in MRS) in the given time signal $\{c_n\}$ modelled by (5.11) or (7.10).

The condition (7.51) for determining the rank K of the harmonic inversion problem, inherent in the time signal $\{c_n\}$ from the geometric sequence (7.10), emerged from a practical need to bypass explicit evaluations of high-order Hankel determinants. These determinants are encountered in the conditions (7.11) and (7.12) for the uniqueness of the solution (7.10) of the Kth order homogeneous difference equation (7.7) which is the direct problem counterpart of the harmonic inversion. However, there is more to (7.51) than the sole determination of the rank K, as will be shown in chapters 15 and 18. As a preview, notice that instead of (7.10) we could have considered, as we shall do in chapter 20, an auxiliary sequence of some given numbers $\{\bar{c}_n\}$ defined by

$$\bar{c}_n = \bar{c}_\infty + \sum_{k=1}^{K} d_k e^{-in\omega_k \tau} \qquad \mathrm{Im}(\omega_k) < 0 \tag{7.59}$$

where $\bar{c}_\infty$ is the limiting value or the equilibrium value attained when all the transients, i.e. the evanescent modes $\{d_k \exp(-in\omega_k \tau)\}$ have decayed to zero as $n\tau \longrightarrow \infty$. Obviously

$$\bar{c}_n = c_n \tag{7.60}$$

provided that

$$\bar{c}_\infty = 0 \tag{7.61}$$

where c_n is from (5.11). Here the following important question emerges: which operational form should a mathematical transform $S(\bar{c}_n)$ have in order to secure that the base $\bar{c}_\infty$ will be obtained exactly, when the sequence $\{\bar{c}_n\}$ is subjected to $S(\bar{c}_n)$

$$S(\bar{c}_n) = \bar{c}_\infty \qquad (7.62)$$

where all the K transients $\{d_k \exp(-in\omega_k \tau)\}_{k=1}^K$ are filtered out *explicitly* and *exactly* from the signal (7.59), so that one could arrive straight at the equilibrium value $\bar{c}_\infty$? The exact answer to this question is: the sought transform is the Shanks transform $S(\bar{c}_n) = e_K(\bar{c}_n)$ as will be analysed in chapter 16

$$e_K(\bar{c}_n) = \bar{c}_\infty. \qquad (7.63)$$

In the special case (7.61), when $\bar{c}_\infty = 0$, we know that (7.59) is reduced to (7.10) according to (7.60). In such a circumstance (7.63) becomes $e_K(c_n) = 0$ which is (7.51). Hence, on top of having the capacity to determine the rank K, the condition (7.51) is also capable of representing *the exact filter* for all the K transients $\{d_k u_k^n\}$ in the signal c_n from (5.11). This can be rephrased as follows. Suppose that we are given a sequence of equidistantly sampled entries $\{c_n\}$ stemming from either some theoretical calculations or from experimental measurements such that (5.11) is chosen as a mathematical model to represent these data. Then, as pointed out by Wynn [73], in order to establish the suitability of the model (5.11) for the entries $\{c_n\}$ and, more importantly, to determine the rank K, it merely suffices to apply the Shanks transformation $e_K(c_n)$ to the data $\{c_n\}$ viz $e_K(c_n) = 0$ as in (7.51) [71, 72]. Here, the word 'suitability' used relative to the data $\{c_n\}$ means that the transform $e_K(c_n)$ is able to determine uniquely the number K of the normal mode harmonics in (5.11) and also to extract subsequently all the $2K$ spectral parameters $\{\omega_k, d_k\}_{k=1}^K$ consistent with the data $\{c_n\}$, where $\omega_k = (i/\tau)\ln(u_k)$. Several concrete procedures of the extraction of the parameters $\{\omega_k, d_k\}$ from the data $\{c_n\}$ have been thoroughly analysed in [17, 18].

To conclude this chapter, we recall that the present book has the main focus on the inverse problems or reconstructions. Nevertheless, as this chapter illustrates, the associated direct problems also play a pivotal role in their inverse counterparts. Thus, starting from the difference equation (7.7) with constant coefficients, the solution (7.10) is obtained. This solution will be unique if and only if the two conditions (7.11) and (7.12) are simultaneously fulfilled, in addition to knowing the boundary conditions to the difference equation (7.7). Despite being the critical part of a direct problem, the conditions (7.11) and (7.12) are also identified in this chapter as the entire source of the harmonic inversion problem. Indeed, as shown above, the conditions (7.11) and (7.12) translate directly into the requirement (7.51) or (7.58) for the suitability of the inverse modelling (5.11) to represent the given data $\{c_n\}$ as a linear combination of damped complex exponentials whose number K, as well as the nodal frequencies $\{\omega_k\}$ and the corresponding amplitudes $\{d_k\}$, are completely unknown prior to the analysis.

Chapter 8

The Schrödinger eigenproblem in global spectral analysis

Since the exponentially damped signal (5.11) is obtained from the matrix element $(\Phi_0|\hat{U}(\tau)|\Phi_0)$, as is clear from (5.7), it follows that the spectral parameters $\{\omega_k, d_k\}$ of $\{c_n\}$ can be obtained without any fitting by, e.g. diagonalizing the evolution operator $\hat{U}(\tau)$ via (5.8) or generally

$$\hat{U}^s(\tau)|\Upsilon_k) = u_k^s|\Upsilon_k) \qquad \hat{U}(\tau) = e^{-i\hat{\Omega}\tau} \tag{8.1}$$

where $\hat{\Omega}$ is the 'Hamiltonian' of the system and s is any finite integer ($s = \ldots, -2, -1, 0, 1, 2, \ldots$) [108]–[115]. The eigenvalue problem (8.1) of the sth power of the evolution operator $\hat{U}$ is obtained from (5.8) for $f(\hat{U}) = \hat{U}^s(\tau)$. In diagonalizations, one does not necessarily need the explicit knowledge of the operator $\hat{U}(\tau)$ itself, but only its matrix elements $(\xi_m|\hat{U}(\tau)|\xi_n)$ are requested. Here, $\{|\xi_n)\}$ is a suitably selected complete set of the expansion functions $\{|\xi_n)\}$ that form a basis which does not need to be orthonormalized. For example, the collection of functions $\{|\Phi_n)\}_{n=0}^{M-1}$ represents one such basis which is called the Schrödinger or Krylov basis[1]. The two different 'orbitals' $|\Phi_n)$ and $|\Phi_m)$ are not mutually orthogonal, so that their overlap is generally non-zero [229]

$$S_{n,m} \equiv S_{nm} = (\Phi_m|\Phi_n) \neq c_0 \delta_{n,m} \qquad |\Phi_n) = \hat{U}^n|\Phi_0). \tag{8.2}$$

Inserting $|\Phi_n)$ and $(\Phi_m|$ into (8.2) and using the symmetry (3.3) of the scalar product, we have

$$S_{n,m} = c_{n+m}. \tag{8.3}$$

---

[1] The Schrödinger basis set $\{|\Phi_n)\} = \{\hat{U}^n|\Phi_0)\}$ is not orthogonal. Here, $|\Phi_0)$ is the initial state of the studied system. As in chapter 5, we recall that the set $\{|\Phi_n)\}$ is also known as the Krylov basis. As mentioned, the reason for which we adhere to the nomenclature 'the Schrödinger basis' for $\{|\Phi_n)\}$ is that the continuous function $|\Phi(t))$ satisfies the time-dependent Schrödinger equation $(i\partial/\partial t)|\Phi(t)) = \hat{\Omega}|\Phi(t))$. The state $|\Phi_n)$ is obtained from $|\Phi(t))$ by discretization $t = n\tau$, so that $|\Phi(n\tau)) \equiv |\Phi_n)$ where τ is the sampling time.

The Schrödinger eigenproblem in global spectral analysis 57

In general, we have $c_{n+m} \neq \delta_{n,m}$, so that indeed the states $|\Phi_n)$ and $|\Phi_m)$ are non-orthogonal for $n \neq m$ as stated in (8.2). Given N signal points $\{c_n\}$, the sum of the maximal values of indices n and m in (8.3) is $n_{\max} + m_{\max} = 2M - 2$. Thus, the upper limit M of the available Schrödinger states $\{|\Phi_n)\}$ will satisfy the relation

$$2M - 1 = N. \tag{8.4}$$

Therefore, the overlap matrix $\mathbf{S}$ of the dimension $M \times M$ is given by

$$\mathbf{S} = \{S_{n,m}\} \equiv \{S_{nm}\} = \{c_{n+m}\}_{n,m=0}^{M-1}. \tag{8.5}$$

The non-orthogonality of the basis $\{|\Phi_n)\}$ implies that the decomposition of the unity operator will not be as simple as in (3.6), since the inverse matrix $\mathbf{S}^{-1}$ must be involved via [97, 229]

$$\hat{1} = \sum_{k=1}^{K} \sum_{k'=1}^{K} |\Phi_k) S_{kk'}^{-1} (\Phi_{k'}|. \tag{8.6}$$

We assume that the Schrödinger basis $\{|\Phi_n)\}$ is locally complete and this would permit a development of the eigenfunction $|\Upsilon_k)$ of $\hat{U}(\tau)$ from (8.1), as follows

$$|\Upsilon_k) = \sum_{n=0}^{M-1} A_{n,k} |\Phi_n) \tag{8.7}$$

where M is given in (8.4). The expansion coefficients $\{A_{n,k}\}$ from (8.7) are the elements of the column matrix $\mathbf{A}_k = \{A_{n,k}\} \equiv \{A_{nk}\}$. Here, the subscript k in the column vector $\mathbf{A}_k$ should not be interpreted as the dimension, but rather as the counting, i.e. the running index. We insert this state vector into (8.1) which is afterwards projected onto the state $(\Phi_m|$ to yield

$$\sum_{n=0}^{M-1} U_{n,m}^{(s)} A_{n,k} = u_k^s \sum_{n=0}^{M-1} S_{n,m} A_{n,k}. \tag{8.8}$$

The matrix element $U_{n,m}^{(s)}$ is associated with the evolution operator raised to the power s such that

$$U_{n,m}^{(s)} \equiv U_{nm}^{(s)} = (\Phi_m | \hat{U}^s(\tau) | \Phi_n) \tag{8.9}$$

where $U_{n,m}^{(0)} \equiv S_{n,m}$ and $U_{n,m}^{(1)} \equiv U_{n,m}$ as in (5.13). Then the system of linear equations (8.8) can be compactly written in its corresponding matrix representation

$$\mathbf{U}^{(s)} \mathbf{A}_k = u_k^s \mathbf{S} \mathbf{A}_k \qquad \mathbf{U}^{(s)} = \{U_{n,m}^{(s)}\}_{n,m=0}^{M-1}. \tag{8.10}$$

The expansion column matrix $\mathbf{A}_k$ from (8.10) is the eigenvector of $\mathbf{U}^{(s)}$ and the elements $\{S_{nm}\}$ of the overlap matrix $\mathbf{S}$ are given in (8.3). The obtained expression (8.10) is not an ordinary, but rather a generalized eigenvalue problem involving the overlap matrix **S**, due to the mentioned lack of orthogonality of the Schrödinger basis functions $|\Phi_m\rangle$ and $|\Phi_n\rangle$ for $n \neq m$. As a direct consequence of this latter fact, instead of (8.1), one solves a more difficult generalized eigenvalue problem involving the overlap matrix **S** which is a bottleneck of every non-orthogonal basis. In fact, in such a case, one solves two eigenvalue problems in succession, since diagonalization of **S** precedes that of **U**. Each of these two diagonalizations customarily leads to spurious eigenvalues. This problem could be considerably mitigated if an orthogonal basis set is used from the onset, which deals only with ordinary eigenvalue problems where the matrix **S** is not present at all. Such a basis has been thoroughly analysed in [17] using the Lanczos recursive algorithm of wave packet propagations. In the Schrödinger basis $\{|\Phi_n\rangle\}$, the elements of matrix $\mathbf{U}^{(s)}$ are given by (5.13), i.e.

$$U_{n,m}^{(s)} \equiv (\Phi_m|\hat{U}^s(\tau)|\Phi_n) = (\Phi_0|\hat{U}^{n+m+s}(\tau)|\Phi_0) = c_{n+m+s}$$
$$U_{n,m}^{(s)} = c_{n+m+s} \qquad (8.11)$$

where (5.6), (5.9) and (5.12) are used. Hence, the matrix element of the sth power of the evolution operator $\hat{U}(\tau)$ taken over two general Schrödinger states $|\Phi_n\rangle$ and $(\Phi_m|$ is reduced to only one single value of the auto-correlation function or the time signal $c_{n+m+s} = C_{n+m+s}$. Obviously, this result also includes the overlap matrix **S** which is obtained for $s = 0$ as a special case of $\mathbf{U}^{(s)}$. Likewise, the evolution matrix $\mathbf{U} = \mathbf{U}^{(1)}$ is obtained for $s = 1$ and it reads as

$$\mathbf{U} = \{U_{n,m}\} \equiv \{U_{nm}\} = \{c_{n+m+1}\}_{n,m=0}^{M-1}. \qquad (8.12)$$

Once the whole set $\{u_k, A_{n,k}\}$ is obtained by solving the generalized eigenvalue problem (8.10), the eigenfrequencies are deduced from the relation

$$\omega_k = i\tau^{-1} \ln(u_k) \qquad (8.13)$$

where the principal branch should be taken from the multi-valued complex logarithm. Throughout this book we shall adhere to the Landau–Lifshitz convention [144] according to which the correct branch of a multi-valued function $f(z)$ is obtained by selecting the least value of the argument $\mathrm{Arg}(f(z)) = \tan^{-1}([\mathrm{Im}\{f(z)\}]/[\mathrm{Re}\{f(z)\}])$. The residues $\{d_k\}$ associated with the eigenfrequencies $\{\omega_k\}$ are calculated by inserting the expansion (8.7) for $|\Upsilon_k\rangle$ into (3.10) and using (5.9) as well as (5.12) with the result

$$d_k = \left(\sum_{n=0}^{M-1} c_n A_{n,k}\right)^2. \qquad (8.14)$$

The above procedure constitutes a computational tool for obtaining the spectrum of the evolution matrix $\mathbf{U}_M = \{c_{n+m+1}\}_{n,m=0}^{M-1}$ through diagonalization in the Schrödinger basis $\{|\Phi_n\rangle\}$. This procedure works optimally if the signal length

N is not too long, say $N \leq N_{\text{cut}} \approx 300$, in which case the problem of ill-conditioning in solving the generalized eigenvalue problem (8.10) can be kept under control [108]–[115]. Using the sequence $\{|\Phi_n)\}$ as a basis set makes the data matrix **U** full, since these Schrödinger states are delocalized. The term 'delocalized' means that the set $\{|\Phi_n)\}$ is totally independent of the selected frequency window $[\omega_{\min}, \omega_{\max}]$ from which the eigenvalues $\{u_k\}$ are supposed to be extracted. The basis set $\{|\Phi_n)\}$ is not practical for large N [108]–[115]. An attempt to extend the analysis to an arbitrarily large N, brings us to the topic of the so-called dimensionality reduction problem which is analysed in the next two chapters.

Chapter 9

Dimensionality reduction in the frequency domain

The advantage of quantum-mechanical signal processing is in a direct reliance upon the dynamics of the examined system whose time evolution is described by a first-order differential equation, which is the Schrödinger equation (4.1) for the continuous or analog state vector $|\Phi(t)\rangle$. As is well-known, linearity of the Schrödinger ansatz (4.1) implies that any sum of the states $\{|\Phi(t)\rangle\}$ with constant coefficients also satisfies the same equation. Hence the flexibility of this class of methods operating with state vectors that permit changes, so that one can switch from one basis to another without altering the sought eigensolutions. If the signal length N is not too large, say of the order of $N_{\text{cut}} \leq 300$, the Schrödinger basis set $\{|\Phi_n\rangle\}$ from (8.2) should suffice. However, this is not true any longer for large values of N, in which case one can resort to the technique of windowing as implemented by the FD [108]–[112] in the frequency domain or by the DSD [115] in the time domain. In the FD, this technique of changing the basis amounts to using the Schrödinger–Fourier or Krylov–Fourier basis set $\{|\Psi_j\rangle\}_{j=1}^{J}$ which is given as a linear combination of $|\Phi_n\rangle$ with the preassigned Fourier-type coefficients

$$|\Psi_j\rangle = \sum_{n=0}^{M-1} e^{in\varphi_j} |\Phi_n\rangle \qquad (1 \leq j \leq J). \qquad (9.1)$$

Here, the elements φ_j of the set $\{\varphi_j\}_{j=1}^{J}$ represent the equidistantly spaced grid points on the unit circle in the complex dimensionless angular frequency plane $\varphi = \omega\tau$. In principle, the frequency φ_j can be chosen freely in the interval $[\varphi_{j_{\min}}, \varphi_{j_{\max}}]$ or it can also be selected from the Fourier grid $\varphi_j = \tilde{\omega}_j \tau$ where $\tilde{\omega}_j = 2\pi j/(J\tau)$. Assuming that the basis $\{|\Psi_j\rangle\}$ is locally complete, i.e.

$$\sum_{j=1}^{J} |\Psi_j\rangle(\Psi_j| = 1 \qquad (9.2)$$

we can expand the full wavefunction $|\Upsilon_k\rangle$ according to

$$|\Upsilon_k\rangle = \sum_{j=1}^{J} B_{j,k}|\Psi_j\rangle \qquad (9.3)$$

where $\{B_{j,k}\}$ are the expansion coefficients. Inserting (9.3) into (8.1) and projecting the obtained equation onto the state $\langle\Psi_{j'}|$ will yield the following generalized eigenproblem

$$\mathbf{U}^{(s)}\mathbf{B}_k = u_k^s \mathbf{S}\mathbf{B}_k \qquad (9.4)$$

with $\mathbf{B}_k = \{B_{j,k}\}$ being a column matrix and where the matrix $\mathbf{U}^{(s)}$ has the elements $\mathrm{U}^{(s)}_{j,j'}$ as[1]

$$\mathrm{U}^{(s)}_{j,j'} \equiv \mathrm{U}^{(s)}(\varphi_j, \varphi_{j'}) = \langle\Psi_{j'}|\hat{U}^s(\tau)|\Psi_j\rangle. \qquad (9.5)$$

Inserting $|\Psi_j\rangle$ and $\langle\Psi_{j'}|$ from (9.1) into (9.5), we have

$$\mathrm{U}^{(s)}_{j,j'} = \sum_{n=0}^{M-1}\sum_{n'=0}^{M-1} c_{n+n'+s} z_j^{-n} z_{j'}^{-n'} \qquad z_m = e^{-i\varphi_m} \qquad (9.6)$$

where the relation $2M - 1 = N$ from (8.4) is used. Here, one of the two sums can be carried out analytically as in [108, 109] with the following pair of the results, for $z_{j'} \neq z_j$

$$\mathrm{U}^{(s)}_{j,j'} = \frac{1}{z_j - z_{j'}} \Bigg\{ z_j \sum_{n=0}^{M-1} c_{n+s} z_{j'}^{-n} - z_{j'} \sum_{n=0}^{M-1} c_{n+s} z_j^{-n}$$
$$- z_j^{-(M-1)} \sum_{n=M}^{2M-2} c_{n+s} z_{j'}^{M-n} + z_{j'}^{-(M-1)} \sum_{n=M}^{2M-2} c_{n+s} z_j^{M-n} \Bigg\} \qquad (9.7)$$

and for $z_{j'} = z_j$

$$\mathrm{U}^{(s)}_{j,j} = \sum_{n=0}^{2M-2} (M - |M - 1 - n|) c_{n+s} z_j^{-n}. \qquad (9.8)$$

An alternative expression has been derived in [18] for $\mathrm{U}^{(s)}_{j,j'}$ which is automatically valid for both cases $z_{j'} \neq z_j$ and $z_{j'} = z_j$. This is accomplished by introducing an auxiliary bivariate function as

$$\tilde{\mathrm{U}}^{(s)}_{j,j'} = \sum_{n=0}^{M-1}\sum_{n'=0}^{M-1} \tilde{c}_{n+s,n'} z_j^{-n} z_{j'}^{-n'} \qquad (9.9)$$

---

[1] No confusion should exist when, in an attempt to avoid introducing too many different notations, we use the label $\mathrm{U}^{(s)}_{j,j'}$ for the elements of the evolution matrix in the Schrödinger–Fourier basis $\{|\Psi_j\rangle\}$ as the same symbol $\mathrm{U}^{(s)}_{n,n'}$ which is also employed for the Schrödinger basis $\{|\Phi_n\rangle\}$.

which coincides with $\tilde{U}^{(s)}_{j,j'}$ for $\tilde{c}_{n+s,n'} \equiv c_{n+n'+s}$. Using the Kronecker δ-symbol (3.7), we can insert an additional sum over m into the rhs of (9.9)

$$\tilde{U}^{(s)}_{j,j'} = \sum_{m=0}^{2M-2} \sum_{n=0}^{M-1} \sum_{n'=0}^{M-1} \tilde{c}_{n+s,n'} z_j^{-n} z_{j'}^{-n'} \delta_{n+n',m}. \tag{9.10}$$

Here, we set $n' = m - n$, so that

$$\tilde{U}^{(s)}_{j,j'} = \sum_{m=0}^{2M-2} \sum_{\substack{n+n'=m \\ n+n'=0 (n,n' \geq 0)}} \tilde{c}_{n+s,n'} z_j^{-n} z_{j'}^{-n'} = \sum_{m=0}^{2M-2} \sum_{n=n_1}^{n_2} \tilde{c}_{n+s,m-n} z_j^{-n} z_{j'}^{n-m}$$

$$\tilde{U}^{(s)}_{j,j'} = \sum_{m=0}^{2M-2} z_{j'}^{-m} \sum_{n=n_1}^{n_2} \tilde{c}_{n+s,m-n} \xi^n \tag{9.11}$$

$$\xi = \frac{z_{j'}}{z_j} \qquad n_1 = \max\{0, m-M+1\} \qquad n_2 = \min\{M-1, m\}. \tag{9.12}$$

To retrieve the quantity $U^{(s)}_{j,j'}$ from (9.6) we need to put $\tilde{c}_{n+s,n'} = c_{n+n'+s}$ in (9.11) evaluated at $n' = m - n$ which eliminates the index n' from each signal point $[\tilde{c}_{n+s,n'}]_{n'=m-n} = \tilde{c}_{n+s,m-n} = c_{n+(m-n)+s} = c_{m+s}$ and thus

$$U^{(s)}_{j,j'} = \sum_{m=0}^{2M-2} c_{m+s} z_{j'}^{-m} g_m(\xi) \tag{9.13}$$

where $g_m(\xi)$ is a geometric progression given in terms of the Dirichlet kernel $D_n(z)$ [64]

$$g_m(\xi) = \sum_{n=n_1}^{n_2} \xi^n = \xi^{n_1} \left\{ \frac{1 - \xi^{n_2-n_1+1}}{1-\xi} \right\}$$

$$= \xi^{(n_1+n_2)/2} \left\{ \frac{\xi^{-(n_2-n_1+1)/2} - \xi^{(n_2-n_1+1)/2}}{\xi^{-1/2} - \xi^{1/2}} \right\}$$

$$= e^{i(n_1+n_2)(\varphi_j - \varphi_{j'})/2} D_{n_2-n_1+1}(\varphi_j - \varphi_{j'})$$

$$g_m(\xi) \equiv \sum_{n=n_1}^{n_2} \xi^n = e^{i(n_1+n_2)(\varphi_j - \varphi_{j'})/2} D_{n_2-n_1+1}(\varphi_j - \varphi_{j'}) \tag{9.14}$$

$$D_n(z) = \frac{\sin(nz/2)}{\sin(z/2)}. \tag{9.15}$$

Using (9.12), it follows

$$n_2 - n_1 = \min\{M-1, m\} - \max\{0, m-M+1\}$$

$$= \begin{cases} m & m \in [0, M-1] \\ 2M-2-m & m \in [M, 2M-2]. \end{cases} \tag{9.16}$$

The rhs of this equation can also be written in a more condensed form as $M - 1 - |M - 1 - m|$ which encompasses both subintervals $[0, M - 1]$ and $[M, 2M - 2]$, so that

$$n_2 - n_1 = M - 1 - |M - 1 - m| \qquad m \in [0, 2M - 2]. \tag{9.17}$$

Similarly, it is immediately apparent from (9.12) that $n_1 + n_2$ reduces to m irrespective of whether $m \in [0, M - 1]$ or $m \in [M, 2M - 2]$

$$n_1 + n_2 = m \qquad m \in [0, 2M - 2]. \tag{9.18}$$

Finally, inserting (9.14) into (9.13) and using (9.17) and (9.18), we arrive at a new and compact expression for the matrix element $U_{j,j'}^{(s)}$

$$U_{j,j'}^{(s)} = \sum_{n=0}^{N-1} c_{n+s} e^{in(\varphi_j + \varphi_{j'})/2} D_{M-|M-n-1|}(\varphi_j - \varphi_{j'}) = U_{j',j}^{(s)} \tag{9.19}$$

where the upper summation limit $2M - 2$ is identified as $N - 1$ following (8.4). Thus, unlike (9.7), the result (9.19) for the matrix element $U_{j,j'}^{(s)}$ is valid for $\varphi_{j'} \neq \varphi_j$ and $\varphi_{j'} = \varphi_j$. The final formula (9.19) is consistent with the fact that a purely rectangular time signal $c_n = \theta(N-1-n)$ has the discrete Fourier transform given by $D_N(\omega\tau) \exp(i\omega\tau[N - 1]/2)$ [64]. Here, $\theta(m)$ is the Heaviside θ-step function [132]

$$\theta(x - y) = \begin{cases} 1 & x > y \\ 0 & x < y. \end{cases} \tag{9.20}$$

The form (9.7) of the matrix element $U_{j,j'}^{(s)}$ from [108, 109] also exhibits a behaviour of the type of the sinc-function $\mathrm{sinc}(z) = z^{-1} \sin(z)$. We have verified that both (9.7) and (9.19) yield the same numerical results for the standard harmonic inversion problem solved by the FD method. As an analytical check, we used (9.19) to compute the matrix elements $U_{j,j'}^{(s)}$ for $M = 3$ which gives

$$U_{j,j'}^{(s)} = c_s + c_{s+1}(e^{i\varphi_j} + e^{i\varphi_{j'}}) + c_{s+2}[e^{2i\varphi_j} + e^{i(\varphi_j + \varphi_{j'})} + e^{2i\varphi_{j'}}]$$
$$+ c_{s+3}[e^{i(\varphi_j + 2\varphi_{j'})} + e^{i(\varphi_{j'} + 2\varphi_j)}]. \tag{9.21}$$

For $M = 3$ the same analytical result (9.21) is verified to follow also from (9.7) as well as from the double sum (9.6), which defines the matrix element $U_{j,j'}^{(s)}$. More generally, it is possible to show that (9.7) and (9.8) can also be analytically reduced to (9.19). To achieve this explicitly, we first introduce the index change $n' = n - M$ in the last two sums of (9.7) to write

$$U_{j,j'}^{(s)} = \sum_{n=0}^{M-1} \left[c_{n+s} \frac{1 - \zeta^{n+1}}{1 - \zeta} z_j^{-n} + c_{n+s+M} \frac{1 - \zeta^{M-n-1}}{1 - \zeta} z_j^{1-M} z_{j'}^{-n-1} \right] \tag{9.22}$$

where $\zeta = 1/\xi = z_j/z_{j'}$. The two quotients in the round brackets in (9.22) are recognized as the results of the geometric progression (9.14) viz

$$\sum_{n=0}^{M-1} \zeta^n = \frac{1-\zeta^M}{1-\zeta} = e^{i(M-1)(\varphi_{j'}-\varphi_j)/2} D_M(\varphi_j - \varphi_{j'}) \qquad (9.23)$$

and this reduces (9.22) to the form

$$U_{j,j'}^{(s)} = \sum_{n=0}^{M-1} [c_{n+s} D_{n+1}(\varphi_j - \varphi_{j'}) e^{in(\varphi_j+\varphi_{j'})/2}$$
$$+ c_{n+s+M} D_{M-n-1}(\varphi_j - \varphi_{j'}) e^{i(M+n)(\varphi_j+\varphi_{j'})/2}]. \qquad (9.24)$$

The starting expression (9.7) was valid only for $z_{j'} \neq z_j$. However, the formulae (9.22) and (9.24), that are derived from (9.7) are well defined for both $z_{j'} \neq z_j$ and $z_{j'} = z_j$. In (9.24) we can use the property $D_M(0) = M$, so that for $j' = j$ or for $z_{j'} = z_j$, i.e. $\varphi_{j'} = \varphi_j$, we have

$$U_{j,j}^{(s)} = \sum_{n=0}^{M-1} [(n+1) c_{n+s} e^{in\varphi_j} + (M-n-1) c_{n+s+M} e^{i(M+n)\varphi_j}]. \qquad (9.25)$$

Due to the identity $M - n - 1 = |M - n - 1|$ for $n \leq M - 1$ which is the case in (9.24), we can write $D_{n+1}(\varphi_j - \varphi_{j'}) = D_{M-\{M-n-1\}}(\varphi_j - \varphi_{j'}) = D_{M-|M-n-1|}(\varphi_j - \varphi_{j'})$, so that

$$U_{j,j'}^{(s)} = \sum_{n=0}^{M-1} c_{n+s} D_{M-|M-n-1|}(\varphi_j - \varphi_{j'}) e^{in(\varphi_j+\varphi_{j'})/2} + W_{j,j'}^{(s)} \qquad (9.26)$$

$$W_{j,j'}^{(s)} = \sum_{n=0}^{M-1} c_{n+s+M} D_{M-n-1}(\varphi_j - \varphi_{j'}) e^{i(M+n)(\varphi_j+\varphi_{j'})/2}. \qquad (9.27)$$

Changing the summation index in (9.27) according to $n' = n + M$, it follows

$$W_{j,j'}^{(s)} = \sum_{n=M}^{2M-1} c_{n+s} D_{2M-n-1}(\varphi_j - \varphi_{j'}) e^{in(\varphi_j+\varphi_{j'})/2}$$
$$= \sum_{n=M}^{2M-1} c_{n+s} D_{M-|M-n-1|}(\varphi_j - \varphi_{j'}) e^{in(\varphi_j+\varphi_{j'})/2} \qquad (9.28)$$

where $2M - n - 1 = M - |M - n - 1|$ for $n \in [M, 2M - 1]$. Now (9.28) can be simplified as

$$W_{j,j'}^{(s)} = \sum_{n=M}^{2M-2} c_{n+s} D_{M-|M-n-1|}(\varphi_j - \varphi_{j'}) e^{in(\varphi_j+\varphi_{j'})/2} \qquad (9.29)$$

where the upper summation limit $2M-1$ is reduced to $2M-2$, since the last term for $n=2M-1$ has the overall multiplicative function $D_{(2M-1)-(2M-1)}(\varphi_j + \varphi_{j'}) = D_0(\varphi_j + \varphi_{j'})$ which is zero due to the property $D_0(z) = 0$. Inserting the result (9.28) for the second term $W^{(s)}_{j,j'}$ into (9.26) yields

$$U^{(s)}_{j,j'} = \sum_{n=0}^{M-1} c_{n+s} D_{M-|M-n-1|}(\varphi_j - \varphi_{j'}) e^{in(\varphi_j+\varphi_{j'})/2}$$
$$+ \sum_{n=M}^{2M-2} c_{n+s} D_{M-|M-n-1|}(\varphi_j - \varphi_{j'}) e^{in(\varphi_j+\varphi_{j'})/2}$$
$$= \sum_{n=0}^{2M-2} c_{n+s} D_{M-|M-n-1|}(\varphi_j - \varphi_{j'}) e^{in(\varphi_j+\varphi_{j'})/2}. \qquad (9.30)$$

If in the third equation of (9.30), we use $2M-2 = N-1$ from (8.4), then we obtain the result

$$U^{(s)}_{j,j'} = \sum_{n=0}^{N-1} c_{n+s} e^{in(\varphi_j+\varphi_{j'})/2} D_{M-|M-n-1|}(\varphi_j - \varphi_{j'}) \qquad \text{(QED)} \quad (9.31)$$

in agreement with the previously derived formula from (9.19) and this was set up to be proven.

According to the result (9.19), the matrix elements $U^{(s)}_{j,j'}$ are diagonally dominated. The function $D_n(z) = [\sin(nz/2)]/\sin(z/2)$ from (9.19) is the Dirichlet kernel [64] defined in (9.15). This function is also the Gegenbauer polynomial or the Chebyshev polynomial of the second kind both in the variable $\cos z$ [180]. The Dirichlet kernel $D_n(z-z')$ peaks at $z \approx z'$ with the maximal height equal to n. Away from this central peak, i.e. the main lobe, the function $D_n(z-z')$ decreases rapidly in an oscillatory manner. Given these features of the Dirichlet kernel, it is clear that the net effect of constructing the Fourier sum (9.1) from the Schrödinger states $\{|\Phi_n\rangle\}$ is equivalent to the generation of the well-localized basis set $\{|\Psi_j\rangle\}$. With such a narrow-band basis set at hand, one can achieve an efficient diagonalization of the evolution matrix $\mathbf{U}$ of an arbitrary size. A linear combination of such Dirichlet kernels weighted with the phase modulated c_n as in (9.19) is found to produce a considerable cancellation, yielding negligibly small values for off-diagonal matrix elements $U^{(s)}_{j,j'}$ for $j' \neq j$. The presence of the Dirichlet kernel in (9.19) ensures that the diagonal elements of matrix $\mathbf{U}^{(s)}$ are much larger than the off-diagonal ones, provided that

$$(M-1)|\varphi_j - \varphi_{j'}| \gg 1. \qquad (9.32)$$

Therefore, it follows from (9.19) that the merit of the Schrödinger–Fourier basis set (9.1) is in making the matrix elements of $\mathbf{U}$ diagonally dominated with rapidly decaying off-diagonal elements. It is this structure which underlies the FD method and causes a significant reduction in the size of the original large matrix **U**. The appearance of the damped sinusoidal behaviour of the matrix elements (9.19), similar to the so-called sinc-function $z^{-1}\sin z$, indicates the presence of a rectangular or a box-type filter [109]. This filter is indeed built in (9.1), which is precisely the vector valued DFT of the Schrödinger functions $\{|\Phi_n\rangle\}$ when, e.g. the Fourier grid $\varphi_j = \tilde{\omega}_j \tau$ is selected according to $\tilde{\omega}_j = 2\pi j/(J\tau)$ $(1 \leq j \leq J)$. This DFT is a band-limited Fourier sequence [17, 64], since the angle φ_j belongs to a preselected interval $\varphi_j \in [\varphi_{j\min}, \varphi_{j\max}]$. The purpose of the introduction of this rectangular filter is to create a basis which is dependent upon the chosen frequency range $[\omega_{\min}, \omega_{\max}]$. The resulting basis $\{|\Psi_j\rangle\}$ effectively filters out the Fourier frequencies that do not belong to the window of interest $[\omega_{\min}, \omega_{\max}]$. As a consequence, the created localized functions $\{|\Psi_j\rangle\}$ are expected to have a negligible overlap with the wave packets $\hat{U}^s(\tau)|\Psi_{j'}\rangle$ for those angles φ_j that lie outside the diagonally located strip $[\varphi_{j\min}, \varphi_{j\max}]$. In other words, the matrix $\mathbf{U}^{(s)} = \{U_{j,j'}^{(s)}\}$ from (9.19) becomes block-diagonal [176]. Hence the needed reduction of the original dimension of the problem. Having obtained the solution $\{u_k, \mathbf{B}_k\}$ of (9.4) in the basis $\{|\Psi_j\rangle\}$, the residue d_k follows from $d_k = (\Upsilon_k|\Phi_0)^2$ as in (3.10) using $|\Upsilon_k\rangle$ from (9.3). Thus (9.1) leads to the following result for the residue d_k in the FD method [108, 109]

$$d_k = \left(\sum_{n=0}^{M-1} c_n \tilde{B}_{n,k}\right)^2 \qquad \tilde{B}_{n,k} = \sum_{j=1}^{J} B_{j,k} e^{-in\varphi_j}. \qquad (9.33)$$

If φ_j are chosen to coincide with the corresponding Fourier grid points $\tilde{\omega}_j \tau$, then $\{\tilde{B}_{n,k}\}$ will represent the DFT of $\{B_{j,k}\}$. According to (8.4), the formula (9.33) for the residue d_k in the FD method includes the data $\{c_n\}$ from one half of the full signal length $M - 1 = (N-1)/2$. This occurrence is found in the FD to lead to insufficiently accurate numerical values for the elements of the set $\{d_k\}$ [109]. To achieve the needed accuracy for $\{d_k\}$ that would match the precision of the corresponding values of $\{u_k\}$ the following averaging procedure is required in the FD [109]

$$d_k = \left[\frac{2\sin\omega_k}{(2M+1)\sin\omega_k + \sin(2M+1)\omega_k} \sum_{j=1}^{J} B_{j,k} U^{(0)}(\varphi_j, \omega_k)\right]^2. \qquad (9.34)$$

By contrast, working merely with the Schrödinger basis $\{|\Phi_n\rangle\}$, the machine accuracy in the DSD method [115] for noiseless problems is achievable without any averaging in (8.14) when the so-called band-limited decimated (bld) signal $\{c_n^{\text{bld}}\}$ is used in place of $\{c_n\}$, as in chapter 10.

Chapter 10

Dimensionality reduction in the time domain

Windowing in the FD is done in the frequency domain in the course of switching from the basis set $\{|\Phi_n)\}$ to $\{|\Psi_j)\}$ while diagonalizing the matrix $\mathbf{U}$. Therefore, such a windowing is typical of the FD and cannot be used in other methods. An alternative windowing known as the band-limited decimation [113, 115] is performed directly on the original long signal $\{c_n\}$ of the length N leading to a short signal to which any desired method could be applied for subsequent data processing [115]. The band-limited decimation can be introduced as follows. We are given a discrete time signal $\{c_n\}(0 \leq n \leq N-1)$ of length N digitized using equidistant sampling with rate τ. To initialize windowing in the time domain, the signal $\{c_n\}$ is first subjected to the FFT. This yields a low resolution spectrum, since N is generally insufficiently long for the FFT to resolve dense nodal frequencies $\{\omega_k\}$ from $\{c_n\}$. By construction, the FFT is defined only at the Fourier grid points $\tilde{\omega}_k = 2\pi k/(N\tau)$. We take these latter points as the window end points and we subdivide the whole Fourier spectrum into M intervals/windows. We make sure that each of these windows contains at most, e.g. 300 Fourier grid points in order to diminish ill-conditioning of the subsequent processing stage by a selected high-resolution parametric estimator. For the purpose of scanning and checking the validity of the procedure in the entire Nyquist bandwidth $[-\pi/\tau, +\pi/\tau]$, we compute the Fourier spectra $\{F_k^{\text{bld}}\}$ separately for each of the M windows by applying the FFT to the so-called 'band-limited decimated' (bld) signals $\{c_n^{\text{bld}}\}(0 \leq n \leq N_d - 1)$ of a shorter length $N_d = [N/M]$, where $[x]$ is the integer part of the real number x. By construction, in any of the windows $[\omega_{\min}, \omega_{\max}]$, the resulting FFT spectra $\{F_k^{\text{bld}}\}$ created from $\{c_n^{\text{bld}}\}$ are identical to the corresponding Fourier spectra $\{F_k\}$ obtained from the original signal $\{c_n\}$. This constitutes the critical feature of the band-limited decimation which is the preservation of the full informational content in each of the M individual windows. This key property of the band-limited decimation is radically different from a straight-forward decimation which always leads to a loss of information. Moreover, as we shall explicitly demonstrate, the original signal $\{c_n\}$ is *not decimated at all* by the band-limited decimation. This is because each member from the set $\{c_n^{\text{bld}}\}$ will be shown to contain the intact assembly of all the N original signal points $\{c_n\}$ ($0 \leq n \leq N-1$) and the decimation is restricted exclusively to the elements of the matrix which transforms $\{c_n\}$ to $\{c_n^{\text{bld}}\}$.

The new signals $\{c_n^{\text{bld}}\}$ are generated separately in every window by setting to zero the FFT spectra outside the given frequency range $[\omega_{\min}, \omega_{\max}]$. Simultaneously, we shift the unaltered spectral remainders within the window $[\omega_{\min}, \omega_{\max}]$ in order to relocate them symmetrically about the frequency origin $\omega = 0$. This procedure yields a spectrum centred about $\omega = 0$ with the bandwidth $2\pi/(M\tau)$. In this way, considering $\omega_0 = (\omega_{\min} + \omega_{\max})/2$ as the centre of the window, we subtract ω_0 from every frequency ω belonging to the actual window. The results are the usual band-limited (bl) spectra that are subsequently subjected to the inverse FFT to arrive in each window at the band-limited signals $\{c_n^{\text{bl}}\}$ ($0 \leq n \leq N-1$) of the same length N as the original data $\{c_n\}$. But the new bandwidth is reduced M times from that of the original signal. This circumstance allows resampling $\{c_n^{\text{bl}}\}$ with M times larger sampling time $\tau_d = M\tau$ over the same total acquisition time T used for c_n, i.e. $T = N_d \tau_d = N\tau$. Hence the problem has been reduced to the one of processing a signal of a greatly shortened effective length $N_d \ll N$ [113, 115].

We re-emphasize that the whole procedure of the band-limited decimation leads to no loss of information in any of the selected windows. This technique gives M theoretically equivalent bld-signals $\{c_n^{\text{bld}}\}$ of the same short length $N_d = [N/M]$. Such signals can be arranged as the set $\{c_\ell^{\text{bl}}, c_{\ell+M}^{\text{bl}}, c_{\ell+2M}^{\text{bl}}, \ldots\}$ ($0 \leq \ell \leq M-1$). When noiseless input data are used, both the original set $\{c_n\}$ and the above sequence of M bld-signals give the same spectra to within machine accuracy in, e.g. the three novel high-resolution parametric processors introduced in [115]. Such an equivalence among these M bld-signals ceases to exist for noise corrupted data $\{c_n\}$. However, the genuine structures are expected to be approximately repeated in each M bld-spectra. Moreover, other peaks that could exhibit more noticeable fluctuations when passing from one to another of the M bld-spectra are naturally interpreted as the unstable structures and, as such, they are likely to be connected with noise. Hence, the availability of M subsignals can turn into an advantage and provide one of the potential ways to identify the noise level in corrupted signals and spectra.

To summarize the above-outlined description, we shall now give the main working prescription for the application of the band-limited decimation. The final result of this preprocessing via the successive applications of the direct and inverse FFT is a substantially shorter ($N_d \ll N$) band-limited decimated signal $\{c_n^{\text{bld}}\}$ which can afterwards be subjected to any selected estimator for the parametric or non-parametric spectral analysis [115]. Then this prescription runs as follows:

(i) Subdivide the entire Nyquist interval $[-\pi/\tau, +\pi/\tau]$ into a sequence of M smaller windows $[\omega_{\min}, \omega_{\max}]$ with τ being the sampling time increment of the original signal c_n, where $\omega_{\min}$ and $\omega_{\max}$ are the lower and upper limits of the selected frequency window. Note that here we work exclusively with the Fourier grid $\tilde{\omega}_k = 2\pi k/T$, so that the frequency window of interest is equivalent to the k-interval $[k_{\min}, k_{\max}]$ with $\tilde{\omega}_k = 2\pi k/(N\tau)$

$$k_{\min} = \left[\frac{T}{2\pi}\omega_{\min}\right] \quad k_{\max} = \left[\frac{T}{2\pi}\omega_{\max}\right] \tag{10.1}$$

where the symbol $[n/2]$ represents the largest integer in the rational number $n/2$.

(ii) Obtain the whole FFT spectrum, as an array $\{F_k\}$ of the length N from the original signal $\{c_n\} (0 \le n \le N-1)$, or symbolically $F_k = \text{FFT}(c_n)$

$$F_k = \text{FFT}(c_n) \equiv \frac{1}{N}\sum_{n=0}^{N-1} c_n e^{2i\pi nk/N}. \tag{10.2}$$

(iii) Create the band-limited FFT spectrum $\{F_k^{\text{bl}}\}$ of the length N through the replacement of all the F_k terms outside the window $[k_{\min}, k_{\max}]$ by zeros, without altering the remaining elements F_k inside the window

$$F_k^{\text{bl}} = F_k \, \Pi(k_{\min}, k_{\max}). \tag{10.3}$$

Here $\Pi(k_{\min}, k_{\max}) = \text{rec}(k_{\min}, k_{\max})$ is the Heaviside-type step function, or the so-called rectangular or rec-function[1], which is unity for $k \in [k_{\min}, k_{\max}]$ and zero elsewhere [64]

$$\Pi(k_{\min}, k_{\max}) \equiv \text{rec}(k_{\min}, k_{\max}) = \begin{cases} 1 & k \in [k_{\min}, k_{\max}] \\ 0 & k \notin [k_{\min}, k_{\max}]. \end{cases} \tag{10.4}$$

(iv) Shift the spectrum inside the window by frequency ω_0. In this way all the spectral points F_k within the window $k \in [k_{\min}, k_{\max}]$ are scaled by the central index k_0

$$k_0 = \left[\frac{T}{2\pi}\omega_0\right] = \left[\frac{k_{\min}+k_{\max}}{2}\right] \quad \omega_0 = \frac{\omega_{\min}+\omega_{\max}}{2}. \tag{10.5}$$

In other words, the shift $k \to k - k_0$ redistributes the grid points symmetrically around the origin $k = 0$. This is compensated afterwards by adding ω_0 to every obtained natural frequency ω_k of the signal. In other words, the sequence $\{F_k^{\text{bl}}\} (0 \le k \le N-1)$ is now centred at $k = k_0$. Then after step (iv) we are left with the set $\{F_{k-k_0}^{\text{bl}}\} (0 \le k \le N-1)$.

---

[1] The rec-function is similar to the conventional definition of the jump function (9.20).

(v) Construct a generally complex valued band-limited signal $\{c_n^{bl}\}(0 \le n \le N-1)$ of length N by subjecting $\{F_{k-k_0}^{bl}\}$ from step (iv) to the inverse FFT, i.e. FFT^{-1}

$$c_n^{bl} = \text{FFT}^{-1}(F_{k-k_0}^{bl}) = \sum_{k=0}^{N-1} F_{k-k_0} e^{-2\pi i n k/N}. \tag{10.6}$$

(vi) Generate the band-limited decimated signal $\{c_n^{bld}\}$ of length N_d with $N_d = [N/M]$ by decimating via skipping the signal points from $\{c_n^{bl}\}$ along the whole sequences of length M. The new sampling rate τ_d of $\{c_n^{bld}\}$ is M times larger than $\{c_n\}$ or $\{c_n^{bl}\}$, i.e. $\tau_d = M\tau$, but the total acquisition time T remains unaltered $T = N\tau = N_d \tau_d$

$$c_n^{bld}(n = 0, 1, 2, \ldots, N_d - 1) = c_n^{bl}(n = 0, M, 2M, \ldots, N-1). \tag{10.7}$$

The Nyquist intervals of the original and band-limited decimated signals are $\Delta\omega = 1/(2\tau)$ and $\Delta\omega_d = 1/(2\tau_d)$, respectively. This enables a reduction of the signal length of c_n^{bld} by the factor $\Delta\omega/\Delta\omega_d = \Delta\tau_d/\Delta\tau = M$. The whole sequence of steps (i)–(vi) of the band-limited decimation is symbolized as

$$c_n^{bld} = \left\{ \text{FFT}^{-1} \left([\text{FFT}(c_n)] \Pi(k_{\min}, k_{\max}) \right]_{k \to k-k_0}^{\text{shifting}} \right) \right\}_{N \to N_d}^{\text{decimation}}. \tag{10.8}$$

The explicit recipe in (10.8) is useful, since it provides a direct way to arrive at the analytical result for $\{c_n^{bld}\}$ in the case of any time signal $\{c_n\}$. Thus, from the preparatory step (i) of partitioning the Nyquist range into M equal windows, we proceed to step (ii) by computing the complete Fourier spectrum of the original signal (in general, the lengths of all the windows may be different)

$$F_k = \frac{1}{N} \sum_{n=0}^{N-1} c_n e^{2\pi i n k/N} \qquad 0 \le k \le N - 1. \tag{10.9}$$

Then the band-limited Fourier spectrum is created according to step (iii) via

$$F_k^{bl} = F_k \Pi(k_{\min}, k_{\max}) = \begin{cases} F_k & k \in [k_{\min}, k_{\max}] \\ 0 & k \notin [k_{\min}, k_{\max}]. \end{cases} \tag{10.10}$$

In step (iv) and (v), every k from the set F_k^{bl} with $k \in [k_{\min}, k_{\max}]$ is shifted by k_0. This is followed by the application of the inverse band-limited Fourier transform which then leads to the band-limited time signal c_n^{bl} viz

$$c_n^{bl} = \sum_{k=k_{\min}}^{k_{\max}} F_k^{bl} e^{-2i\pi n(k-k_0)/N}. \tag{10.11}$$

Dimensionality reduction in the time domain

By inserting (10.10) into (10.11) and using the geometric progression (9.14), it follows

$$c_n^{bl} = \frac{1}{N} \sum_{n'=0}^{N-1} c_{n'} e^{2i\pi n k_0/N} \left\{ \sum_{k=k_{min}}^{k_{max}} e^{2i\pi k(n'-n)/N} \right\}$$

$$= \frac{1}{N} \sum_{n'=0}^{N-1} c_{n'} e^{2i\pi n k_0/N} \left\{ e^{2i\pi k_0(n'-n)/N} D_{1+\Delta k}\left(2\pi \frac{n'-n}{N}\right) \right\}$$

where $D_m(x)$ is the Dirichlet kernel from (9.15) and

$$\Delta k = k_{max} - k_{min}. \qquad (10.12)$$

The length Δk of the interval $[k_{min}, k_{max}]$ is also equal to $N_d - 1 = [N/M] - 1$, so that the band-limited time signal c_n^{bl} can be written as

$$c_n^{bl} = \frac{1}{N} \sum_{n'=0}^{N-1} c_{n'} e^{2\pi i n' k_0/N} D_{N_d}\left(2\pi \frac{n'-n}{N}\right)$$

$$= \frac{1}{N} \sum_{n'=0}^{N-1} c_{n'} e^{2\pi i n' k_0/N} \frac{\sin(\pi N_d[n'-n]/N)}{\sin(\pi[n'-n]/N)}$$

$$= \frac{1}{N} \sum_{n'=0}^{N-1} c_{n'} e^{2\pi i n' k_0/N} \frac{\sin(\pi[n'-n]/M)}{\sin(\pi[n'-n]/N)} \qquad (10.13)$$

$$N_d = 1 + \Delta k = \left[\frac{N}{M}\right] \quad (0 \leq n \leq N-1). \qquad (10.14)$$

Finally, the band-limited decimated signal c_n^{bld} is obtained by decimating c_n^{bl} from which we retain only every Mth term

$$c_n^{bld} = \frac{1}{N} \sum_{n'=0}^{N-1} c_{n'} e^{2\pi i n' k_0/N} \frac{\sin(\pi[n'-n]/M)}{\sin(\pi[n'-n]/N)} \quad (0 \leq n \leq N_d - 1). \qquad (10.15)$$

According to (10.15), the decimation of the band-limited signal (10.13) takes place exclusively in the Dirichlet kernel $D_{N_d}(2\pi[n'-n]/N) = \{\sin(\pi[n'-n]/M)\}/\sin(\pi[n'-n]/N)$. In other words, the original set of signal points $\{c_{n'}\}_{n'=0}^{N-1}$ is not decimated at all, i.e. it is used as intact in (10.15). Notice that the complex exponential $\exp(2\pi i n' k_0/N)$ from (10.13) and (10.15) comes from shifting $k \to k - k_0$ prescribed by step (iv). Due to this latter scaling, every frequency ω_k extracted from c_n^{bld} by a given parametric method must also be shifted according to $\omega_k \to \omega_k + \omega_0$ where ω_0 is from (10.5). It should be observed that the term $\exp(2\pi i n' k_0/N)$ is the only place where the specific window $[\omega_{min}, \omega_{max}]$ is present explicitly in (10.15). As mentioned before, the

Dirichlet kernel $D_n(z)$ from (9.15) is related to the Chebyshev polynomial of the second kind and, therefore, a recursive computation is possible. Thus, using (3.7) we have [176, 180]

$$D_{n+1}(z) = (2 - \delta_{n,0})z D_n(z) - D_{n-1}(z) \qquad (10.16)$$

with the initializations $D_{-1}(z) = 0$ and $D_0(z) = 1$. The advantage of the bld-windowing is in having to deal only with a signal $\{c_n^{bld}\}$ of short length, $N_d \ll N$ (see figure 1 in [115] for an illustration). Unlike windowing in the FD [108, 109], the band-limited decimated signal $\{c_n^{bld}\}$ can be subjected to any desired method for data processing. For example, if the signal $\{c_n^{bld}\}$ is used instead of $\{c_n\}$ for obtaining the solutions $\{u_k, A_{n,k}\}$ of the generalized eigenvalue problem (8.10), then the method called the decimated signal diagonalization (DSD) would emerge [113, 115]. The same substitution of signal $\{c_n\}$ by $\{c_n^{bld}\}$ within the PA and LP would result in the decimated Padé approximant (DPA) and decimated linear predictor (DLP), respectively [115].

Chapter 11

The basic features of the Padé approximant (PA)

The standard PA is an abundantly studied subject in the theory of rational approximations [65]–[92]. This method is called the z-transform in engineering literature on signal processing. Also, as mentioned, the acronym ARMA is used for the PA in mathematical statistics and in probability theory dealing with stochastic phenomena. Of course, PA/ARMA is applicable to both deterministic and stochastic signals. The PA is the most prominent member from the whole family of the known nonlinear sequence-to-sequence transformations. There exists a number of more powerful nonlinear transformations than the PA [86, 89, 130, 230]. Yet, the PA continues to be the most frequently employed rational approximation not only in physics, but also in other basic and applied sciences including engineering [118]–[123]. The main reason is that the PA often outperforms its competitors in robustness and simplicity. The overriding rationale for a firm establishment of the PA in physics is its mathematical equivalence with a number of the leading methods in quantum mechanics, e.g. Born's perturbation expansions, finite-rank expansions with separable potential expansions, the Schwinger variational principles, the Green functions, the Fredholm determinants, etc [231]–[235].

The Weierstrass theorem [236] states that, in principle, every continuous function in a given domain can be approximated to any prescribed accuracy by a polynomial. However, in practice many functions possess singularities in some parts of their definition domains, so that approximations other than polynomials are required. One of the possibilities is to use the rational polynomial

$$f(z) \approx \frac{A_L(z)}{B_K(z)}. \tag{11.1}$$

Here, $A_L(z)$ and $B_K(z)$ are usual polynomials of degrees L and K, respectively

$$A_L(z) = \sum_{\ell=0}^{L} a_\ell z^\ell \qquad B_K(z) = \sum_{k=0}^{K} b_k z^k \tag{11.2}$$

where, in general, the variable z and the coefficients $\{a_\ell, b_k\}$ are complex valued. The rational polynomial $A_L(z)/B_K(z)$ from (11.1) forms a two-dimensional $L \times K$ table, called the Padé table, which represents a whole set of functions of varying degrees L and K. The original concept of the Padé table is not due to Padé, but rather to Frobenius [66] who in 1879 developed the basic algorithmic aspects of the theory. However, prior to Frobenius, quotients of two polynomials were studied in depth by Prony in 1797, then by Cauchy in 1821 and afterwards by Jacobi in 1845 [65]. More specifically, Frobenius [66] established the theory of what was later known as the normal Padé table, which has all distinct elements. The work of Padé [67] came later in 1892 when he expanded the Frobenius concept to encompass certain cases of the so-called abnormal tables in which some of the elements could be equal to each other. To understand the concrete meaning of the relation (11.1), we employ the standard symbol $[L/K]_f(z)$ to denote the PA for a given function $f(z)$

$$[L/K]_f(z) \equiv \frac{A_L(z)}{B_K(z)} \qquad K \geq L. \qquad (11.3)$$

This definition of the PA is valid provided that the following three conditions are satisfied: (a) the numerator and denominator polynomials $A_L(z)$ and $B_K(z)$ have no common divisors except for a possible constant term, (b) $B_K(z) \neq 0$ and (c) the Maclaurin series development of $f(z)$

$$f(z) = \sum_{n=0}^{\infty} c_n z^n \qquad (11.4)$$

and the Maclaurin expansion of $[L/K]_f(z)$ in powers of z agree term-by-term with each other through all orders including z^n ($n \leq L + K$). The expansion coefficients $\{c_n\}$ in (11.4) are supposed to be known real or complex numbers, but they do not need to be identified with the auto-correlation functions or signal points. The condition (c) can be equivalently stated as

$$f(z) - \frac{A_L(z)}{B_K(z)} = O(z^{L+K+1}). \qquad (11.5)$$

Here, the appearance of the algebraic O-symbol indicates that the rhs of (11.5) represents a power series in z containing the terms $\{z^{L+K+m}\}$ for $m \geq 1$. In other words, by definition, the PA for the function $f(z)$ from (11.4) is given by (11.3), which assumes that the *equality*

$$f(z) = \frac{A_L(z)}{B_K(z)} \qquad (11.6)$$

is valid through all orders z^{L+K}, i.e. that every coefficient of the higher terms z^{L+K+m} with $m \geq 1$ is automatically taken as zero. Of great importance in every application of the PA is its uniqueness which means that there is one and only one rational polynomial $A_L(z)/B_K(z)$ of the degree (L, K) for the series (11.4) of the function $f(z)$. Padé assumed that both c_n and z can take any real or complex finite values with the exception $z = 0$ and $z = 1$ as in (11.4). In spectral analysis, the elements of the set $\{c_n\}$ usually represent time signals or auto-correlation functions, and the series in (11.4) is related to the DFT from (5.3) for an infinite sequence ($N = \infty$). Convergence of the PA, which actually represents an *analytical continuation* [28, 72, 75] of $f(z)$ is usually faster than the original series (11.4). In addition, the PA can resum wildly divergent series for $f(z)$ even with the zero convergence radius [88] or if the coefficients c_n grow as exponentials [237] or factorials [85, 86, 89] when n increases, etc. Here, the term 'resum' or 'resummation' is used as a compound word which means acceleration of slowly convergent series and/or induced, i.e. 'forced' convergence of divergent series.

The prescription $f(z) \approx A_L(z)/B_K(z)$ from (11.1) indicates itself that the unique univariate PA to a given series representation (11.4) of $f(z)$ has a richer mathematical structure than the corresponding single polynomial ansatz $f(z) \approx A_{L+K}(z)$. The zeros of the numerator polynomial $A_L(z)$ give the L roots of the equation $f(z) = 0$. Similarly, the zeros of the denominator polynomial $B_K(z)$ are simultaneously the K poles of the function $f(z)$. For this reason, the Padé ansatz (11.6) is also known as the 'all-pole-all-zero' approximation to $f(z)$ [58, 61, 64, 127]. There are two special cases of the PA and they are called the 'all-zero' and the 'all-pole' approximations [58, 61]. The 'all-zero' approximation, which is the MA model is obtained by simply setting the denominator polynomial $B_K(z)$ to a constant, say unity $B_K(z) = 1$. This leads to the ordinary polynomial approximation $f(z) \approx A_L(z)$. The 'all-pole' approximation, which is the AR model, follows from the requirement that the numerator polynomial $A_L(z)$ is a constant, e.g. unity $A_L(z) = 1$ yielding the approximation $f(z) \approx 1/B_K(z)$. The name 'auto-regressive' is plausible in studies of stationary phenomena where the coefficients $\{c_n\}$ from (11.4) represent the auto-correlation functions $\{C_n\}$ from (5.9). Here, every element $C_n = c_n$ of the set $\{c_n\}$ can be found by *regressing* to the lower values of the subscript n via a linear combination of all points $\{c_m\}$ for $m < n$. If the variable z of function $f(z)$ is random, the AR approximation is found to coincide with the maximum entropy method [58, 61, 106]. For example, in speech processing, one of the most commonly used spectral estimators is the AR model, which is also called the linear predictive coding (LPC) [124]–[128]. This is appropriate for non-nasal voiced speech sounds for which the transfer function of the vocal tract has no zeros [124, 125]. However, unvoiced and nasal (fricatives) sounds lead to spectra with the appearance of valleys, i.e. zeros or dips called anti-resonances. In such cases, the 'all-pole' character of AR/LPC is the cause for serious processing limitations. This may be partially responsible for the perceived difference between real speech and its best synthetic counterpart. One of the adequate candidates that naturally fills in this gap in speech processing is the 'all-pole-all-zero' or ARMA, i.e. PA. This has been illustrated in the literature on speech processing [127] using a vocal-tract transfer function as a rational polynomial over a fixed time interval. Such a function has a finite number of zeros and poles that are well described by the numerator and denominator of the parametric PA/ARMA. Industrial companies in telephone communications are also interested in robust processing via PA/ARMA, i.e. the pole-zero modelling for applications in, e.g. effective speech coding, voice pattern recognition and verification, etc.

The quotient of two polynomials $A_L(z)/B_K(z)$ is more powerful than an ordinary single polynomial approximation to a given function $f(z)$. This is because the ratio $A_L(z)/B_K(z)$ explicitly contains $L+K$ 'adjustable parameters' even though only the polynomial $A_L(z)$ and $B_K(z)$ of degrees L and K, respectively, are computed [238]. By contrast, the request for having the same number $L+K$ of 'free parameters' in the single polynomial approximation $f(z) \approx A_{L+K}(z)$ can only be fulfilled by increasing the degree of the generated polynomial. In many practical circumstances, the rational approximation $f(z) \approx A_L(z)/B_K(z)$ will possess an error which is smaller than in the case of any single polynomial approximation $f(z) \approx A_{L+K}(z)$, despite the same number of the input data $\{c_n\}$ used by both methods applied to (11.4). Not only that rational polynomials $A_L(z)/B_K(z)$ can be used for functions that are less efficiently approximated by ordinary polynomials but, even more importantly, the ansatz (11.6) may also succeed where the single polynomial approximation fails completely [238]. This is the case with, e.g. discontinuous and/or singular functions that cannot be successfully treated by single polynomials in, e.g. the MA model, etc. However, the rational polynomial approximation may well be adequate here too, since the zeros of the denominator polynomial $B_K(z)$ could mimic closely the singularities of the studied function $f(z)$. Furthermore, in the most frequent applications, a given function $f(z)$ is required to be approximated over a very large interval (possibly infinite) of variable z. In such cases, separate single polynomial approximations are customarily constructed, since ordinarily one polynomial cannot cover all the regions of interest of the variable z. By comparison, a single approximation $f(z) \approx A_L(z)/B_K(z)$ is usually valid throughout the entire interval of the values of z under investigation. The PA accomplishes both *interpolation* and *extrapolation*. Moreover, a sequence of branch point singularities of the function $f(z)$ can be approximately described by the poles of the quotient $A_L(z)/B_K(z)$ [107]. Furthermore, if some specific cuts or branch point singularities of a given function $f(z)$ are desired to be incorporated into the corresponding PA from the onset, one could appropriately redefine the variable z. For example, if the studied function $f(z)$ has the square root branch points, the PA could be taken as a rational approximation in the variable $z^{1/2}$, i.e. $f(z^{1/2}) = A_L(z^{1/2})/B_K(z^{1/2})$. Here, the polynomials $A_L(z^{1/2})$ and $B_K(z^{1/2})$ can be determined by equating the coefficients of the fractional powers of the variable z in the Maclaurin expansion of functions $f(z^{1/2})$ and $A_L(z^{1/2})/B_K(z^{1/2})$ much in the same fashion as customarily achieved with z raised to integer powers [90]. Also logarithmic branch point singularities could

The basic features of the Padé approximant (PA) 77

be incorporated from the onset into the development of the PA, as has recently been shown in [92].

A computation of $[L/K]_f(z)$ requires at least $L + K$ coefficients $\{c_n\}$ which are available from (11.4). From the two-dimensional $L \times K$ Padé table, one usually selects the 'paradiagonal' element $L = K - 1$ yielding the so-called paradiagonal PA, whereas the diagonal PA is given for $L = K$. The unknown coefficients $\{a_\ell, b_k\}$ can be obtained in several ways. If we ignore the power function z^{L+K+1} and all the higher-degree terms, the equality $f(z)B_K(z) = A_L(z)$ becomes exact as in (11.6). Multiplying (11.6) by the denominator polynomial $B_K(z) \neq 0$ one obtains the relation

$$B_K(z) \sum_{n=0}^{L+K} c_n z^n = A_L(z). \tag{11.7}$$

Substituting the power series representations for $A_L(z)$ and $B_K(z)$ from (11.2) into (11.7), we have

$$(c_0 + c_1 z + c_2 z^2 + \cdots + c_{L+K} z^{L+K})(b_0 + b_1 z + b_2 z^2 + \cdots + b_K z^K)$$
$$= (a_0 + a_1 z + a_2 z^2 + \cdots + a_L z^L)$$
$$+ (0 \cdot z^{L+1} + 0 \cdot z^{L+2} + \cdots + 0 \cdot z^{L+K}).$$

Here, we compare like powers of z to arrive at two systems of linear equations for the coefficients $\{a_\ell\}$ and $\{b_k\}$ of the polynomials $A_L(z)$ and $B_K(z)$, respectively [83, 107]

$$\left.\begin{aligned}
c_{L+1}b_0 + c_L b_1 + \cdots + c_{L-K+1}b_K &= 0 \\
c_{L+2}b_0 + c_{L+1}b_1 + \cdots + c_{L-K+2}b_K &= 0 \\
c_{L+3}b_0 + c_{L+2}b_1 + \cdots + c_{L-K+3}b_K &= 0 \\
&\vdots \\
c_{L+K}b_0 + c_{L+K-1}b_1 + \cdots + c_L b_K &= 0
\end{aligned}\right\} K \text{ linear equations} \tag{11.8}$$

$$\left.\begin{aligned}
c_0 b_0 &= a_0 \\
c_1 b_0 + c_0 b_1 &= a_1 \\
c_2 b_0 + c_1 b_1 + c_0 b_2 &= a_2 \\
&\vdots \\
c_L b_0 + c_{L-1} b_1 + \cdots + c_{L-K} b_K &= a_L
\end{aligned}\right\} L+1 \text{ linear equations.} \tag{11.9}$$

The first system (11.8) of linear equations can succinctly be rewritten as

$$c_\ell = -\frac{1}{b_0} \sum_{r=1}^{K} b_r c_{\ell-r} \qquad (\ell \geq K) \qquad b_0 \neq 0 \tag{11.10}$$

where the index ℓ in $c_{\ell-r}$ must be non-negative and this is the case for $\ell \geq K$. The system of linear equations in (11.8) coincides with the defining equations for the predicting coefficients $\{b_r\}$ in the LP from signal processing [61]. The system (11.9) of linear equations for the coefficients $\{a_\ell\}$ is defined solely via $\{c_n\}$ and $\{b_k\}$. Thus, such a system represents an *explicit* analytical expression, once the coefficients $\{b_k\}$ are found from (11.8) for the given set $\{c_n\}$

$$a_\ell = \sum_{r=0}^{\ell} b_r c_{\ell-r} \qquad (\ell = 0, 1, 2, \ldots, L). \tag{11.11}$$

The system (11.8) of linear equations for the coefficients $\{b_k\}$ is mathematically ill-conditioned and, as such, it should not be used in practice for large K [82, 107]. Recently [115], this system has been solved successfully for smaller values of $K \leq 300$ by using the singular value decomposition (SVD) and the QZ algorithm [176, 178]. Presently, we solve (11.8) within a few minutes by employing the SVD for $K \leq 1024$ (see chapter 73). Alternatively, the system of equations originating from the condition (11.7) can be solved by using other competitive methods, e.g. the algorithm of Shanks [71, 72] or Wynn [73] and Longman [80] as discussed in chapter 18. The Shanks transform is an extension of the well-known Δ^2-extrapolation of Aitken [70, 176]. The Wynn ε-algorithm is a recursive way of evaluating the Shanks transform. The Aitken, Shanks and Wynn sequence-to-sequence mappings are all related directly to the Padé quotient $A_L(z)/B_K(z)$ as a whole, but they do not extract explicitly the polynomials $A_L(z)$ and $B_K(z)$. By contrast, the algorithm from the Padé–Lanczos approximant (PLA) [17] obtains explicitly the coefficients $\{a_\ell\}$ and $\{b_k\}$ from (11.7). Therefore, from the signal processing viewpoint, the original algorithms of Aitken, Shanks and Wynn belong to the category of non-parametric methods, since they can provide directly only the shape of the considered spectrum[1]. We say that these three algorithms are of the type of 'transforms' in the spirit of the Fourier transform. However, the PLA can play a twofold role of being a non-parametric and parametric estimator of spectra. In the non-parametric case, only the quotient $A_L(z)/B_K(z)$ is computed at a pre-assigned range of the variable z. In the parametric case, the explicit knowledge of the polynomials $A_L(z)$ and $B_K(z)$ permits obtaining a spectral representation as the sum of the Heaviside [132, 141] partial fractions for simple poles, i.e. non-degenerate roots ($z_k \neq z_{k'}, k' \neq k$) of the denominator polynomial $B_K(z)$ (the expansion theorem)[2]:

$$\frac{A_L(z)}{B_K(z)} = \sum_{k=1}^{K} \frac{\rho_k}{z - z_k}. \tag{11.12}$$

---

[1] Methods predicting only the shape of a spectrum are spectral estimators as opposed to parameter estimators.

[2] Those roots of a given polynomial that have their multiplicity greater than one are hereafter referred to as degenerate (multiple, confluent, coincident) roots as opposed to simple roots whose multiplicity is equal to one.

The basic features of the Padé approximant (PA)

Here, the elements of the set $\{\rho_k\}$ are the residues of $A_L(z)/B_K(z)$ taken at the non-degenerate roots $\{z_k\}$ of the characteristic equation $B_K(z_k) = 0$

$$\rho_k = \frac{A_L(z_k)}{B'_K(z_k)} \quad (11.13)$$

where $B'_K(z)$ is the first derivative $B'_K(z) \equiv B_{K,1}(z) = (d/dz)B_K(z)$. If some J roots of $B_K(z)$ from the whole set $\{z_k\}(1 \leq k \leq K)$ are degenerate, meaning that the jth root z_j has multiplicity m_j such that $m_1 + m_2 + \cdots + m_J = K$, then the Heaviside partial fractions from (11.12) for the quotient $A_L(z)/B_K(z)$ must be modified according to [141]

$$\frac{A_L(z)}{B(z)} = \sum_{k=1}^{J}\sum_{j=1}^{m_k} \frac{\rho_{k,j}}{(z-z_k)^j} \quad \rho_{k,j} = \frac{A_L(z_k)}{B_{K,j}(z_k)} \quad (11.14)$$

where $B_{K,j}(z)$ is the jth derivative of $B_K(z)$, i.e. $B_{K,j}(z) = (d/dz)^j B(z)$. Using (11.2), we have

$$B_{K,j}(z) \equiv \left(\frac{d}{dz}\right)^j B_K(z) = \sum_{k=0}^{K} k_j b_k z^{k-j} \quad k_j = \frac{k!}{(k-j)!}. \quad (11.15)$$

The parametric model of the PA from (11.14) is for non-Lorentzian spectra with overlapping resonances. The Fourier integral of (11.14) is the time signal [141]

$$c(t) = \sum_{k=1}^{J}\sum_{j=1}^{m_k} \rho_{k,j} t^{j-1} e^{-iz_k t}. \quad (11.16)$$

The conventional PA is not a method without pitfalls and neither are all other parametric estimators including the FD, DSD, DLP, DPA, etc. The main common difficulty of every existing parametric estimator is instability due to emergence of spurious/extraneous eigenvalues of the evolution matrix **U** in, e.g. the FD or the DSD, or due to unphysical roots of the characteristic polynomial $B_K(z)$ in the PA, DLP, DPA, etc. The FD [108, 109] and the DSD [115] use the same procedure to identify and subsequently handle spurious roots. The diagonal DPA is mathematically equivalent to DSD and DLP. As mentioned, in the case of noiseless signals, these three latter methods have been shown to give identical numerical results for the pair $\{u_k, d_k\}$ to within machine accuracy by using entirely different computational tools [115]. Furthermore, when the difference of the degrees of the numerator and denominator polynomials is equal to $s + 1(s = 0, 1, 2, \ldots)$, the ensuing paradiagonal DPA is found to coincide with the DSD based upon the evolution operator raised to the power s. Hence, the same procedure for regularizing spurious roots from the FD and the DSD can also be used in the DPA. We compare the diagonal and several paradiagonal DPAs from the Padé table [107]. In doing so, we observe that some roots do not change noticeably, whereas others alter considerably when passing from diagonal to paradiagonal DPAs. Then the former stable quantities are considered as genuine or physical roots, whereas the latter unstable ones are viewed as spurious or unphysical roots. Once identified, the spurious roots whose imaginary parts are positive are reflected, i.e. subjected to the transformation $\text{Im}(\omega_k) \longrightarrow -\text{Im}(\omega_k)$. The latter transformation is the standard root reflection which is frequently used in signal processing [61]–[64]. An alternative procedure called the *constrained root reflection* is outlined in chapter 13 within the PA for an analytical identification and regularization of spurious roots with the simultaneous preservation of the phase-minimum and uniqueness of the power spectra.

Chapter 12

Gaussian quadratures and the Padé approximant

Here we shall analyse the essence of a numerical integration within the realm of the PA. Consider the following one-dimensional real integral of a univariate function $f(x)$ of real variable x in the closed interval $[a, b]$

$$I = \int_a^b dx\, w(x) f(x) \tag{12.1}$$

where $w(x)$ is a given real weight function which is non-negative for $x \in [a, b]$. The standard quadrature rule for this is the prescription[1]

$$\int_a^b dx\, w(x) f(x) \approx \sum_{k=1}^K w_k f(x_k) + \mathcal{R}_K(f) \tag{12.2}$$

where $\mathcal{R}_K(f)$ is the remainder which estimates the error of the integration in (12.1). The set $\{x_k\}$ in (12.2) represents the K distinct fixed points that are the real zeros of an auxiliary, pre-assigned polynomial $B_K(x)$

$$B_K(x_k) = 0 \quad (1 \leq k \leq K) \tag{12.3}$$

and $\{w_k\}$ are the corresponding positive weights. The approximate formula (12.1) can be made exact, i.e. $\mathcal{R}_K(f) = 0$ if $f(x)$ is a polynomial of order $2K - 1$, since $B_K(x)$ is a prescribed polynomial of order K. These pre-assigned polynomials $B_K(x)$ are usually the classical polynomials [176, 180] in which case the resulting rules are called Gauss–Legendre, Gauss–Laguerre or Gauss–Hermite, etc [180]. Their importance is in guaranteeing that they all converge with increasing order K and the invoked error $\mathcal{R}_K(f)$ can be computed from the estimated formula incorporating the $2K$th derivative of $f(x)$ at some point $\xi \in [a, b]$. Of course,

---
[1] Gaussian quadrature rules for certain classes of complex valued weight functions $w(x)$ are also possible to construct [239].

if $f(x)$ is not a polynomial, the rule (12.1) is approximate and the quality of integration, both regarding the convergence rate and the estimated error, is dependent upon the nature of the integrated function. The $2K$th derivative of $f(x)$ needed to estimate the error in $\mathcal{R}_K(f)$ might be difficult to obtain even for some complicated analytical functions. If $f(x)$ is only known through its tabular values, such as those acquired in measurements, then high-order numerical differentiation is not recommended, as it is prone to severe round-off errors. An alternative is to estimate the error through quadratures of the similar type as the original integral (see Appendix B). The standard Cauchy contour integrals are well suited for this purpose. Additionally, they will facilitate a formulation of the numerical integration (12.1) as a spectral problem.

Let us now replace the real variable x by the complex variable z and introduce contour integrals. For this purpose we choose a closed contour C which in its interior includes the whole interval $[a, b]$ of the Re z-axis along with the entire set $\{x_k\}$. Both functions $f(z)$ and $w(z)$ are assumed to be analytic[2] within the contour C. Due to these circumstances, we can write the identity

$$f(y) = \frac{1}{2\pi i} \oint_C dz \frac{f(z)}{y-z}. \tag{12.4}$$

Let us now define the real Hilbert transform of a function $g(x)$ which is assumed to be analytic within $x \in [a, b]$ by the following Riemann integral

$$(\tilde{H}g)(x) = \mathcal{P} \int_a^b dy \frac{g(y)}{x-y} \tag{12.5}$$

where the symbol $\mathcal{P}$ denotes the principal Cauchy value. If an integrand $F(x)$ is singular at $x = c \in [a, b]$, we have [240]

$$\mathcal{P} \int_a^b F(x)dx = \lim_{\epsilon \to 0} \left[\int_a^{c-\epsilon} dx\, F(x) + \int_{c+\epsilon}^b dx\, F(x) \right]. \tag{12.6}$$

Next we introduce two polynomials $A_{K-1}(z)$ and $\tilde{A}_{K-1}(z)$ of the degree $K-1$ such that both of them are analytic within the contour C as the finite Hilbert transforms

$$A_{K-1}(z) = \mathcal{P} \int_a^b dy\, w(y) \frac{B_K(y)}{z-y} \tag{12.7}$$

$$\tilde{A}_{K-1}(z) = \int_a^b dy\, w(y) \frac{B_K(z) - B_K(y)}{z-y}. \tag{12.8}$$

In (12.8) the principal value symbol $\mathcal{P}$ is unnecessary, since the integrand is a regular function at $z = y$, due to the relation $[B_K(z) - B_K(y)]/(z-y) = B'_K(y)$

---

[2] Function $B_K(z)$ being a polynomial is automatically an analytic function.

at $z = y$ with $B'_K(z) = dB_K(z)/dz$. Further, let $R(z)$ and $\tilde{R}(z)$ denote two rational functions via the following PA quotients

$$R(z) = \frac{A_{K-1}(z)}{B_K(z)} \qquad \tilde{R}(z) = \frac{\tilde{A}_{K-1}(z)}{B_K(z)}. \qquad (12.9)$$

The functions $R(z)$ as well as $\tilde{R}(z)$ are analytic within C and they share the common set of roots $\{x_k\}$ from $B_K(z)$ in their denominators, so that their residues can be calculated as $w_k = \lim_{x \to x_k}(x - x_k)R(z)$ and $\tilde{w}_k = \lim_{x \to x_k}(x - x_k)\tilde{R}(z)$, with the result for w_k in the form of the so-called Christoffel coefficients [176]

$$w_k = \frac{A_{K-1}(x_k)}{B'_K(x_k)} = \frac{1}{B'_K(x_k)} \mathcal{P} \int_a^b dy\, w(y) \frac{B_K(y)}{x_k - y}. \qquad (12.10)$$

Accounting for the characteristic equation $B_K(x_k) = 0$ according to (12.3), we also have

$$\tilde{w}_k = -w_k. \qquad (12.11)$$

The polynomial $A_{K-1}(z)$ is defined on the complex z-plane, which is cut on the segment $[a, b]$ of the real axis $\text{Re}(z)$. In the residue w_k from (12.10) the value $A_{K-1}(x_k)$ is required, where $x_k \in [a, b]$ and this may cause concern for the definition of w_k. However, no problem arises here since for $x \in (a, b)$ two different functions can be defined, $A_{K-1}(x + i\epsilon)$ and $A_{K-1}(x - i\epsilon)$, such that

$$\lim_{\epsilon \to 0} A_{K-1}(x_k - i\epsilon) = \lim_{\epsilon \to 0} A_{K-1}(x_k + i\epsilon) = A_{K-1}(x_k) \qquad (12.12)$$

which defines properly the needed value $A_{K-1}(x_k)$. Then, we have the feature $\lim_{\epsilon \to 0}[A_{K-1}(x_k - i\epsilon) - A_{K-1}(x_k + i\epsilon)] = 0$ at the real zero $x = x_k$ of $B_K(x)$, and this follows from

$$\lim_{\epsilon \to 0}[A_{K-1}(x - i\epsilon) - A_{K-1}(x + i\epsilon)] = \frac{1}{2\pi i}\oint_C dz\, w(z) \frac{B_K(z)}{x - z} = w(x)B_K(x). \qquad (12.13)$$

The rhs of (12.13) is equal to zero at $x = x_k \in [a, b]$ on the account of (12.3). Associated with the pair of functions $f(z)$ and $\tilde{R}(z)$ we consider the following contour integral

$$\tilde{I} = \frac{1}{2\pi i}\oint_C dz\, \tilde{R}(z) f(z). \qquad (12.14)$$

Since $f(z)$ is analytic within the contour C, the only singularities in the integrand $\tilde{R}(z) f(z)$ are the K simple poles $\{x_k\}$ of $\tilde{R}(z)$. Therefore the integral on the rhs of (12.14) can be readily calculated by the Cauchy residue theorem and the result for $\tilde{I}$ is

$$\tilde{I} = \sum_{k=1}^{K} w_k f(x_k). \qquad (12.15)$$

Replacing the result $\sum_{k=1}^{K} w_k f(x_k)$ for $\tilde{I}$ in (12.14) and inserting $\tilde{R}(z)$ from (12.9) into the rhs of (12.14), we have the sequel

$$\sum_{k=1}^{K} w_k f(x_k) = \frac{1}{2\pi i} \oint_C dz\, \tilde{R}(z) f(z) = \frac{1}{2\pi i} \oint_C dz\, \frac{\tilde{A}_{K-1}(z)}{B_K(z)} f(z)$$

$$= \frac{1}{2\pi i} \oint_C dz \left\{ \int_a^b dy\, w(y) \frac{B_K(z) - B_K(y)}{z - y} \right\} \frac{f(z)}{B_K(z)}$$

$$= \frac{1}{2\pi i} \oint_C dz \left\{ P \int_a^b dy\, \frac{w(y)}{z-y} \right\} f(z)$$

$$- \frac{1}{2\pi i} \oint_C dz \left\{ P \int_a^b dy\, w(y) \frac{B_K(y)}{z-y} \right\} \frac{f(z)}{B_K(z)}$$

$$= \int_a^b dy\, w(y) \left\{ \frac{1}{2\pi i} \oint_C dz\, \frac{f(z)}{z-y} \right\}$$

$$- \frac{1}{2\pi i} \oint_C dz\, \frac{A_{K-1}(z)}{B_K(z)} f(z) = \int_a^b dy\, w(y) f(y)$$

$$- \frac{1}{2\pi i} \oint_C dz\, R(z) f(z) = I - \frac{1}{2\pi i} \oint_C dz\, R(z) f(z) \quad (12.16)$$

where (12.4) is used, so that after rewriting the first and the last line of (12.16), we have [241]

$$I \equiv \int_a^b dx\, w(x) f(x) = \sum_{k=1}^{K} w_k f(x_k) + \mathcal{R}_K(f) \quad (12.17)$$

$$\mathcal{R}_K(f) = \frac{1}{2\pi i} \oint_C dz\, R(z) f(z). \quad (12.18)$$

Here, the error term $\mathcal{R}_K(f)$ from (12.1) is now identified as the contour integral of the product of the given function $f(z)$ and the PA given by $R(z)$ from (12.9). The Padé quadrature rule (12.17) determines the pairs $\{x_k, w_k\}$ through the standard PA procedure, consisting of setting up the quotient of two polynomials $R(z) = A_{K-1}(z)/B_K(z)$ as in (12.9) with the subsequent rooting $B_K(x_k) = 0$ according to (12.3) and, finally, determining the corresponding residues via the explicit expression $w_k = A_{K-1}(x_k)/B'(x_k)$, as in (12.10). The error committed by this procedure can be assessed by evaluating numerically the remainder $\mathcal{R}_K(f)$ from the contour integral (12.18) in the complex z-plane over the integrand $R(z) f(z)$. Alternatively, a real variable form can be derived for the error term involving the $2K$th derivative of the given function $f(x)$. In the above procedure, only the denominator polynomial $B_K(z)$ is left free, whereas the numerator polynomial $A_{K-1}(z)$ is fixed by the definition (12.7). One can pre-assign $B_K(z)$ by selecting, e.g. one of the known classical polynomials [180]. In this case, $A_{K-1}(z)$ is automatically known from the set of the existing polynomials of the second kind for each of the mentioned $B_K(z)$. There are many useful asymptotic expressions for the ratio $A_{K-1}(z)/B_K(z) = R(z)$ of such polynomials at, e.g. large values of z or for high orders K that could facilitate adequate evaluations of the estimates of the error term $\mathcal{R}_K(f)$ in the integral (12.18) [241]. It should be pointed out that the error analysis within the PA, or equivalently, the Gaussian quadrature rule can also be systematically carried out using the so-called Stieltjes and Geronimus polynomials as has been done in, e.g. [242].

Chapter 13

Padé–Schur approximant (PSA) with no spurious roots

In this chapter we shall examine the stability problem of the standard PA with the purpose of unambiguously identifying both the genuine and spurious roots by the analytical procedure of Holtz and Leondes [117]. Once we disentangle the unphysical from the physical roots, we will be in a position to design a stable PA, which is free from any spurious roots stemming from either the theoretical processing or the input signal points $\{c_n\}$. In this way some of the noise from the raw data $\{c_n\}$ will also be regularized. Noise poles can be on either side of the unit circle appearing as mixed with or separated from the genuine poles. When they are on the 'wrong' side of the unit circle, noise poles will be regularized along with the like portion of the genuine harmonics from the input data $\{c_n\}$. Here, regularization does not mean elimination of spurious roots, but rather their reflection to the physical side of the unit circle, so that all the harmonics from $\{c_n\}$ are exponentially damped at large times. Most previous studies on signal processing tried to reduce or even eliminate noise with various devices. This could be a highly risky undertaking, since noise is mostly random, i.e. intrinsically unknown and immersed into the physical signal. Moreover, some of the spurious resonances might well be physical resonances, but still have their frequencies with the wrong signs of the imaginary parts $\{\text{Im}(\omega_k)\}$ because, e.g. the corresponding excited transients could not decay during the given acquisition time T. These physical transients that appear disguised as noise are weak, i.e. sensitive to any perturbation and, as such, they could easily evade any of the existing stability tests ending up as being erroneously discarded. Hence, it is more appropriate to process noise as the constituent part of the measured/raw time signal $\{c_n\}$ rather than trying to suppress it from the input data. In other words, we intentionally do not want to discard spurious resonances. To achieve this goal, we shall treat the 'noisy' and the genuine content from $\{c_n\}$ on the same footing, discarding nothing, but still encountering only the exponentially damped harmonics. The general framework for this task can be a typical rational polynomial $F_{L,K}(u^{-1})$

of the form

$$F_{L,K}(u^{-1}) = \frac{A_L(u^{-1})}{B_K(u^{-1})} \qquad F_{L,K}(u^{-1}) \approx g(u^{-1}) \equiv \sum_{n=0}^{\infty} c_n u^{-n} \qquad (13.1)$$

with u being a complex variable $u = \exp(-i\omega\tau)$, as in (7.3) where the frequency ω and the sampling time τ are both real. Here, we define $F_{L,K}(u^{-1})$ as the PA, i.e. $[L/K]_g(u^{-1})$ for the function $g(u^{-1})$ which is given by its series development in powers of u^{-1}. The numerator $A_L(u^{-1})$ and denominator $B_K(u^{-1})$ polynomials in (13.1) are the ordinary polynomials of the degrees L and K in the complex variable u^{-1}, i.e. the truncated series

$$A_L(u^{-1}) = \sum_{\ell=0}^{L} a_\ell u^{-\ell} \qquad B_K(u^{-1}) = \sum_{k=0}^{K} b_k u^{-k}. \qquad (13.2)$$

All the algorithms that exist in the current literature for computation of the PA yield spurious roots for arbitrary degrees L and K or for the 'paradiagonal' $L = K - 1$ or the diagonal $L = K$ variant. Therefore, construction of a stable PA without any spurious or extraneous roots from the onset in either $A_L(u^{-1})$ or $B_K(u^{-1})$ is one of the most important themes to study. For simplicity of illustration, the analysis in this chapter will be given for real expansion coefficients $\{c_n\}$ alone[1]. In such a case, the generated polynomial coefficients $\{a_\ell, b_k\}$ will also be real, and in this chapter, they will be assumed to be precomputed via any of the existing algorithms. As mentioned in chapter 11, from a theoretical viewpoint, the polynomials $A_L(u^{-1})$ and $B_K(u^{-1})$ are unique for a given series (13.1). In practice, however, uniqueness might not be preserved because the numerator polynomial $A_L(u^{-1})$ could also possess spurious roots. Likewise, such extraneous roots of the denominator polynomial $B_K(u^{-1})$ routinely appear in applications leading to considerable instability in (13.1). Since the variable of the numerator and denominator polynomials is u^{-1}, these spurious or ghost roots of both $A_L(u^{-1})$ and $B_K(u^{-1})$ have their magnitudes, i.e. the absolute values less than unity (inside the unit circle)[2]. In particular, if the expansion coefficients $\{c_n\}$ in (13.1) are auto-correlation functions or signal points, the spurious roots of the denominator polynomial $B_K(u^{-1})$ will cause the corresponding harmonics to increase indefinitely with the augmented time. This is because all the spurious roots $\{u_k^{-1}\} = \{\exp(i\omega_k \tau)\}$ from $B_K(u^{-1})$ are those roots for which the imaginary parts of the characteristic frequencies are negative ($\text{Im}(\omega_k) < 0$), as follows from the complex exponential

---

[1] The extension of the analysis from this chapter to complex $\{c_n\}$ is given in [142].
[2] Had we decided to work with the variable u instead of u^{-1}, the corresponding PA would be stable only if the associated numerator and denominator polynomials in u possess the roots with the absolute values that are less than unity (inside the unit circle). In this case, the spurious roots are those roots of polynomials $A_L(u)$ and $B_K(u)$ that have their absolute values greater than unity (outside the unit circle) [243].

exp $(i\omega_k \tau)$ in c_n

$$c_n = \sum_{k=1}^{K} d_k u_k^{-n} \qquad u_k^{-1} = e^{i\omega_k \tau} = e^{i\tau \operatorname{Re}(\omega_k) - \tau \operatorname{Im}(\omega_k)} \qquad \tau > 0. \qquad (13.3)$$

Due to these spurious roots, the corresponding transients of the signal c_n from (13.3) will never decay, and this is an unphysical effect, which must be regularized[3]. It would be important to have a procedure which could guarantee the emergence of only stable or Schur polynomials [56] $A_L(u^{-1})$ and $B_K(u^{-1})$ with all their roots outside the unit circles in, e.g. the Padé power spectrum

$$|F_{L,K}(u^{-1})|^2 = \left| \frac{A_L(u^{-1})}{B_K(u^{-1})} \right|^2 \qquad (13.4)$$

or in the Padé magnitude spectrum $|F_{L,K}(u^{-1})|$. The net effect of instabilities of the denominator polynomial $B_K(u^{-1})$ in the power spectrum $|F_{L,K}(u^{-1})|^2$ is the emergence of artificial spikes that might be unphysical resonances. Moreover, these artefacts are excessively sensitive to any changes in the procedure such as reduction of the signal length, enlargement or shortening of the studied frequency interval, etc. Such a circumstance leads to an overall unacceptable sensitivity of the ensuing PA to changes of the external parameters, even for variations that are otherwise expected to produce minor effects upon the generated spectrum. Similar spikes in the spectrum do not follow from the spurious zeros of $A_L(u^{-1})$, since this is the numerator polynomial. Nevertheless, the spurious zeros of $A_L(u^{-1})$ are also unwanted, since they can lead to unphysical valleys, i.e. anti-resonances in the spectrum yielding loss of uniqueness of the PA. The first idea which comes to mind, as often implemented in signal processing within most of the existing parametric methods [64, 108, 109, 115], is to employ the usual root reflection to simply replace $\operatorname{Im}(\omega_k)$ by $-\operatorname{Im}(\omega_k)$ whenever $\operatorname{Im}(\omega_k) < 0$ in u_k^{-1}. This amounts to relocating the roots of $B_K(u^{-1})$ from inside to outside the unit circle. This root reflection replaces the exponentially diverging terms from the time signal (13.3) by the exponentially converging, i.e. damped or attenuated fundamental harmonics.

The drawback of the usual root reflection is a loss of information, since this kind of regularization alters the spectrum in such a way that the original signal cannot be reconstructed any longer within any given parametric estimator. Hence, an alternative solution to the problem of spurious roots is sought, such that invariance of magnitude or power spectra is assured under a regularization of spurious roots [117]. One such solution will be analysed in this chapter. We shall

---

[3] As opposed to the c_n from (5.11) given in terms of $\{u_k^n\}$, the c_n from (13.3) is defined via $\{u_k^{-n}\}$ since the PA from (13.1) is introduced using the variable u^{-1} rather than u.

first write the polynomials $A_L(u^{-1})$ and $B_K(u^{-1})$ in their complex polar forms

$$A_L(u^{-1}) = |A_L(u^{-1})|e^{i\theta_A} \qquad B_K(u^{-1}) = |B_K(u^{-1})|e^{i\theta_B} \qquad (13.5)$$
$$|A_L(u^{-1})| = \{[\text{Re}\{A_L(u^{-1})\}]^2 + [\text{Im}\{A_L(u^{-1})\}]^2\}^{1/2} \qquad (13.6)$$
$$|B_K(u^{-1})| = \{[\text{Re}\{B_K(u^{-1})\}]^2 + [\text{Im}\{B_K(u^{-1})\}]^2\}^{1/2} \qquad (13.7)$$
$$\theta_A = \tan^{-1}\frac{\text{Im}\{A_L(u^{-1})\}}{\text{Re}\{A_L(u^{-1})\}} \qquad \theta_B = \tan^{-1}\frac{\text{Im}\{B_K(u^{-1})\}}{\text{Re}\{B_K(u^{-1})\}}. \qquad (13.8)$$

The real/imaginary parts of the numerator/denominator polynomials are

$$\text{Re}\{A_L(u^{-1})\} = \sum_{\ell=0}^{L} a_\ell \cos(\ell\omega\tau) \qquad \text{Im}\{A_L(u^{-1})\} = \sum_{\ell=0}^{L} a_\ell \sin(\ell\omega\tau) \quad (13.9)$$

$$\text{Re}\{B_K(u^{-1})\} = \sum_{k=0}^{K} b_k \cos(k\omega\tau) \qquad \text{Im}\{B_K(u^{-1})\} = \sum_{k=0}^{K} b_k \sin(k\omega\tau).$$
$$(13.10)$$

The phases θ_A and θ_B can be obtained explicitly from the following expressions

$$\theta_A = \tan^{-1}\frac{\sum_{\ell=0}^{L} a_\ell \sin(\ell\omega\tau)}{\sum_{\ell=0}^{L} a_\ell \cos(\ell\omega\tau)} \qquad \theta_B = \tan^{-1}\frac{\sum_{k=0}^{K} b_k \sin(k\omega\tau)}{\sum_{k=0}^{K} b_k \cos(k\omega\tau)} \qquad (13.11)$$

whereas the squared magnitudes $|A_L(u^{-1})|^2$ and $|B_K(u^{-1})|^2$ are given by

$$|A_L(u^{-1})|^2 = \sum_{\ell=0}^{L}\sum_{\ell'=0}^{L} a_\ell a_{\ell'} \cos([\ell-\ell']\omega\tau) = \sum_{\ell=0}^{L}\sum_{\ell'=0}^{L} a_\ell a_{\ell'} T_{\ell-\ell'}(\omega\tau)$$
$$(13.12)$$

$$|B_K(u^{-1})|^2 = \sum_{k=0}^{K}\sum_{k'=0}^{K} b_k b_{k'} \cos([k-k']\omega\tau) = \sum_{k=0}^{K}\sum_{k'=0}^{K} b_k b_{k'} T_{k-k'}(\omega\tau)$$
$$(13.13)$$

with $T_n(x)$ being the Chebyshev polynomial [180]

$$T_n(x) = \cos(nx). \qquad (13.14)$$

The double sums in the results (13.12) and (13.13) can be reduced to the corresponding single sums as follows. Using the symmetry property $\cos(-z) =$

cos z, we first write

$$\sum_{\ell=0}^{L}\sum_{\ell'=0}^{L} a_\ell a_{\ell'} T_{\ell-\ell'}(\omega\tau) = \sum_{\ell=0}^{L}\sum_{\ell'=0}^{L} a_\ell a_{\ell'} T_{|\ell-\ell'|}(\omega\tau)$$

$$= \sum_{r=0}^{L}\left[\sum_{\ell=0}^{L}\sum_{\ell'=0}^{L} a_\ell a_{\ell'} T_{|\ell-\ell'|}(\omega\tau)\right]\delta_{r,|\ell-\ell'|}$$

$$= \sum_{r=0}^{L}\left[\sum_{\ell=0}^{L}\sum_{\ell'=0}^{L} a_\ell a_{\ell'} \delta_{r,|\ell-\ell'|}\right] T_r(\omega\tau)$$

$$\equiv \sum_{r=0}^{L} \tilde{a}_r T_r(\omega\tau)$$

with the new coefficients $\{\tilde{a}_r\}$ defined as

$$\tilde{a}_r = \sum_{\ell=0}^{L}\sum_{\ell'=0}^{L} a_\ell a_{\ell'} \delta_{r,|\ell-\ell'|}$$

$$= \sum_{\ell=0}^{L} a_\ell \left[\sum_{\ell'=0}^{\ell} a_{\ell'} \delta_{r,\ell-\ell'} + \sum_{\ell'=\ell+1}^{L} a_{\ell'} \delta_{r,\ell'-\ell}\right] = \sum_{\ell=0}^{L} a_\ell (a_{\ell-r} + a_{\ell+r})$$

$$= \sum_{\ell'=-r}^{L-r} a_{\ell'} a_{\ell'+r} + \sum_{\ell=0}^{L} a_\ell a_{\ell+r}$$

$$= \sum_{\ell'=0}^{L-r} a_{\ell'} a_{\ell'+r} + \sum_{\ell=0}^{L-r} a_\ell a_{\ell+r} = 2\sum_{\ell=0}^{L-r} a_\ell a_{\ell+r} \qquad r \neq 0.$$

In the fourth line of this sequel of equations, the index ℓ' in the first sum is redefined to start from zero rather from $-r$, since $a_{-|\ell'|} = 0$ for any integer $\ell' \neq 0$. Moreover, in the same fourth line, the upper summation index $\ell = L$ in the second sum is reduced to $\ell = L - r$ because we have $a_{L+r} = 0$ for $r > 0$. In the special case $r = 0$, it follows

$$\sum_{\ell=0}^{L}\sum_{\ell'=0}^{L} a_\ell a_{\ell'} \delta_{0,|\ell-\ell'|} = \sum_{\ell=0}^{L}\sum_{\ell'=0}^{L} a_\ell a_{\ell'} \delta_{0,\ell-\ell'} = \sum_{\ell=0}^{L} a_\ell^2.$$

Padé–Schur approximant (PSA) with no spurious roots

Of course, the same type of argument is also valid for (13.13). Thus, we shall have the new forms of (13.12) and (13.13)

$$|A_L(u^{-1})|^2 = \sum_{\ell=0}^{L} \tilde{a}_\ell T_\ell(\omega\tau) \qquad |B_K(u^{-1})|^2 = \sum_{k=0}^{K} \tilde{b}_k T_k(\omega\tau) \quad (13.15)$$

$$\tilde{a}_\ell = 2\sum_{r=0}^{L-\ell} a_r a_{r+\ell} \quad (\ell \neq 0) \qquad \tilde{a}_0 = \sum_{r=0}^{L} a_r^2 \quad (13.16)$$

$$\tilde{b}_k = 2\sum_{s=0}^{K-k} b_s b_{s+k} \quad (k \neq 0) \qquad \tilde{b}_0 = \sum_{s=0}^{K} b_s^2. \quad (13.17)$$

As a check, the explicit computation from either (13.12) or (13.15) leads to the following results

$$|A_0(u^{-1})|^2 = a_0^2 \qquad |A_1(u^{-1})|^2 = a_0^2 + a_1^2 + 2a_0 a_1 \cos\omega\tau \quad (13.18)$$

$$|A_2(u^{-1})|^2 = a_0^2 + a_1^2 + a_2^2 + 2(a_0 a_1 + a_1 a_2)\cos\omega\tau + 2a_0 a_2 \cos 2\omega\tau \quad (13.19)$$

$$|A_3(u^{-1})|^2 = a_0^2 + a_1^2 + a_2^2 + a_3^2 + 2(a_0 a_1 + a_1 a_2 + a_2 a_3)\cos\omega\tau$$
$$+ 2(a_0 a_2 + a_1 a_3)\cos 2\omega\tau + 2a_0 a_3 \cos 3\omega\tau \quad (13.20)$$

where the variable u^{-1} is implicitly present via

$$\cos\omega\tau = \frac{u + u^{-1}}{2} = \frac{1 + u^{-2}}{2u^{-1}}. \quad (13.21)$$

Thus using (13.4) and (13.15) we can express the Padé power spectrum $|F_{L,K}(u^{-1})|^2$ as the Padé–Chebyshev approximant (PCA) [244] with the different polynomial coefficients $\{\tilde{a}_\ell, \tilde{b}_k\}$ relative to the original equation (13.1)

$$|F_{L,K}(u^{-1})|^2 = \frac{\sum_{\ell=0}^{L} \tilde{a}_\ell T_\ell(\omega\tau)}{\sum_{k=0}^{K} \tilde{b}_k T_k(\omega\tau)} \equiv T_{L,K}^{PCA}(\cos\omega\tau). \quad (13.22)$$

In (13.22) we introduced a real valued positive definite polynomial quotient $T_{L,K}^{PCA}(\cos\omega\tau)$, which is hereafter called the rational Chebyshev polynomial and this is also the PCA. According to (13.16) and (13.17), the new numerator and denominator polynomial coefficients $\{\tilde{a}_\ell, \tilde{b}_k\}$ are related to the original parameters $\{a_\ell, b_k\}$ through the discrete auto-covariance type expressions. At the same time, both the numerator and denominator polynomials in (13.22) are of the form of the cosine Fourier sums taken at any frequency ω and not just at the Fourier grid $\omega = \tilde{\omega}_k = 2\pi k/T$. Having obtained the power spectrum $|F_{L,K}(u^{-1})|^2$ in (13.22) and using the angles (13.11), the complex rational polynomial $F_{L,K}(u^{-1})$ can be constructed as

$$F_{L,K}(u^{-1}) = \sqrt{T_{L,K}^{PCA}(\cos\omega\tau)}\, e^{i\theta_{AB}(\omega\tau)} \quad (13.23)$$

where $\theta_{AB}(\omega\tau) = \theta_A - \theta_B$. The obtained result (13.23) exhibits the property of periodicity of the complex Padé spectrum $F_{L,K}(u)$. This is because both the magnitude $|F_{L,K}(u)|$ and the angle $\theta_{AB}(\omega\tau)$ are the ratios of linear combinations of sines and cosines of a multiple of the real angle $\omega\tau$. These latter trigonometric functions are periodic with the period equal to $2\pi/\tau$ such that $\cos(\omega\tau + 2\pi m) = \cos(\omega\tau)$ and $\sin(\omega\tau + 2\pi m) = \sin(\omega\tau)$ where $m = 0, \pm 1, \pm 2, ...$, and thus

$$F_{L,K}(e^{i\tau[\omega+2\pi m/\tau]}) = F_{L,K}(e^{i\omega\tau}) \qquad (m = 0, \pm 1, \pm 2, \ldots). \qquad (13.24)$$

Such periodicity is important in practice since it reduces the overall computational effort. In particular, one needs to perform the spectral analysis only in the interval $\omega \in [0, \pi/\tau]$. When this is followed by the use of (13.24), the spectral information about the entire Nyquist range $-\pi/\tau \le \omega \le +\pi/\tau$ will become available with one half of the computational time.

Next, it will prove essential to convert the linear combination of the Chebyshev polynomials from (13.22) into ordinary polynomials in variable $\cos(\omega\tau)$. With this goal, let us consider the following general expression

$$\mathcal{T}_N = \sum_{n=0}^{N-1} a_n T_n(\omega\tau) \qquad (13.25)$$

where the set $\{a_n\}_{n=0}^{N-1}$ of any finite constants (real or complex) a_n is assumed to be known. The rhs of (13.25) is the cosine discrete Fourier transform if ω is on the Fourier grid [64, 245]. To express $\mathcal{T}_N$ as a polynomial in $\cos(\omega\tau)$, one can use the power series representation for $T_n(\cos\omega\tau)$ [180]

$$T_n(x) = \sum_{m=0}^{[n/2]} t_{n,m} x^{n-2m} \qquad x = \cos\omega\tau \qquad (13.26)$$

where $[n/2]$ is the largest integer part of $n/2$, as in (10.1) and

$$t_{n,m} = (-1)^m \frac{n}{m!} \frac{(n-m-1)!}{(n-2m)!} 2^{n-2m-1} \qquad t_{0,0} = 1. \qquad (13.27)$$

Despite the availability of this analytical expression for the coefficients $\{t_{n,m}\}$, it is desirable to have an alternative procedure which avoids factorials altogether. This can be done using the recurrence

$$T_{n+1}(x) = \xi_n x T_n(x) - T_{n-1}(x) \qquad T_{-1}(x) = 0 \qquad T_0(x) = 1 \quad (13.28)$$

where $\xi_n = 1$ for $n = 0$ and $\xi_n = 2$ for $n > 0$ as in (10.16). We then write

$$T_n(x) = \sum_{m=0}^{n} \gamma_{n,n-m} x^m. \qquad (13.29)$$

Padé–Schur approximant (PSA) with no spurious roots 93

The expansion coefficients $\{\gamma_{n,n-m}\}$ are obtained by inserting (13.29) into (13.28) and comparing the terms multiplying like powers of x. This procedure yields $\gamma_{(n+1),(n+1)-m} = 2\gamma_{n,n-(m-1)} - \gamma_{(n-1),(n-1)-m}$, so that

$$\gamma_{n+1,n+1-m} = 2\gamma_{n,n+1-m} - \gamma_{n-1,n-1-m} \quad (13.30)$$

$$\gamma_{0,0} = 1 \quad \gamma_{1,0} = 1 \quad \gamma_{n,m} = 0 \quad (m > n) \quad \gamma_{n,-|m|} = 0 \quad (m \geq 1). \quad (13.31)$$

Hence, all the expansion coefficients $\{\gamma_{n,n-m}\}$ are integers implying that the highest possible accuracy and stability of the recursion (13.30) are achievable. If we put the particular value of the index $m = n + 1$ into the recursion (13.30), it follows that $\gamma_{n+1,0} = 2\gamma_{n,0} - \gamma_{n-1,-2}$ and

$$\gamma_{n+1,0} = 2\gamma_{n,0} \quad (13.32)$$

where $\gamma_{n-1,-2} = 0$ is used according to the initialization (13.31). The simplified relationship (13.32) can be solved explicitly by recurring downward to $n = 0$ giving $\gamma_{n+1,0} = 2\gamma_{n,0} = 2^2\gamma_{n-1,0} = 2^3\gamma_{n-2,0} = \cdots = 2^n\gamma_{1,0}$ and

$$\gamma_{n,0} = 2^{n-1} \quad (13.33)$$

where the relation $\gamma_{0,0} = 1$ from (13.31) is employed. As a check, we calculated the first few expansion coefficients using the recursion (13.30) with the results

$$\gamma_{1,1} = 0 \quad \gamma_{2,2} = -1 \quad \gamma_{2,1} = 0 \quad \gamma_{2,0} = 2 \quad (13.34)$$

$$\gamma_{3,3} = 0 \quad \gamma_{3,2} = -3 \quad \gamma_{3,1} = 0 \quad \gamma_{3,0} = 4. \quad (13.35)$$

Substituting the values (13.34) and (13.35) into (13.29), we find

$$T_0(x) = 1 \quad T_1(x) = x \quad T_2(x) = 2x^2 - 1 \quad T_3(x) = 4x^3 - 3x \quad (13.36)$$

in agreement with the well-known expressions for the Chebyshev polynomials $T_n(x)$ [180]. Inserting the expansion (13.29) into the sum (13.25), the final result for $\mathcal{T}_N$ is obtained in the form

$$\mathcal{T}_N = \sum_{n=0}^{N-1} b_{n,N-1} \cos^n \omega\tau \quad (13.37)$$

$$b_{n,N-1} = \sum_{m=0}^{N-n-1} a_{n+m}\gamma_{n+m,m} = \sum_{m=n}^{N-1} a_m \gamma_{m,m-n} \quad (13.38)$$

$$b_{N-1,N-1} = 2^{N-2} a_{N-1}. \quad (13.39)$$

For verification purposes, we made use of (13.38) for several values of N and obtained the explicit formulae for the coefficients $\{b_{n,N}\}$

$$b_{0,0} = a_0 \quad b_{0,1} = a_0 \quad b_{1,1} = a_1 \quad (13.40)$$

$$b_{0,2} = a_0 - a_2 \quad b_{1,2} = a_1 \quad b_{2,2} = 2a_2 \quad (13.41)$$

$$b_{0,3} = a_0 - a_2 \quad b_{1,3} = a_1 - 3a_3 \quad b_{2,3} = 2a_2 \quad b_{3,3} = 4a_3. \quad (13.42)$$

Substitution of these expressions into the sum rule (13.37) yields the following values of $\mathcal{T}_N$

$$\mathcal{T}_1 = a_0 \qquad \mathcal{T}_2 = a_0 + a_1 \cos\omega\tau + 2a_2 \cos^2\omega\tau \qquad (13.43)$$
$$\mathcal{T}_3 = (a_0 - a_2) + a_1 \cos\omega\tau + 2a_2 \cos^2\omega\tau \qquad (13.44)$$
$$\mathcal{T}_4 = (a_0 - a_2) + (a_1 - 3a_3)\cos\omega\tau + 2a_2 \cos^2\omega\tau + 4a_3 \cos^3\omega\tau. \qquad (13.45)$$

Inserting the Chebyshev polynomials (13.36) into (13.25) we obtain the same results as those stated in (13.43)–(13.45). With the result (13.37) at hand, we have the desired expressions for the two Chebyshev sums from (13.22)

$$\sum_{\ell=0}^{L} \tilde{a}_\ell T_\ell(\omega\tau) = \sum_{\ell=0}^{L} \tilde{p}_\ell \cos^\ell \omega\tau \qquad \tilde{p}_\ell = \sum_{r=0}^{L-\ell} \tilde{a}_r \gamma_{\ell+r,r} \qquad (13.46)$$

$$\sum_{k=0}^{K} \tilde{b}_k T_k(\omega\tau) = \sum_{k=0}^{K} \tilde{q}_k \cos^k \omega\tau \qquad \tilde{q}_k = \sum_{s=0}^{K-k} \tilde{b}_s \gamma_{k+s,s}. \qquad (13.47)$$

Of course, both sets of the new polynomial coefficients $\{\tilde{p}_\ell, \tilde{q}_k\}$ from (13.46) and (13.47) are real due to the assumed realness of the original coefficients $\{a_\ell, b_k\}$ from (13.2). The representations (13.46) and (13.47) permit writing (13.22) in the form of the usual PA for the associated power spectrum

$$T_{L,K}^{\mathrm{PCA}}(\cos\omega\tau) = \frac{\tilde{P}_L(\cos\omega\tau)}{\tilde{Q}_K(\cos\omega\tau)} \qquad (13.48)$$

$$\tilde{P}_L(\cos\omega\tau) = \sum_{\ell=0}^{L} \tilde{p}_\ell \cos^\ell\omega\tau \qquad \tilde{Q}_K(\cos\omega\tau) = \sum_{k=0}^{K} \tilde{q}_k \cos^k\omega\tau. \qquad (13.49)$$

The numerator and denominator polynomials $\tilde{P}_L(\cos\omega\tau)$ and $\tilde{Q}_K(\cos\omega\tau)$ are real, since they are both defined in terms of the powers of the real variable $\cos\omega\tau$ with the real coefficients $\{\tilde{p}_\ell, \tilde{q}_k\}$. Let r_ℓ and s_k denote, respectively, the roots of the polynomials $\tilde{P}_L(\cos\omega\tau)$ and $\tilde{Q}_K(\cos\omega\tau)$ in (13.49). Then setting $\eta_{L,K} = \tilde{p}_L/\tilde{q}_K$, the Padé spectral density follows immediately

$$T_{L,K}^{\mathrm{PCA}}(\cos\omega\tau) = \eta_{L,K} \frac{\prod_{\ell=1}^{L}(\cos\omega\tau - r_\ell)}{\prod_{k=1}^{K}(\cos\omega\tau - s_k)}. \qquad (13.50)$$

For an efficient identification and subsequent regularization of spurious roots from (13.50), it is now convenient to return to the explicit complex variable u^{-1} by using (13.21). We then have

$$\cos\omega\tau - r_\ell = \frac{u^{-2} - 2r_\ell u^{-1} + 1}{2u^{-1}} \qquad \cos\omega\tau - s_k = \frac{u^{-2} - 2s_k u^{-1} + 1}{2u^{-1}}$$

$$T_{L,K}^{\mathrm{PCA}}\left(\frac{u + u^{-1}}{2}\right) = \eta_{L,K}(2u^{-1})^{K-L} \frac{\prod_{\ell=1}^{L}(u^{-2} - 2r_\ell u^{-1} + 1)}{\prod_{k=1}^{K}(u^{-2} - 2s_k u^{-1} + 1)}$$

$$\equiv R_{L,K}^{\mathrm{PCA}}(u^{-1}) \qquad (13.51)$$

Padé–Schur approximant (PSA) with no spurious roots

where $R_{L,K}^{PCA}(u^{-1})$ is used to relabel $T_{L,K}^{PCA}(\cos\omega\tau)$ from (13.22) as follows

$$T_{L,K}^{PCA}(\cos\omega\tau) = T_{L,K}^{PCA}\left(\frac{u+u^{-1}}{2}\right) = T_{L,K}^{PCA}\left(\frac{1+u^{-2}}{2u^{-1}}\right) \equiv R_{L,K}^{PCA}(u^{-1}). \tag{13.52}$$

Realness of the coefficients $\{\tilde{p}_\ell\}$ appearing in the polynomial $\tilde{P}_L(\cos\omega\tau)$ from (13.49) implies that the roots $\{r_\ell\}$ will form two groups. One group is comprised of real roots only, whereas the other group contains exclusively complex roots that must appear as pairs of complex numbers and their complex conjugate counterparts in order to maintain the realness of $\tilde{P}_L(\cos\omega\tau)$. The same is true for the roots $\{s_k\}$, since the coefficients $\{\tilde{q}_k\}$ of the polynomial $\tilde{Q}_K(\cos\omega\tau)$ in (13.49) are also real. In other words, for complex roots $\{r_\ell, s_k\}$ of the polynomials $\{\tilde{P}_L(\cos\omega\tau), \tilde{Q}_K(\cos\omega\tau)\}$, respectively, the spectral representation from (13.51) must contain the pair of the following products

$$(u^{-2} - 2r_\ell u^{-1} + 1)(u^{-2} - 2r_\ell^* u^{-1} + 1)$$
$$= \frac{1}{u^4} - \frac{4\tilde{r}_\ell}{u^3} + 2\frac{1+2|r_\ell|^2}{u^2} - \frac{4\tilde{r}_\ell}{u} + 1 \tag{13.53}$$
$$(u^{-2} - 2s_k u^{-1} + 1)(u^{-2} - 2s_k^* u^{-1} + 1)$$
$$= \frac{1}{u^4} - \frac{4\tilde{s}_k}{u^3} + 2\frac{1+2|s_k|^2}{u^2} - \frac{4\tilde{s}_k}{u} + 1 \tag{13.54}$$

where $\tilde{r}_\ell \equiv \text{Re}(r_\ell)$ and $\tilde{s}_k \equiv \text{Re}(s_k)$. Let the roots of the quadratic equations

$$u^{-2} - 2r_\ell u^{-1} + 1 = 0 \qquad u^{-2} - 2s_k u^{-1} + 1 = 0 \tag{13.55}$$

be denoted by $\rho_{\ell,1/2}$ and $\sigma_{k,1/2}$, respectively

$$\rho_{\ell,1/2} = r_\ell \pm \sqrt{r_\ell^2 - 1} \qquad \sigma_{k,1/2} = s_k \pm \sqrt{s_k^2 - 1} \tag{13.56}$$

by means of which (13.50) is cast into the form

$$T_{L,K}^{PCA}(\cos\omega\tau) = R_{L,K}^{PCA}(u^{-1}) = \eta_{L,K}(2u^{-1})^{K-L}$$
$$\times \frac{\prod_{\ell=1}^{L}(u^{-1} - \rho_{\ell,1})(u^{-1} - \rho_{\ell,2})}{\prod_{k=1}^{K}(u^{-1} - \sigma_{k,1})(u^{-1} - \sigma_{k,2})} \tag{13.57}$$

where the following factorizations are used

$$u^{-2} - 2r_\ell u^{-1} + 1 = (u^{-1} - \rho_{\ell,1})(u^{-1} - \rho_{\ell,2}) \tag{13.58}$$
$$u^{-2} - 2s_k u^{-1} + 1 = (u^{-1} - \sigma_{k,1})(u^{-1} - \sigma_{k,2}). \tag{13.59}$$

The roots $\{\rho_{\ell,1/2}, \sigma_{k,1/2}\}$ are grouped together according to whether $\{r_\ell, s_k\}$ are real or complex. For example, when $\{r_\ell\}$ are real, the corresponding roots $\{\rho_{\ell,1/2}\}$

can be real or complex, depending upon whether $r_\ell > 1$ or $r_\ell < 1$ as is clear from (13.56). If they are complex, they must be always accompanied by their complex conjugates to preserve realness of the rhs of (13.57). On the other hand, if $\{r_\ell\}$ are complex, then the products $(u^{-2} - 2r_\ell u^{-1} + 1)(u^{-2} - 2r_\ell^* u^{-1} + 1)$ must systematically emerge, so that the rhs of (13.50) is real. Such products lead to polynomials with real coefficients, as is obvious from (13.53) and, therefore, also for complex $\{r_\ell\}$, the complex roots $\{\rho_{\ell,1/2}\}$ must always be accompanied by their complex conjugate counterparts

$$(u^{-2} - 2r_\ell u^{-1} + 1)(u^{-2} - 2r_\ell^* u^{-1} + 1)$$
$$= [(u^{-1} - \rho_{\ell,1})(u^{-1} - \rho_{\ell,2})][(u^{-1} - \rho_{\ell,1}^*)(u^{-1} - \rho_{\ell,2}^*)]$$
$$= [(u^{-1} - \rho_{\ell,1})(u^{-1} - \rho_{\ell,1}^*)][(u^{-1} - \rho_{\ell,2})(u^{-1} - \rho_{\ell,2}^*)].$$

Of course, the same argument is also valid for roots $\{s_k\}$ and $\{\sigma_k\}$. In any case, for real or complex $\{r_\ell, s_k\}$ the fact that the free terms in both quadratic equations in (13.55) are equal to unity implies that the root $\rho_{\ell,2}$ must be the inverse of $\rho_{\ell,1}$ and likewise for $\{\sigma_{k,1}, \sigma_{k,2}\}$

$$\rho_{\ell,1}\rho_{\ell,2} = 1 \qquad \sigma_{k,1}\sigma_{k,2} = 1. \tag{13.60}$$

Let us now write the roots $\{\rho_{\ell,1}, \sigma_{k,1}\}$ in their corresponding polar forms

$$\rho_{\ell,1/2} = \rho_\ell^{\pm 1} e^{\pm i\phi_\ell} \qquad \sigma_{k,1/2} = \sigma_k^{\pm 1} e^{\pm i\varphi_k} \tag{13.61}$$

where the magnitudes $\{\rho_\ell, \sigma_k\} = \{|\rho_{\ell,1}|, |\sigma_{k,1}|\}$ and phases $\{\phi_\ell, \varphi_k\}$ are

$$\rho_\ell = \max\left\{\left|r_\ell + \sqrt{r_\ell^2 - 1}\right|, \left|r_\ell - \sqrt{r_\ell^2 - 1}\right|\right\}$$

$$\phi_\ell \equiv \mathrm{Arg}(\rho_{\ell,1}) = \tan^{-1}\frac{\mathrm{Im}(\rho_{\ell,1})}{\mathrm{Re}(\rho_{\ell,1})}$$

$$\sigma_k = \max\left\{\left|s_k + \sqrt{s_k^2 - 1}\right|, \left|s_k - \sqrt{s_k^2 - 1}\right|\right\}$$

$$\varphi_k \equiv \mathrm{Arg}(\sigma_{k,1}) = \tan^{-1}\frac{\mathrm{Im}(\sigma_{k,1})}{\mathrm{Re}(\sigma_{k,1})}.$$

The magnitudes $\{\rho_\ell, \sigma_k\}$ are constant in their values that will be always greater than unity along any ellipse with the points ± 1 as foci [117, 241], provided that the signs of the square roots $\sqrt{r_\ell^2 - 1}$ and $\sqrt{s_k^2 - 1}$ are correctly chosen according to the Landau–Lifshitz convention [144], as stated earlier in relation to (8.13)

$$\rho_\ell > 1 \qquad \sigma_k > 1 \qquad (\ell = 1, 2, 3, \ldots, L \quad \text{and} \quad k = 1, 2, 3, \ldots, K). \tag{13.62}$$

Recall that an ellipse can be described by the equation $|z + \sqrt{z^2 - 1}| = R$ where R is a constant. For any real number a the following identity holds [176],

$$\sqrt{2(1 + ia)} = \sqrt{\sqrt{1 + a^2} + 1} + i\,\mathrm{sgn}(a)\sqrt{\sqrt{1 + a^2} - 1} \tag{13.63}$$

Padé–Schur approximant (PSA) with no spurious roots

where sgn symbolizes the sign function $\text{sgn}(a) = |a|/a$ which is equal to 1 or -1 for $a > 0$ or $a < 0$ and 0 for $a = 0$. Now the PCA from (13.57) can be written as

$$T_{L,K}^{\text{PCA}}(\cos\omega\tau) = R_{L,K}^{\text{PCA}}(u^{-1}) = \eta_{L,K}(2u^{-1})^{K-L}$$
$$\times \frac{\prod_{\ell=1}^{L}(u^{-1} - \rho_\ell e^{i\phi_\ell})(u^{-1} - \rho_\ell^{-1} e^{-i\phi_\ell})}{\prod_{k=1}^{K}(u^{-1} - \sigma_k e^{i\varphi_k})(u^{-1} - \sigma_k^{-1} e^{-i\varphi_k})}. \quad (13.64)$$

The roots $\{\rho_{\ell,2}\}$ and $\{\sigma_{k,2}\}$ of the numerator and denominator polynomials in (13.64) are spurious, i.e. unphysical, since they lie inside the unit circles ($|\rho_{\ell,2}| = \rho_\ell^{-1} < 1$, $|\sigma_{k,2}| = \sigma_k^{-1} < 1$). However, the roots $\{\rho_{\ell,1}\}$ and $\{\sigma_{k,1}\}$ of the numerator and denominator polynomials in (13.64) are genuine/physical, since they are located outside the unit circles ($|\rho_{\ell,1}| = \rho_\ell > 1$, $|\sigma_{k,1}| = \sigma_k > 1$). This perfect separation of genuine and spurious roots is due to the two relations from (13.60). Thus

The roots of PCA [equation (13.64)]:

$$\begin{cases} \text{genuine (outside the unit circle)} \begin{cases} \rho_\ell e^{i\phi_\ell} \text{ [numerator]} \\ \sigma_k e^{i\varphi_k} \text{ [denominator]} \end{cases} \\ \text{spurious (inside the unit circle)} \begin{cases} \rho_\ell^{-1} e^{-i\phi_\ell} \text{ [numerator]} \\ \sigma_k^{-1} e^{-i\varphi_k} \text{ [denominator]}. \end{cases} \end{cases} \quad (13.65)$$

The extraneous or unphysical roots must be regularized because they lead to various artefacts such as spectral distortions, spikes, etc. The goal is to accomplish such a regularization of spurious roots without altering the computed value of the power spectrum $T_{L,K}^{\text{PCA}}(\cos\omega\tau)$ in the original variable $\cos\omega\tau$ [58, 117, 125]. To this end, we shall use a procedure (hereafter termed the constrained root reflection) only for those product terms in (13.64) that are associated with spurious roots. The constrained root reflection is a functional C_* which acts only upon the spurious roots that appear in the power spectrum from (13.64). This is done in two steps by first using the following 'inner prescriptions'

$$C_*\{(u^{-1} - \rho_\ell e^{i\phi_\ell})(u^{-1} - \rho_\ell^{-1} e^{-i\phi_\ell}\}$$
$$= (u^{-1} - \rho_\ell e^{i\phi_\ell}) C_*(u^{-1} - \rho_\ell^{-1} e^{-i\phi_\ell})$$
$$= (u^{-1} - \rho_\ell e^{i\phi_\ell})(u^{-1} - \rho_\ell^{-1} e^{-i\phi_\ell})^*$$
$$= (u^{-1} - \rho_\ell e^{i\phi_\ell})(u - \rho_\ell^{-1} e^{i\phi_\ell})$$
$$= -u\rho_\ell^{-1} e^{i\phi_\ell}(u^{-1} - \rho_\ell e^{i\phi_\ell})(u^{-1} - \rho_\ell e^{-i\phi_\ell})$$
$$= -\frac{u}{\rho_{\ell,1}^*}(u^{-1} - \rho_\ell e^{i\phi_\ell})(u^{-1} - \rho_\ell e^{-i\phi_\ell}) \quad (13.66)$$
$$C_*\{(u^{-1} - \sigma_k e^{i\varphi_k})(u^{-1} - \sigma_k^{-1} e^{-i\varphi_k}\}$$
$$= (u^{-1} - \sigma_k e^{i\varphi_k}) C_*(u^{-1} - \sigma_k^{-1} e^{-i\varphi_k})$$

$$= (u^{-1} - \sigma_k e^{i\varphi_k})(u^{-1} - \sigma_k^{-1} e^{-i\varphi_k})^*$$
$$= (u^{-1} - \sigma_k e^{i\varphi_k})(u - \sigma_k^{-1} e^{i\varphi_k})$$
$$= -u\sigma_k^{-1} e^{i\varphi_k}(u^{-1} - \sigma_k e^{i\varphi_k})(u^{-1} - \sigma_k e^{-i\varphi_k})$$
$$= -\frac{u}{\sigma_{k,1}^*}(u^{-1} - \sigma_k e^{i\varphi_k})(u^{-1} - \sigma_k e^{-i\varphi_k}) \qquad (13.67)$$

and by subsequently employing the 'outer prescription'

$$|C_*|F_{L,K}(u^{-1})|^2| = \left| C_* \left| \frac{A_L(u^{-1})}{B_K(u^{-1})} \right|^2 \right| = |C_* T_{L,K}^{\text{PCA}}(\cos \omega \tau)|$$

$$= |C_* R_{L,K}^{\text{PCA}}(u^{-1})| = \left| C_* \frac{\tilde{P}_L(\cos \omega \tau)}{\tilde{Q}_K(\cos \omega \tau)} \right|$$

$$\equiv \left| \left\{ \frac{\tilde{P}_L(\cos \omega \tau)}{\tilde{Q}_K(\cos \omega \tau)} \right\}_{\text{reg}} \right|. \qquad (13.68)$$

Here, the subscript 'reg' stands for the 'regularization' to indicate that the application of functional C_* to $\tilde{P}_L(\cos \omega \tau) = \tilde{P}_L([1 + u^{-2}]/[2u^{-1}])$ and $\tilde{Q}_K(\cos \omega \tau) = \tilde{Q}_K([1 + u^{-2}]/[2u^{-1}])$ have been carried out using the inner prescriptions from (13.66) and (13.67), respectively. In the first line $|C_*|F_{L,K}(u^{-1})|^2| = |C_*|A_L(u^{-1})/B_K(u^{-1})|^2|$ of the outer prescription (13.68), the functional C_* acts like the unity operator. This part of (13.68) is written simply to remind us of the origin of the remainder of (13.68) which reads as, $T_{L,K}^{\text{PCA}}(\cos \omega \tau) = \tilde{P}_L(\cos \omega \tau)/\tilde{Q}_K(\cos \omega \tau) = R_{L,K}^{\text{PCA}}(u^{-1})$. Functional C_* is set into action only after the emergence of the rational polynomial in variable $\cos \omega \tau$, namely $T_{L,K}^{\text{PCA}}(\cos \omega \tau) = \tilde{P}_L(\cos \omega \tau)/\tilde{Q}_K(\cos \omega \tau)$ in which case the inner prescriptions (13.66) and (13.67) are followed. If C_* is omitted from $|C_*|F_{L,K}(u^{-1})|^2|$ in (13.68), the extra absolute value would be superfluous in the positive definite power spectrum $|\{|F_{L,K}(u^{-1})|^2\}| = |F_{L,K}(u^{-1})|^2$. However, C_* is a mapping from a space of real positive functions to complex valued functions which are produced by the reflection of the spurious roots from inside to outside the unit circle. Therefore, the absolute value must be invoked in (13.68) to restore the expected positive definiteness of the root-reflected/regularized power spectrum. According to (13.64), the spurious roots are identified in (13.66) and (13.67) via $\rho_\ell^{-1} \exp(-i\phi_\ell)$ and $\sigma_k^{-1} \exp(-i\varphi_k)$, since $\rho_\ell > 1$ and $\sigma_k > 1$ as in (13.62). Thus, to regularize these unphysical roots, the constrained root reflection simply replaces the terms $\{(u^{-1} - \rho_{\ell,2}), (u^{-1} - \sigma_{k,2})\}$ by $\{(u^{-1} - \rho_{\ell,2})^*, (u^{-1} - \sigma_{k,2})^*\} = \{(u - \rho_{\ell,2}^*), (u - \sigma_{k,2}^*)\}$, respectively. The final result of the constrained root reflection is the following transformation of (13.64)

$$|C_* R_{L,K}^{PCA}(u^{-1})| \equiv \mathcal{R}_{L,K}^{PSA}(u^{-1}) \tag{13.69}$$

$$\mathcal{R}_{L,K}^{PSA}(u^{-1}) = \left| \eta_{L,K}(-2u^{-2})^{K-L} \left(\frac{\sigma}{\rho}\right)^* \right.$$

$$\left. \times \frac{\prod_{\ell=1}^{L}(u^{-1} - \rho_\ell e^{i\phi_\ell})(u^{-1} - \rho_\ell e^{-i\phi_\ell})}{\prod_{k=1}^{K}(u^{-1} - \sigma_k e^{i\varphi_k})(u^{-1} - \sigma_k e^{-i\varphi_k})} \right| \tag{13.70}$$

$$\rho = \prod_{\ell=1}^{L} \rho_{\ell,1} \qquad \sigma = \prod_{k=1}^{K} \sigma_{k,1}. \tag{13.71}$$

Here, the quantity $\mathcal{R}_{L,K}^{PSA}(u^{-1})$ is the positive definite power spectrum in the ensuing Padé–Schur approximant (PSA) with the key property[4]

$$\mathcal{R}_{L,K}^{PSA}(u^{-1}) = T_{L,K}^{PCA}(\cos \omega \tau). \tag{13.72}$$

Therefore, by subjecting the PCA, i.e. the rational polynomial $R_{L,K}^{PCA}(u^{-1})$ to the constrained root reflection, we obtain the result (13.72) which is the definition of the PSA. The expression (13.72) shows that the power spectrum in the PSA is invariant under the constrained root reflection. Likewise, the same conclusion holds true also for the corresponding magnitude spectrum in the PSA. Hence, the final working formula (13.72) for the power spectrum computed by the PSA contains no spurious roots at all and this is true for both the numerator and the denominator polynomials

$$\begin{cases} \text{The roots of PSA [equation (13.72)]:} \\ \text{genuine (outside the unit circle)} \begin{cases} \rho_\ell e^{i\phi_\ell} \text{ [numerator]} \\ \sigma_k e^{i\varphi_k} \text{ [denominator]} \end{cases} \\ \text{spurious (inside the unit circle)} \begin{cases} \text{none [numerator]} \\ \text{none [denominator]}. \end{cases} \end{cases} \tag{13.73}$$

This proves the sought stability of the PSA as opposed to the unstable PCA from (13.64) and (13.65).

Thus, the starting expression (13.1) of the PA, i.e. $F_{L,K}(u^{-1})$ is a rational complex polynomial in the variable $u^{-1} = \exp(i\omega\tau)$ for real $\omega\tau$. This circumstance permits a subsequent writing of the power spectrum $|F_{L,K}(u^{-1})|^2$ as a real positive definite rational Chebyshev polynomial via (13.48), i.e. $T_{L,K}^{PCA}(\cos \omega\tau)$ in the variable $\cos(\omega\tau)$. From this latter step, the relation (13.21) is used to obtain the power spectrum as a rational real polynomial (13.64), i.e.

---

[4] Despite the appearance of the absolute value rather than its square in (13.69), the transformed function $\mathcal{R}_{L,K}^{PSA}(u)$ does represent the power and *not* the magnitude spectrum. This is because $\mathcal{R}_{L,K}^{PSA}(u)$ originates from (13.48) which is the power spectrum $T_{L,K}^{PCA}(\cos \omega\tau)$.

$R_{L,K}^{PCA}(u^{-1})$ in the original complex variable u^{-1}. Such a rational polynomial facilitates an easy identification of the spurious roots of the PCA as given in (13.65). According to the scheme (13.65), these extraneous roots are located inside the unit circles due to the relations $\rho_\ell > 1$ and $\sigma_k > 1$. Here, σ_ℓ and ρ_k are the magnitudes of the physical roots $\{\rho_{\ell,1}, \sigma_{k,1}\}$ of the numerator and denominator polynomial, respectively, of the power spectrum $R_{L,K}^{PCA}(u^{-1})$. Afterwards, the spurious roots are subjected to the constrained root reflection, symbolized through the application of the constrained complex conjugation C_* from (13.66) and (13.67) to yield a real valued rational polynomial in variable u^{-1}, i.e. $|C_* R_{L,K}^{PCA}(u^{-1})| = \mathcal{R}_{L,K}^{PSA}(u^{-1})$ as given by (13.70). The sought result $\mathcal{R}_{L,K}^{PSA}(u^{-1})$ is a non-negative power spectrum which coincides with the original positive definite spectrum $T_{L,K}^{PCA}(\cos\omega\tau)$ in the variable $\cos\omega\tau$ according to (13.72). The scheme for the regularized roots in the PSA is displayed in (13.73). The advantage of the whole procedure is that it transforms an unstable power spectrum from the PCA to a stable power spectrum in the PSA. This establishes the PSA, which has no spurious roots at all in either the numerator and denominator polynomials for any studied spectrum. Thus the following theorem can be formulated for regularization of all spurious poles and zeros with the simultaneous preservation of the phase-minimum magnitude and power spectra in the original variable $\cos\omega\tau$ [117]:

For any positive definite Padé power spectrum $A(\cos\omega\tau)/B(\cos\omega\tau)$ in the variable $\cos\omega\tau$ there exists a rational polynomial $C(u^{-1})/D(u^{-1})$ in $u^{-1} = \exp(i\omega\tau)$ having its magnitude identical to the original polynomial quotient $|C(u^{-1})/D(u^{-1})| = A(\cos\omega\tau)/B(\cos\omega\tau)$. The constrained root reflection yields the rational Schur (stable) polynomial $|C(u^{-1})/D(u^{-1})|$ without any spurious zeros *and* poles. The zeros of $C(u^{-1})$ and $D(u^{-1})$ have their absolute values greater than unity, as they are all located in the convergent region (the stability domain), i.e. outside the unit circle. The ensuing Padé–Schur approximant $|C(u^{-1})/D(u^{-1})|$ represents the power spectrum in the variable u^{-1} with the preserved phase-minimum and uniqueness. This is due to the invariance of the two power spectra $A(\cos\omega\tau)/B(\cos\omega\tau) = |C(u^{-1})/D(u^{-1})|$ under the passage from the original variable $\cos\omega\tau$ to the variable u^{-1}.

Artefacts and spikes in the computed spectrum are caused by the spurious zeros from the denominator $B_K(u^{-1})$ of the quotient $A_L(u^{-1})/B_K(u^{-1})$ and not from the numerator $A_L(u^{-1})$. Thus, for e.g. designing a stable digital filter one could think that it would suffice to regularize only the denominator polynomial $B_K(u^{-1})$ in $A_L(u^{-1})/B_K(u^{-1})$ and leave the numerator polynomial $A_L(u^{-1})$ unregularized as in [117]. However, a general analysis for designing a stable processor, which preserves the uniqueness *and* the phase-minimum of the PA, requires that *both* the numerator $A_L(u^{-1})$ and denominator $B_K(u^{-1})$ of the rational polynomials $A_L(u^{-1})/B_K(u^{-1})$ are simultaneously regularized by the same procedure. In such a pursuit, extending [117] we supplied the proof of the stated theorem for regularization of both the numerator and denominator of the PA in $A_L(u^{-1})/B_K(u^{-1})$. The resulting method, having only stable zeros in the stable numerator and denominator polynomials of the rational polynomial $A_L(u^{-1})/B_K(u^{-1})$ is known as the Padé–Schur approximant [18]. This is the PSA which regularizes all the encountered spurious zeros and poles that enter the analysis from the noise-corrupted input data. In other words, the initial purpose of the outlined analysis is not to eliminate the noise, which is otherwise indiscernable from, e.g. experimentally measured time signals $\{c_n\}$, but rather to process it along with the genuine physical content of the input data. Unlike the usual i.e the unconstrained root reflection, it then becomes clear that the constrained root reflection loses no information. The standard root reflection operates directly on the original spectrum $A_L(u^{-1})/B_K(u^{-1})$ by reversing the wrong signs $\text{Im}(\omega_k) < 0$ of the imaginary parts of the eigenfrequencies from the roots $\{u_k^{-1}\} = \{\exp(i\omega_k\tau)\}$ of the polynomial $B_K(u^{-1})$ in the variable u^{-1}. This should be distinguished from the constrained root reflection which does not operate at all on the roots $\{u_k^{-1}\}$ of the original polynomials from $A_L(u^{-1})/B_K(u^{-1})$. Instead, the constrained root reflection $\{(u^{-1} - \rho_{\ell,2}), (u^{-1} - \sigma_{k,2})\} \longrightarrow \{(u^{-1} - \rho_{\ell,2})^*, (u^{-1} - \sigma_{k,2})^*\} = \{-(u/\rho_{\ell,1}^*)(u^{-1} - \rho_{\ell,1}^*), -(u/\rho_{\ell,1}^*)(u^{-1} - \sigma_{k,1}^*)\}$ converts the unequivocally identified spurious roots $\{\rho_{\ell,2}, \sigma_{k,2}\}$ into the physical roots $\{\rho_{\ell,1}, \sigma_{k,1}\}$ in the polynomials $\{\tilde{P}_L(\cos\omega\tau), \tilde{Q}_K(\cos\omega\tau)\}$ from the Padé power spectrum $|A_L(u^{-1})/B_K(u^{-1})|^2 = \tilde{P}_L(\cos\omega\tau)/\tilde{Q}_K(\cos\omega\tau)$ [142].

Chapter 14

Spectral analysis and systems of inhomogeneous linear equations

While studying a Lorentzian spectrum, a sequence of time signal points $\{c_n\}_{n=0}^{N-1}$ is available and the task is to solve the inverse problem by reconstructing the spectral harmonics with their complex amplitudes and frequencies as the building constituents of $\{c_n\}$ given by the sum (5.11) of damped exponentials. The representation (5.11), which is otherwise a geometric sequence, is also known as the harmonic inversion problem [17, 108, 109, 165]. By definition, a geometric sequence is the one whose differences of the successive terms are always equal to each other. There are many apparently different, but often mathematically equivalent methods for solving the inverse problem (5.11) and most of them resort to a system of linear equations in some important computational stages of the analysis. For example, the DPA [115] solves such a system while obtaining the coefficients of the denominator polynomial, which is equal to the secular or characteristic polynomial. The same system is also encountered in the DLP [115] which solves an additional system of linear equations for the amplitudes $\{d_k\}$. In the FD [108, 109] and the DSD [115] one solves a generalized eigenvalue problem (8.10) of the so-called data matrix which is the Hankel matrix $\mathbf{H}_n(c_s) = \{c_{j+k+s}\}_{j,k=0}^{n-1}$ from (5.14) with signal points (5.11) as its elements. Of course, in general, the elements of the Hankel matrix (5.14) are some arbitrary real or complex numbers $\{c_{j+k+s}\}$ that do not need to coincide with the signal points from (5.11). The Hankel matrix $\mathbf{H}_n(c_s)$ is identical to the sth power of the non-Hermitean evolution or relaxation matrix $\mathbf{U}^{(s)} = \{U_{nm}^{(s)}\}$ from (8.10) as stated in (5.14). The generalized eigenvalue problem (8.10) of the matrix $\mathbf{U}^{(s)} = \mathbf{H}_n(c_s)$ in the FD and the DSD is solved by resorting to a system of linear equations. Hence, indeed in solving the harmonic inversion problem, one very often encounters a system of linear inhomogeneous equations. But, here, there ought to be a degree of reciprocity suggesting that also, while solving a system of linear inhomogeneous equations *per se* by certain methods, one should arrive at the harmonic inversion problem. This is indeed true, as shown in chapter 15 by using the Schmidt [71] variant of the Gauss–Seidel [223] iterative method of successive approximations for the exact solution of the system of linear inhomogeneous equations.

In practice, it appears that one often tries to reduce the computational part of the studied problem to a system of linear homogeneous or inhomogeneous equations. This is understandable in view of the fact that major efforts in building the existing mathematical software [176], have thus far been invested in supplying powerful algorithms for solutions of systems of linear equations. This latter type of algorithms undoubtedly belongs to the top category of the most accurate, robust and efficient algorithms in the leading libraries on linear algebra [176]. The way of solving a system of linear equations via, e.g. the Gauss–Seidel [223] method and the like is no longer in routine use for extensive numerical purposes, due to their performance with comparatively low efficiency relative to the modern software of the LU (lower-upper) decomposition types [176], etc. However, concentrating exclusively on aspects of computational efficiency might occasionally miss the opportunities offered by some of the alternative techniques, particularly when it comes to connecting, by means of analytical methods, certain apparently disjoint strategies of computations. In the next chapter we assess the usefulness of this methodology. This is illustrated not only by elucidating a fundamental link between a system of linear inhomogeneous equations, harmonic inversion problem and nonlinear acceleration of convergent/divergent series/sequences, but also in providing a very useful computational tool for spectral analysis. The exact iterative solution of a system of linear equations will be found within the Gauss–Seidel methodology [223] in a versatile setting of the Schmidt relaxations [71].

Chapter 15

Exact iterative solution of a system of linear equations

Consider a system of inhomogeneous linear equations of the following general matrix form

$$\mathbf{A}\mathbf{x} = \mathbf{b} \tag{15.1}$$

$$\left.\begin{array}{l} a_{11}x_1 + a_{12}x_2 + a_{13}x_3 + \cdots + a_{1m}x_m = b_1 \\ a_{21}x_1 + a_{22}x_2 + a_{23}x_3 + \cdots + a_{2m}x_m = b_2 \\ a_{31}x_1 + a_{32}x_2 + a_{33}x_3 + \cdots + a_{3m}x_m = b_3 \\ \quad\vdots \\ a_{m1}x_1 + a_{m2}x_2 + a_{m3}x_3 + \cdots + a_{mm}x_m = b_m \end{array}\right\} \tag{15.2}$$

where $\mathbf{x} = \{x_k\}_{k=1}^m$ is the sought column vector $\{x_1, x_2, x_3, \ldots, x_m\}$. The matrix $\mathbf{A}$ is the given $m \times m$ matrix $\mathbf{A} = \{a_{jk}\}$ and $\mathbf{b} = \{b_k\}$ is the column vector with the known coefficients $\{b_1, b_2, b_2, \ldots, b_m\}$. The exact and unique solution of (15.2) for the unknown x_r is given by the Cramer rule [228]

$$x_r = \frac{D_r}{D} \tag{15.3}$$

provided that

$$D \neq 0 \tag{15.4}$$

where D is the determinant of the system (15.2)

$$D = \begin{vmatrix} a_{11} & a_{12} & a_{13} & \cdots & a_{1m} \\ a_{21} & a_{22} & a_{23} & \cdots & a_{2m} \\ a_{31} & a_{32} & a_{33} & \cdots & a_{3m} \\ \vdots & \vdots & \vdots & \ddots & \vdots \\ a_{m1} & a_{m2} & a_{m3} & \cdots & a_{mm} \end{vmatrix} \qquad (15.5)$$

$$D_r = \begin{vmatrix} a_{11} & a_{12} & a_{13} & \cdots & a_{1,r-1} & b_1 & a_{1,r+1} & \cdots & a_{1m} \\ a_{21} & a_{22} & a_{23} & \cdots & a_{2,r-1} & b_2 & a_{2,r+1} & \cdots & a_{2m} \\ a_{31} & a_{32} & a_{33} & \cdots & a_{3,r-1} & b_3 & a_{3,r+1} & \cdots & a_{3m} \\ \vdots & \vdots & \vdots & \vdots & \vdots & \vdots & \vdots & \ddots & \vdots \\ a_{m1} & a_{m2} & a_{m3} & \cdots & a_{m,r-1} & b_m & a_{m,r+1} & \cdots & a_{mm} \end{vmatrix}. \qquad (15.6)$$

The *exact* solution **x** of this problem can also be found by using the Schmidt [71] iterative procedure of successive approximations/relaxations as reminiscent of the Gauss–Seidel [223] elimination technique. This method permits a natural introduction of the base-transient concept through the appearance of the same type of geometric sequences that are also encountered in the harmonic inversion problem for signal processing and in a nonlinear acceleration of series [71, 72].

In every iterative method one should have an initial approximation of the sought solution. Let the vector $\mathbf{x}_0 = \{x_{r,0}\}$ be such a starting guess. An approximation of the exact solution $\mathbf{x} = \{x_r\}$ obtained after n successive iterations will be denoted by $\mathbf{x}_n = \{x_{r,n}\}$ and this will be called the nth-order iterate. Then the successive approximations are obtained as follows. We first determine the first-order iterates $\{x_{r,1}\}$ of the corresponding exact components $\{x_r\}$. Thus in order to obtain, e.g. the iterate $x_{1,1}$ we first replace the set $\{x_1, x_2, x_3, \ldots, x_m\}$ by $\{x_{1,1}, x_{2,0}, x_{3,0}, \ldots, x_{m,0}\}$, respectively[1]. We do this only in the first equation of the sequel (15.2)

$$a_{11}x_{1,1} + a_{12}x_{2,0} + a_{13}x_{3,0} + \cdots + a_{1m}x_{m,0} = b_1 \qquad (15.7)$$

from which the unknown $x_{1,1}$ is determined. In the second step of the iteration we compute the first-order iterate $x_{2,1}$ linked to the component x_2 of **x**. To this end, only the second equation in (15.2) is used, where the vector $\{x_1, x_2, x_3, \ldots, x_m\}$ is replaced by $\{x_{1,1}, x_{2,1}, x_{3,0}, \ldots, x_{m,0}\}$, respectively

$$a_{21}x_{1,1} + a_{22}x_{2,1} + a_{23}x_{3,0} + \cdots + a_{2m}x_{m,0} = b_2. \qquad (15.8)$$

This yields $x_{2,1}$ since $x_{1,1}$ is already known from (15.7). Notice that the number of the zeroth-order estimates $\{x_{r,0}\}$, i.e. the initial guesses that are present in (15.7) and (15.8), is equal to $m-1$ and $m-2$, respectively. The general pattern now becomes clear when one progresses further down to the end of

---

[1] Notice that in the string $\{x_{1,1}, x_{2,0}, x_{3,0}, \ldots, x_{m,0}\}$ all the entries except the first element are equalized with their initial values $\{x_{n,0}\}(n \geq 2)$.

the sequel (15.2) by using only one equation at a time and taking advantage of the previously obtained first-order iterates. This process simultaneously and progressively reduces the number of the zeroth-order iterates $\{x_{r,0}\}$ from each of the equations in (15.2). Thus, in the end, while obtaining the iterate $x_{m,1}$ all the iterates $\{x_{1,1}, x_{2,1}, x_{3,1}, \ldots, x_{m-1,1}\}$ have become known. Therefore, in the last equation in (15.2) we replace $\{x_1, x_2, x_3, \ldots, x_m\}$ by $\{x_{1,1}, x_{2,1}, x_{3,1}, \ldots, x_{m,1}\}$, respectively, with no zeroth-order iterate encountered

$$a_{m1}x_{1,1} + a_{m2}x_{2,1} + a_{m3}x_{3,1} + \cdots + a_{m,m-1}x_{m-1,1} + a_{mm}x_{m,1} = b_m. \quad (15.9)$$

This equation gives the unknown $x_{m,1}$ in terms of the previously obtained iterates from the string $\{x_{1,1}, x_{2,1}, x_{3,1}, \ldots, x_{m-1,1}\}$. Hence, the outlined first-order iteration produces the vector $\mathbf{x}_1 = \{x_{r,1}\}$ as the *exact* solution of the following system of equations

$$\left.\begin{aligned}
a_{11}x_{1,1} + a_{12}x_{2,0} + a_{13}x_{3,0} + \cdots + a_{1m}x_{m,0} &= b_1 \\
a_{21}x_{1,1} + a_{22}x_{2,1} + a_{23}x_{3,0} + \cdots + a_{2m}x_{m,0} &= b_2 \\
a_{31}x_{1,1} + a_{32}x_{2,1} + a_{33}x_{3,1} + \cdots + a_{3m}x_{m,0} &= b_3 \\
&\vdots \\
a_{m1}x_{1,1} + a_{m2}x_{2,1} + a_{m3}x_{3,1} + \cdots + a_{mm}x_{m,1} &= b_m
\end{aligned}\right\}. \quad (15.10)$$

In the original system of simultaneous linear equations (15.2) all the equations are coupled. However, no such coupling exists in the auxiliary system of linear equations (15.10) where each equation is solvable exactly in succession, one after the other in the spirit of the standard Gauss–Seidel method [223].

To get the second-order iterate $\mathbf{x}_2 = \{x_{r,2}\}$ we return to the original system of equations (15.2) and use the obtained vector $\mathbf{x}_1 = \{x_{r,1}\}$ as the new initial values. We then repeat exactly the same procedure as done with the initial values $\mathbf{x}_0 = \{x_{r,0}\}$. The higher-order iterates are obtained in a similar way. Thus, e.g. the nth-order iterate $\mathbf{x}_n = \{x_{r,n}\}$ which approximates the sought vector $\mathbf{x}$ after $n + 1$ successive steps, will actually represent the exact solution of the following system

$$\left.\begin{aligned}
a_{11}x_{1,n+1} + a_{12}x_{2,n} + a_{13}x_{3,n} + \cdots + a_{1m}x_{m,n} &= b_1 \\
a_{21}x_{1,n+1} + a_{22}x_{2,n+1} + a_{23}x_{3,n} + \cdots + a_{2m}x_{m,n} &= b_2 \\
a_{31}x_{1,n+1} + a_{32}x_{2,n+1} + a_{33}x_{3,n+1} + \cdots + a_{3m}x_{m,n} &= b_3 \\
&\vdots \\
a_{m1}x_{1,n+1} + a_{m2}x_{2,n+1} + a_{m3}x_{3,n+1} + \cdots + a_{mm}x_{m,n+1} &= b_m
\end{aligned}\right\}. \quad (15.11)$$

To solve this system of m simultaneous difference linear equations we need to homogenize the second subscripts in all the iterates. In other words, we want to map the set $\{x_{r,n+1}\}$ onto $\{x_{r,n}\}$. This can be done through the usage of the displacement or shift-operator or step-operator $\hat{E}$, which acts on the second subscript only

$$\hat{E}x_{m,n} = x_{m,n+1} \qquad (n, m = 0, 1, 2, \ldots). \tag{15.12}$$

Such a shifting can be repeated any number of times, so that the ℓth fold application of the operator $\hat{E}$ onto the iterate $x_{m,n}$ leads to the $(n + \ell)$th iterate $x_{m,n+\ell}$

$$\hat{E}^\ell x_{m,n} = x_{m,n+\ell}. \tag{15.13}$$

In a special case, by operating $\hat{E}^\ell$ on any constant γ we would obtain the same constant

$$\hat{E}^\ell \gamma = \gamma \qquad (\ell = 0, 1, 2, \ldots). \tag{15.14}$$

With this device, we can rewrite (15.11) in a form which is uniform in the second subscripts of the iterates, so that only the unknown set $\{x_{r,n}\}$ will appear

$$\left.\begin{aligned}
a_{11}\hat{E}x_{1,n} + a_{12}x_{2,n} + a_{13}x_{3,n} + \cdots + a_{1m}x_{m,n} &= b_1 \\
a_{21}\hat{E}x_{1,n} + a_{22}\hat{E}x_{2,n} + a_{23}x_{3,n} + \cdots + a_{2m}x_{m,n} &= b_2 \\
a_{31}\hat{E}x_{1,n} + a_{32}\hat{E}x_{2,n} + a_{33}\hat{E}x_{3,n} + \cdots + a_{3m}x_{m,n} &= b_3 \\
&\vdots \\
a_{m1}\hat{E}x_{1,n} + a_{m2}\hat{E}x_{2,n} + a_{m3}\hat{E}x_{3,n} + \cdots + a_{mm}\hat{E}x_{m,n} &= b_m
\end{aligned}\right\}. \tag{15.15}$$

With (15.14), the system of difference equations (15.11) is reduced to a system of ordinary equations (15.15) with the operator-valued coefficients $a_{j,k}\hat{E}^r$ $(r = 0, 1)$. Therefore, the new system (15.15) can be formally solved by the Cramer rule to obtain the rth unknown $x_{r,n}$ in the operational form

$$D(\hat{E})x_{r,n} = D_r(\hat{E})1 \tag{15.16}$$

$$D(\hat{E}) = \begin{vmatrix} a_{11}\hat{E} & a_{12} & a_{13} & a_{14} & \cdots & a_{1m} \\ a_{21}\hat{E} & a_{22}\hat{E} & a_{23} & a_{24} & \cdots & a_{2m} \\ a_{31}\hat{E} & a_{32}\hat{E} & a_{33}\hat{E} & a_{34} & \cdots & a_{3m} \\ \vdots & \vdots & \vdots & \vdots & \ddots & \vdots \\ a_{m1}\hat{E} & a_{m2}\hat{E} & a_{m3}\hat{E} & a_{m4}\hat{E} & \cdots & a_{mm}\hat{E} \end{vmatrix} \tag{15.17}$$

$$D_r(\hat{E}) = \begin{vmatrix} a_{11}\hat{E} & a_{12} & \cdots & a_{1,r-1} & b_1 & a_{1,r+1} & \cdots & a_{1m} \\ a_{21}\hat{E} & a_{22}\hat{E} & \cdots & a_{2,r-1} & b_2 & a_{2,r+1} & \cdots & a_{2m} \\ a_{31}\hat{E} & a_{32}\hat{E} & \cdots & a_{3,r-1} & b_3 & a_{3,r+1} & \cdots & a_{3m} \\ \vdots & \vdots & \vdots & \vdots & \vdots & \vdots & \ddots & \vdots \\ a_{m1}\hat{E} & a_{m2}\hat{E} & \cdots & a_{m,r-1}\hat{E} & b_m & a_{m,r+1}\hat{E} & \cdots & a_{mm}\hat{E} \end{vmatrix}.$$

$$\tag{15.18}$$

The symbol 1 on the rhs of (15.16), after the determinant $D_r(\hat{E})$, is introduced to indicate that the displacement operator acts as a constant of the unit value. Clearly, the operator valued determinants $D(\hat{E})$ and $D_r(\hat{E})$ coincide with D and D_r for $\hat{E} = \hat{1}$, where D and D_r are given by (15.5) and (15.6), respectively. Using the Cauchy expansion theorem to compute the determinant $D_r(\hat{E})$ from (15.18), we shall encounter the operator $\hat{E}$ raised to powers of positive integers. Since $\hat{E}^\ell$ ($\ell = 0, 1, 2, \ldots$) applied to the unity will give the unity according to (15.14), the symbols $\hat{E}$ and 1 can be omitted from the rhs of (15.16), so that

$$D(\hat{E}) x_{r,n} = D_r \qquad (15.19)$$

where D_r is given in (15.6). Developing the determinant $D(\hat{E})$ from (15.17) via the Cauchy expansion theorem and collecting the terms of like powers in $\hat{E}$, we obtain from (15.19)

$$(v_1 \hat{E} + v_2 \hat{E}^2 + v_3 \hat{E}^3 + \cdots + v_m \hat{E}^m) x_{r,n} = D_r. \qquad (15.20)$$

Here, the first and the last coefficients v_1 and v_m, respectively, have their simple forms given by:

$$v_1 = (-1)^{m+1} a_{m1} \prod_{j=1}^{m-1} a_{j,j+1} \qquad v_m = \prod_{j=1}^{m} a_{jj}. \qquad (15.21)$$

Using (15.13) we can rewrite (15.20) as

$$v_1 x_{r,n+1} + v_2 x_{r,n+2} + v_3 x_{r,n+3} + \cdots + v_m x_{r,n+m} = D_r. \qquad (15.22)$$

This is an mth-order inhomogeneous difference equation with constant coefficients. The solution of (15.22) can be obtained indirectly through the zeros $\{q_k\}_{k=1}^{m-1}$ of the characteristic polynomial equation

$$Q_{m-1}(q_k) = 0 \qquad (1 \leq k \leq m-1) \qquad (15.23)$$
$$Q_{m-1}(\hat{E}) \equiv v_1 + v_2 \hat{E} + v_3 \hat{E}^2 + \cdots + v_m \hat{E}^{m-1}. \qquad (15.24)$$

The general solution of the difference equation (15.22) is given by the following expression [218]

$$x_{r,n} = \frac{D_r}{v_1 + v_2 + v_3 + \cdots + v_m} + \sum_{k=1}^{m-1} a_k^{(r)} q_k^n \qquad (15.25)$$

where $\{a_k^{(r)}\}$ are the 'integration' constants that can be fixed by imposing the given initial, i.e. boundary conditions to (15.22). The sum $v_1 + v_2 + v_3 + \cdots + v_m$ appearing in (15.25) is recognized as the value of the determinant $D(\hat{E})$ evaluated at $\hat{E} = \hat{1}$, so that $D(\hat{1}) = D$ where D is given by (15.5). This implies $v_1 + v_2 + v_3 + \cdots + v_m = D(\hat{1}) = D$, so that we have for (15.25)

$$x_{r,n} = x_r + \sum_{k=1}^{m-1} a_k^{(r)} q_k^n \qquad (15.26)$$

where $x_r = D_r/D$ is the exact rth solution which we are seeking for the original system of equations given in (15.2). The concrete framework of this computational procedure is within the general scope of the Gauss–Siedel [223] methodology, but with a distinct feature exhibited by successive relaxations as developed by Schmidt [71].

The solution (15.26) is valid if all the roots are distinct. When some of the roots $\{q_r\}_{k=1}^{m-1}$ coincide with one another, a modification is required as follows. Assume that the q_ℓth root has the multiplicity $s_\ell (1 \leq \ell \leq m-1)$. This means that there are s_ℓ roots labelled by q_ℓ, so that

$$m_1 + m_2 + m_3 + \cdots + m_\ell = m. \qquad (15.27)$$

If some of the roots are simple, their multiplicity is set to unity, e.g. $m_5 = 1$ for the distinct root q_5. Moreover, the constants $\{a_k^{(r)}\}$ from (15.26) should be relabelled, e.g. $\{a_{k,\ell_k}^{(r)}\}$ to reflect the degeneracy of the kth root. Thus, when degenerate roots are present, the general solution (15.26) becomes

$$\begin{aligned} x_{r,n} = x_r &+ \{a_{1,1}^{(r)} + (n+1)a_{1,2}^{(r)} + \cdots + (n+1)^{m_1-1}a_{1,m_1}^{(r)}\}q_1^n \\ &+ \{a_{2,1}^{(r)} + (n+1)a_{2,2}^{(r)} + \cdots + (n+1)^{m_2-1}a_{2,m_2}^{(r)}\}q_2^n \\ &+ \cdots \\ &+ \{a_{m-1,1}^{(r)} + (n+1)a_{m-1,2}^{(r)} + \cdots + (n+1)^{m_\ell-1-1}a_{\ell-1,m_{\ell-1}}^{(r)}\}q_\ell^n \end{aligned}$$

$$x_{r,n} = x_r + \sum_{k=1}^{\ell} A_{k,n}^{(r)} q_k^n \qquad A_{k,n}^{(r)} = \sum_{j=1}^{m_k} a_{k,j}^{(r)} (n+1)^{j-1}. \qquad (15.28)$$

In the non-degenerate case (15.26), the amplitudes $\{a_k^{(r)}\}$ are independent of n. This is not true any longer in the degenerate case (15.28), where the corresponding amplitudes $\{A_{k,n}^{(r)}\}$ depend upon n.

Admittedly, the exact solution $x_r = x_{r,n} - \sum_{k=1}^{m-1} a_k^{(r)} q_k^n$ from (15.26) looks very different from the Cramer counterpart (15.2). This is due to the appearance of the sum $\sum_{k=1}^{m-1} a_k^{(r)} q_k^n$. Recall that the system of linear equations from (15.2) has the exact and *unique* solution if the condition (15.4) is fulfilled. Therefore, we must have

$$\lim_{n \to \infty} x_{r,n} = x_r \qquad (15.29)$$

and this will be satisfied only if the absolute values of *all* the roots $\{q_k\}_{k=1}^{m-1}$ are less than unity, irrespective of the goodness of the initial guess $\{x_{r,0}\}$ and regardless of whether or not some of the roots $\{q_k\}$ are degenerate

$$|q_k| < 1 \quad (1 \leq k \leq m-1). \tag{15.30}$$

In other words, with the increasing value of integer n all the 'transients' $\sum_{k=1}^{m-1} a_k^{(r)} q_k^n$ must be extinguished if the 'equilibrium' value x_r is to be reached. Therefore, the task is to find a transformation which would be capable of filtering out the whole sum of the evanescent modes, i.e. the transients $\sum_{k=1}^{m-1} a_k^{(r)} q_k^n$ from the iterative solution (15.26). Then such a sum would immediately reduce the sequence $\{x_{r,n}\}$ to its limiting value x_r which is called the 'base' of $\{x_{r,n}\}$. The importance of deriving this transformation is also in the possibility of obtaining not only the limit x_r when the condition (15.30) is satisfied, but also the 'anti-limit' of the sequence $\{x_{r,n}\}$ when some of q_k are greater than unity. In this latter case, the sequence $\{x_{r,n}\}$ is divergent and the concept of anti-limit is conceived in the sense of analytical continuation which extends the convergence radius of the original sequence [28, 72, 75]. Both convergent and divergent sequences $\{x_{r,n}\}$ can be handled within the same procedure if we could filter out all the transients $\sum_{k=1}^{m-1} a_k^{(r)} q_k^n$. Moreover, we are aiming at deriving an alternative expression for the general iterative solution $x_{r,n}$ of the system of equations (15.2) in a form which does not need to consider the multiplicity of the roots, since by the mentioned filtering process the roots $\{q_k\}$ should disappear altogether from the analysis along with the associated constants $\{a_k^{(r)}\}$. To accomplish this, we rewrite the expression (15.26) as m equations

$$x_{r,0} = x_r + a_1^{(r)} + a_2^{(r)} + \cdots a_{m-1}^{(r)}$$
$$x_{r,1} = x_r + a_1^{(r)} q_1 + a_2^{(r)} q_2 + \cdots + a_{m-1}^{(r)} q_{m-1}$$
$$x_{r,2} = x_r + a_1^{(r)} q_1^2 + a_2^{(r)} q_2^2 + \cdots + a_{m-1}^{(r)} q_{m-1}^2$$
$$\vdots$$
$$x_{r,m-1} = x_r + a_1^{(r)} q_1^{m-1} + a_2^{(r)} q_2^{m-1} + \cdots + a_{m-1}^{(r)} q_{m-1}^{m-1}. \tag{15.31}$$

Multiplying these equations for $x_{r,0}, x_{r,1}, x_{r,2}, \ldots, x_{r,m-1}$ by $v_1, v_2, v_3, \ldots, v_m$, respectively, and adding together the m resulting equations, will give

$$\sum_{v=0}^{m-1} v_{v+1} x_{r,v} = (v_1 + v_2 + \cdots + v_m) x_r$$

$$+ \left\{ a_1^{(r)} \sum_{v=0}^{m-1} v_{v+1} q_1^v + a_2^{(r)} \sum_{v=0}^{m-1} v_{v+1} q_2^v \right.$$

$$\left. + \cdots + a_{m-1}^{(r)} \sum_{v=0}^{m-1} v_{v+1} q_{m-1}^v \right\}$$

$$= (v_1 + v_2 + \cdots + v_m)x_r + \sum_{k=1}^{m-1} a_k^{(r)} \sum_{v=0}^{m-1} v_{v+1} q_k^v$$

$$= (v_1 + v_2 + \cdots + v_m)x_r$$

$$+ \sum_{k=1}^{m-1} a_k^{(r)} Q_{m-1}(q_k) = (v_1 + v_2 + \cdots + v_m)x_r \quad (15.32)$$

where (15.23) is used. Therefore, by employing (15.4), i.e. $D \neq 0$ we shall have from (15.32)

$$x_r = \frac{N_r}{D} \qquad N_r = \sum_{v=0}^{m-1} v_{v+1} x_{r,v} \quad (15.33)$$

$$x_r = \frac{v_1 x_{r,0} + v_2 x_{r,1} + v_3 x_{r,2} + \cdots + v_m x_{r,m-1}}{v_1 + v_2 + v_3 + \cdots + v_m}. \quad (15.34)$$

Both (15.3) and (15.33) represent the *exact* solution of the system (15.1) of linear equations. When the condition (15.4) is fulfilled, the system (15.1) possesses the unique solution, so that

$$N_r = D_r \implies x_r = \frac{D_r}{D} \quad (15.35)$$

and this can be verified by an explicit calculation. Hence, by filtering out all the transients, the iterated solution $x_{r,n}$ is reduced to the exact Cramer solution, as is clear from (15.3) and (15.35).

The solution (15.33) is now free from transient $\{a_k q_k^n\}$ and, therefore, the multiplicity of the root q_k of the characteristic equation (15.23) should be irrelevant. Nevertheless, this is not strictly correct without an additional proof, since we arrived at (15.33) by explicitly using only the non-degenerate iterate (15.26). Hence, a calculation similar to (15.31) and (15.32) is also needed for the degenerate iterate (15.28)

$$x_{r,0} = x_r + A_{1,0}^{(r)} + A_{2,0}^{(r)} + \cdots A_{m-1,0}^{(r)}$$
$$x_{r,1} = x_r + A_{1,1}^{(r)} q_1 + A_{2,1}^{(r)} q_2 + \cdots + A_{m-1,1}^{(r)} q_{m-1}$$
$$x_{r,2} = x_r + A_{1,2}^{(r)} q_1^2 + A_{2,2}^{(r)} q_2^2 + \cdots + A_{m-1,2}^{(r)} q_{m-1}^2$$
$$\vdots$$
$$x_{r,m-1} = x_r + A_{1,m-1}^{(r)} q_1^{m-1} + A_{2,m-1}^{(r)} q_2^{m-1} + \cdots + A_{m-1,m-1}^{(r)} q_{m-1}^{m-1}.$$

If we multiply the equations for $x_{r,0}, x_{r,1}, x_{r,2}, \ldots, x_{r,m-1}$ by $v_1, v_2, v_3, \ldots, v_m$, respectively, and add the m ensuing equations, the following counterpart of (15.32) will follow

$$\sum_{v=0}^{m-1} v_{v+1} x_{r,v} = (v_1 + v_2 + \cdots + v_m)x_r + \sum_{k=1}^{m-1} \sum_{v=0}^{m-1} v_{v+1} A_{k,v}^{(r)} q_k^v. \quad (15.36)$$

Using (15.28) together with the identity $q_k^\nu = \exp[\nu \ln(q_k)]$, we have

$$A_{k,\nu}^{(r)} q_k^\nu = \sum_{\ell_k=1}^{m_k} a_{k,\ell_k}^{(r)} \left\{ \left[q_k^{-1} \frac{\partial}{\partial \ln(q_k)} \right]^{\ell_k-1} q_k \right\} q_k^\nu \equiv \sum_{\ell_k=1}^{m_k} a_{k,\ell_k}^{(r)} \mathcal{P}_{k,\ell_k} q_k^\nu. \quad (15.37)$$

Inserting (15.37) into (15.36) and employing (15.23) gives

$$\sum_{\nu=0}^{m-1} v_{\nu+1} x_{r,\nu} = (v_1 + v_2 + \cdots + v_m) x_r + \sum_{k=1}^{m-1} \sum_{\ell_k=1}^{m_k} a_{k,\ell_k}^{(r)} \mathcal{P}_{k,\ell_k} \sum_{\nu=0}^{m-1} v_{k+1} q_k^\nu$$

$$= (v_1 + v_2 + \cdots + v_m) x_r + \sum_{k=1}^{m-1} \sum_{\ell_k=1}^{m_k} a_{k,\ell_k}^{(r)} \mathcal{P}_{k,\ell_k} Q_{m-1}(q_k)$$

$$= (v_1 + v_2 + \cdots + v_m) x_r = D x_r \implies x_r = \frac{\sum_{\nu=0}^{m-1} v_{\nu+1} x_{r,\nu}}{D}. \quad (15.38)$$

The result (15.38) for the degenerate iterate (15.37) is seen to coincide with (15.28) for the non-degenerate (QED). This is expected, since the transients $\{a_k q_k^n\}$ are used in an intermediate stage of the analysis for which the multiplicity of the roots $\{q_k\}$ of $Q_{m-1}(q)$ was necessary to consider. The effected proof of invariance of (15.33) to the possibility of the existence of degenerate roots is important in itself. Namely, because of this proof, we can say now that the method of solving the system (15.2) of linear equations by the described Schmidt method of iterated relaxations [71] does not need at all to consider the case of degenerate quotients $\{q_k\}$ of (15.23).

The next elaboration of the Schmidt method [71] is to write down the explicit form of a general transformation for the exact elimination of all the transients $\sum_{k=1}^{m-1} a_k^{(r)} q_k^n$ from (15.26). To this end, we shall introduce the following characteristic polynomial of degree ν associated with an operator-valued determinant similar to (15.17)

$$Q_\nu(\hat{E}) \equiv \bar{v}_0 + \bar{v}_1 \hat{E} + \bar{v}_2 \hat{E}^2 + \cdots + \bar{v}_\nu \hat{E}^\nu \qquad \nu \leq m. \quad (15.39)$$

With this device it would be possible to stop the iterations at $\nu \leq m$ as soon as the contributions from the roots $\{q_\mu < 1\}(\nu \leq \mu \leq m)$ become negligible

$$Q_\nu(q_k) \equiv \bar{v}_0 + \bar{v}_1 q_k + \bar{v}_2 q_k^2 + \cdots + \bar{v}_\nu q_k^\nu = 0 \qquad (1 \leq k \leq \nu). \quad (15.40)$$

Here, $\bar{v}_j$ are the coefficients from the determinant D obtained by the Cauchy expansion theorem. In such a case, the counterpart of the iterative solution (15.26) reads as

$$x_{r,n} = x_r + \sum_{k=1}^{\nu} a_k^{(r)} q_k^n \qquad \nu \leq m. \quad (15.41)$$

Exact iterative solution of a system of linear equations 113

In principle, we should use different symbols, e.g. $q_k^{(v)}, a_k^{(r,v)}, x_{r,n}^{(v)}, x_r^{(v)}$ instead of the previous labels $q_k, a_k^{(r)}, x_{r,n}, x_r$ employed for $v = m - 1$. This is not done to avoid excessive indexing, but it is understood that all the quantities in (15.41) bear an implicit dependence upon the integer v. In order to eliminate the pair $\{q_k, a_k^{(r)}\}$ from $x_{r,n}$, we first write (15.41) explicitly by varying the integer n in the interval $0 \leq n \leq v$

$$\left.\begin{aligned} x_{r,0} &= x_r + a_1^{(r)} + a_2^{(r)} + a_3^{(r)} + \cdots + a_v^{(r)} \\ x_{r,1} &= x_r + a_1^{(r)} q_1 + a_2^{(r)} q_2 + a_3^{(r)} q_3 + \cdots + a_v^{(r)} q_v \\ x_{r,2} &= x_r + a_1^{(r)} q_1^2 + a_2^{(r)} q_2^2 + a_3^{(r)} q_3^2 + \cdots + a_v^{(r)} q_v^2 \\ &\quad\vdots \\ x_{r,v} &= x_r + a_1^{(r)} q_1^v + a_2^{(r)} q_2^v + a_3^{(r)} q_3^v + \cdots + a_v^{(r)} q_v^v \end{aligned}\right\}. \quad (15.42)$$

Proceeding as before in deriving the result (15.34) we shall now have

$$x_r = \frac{\bar{v}_0 x_{r,0} + \bar{v}_1 x_{r,1} + \bar{v}_2 x_{r,2} + \cdots + \bar{v}_v x_{r,v}}{\bar{v}_0 + \bar{v}_1 + \bar{v}_2 + \cdots + \bar{v}_v}. \quad (15.43)$$

Further, the equations for $x_{r,0}, x_{r,1}, x_{r,2}, \ldots, x_{r,v}$ from the sequel (15.42) are multiplied by the coefficients $\bar{v}_0, \bar{v}_1, \bar{v}_2, \ldots, \bar{v}_v$, respectively. Then, adding up all the v resulting equations and using (15.40) will give

$$\left.\begin{aligned} \bar{v}_0(x_r - x_{r,0}) + \bar{v}_1(x_r - x_{r,1}) + \cdots + \bar{v}_v(x_r - x_{r,v}) &= 0 \\ \bar{v}_0(x_r - x_{r,1}) + \bar{v}_1(x_r - x_{r,2}) + \cdots + \bar{v}_v(x_r - x_{r,v+1}) &= 0 \\ \bar{v}_0(x_r - x_{r,2}) + \bar{v}_1(x_r - x_{r,3}) + \cdots + \bar{v}_v(x_r - x_{r,v+2}) &= 0 \\ &\quad\vdots \\ \bar{v}_0(x_r - x_{r,v}) + \bar{v}_1(x_r - x_{r,v+1}) + \cdots + \bar{v}_v(x_r - x_{r,2v}) &= 0 \end{aligned}\right\}. \quad (15.44)$$

Considering $\bar{v}_j$ as the unknowns, the system (15.44) of homogeneous linear equations will have the non-trivial, i.e. non-zero solution if and only if its determinant is equal to zero

$$\begin{vmatrix} x_r - x_{r,0} & x_r - x_{r,1} & x_r - x_{r,2} & \cdots & x_r - x_{r,v} \\ x_r - x_{r,1} & x_r - x_{r,2} & x_r - x_{r,3} & \cdots & x_r - x_{r,v+1} \\ x_r - x_{r,2} & x_r - x_{r,3} & x_r - x_{r,4} & \cdots & x_r - x_{r,v+2} \\ \vdots & \vdots & \vdots & \ddots & \vdots \\ x_r - x_{r,v} & x_r - x_{r,v+1} & x_r - x_{r,v+2} & \cdots & x_r - x_{r,2v} \end{vmatrix} = 0. \quad (15.45)$$

We can extract x_r from (15.45) by performing subtraction of the adjacent columns on the one hand and a similar subtraction of the neighbouring rows. In other words, we subtract the vth from the $(v+1)$st column, then the $(v-1)$st from the vth column, further the $(v-2)$nd from the $(v-1)$st column and so forth. This is followed by doing precisely the same type of subtraction with the rows. Then by a simple rearrangement, the obtained equation can be solved for x_r. For example, in the case of $v = 1$, we have

$$
\begin{aligned}
0 &= \begin{vmatrix} x_r - x_{r,0} & x_r - x_{r,1} \\ x_r - x_{r,1} & x_r - x_{r,2} \end{vmatrix} \\
&= \begin{vmatrix} (x_r - x_{r,0}) - (x_r - x_{r,1}) & (x_r - x_{r,1}) - (x_r - x_{r,2}) \\ x_r - x_{r,1} & x_r - x_{r,2} \end{vmatrix} \\
&= \begin{vmatrix} x_{r,1} - x_{r,0} & x_{r,2} - x_{r,1} \\ x_r - x_{r,1} & x_r - x_{r,2} \end{vmatrix} \\
&= \begin{vmatrix} (x_{r,1} - x_{r,0}) - (x_{r,2} - x_{r,1}) & x_{r,2} - x_{r,1} \\ (x_r - x_{r,1}) - (x_r - x_{r,2}) & x_r - x_{r,2} \end{vmatrix} \\
&= \begin{vmatrix} 2x_{r,1} - x_{r,0} - x_{r,2} & x_{r,2} - x_{r,1} \\ x_{r,2} - x_{r,1} & x_r - x_{r,2} \end{vmatrix} = (x_{r,2} - x_r)\Delta^2 x_{r,1} - (\Delta x_{r,1})^2
\end{aligned}
$$

$$
\therefore \quad x_r = \frac{x_{r,0} x_{r,2} - x_{r,1}^2}{x_{r,2} - 2x_{r,1} + x_{r,0}} = \frac{x_{r,0} x_{r,2} - x_{r,1}^2}{\Delta^2 x_{r,0}} = \frac{\begin{vmatrix} x_{r,0} & x_{r,1} \\ x_{r,1} & x_{r,2} \end{vmatrix}}{\Delta^2 x_{r,0}} \quad (15.46)
$$

$$
\Delta x_{r,n} = x_{r,n+1} - x_{r,n} \qquad \Delta^2 x_{r,n} = x_{r,n+2} - 2x_{r,n+1} + x_{r,n} \quad (15.47)
$$

where $\Delta^2 x_{r,n} \equiv \Delta x_{r,n+1} - \Delta x_{r,n}$. A similar calculation for $v = 2$ gives

$$
x_r = \frac{\begin{vmatrix} x_{r,0} & x_{r,1} & x_{r,2} \\ x_{r,1} & x_{r,2} & x_{r,3} \\ x_{r,2} & x_{r,3} & x_{r,4} \end{vmatrix}}{\begin{vmatrix} \Delta^2 x_{r,0} & \Delta^2 x_{r,1} \\ \Delta^2 x_{r,1} & \Delta^2 x_{r,2} \end{vmatrix}}. \quad (15.48)
$$

Thus, in the case of any integer $v > 0$, the most general solution of the system (15.1) is given by the vector $\mathbf{x} = \{x_r\}$ with the the following determinantal quotient for the rth element x_r

$$
x_r \equiv e_v(x_{r,0}) = \frac{\begin{vmatrix} x_{r,0} & x_{r,1} & x_{r,2} & \cdots & x_{r,v} \\ x_{r,1} & x_{r,2} & x_{r,3} & \cdots & x_{r,v+1} \\ x_{r,2} & x_{r,3} & x_{r,4} & \cdots & x_{r,v+2} \\ \vdots & \vdots & \vdots & \ddots & \vdots \\ x_{r,v} & x_{r,v+1} & x_{r,v+2} & \cdots & x_{r,2v} \end{vmatrix}}{\begin{vmatrix} \Delta^2 x_{r,0} & \Delta^2 x_{r,1} & \Delta^2 x_{r,2} & \cdots & \Delta^2 x_{r,v-1} \\ \Delta^2 x_{r,1} & \Delta^2 x_{r,2} & \Delta^2 x_{r,3} & \cdots & \Delta^2 x_{r,v} \\ \Delta^2 x_{r,2} & \Delta^2 x_{r,3} & \Delta^2 x_{r,4} & \cdots & \Delta^2 x_{r,v+1} \\ \vdots & \vdots & \vdots & \ddots & \vdots \\ \Delta^2 x_{r,v-1} & \Delta^2 x_{r,v} & \Delta^2 x_{r,v+1} & \cdots & \Delta^2 x_{r,2v-4} \end{vmatrix}} \quad (15.49)
$$

or equivalently

$$x_r \equiv e_\nu(x_{r,0}) = \frac{\begin{vmatrix} x_{r,0} & \Delta x_{r,0} & \Delta^2 x_{r,0} & \cdots & \Delta^\nu x_{r,0} \\ \Delta x_{r,0} & \Delta^2 x_{r,0} & \Delta^3 x_{r,0} & \cdots & \Delta^{\nu+1} x_{r,0} \\ \Delta^2 x_{r,0} & \Delta^3 x_{r,0} & \Delta^4 x_{r,0} & \cdots & \Delta^{\nu+2} x_{r,0} \\ \vdots & \vdots & \vdots & \ddots & \vdots \\ \Delta^\nu x_{r,0} & \Delta^{\nu+1} x_{r,0} & \Delta^{\nu+2} x_{r,0} & \cdots & \Delta^{2\nu} x_{r,0} \end{vmatrix}}{\begin{vmatrix} \Delta^2 x_{r,0} & \Delta^3 x_{r,0} & \Delta^4 x_{r,0} & \cdots & \Delta^{\nu+1} x_{r,0} \\ \Delta^3 x_{r,0} & \Delta^4 x_{r,0} & \Delta^5 x_{r,0} & \cdots & \Delta^{\nu+2} x_{r,0} \\ \Delta^4 x_{r,0} & \Delta^5 x_{r,0} & \Delta^6 x_{r,0} & \cdots & \Delta^{\nu+3} x_{r,0} \\ \vdots & \vdots & \vdots & \ddots & \vdots \\ \Delta^{\nu+1} x_{r,0} & \Delta^{\nu+2} x_{r,0} & \Delta^{\nu+3} x_{r,0} & \cdots & \Delta^{2\nu} x_{r,0} \end{vmatrix}}.$$

(15.50)

The rhs of (15.49) is the sought transform, which eliminates completely all the transients $\sum_{k=1}^{m-1} a_k^{(r)} q_k^n$ by expressing the exact solution x_r as a quotient of two determinants that depend only upon the iterates $\{x_{r,j}\}_{j=0}^{2\nu}$. Using (5.15), the numerator and the denominator in (15.49) are recognized as the Hankel determinants, so that

$$x_r \equiv e_\nu(x_{r,0}) = \frac{H_{\nu+1}(x_{r,0})}{H_\nu(\Delta^2 x_{r,0})} \qquad (15.51)$$

$$H_\nu(\Delta^2 x_{r,0}) \neq 0 \qquad (\nu = 1, 2, 3, \ldots). \qquad (15.52)$$

Likewise (15.50) can also be written via a determinantal quotient as

$$x_r \equiv e_\nu(x_{r,0}) = \frac{H_{\nu+1}(y_{r,0})}{H_\nu(z_{r,0})} \qquad H_\nu(z_{r,0}) \neq 0 \qquad (\nu = 1, 2, 3, \ldots) \quad (15.53)$$

$$y_{r,p} = \Delta^p x_{r,0} \qquad z_{r,p} = \Delta^{p+2} x_{r,0} \qquad (0 \leq p \leq \nu) \quad (15.54)$$

where $y_{r,0} = \Delta^0 x_{r,0} = x_{r,0}$ and $z_{r,0} = \Delta^2 x_{r,0}$. The determinantal form of the transform $e_\nu(x_{r,0})$ has been derived by Schmidt in 1941 [71]. Subsequently, in 1949 and 1955 the same transform was rediscovered by Shanks [72], who was apparently unaware of the previous work of Schmidt [71]. In fact, Shanks [72] addressed the problem of acceleration of convergence of series but, as expected, he explicitly arrived at the Schmidt transform (15.49) when dealing with the same type of geometric sequences as those considered by Schmidt [71] while solving the system of linear equations (15.1). In 1956, while introducing his recursive ε-algorithm with the purpose of bypassing any explicit evaluations of the determinants, Wynn [73] stated that the transform (15.49) has been established by Schmidt [71] and Shanks [72]. Nevertheless, all the subsequent publications refer to (15.49) exclusively as the Shanks transform. We shall do likewise for the reason of continuity in terminology, but the reader should bear in mind that the same transform of Shanks from 1949 [71] was previously reported by Schmidt in 1941 [71]. Of course, the work of Shanks [72] is very important particularly for establishing the relationship between the Shanks transform and the Padé approximant for partial sums, as well as for introducing the iterated Shanks transform and for paving the road for the emergence of other more powerful nonlinear transforms [86, 89, 130, 230].

Moreover, the combined merit of the works of Shanks [72] and Wynn [73] is in recreating an immense interest in the usage of nonlinear transformations for data analysis not only in various branches of mathematics, but also in theoretical physics, chemistry and other research fields. A direct computation of high-order Hankel determinants from (15.51) is a drawback which, however, can be avoided altogether. This is done using, e.g. the known identities of the Hankel determinants, so that [73, 223]

$$\frac{H_{\nu+1}(x_{r,0})}{H_\nu(\Delta^2 x_{r,0})} = \varepsilon^{(2\nu)}(x_{r,n}) \tag{15.55}$$

where $\{\varepsilon^{(\nu)}(x_{r,n})\}$ is the Wynn [73] epsilon sequence satisfying the recursion

$$\varepsilon^{(\nu+1)}(x_{r,n}) = \varepsilon^{(\nu-1)}(x_{r,n+1}) + \frac{1}{\varepsilon^{(\nu)}(x_{r,n+1}) - \varepsilon^{(\nu)}(x_{r,n})} \tag{15.56}$$

$$\varepsilon^{(-1)}(x_{r,n}) = 0 \qquad \varepsilon^{(0)}(x_{r,n}) = x_{r,n}. \tag{15.57}$$

Hence, from the computational viewpoint within the Gauss–Seidel [223] framework, the Schmidt [71] iterative method of successive approximations for solving the inhomogeneous system of linear equations (15.1) is most efficiently done through the Wynn ε-recursion (15.56) [85].

Next, it is essential to determine the domain of applicability of the Shanks transform (15.49). This is done by returning to (15.44) from which both x_r and the set $\{\tilde{v}_j\}_{j=0}^\nu$ should be eliminated yielding

$$H_{\nu+1}(\Delta x_{r,0}) \equiv \begin{vmatrix} \Delta x_{r,0} & \Delta x_{r,1} & \Delta x_{r,2} & \cdots & \Delta x_{r,\nu} \\ \Delta x_{r,1} & \Delta x_{r,2} & \Delta x_{r,3} & \cdots & \Delta x_{r,\nu+1} \\ \Delta x_{r,2} & \Delta x_{r,3} & \Delta x_{r,4} & \cdots & \Delta x_{r,\nu+2} \\ \vdots & \vdots & \vdots & \ddots & \vdots \\ \Delta x_{r,\nu} & \Delta x_{r,\nu+1} & \Delta x_{r,\nu+2} & \cdots & \Delta x_{r,2\nu} \end{vmatrix} = 0 \tag{15.58}$$

where $H_\nu(\Delta x_{r,0})$ is the Hankel determinant. The equation (15.58) is the sought criterion for the applicability of the Shanks transform. The proper interpretation of this finding is that if the condition (15.58) is satisfied, then the nth iterate $x_{r,n}$ of the component x_r of the vector $\mathbf{x} = \{x_r\}$ will be given precisely by the form (15.26). On the other hand, the sequence $\{x_{r,n}\}$ from $x_{r,n} = x_r + \sum_{k=1}^{m-1} a_k^{(r)} q_k^n$ is recognized as the geometric sequence in which the differences of the successive terms are equal. This is precisely what is stated by the

Exact iterative solution of a system of linear equations   117

criterion (15.58). Thus for $\nu = 1$, the following three equivalent expressions follow from (15.46)

$$x_r \equiv e_1(x_{r,0}) = x_{r,0} - \frac{[\Delta x_{r,0}]^2}{\Delta^2 x_{r,0}} \qquad (15.59)$$

$$x_r = x_{r,1} - \frac{\Delta x_{r,0} \Delta x_{r,1}}{\Delta^2 x_{r,0}} \qquad (15.60)$$

$$x_r = x_{r,2} - \frac{[\Delta x_{r,1}]^2}{\Delta^2 x_{r,0}}. \qquad (15.61)$$

If these three equations are added together and the result divided by 3, the outcome is precisely (15.46), as it should be. All the three forms (15.59), (15.60) and (15.61) are the Aitken process [70, 176] in which only the second-order and the squared first-order or the product of the first-order difference operators appear, i.e. $\Delta^2 x_{r,0}$, $(\Delta x_{r,0})^2$, $(\Delta x_{r,1})^2$ or $\Delta x_{r,0} \Delta x_{r,1}$. As will be analysed in chapter 20, the Aitken extrapolation formula [70] from (15.59) and (15.60) or (15.61) represents the exact transform for a geometric sequence with only one transient $\{a_1 q_1^n\} (n = 0, 1, 2, \ldots)$. The case $\nu = 1$ of (15.58) is the determinantal equation

$$\begin{vmatrix} \Delta x_{r,0} & \Delta x_{r,1} \\ \Delta x_{r,1} & \Delta x_{r,2} \end{vmatrix} = 0 \qquad (15.62)$$

from which we see that the result (15.59) is valid if

$$\frac{\Delta x_{r,0}}{\Delta x_{r,1}} = \frac{\Delta x_{r,1}}{\Delta x_{r,2}}. \qquad (15.63)$$

This is the constancy of the differences of the successive terms, which is required in the definition of the geometric sequence. Similarly, for $\nu = 2$, the calculations by means of (15.48) gives

$$x_r \equiv e_2(x_{r,0}) = x_{r,0} - \frac{[\Delta x_{r,0}]^2}{\Delta^2 x_{r,0}}$$
$$- \frac{\{\Delta x_{r,0} \Delta x_{r,2} - [\Delta x_{r,1}]^2\}^2}{\Delta^2 x_{r,0} \{\Delta^2 x_{r,0} \Delta^2 x_{r,2} - [\Delta^2 x_{r,1}]^2\}} \qquad (15.64)$$

$$x_r \equiv e_2(x_{r,0}) = x_{r,0} - \frac{[\Delta x_{r,0}]^2}{\Delta^2 x_{r,0}} - \frac{1}{\Delta^2 x_{r,0}} \frac{\begin{vmatrix} \Delta x_{r,0} & \Delta x_{r,1} \\ \Delta x_{r,1} & \Delta x_{r,2} \end{vmatrix}^2}{\begin{vmatrix} \Delta^2 x_{r,0} & \Delta^2 x_{r,1} \\ \Delta^2 x_{r,1} & \Delta^2 x_{r,2} \end{vmatrix}}. \qquad (15.65)$$

According to chapter 20, this expression for $e_2(x_{r,0})$ is the exact transform for a geometric sequence with two transients $\{a_1 q_1^n, a_2 q_2^n\}$ $(n = 0, 1, 2, \ldots)$. We also considered explicitly several cases with $\nu \geq 3$ from (15.48) and found that the results can be succinctly expressed via the Hankel determinants $H_k(\Delta x_{r,0})$ and $H_k(\Delta^2 x_{r,0})$

$$e_1(x_{r,0}) = x_{r,0} - \frac{[H_1(\Delta x_{r,0})]^2}{H_1(\Delta^2 x_{r,0})}$$

$$e_2(x_{r,0}) = x_{r,0} - \frac{[H_1(\Delta x_{r,0})]^2}{H_1(\Delta^2 x_{r,0})} - \frac{[H_2(\Delta x_{r,0})]^2}{H_1(\Delta^2 x_{r,0}) H_2(\Delta^2 x_{r,0})}$$

$$e_3(x_{r,0}) = x_{r,0} - \frac{[H_1(\Delta x_{r,0})]^2}{H_1(\Delta^2 x_{r,0})} - \frac{[H_2(\Delta x_{r,0})]^2}{H_1(\Delta^2 x_{r,0}) H_2(\Delta^2 x_{r,0})}$$
$$- \frac{[H_3(\Delta x_{r,0})]^2}{H_1(\Delta^2 x_{r,0}) H_2(\Delta^2 x_{r,0}) H_3(\Delta^2 x_{r,0})}. \tag{15.66}$$

From here the general ν-order transform $e_\nu(x_{r,0})$ can be inferred as:

$$e_\nu(x_{r,0}) = x_{r,0} - \sum_{k=1}^{\nu} \frac{[H_k(\Delta x_{r,0})]^2}{\prod_{k'=1}^{k} H_{k'}(\Delta^2 x_{r,0})}. \tag{15.67}$$

If we now consider the simplest case of the sequence $\{x_{r,n}\}$ with only one transient $a_1 q_1^n$, then the transform $e_1(x_{r,0})$ from (15.59) must suffice as the exact transform. However, if we consider temporarily that the pair $\{a_1, q_1\}$ is given, so that the corresponding one-transient sequence $\{a_1 q_1^n\}$ could be sampled at some N points by varying the exponent n in the interval $0 \le n \le N - 1$, then certain numbers will be generated. Assume we subsequently forget that such numbers are of the form $\{x_{r,n}\} = \{a_1 q_1, a_1 q_1^2, a_1 q_1^3, \ldots\}$ and then attempt to apply the second-order transform $e_2(x_{r,0})$ to this numerical sequence, as if it were a generic one[2]. In such a case, we must find that

$$e_2(x_{r,0}) = e_1(x_{r,0}). \tag{15.68}$$

This equality indeed follows from the expression (15.64) for $e_2(x_{r,0})$ where the third term is exactly zero on the account of the criterium $\Delta x_{r,0} \Delta x_{r,2} = (\Delta x_{r,1})^2$ from (15.63) for the applicability of $e_1(x_{r,0})$. Likewise, the implication $e_2(x_{r,0}) = e_1(x_{r,0})$ for a given one-transient sequence is seen more directly from (15.65) using (15.62) as a special case ($\nu = 1$) of (15.58) which is the general condition for the applicability of the νth order transform $e_\nu(x_{r,0})$. In other words, if we successively apply $e_1(A_0)$ and $e_2(A_0)$ to a generic sequence $\{A_n\}$ and find $e_2(A_0) = e_1(A_0)$, this would imply $\Delta A_0 / \Delta A_1 = \Delta A_1 / \Delta A_2$, with $\Delta A_n = A_{n+1} - A_n$ as in (15.63), meaning that $e_1(A_0)$ is the exact transform for $\{A_n\}$, which then must be a geometric sequence $\{A_n\}$ with the general element $A_n = B + \sum_{k=1}^{K} a_k q_k^n$ having only one transient $K = 1$, i.e. $A_n = B + a_1 q_1^n$ where B is a constant.

---

[2] Hereafter, the term 'generic sequence' refers to a general sequence, i.e. a sequence which is not necessarily geometric. And even if we take a geometric sequence $\{a_k q_k^n\}(1 \le k \le K)$, we assume that we do not know the total number K of the transients.

Exact iterative solution of a system of linear equations

More generally, for the sequence $\{x_{r,n}\}$ with ν transients $\{a_k q_k\}$ ($1 \leq k \leq \nu$), the transform $e_\nu(x_{r,0})$ from (15.59) must suffice as the exact transform. Again, we can temporarily conceive the set $\{a_k, q_k\}$ ($1 \leq k \leq \nu$) as known, and then proceed to sample the set $\{a_k q_k^n\}$ ($0 \leq n \leq N-1$) generating the corresponding numerical sequence. Further, we pretend that this latter numerical sequence is generic, i.e. not necessarily of the form $\{x_{r,n}\} = \{a_k q_k^n\}$ and we try subjecting it to the transform $e_{\nu+1}(x_{r,0})$ of order $\nu+1$. Then the results of such an application must be

$$e_{\nu+1}(x_{r,0}) = e_\nu(x_{r,0}). \tag{15.69}$$

This can be shown by expressing (15.67) as the following recursion

$$e_{\nu+1}(x_{r,0}) = e_\nu(x_{r,0}) - \rho_{\nu+1}(x_{r,0}) \tag{15.70}$$

$$\rho_\nu(x_{r,0}) = \frac{[H_\nu(\Delta x_{r,0})]^2}{\prod_{k=1}^\nu H_k(\Delta^2 x_{r,0})}. \tag{15.71}$$

Using (15.52), the applicability condition (15.58) for $e_\nu(x_{r,0})$ becomes

$$\rho_{\nu+1}(x_{r,0}) = 0 \tag{15.72}$$

and this reduces (15.70) to (15.69), as it was set up to prove (QED).

Chapter 16

Relaxation methods and sequence accelerations

The discussed relationship between the geometric sequence (15.26) and the criterion (15.58) for the applicability of the Shanks transform (15.49) is entirely general. In other words, this relationship is independent of the framework from which the geometric sequence (15.26) emerged. More specifically, the numbers $\{x_r\}$ are simply the limits or anti-limits of a given geometric sequence $\{x_{r,n}\}$ as n increases (of course, for a general geometric sequence subscript r should be dropped). This is to say that the Shanks transform (15.49) does not need to be restricted only to the problem of solving the system (15.1) of inhomogeneous linear equations as has originally been done by Schmidt [71]. Quite the contrary, the above analysis can be linked directly to the Shanks sequence accelerations [72] and signal processing [28]. This latter liason will be elaborated in chapter 17.

Within the realm of convergence acceleration of sequencies and series, we can advantageously use the concept of Schmidt relaxations to formulate the following general problem related to (7.59) [71, 72]: find the base B for the given general geometric sequence $\{A_n\}$ with $2K + 1$ real or complex numbers A_n with the known pairs $\{q_k, a_k\}(1 \le k \le K)$ in the form

$$A_n = B + \sum_{k=1}^{K} a_k q_k^n. \tag{16.1}$$

The base B can be calculated *exactly* from the ratio of two Hankel determinants of the type (15.51)

$$B = e_K(A_n) = \frac{H_{K+1}(A_0)}{H_K(\Delta^2 A_0)} \tag{16.2}$$

where $\{q_k\}$ do not need to be restricted to exponentially decaying transients. Moreover, some or all $|q_k|$ could be greater than unity.

The base B is the equilibrium value of $A_n = A_\infty$ attained when all the K transients $\{a_k q_k^n\}$ have decayed in the limit $n \longrightarrow \infty$. Stated equivalently, the constant B is the limit or anti-limit of the sequence (16.1) as $n \longrightarrow \infty$. If this sequence is convergent (all $|q_k|$ less than unity), the Shanks transform $e_K(A_n)$ produces another sequence which also converges to B, but does this usually faster than the original sequence $\{A_n\}$ as n increases. If the sequence (16.1) diverges (some or all $|q_k|$ greater than unity), the Shanks transform $e_K(A_n)$ yields the anti-limit B by forcing the divergent sequence to converge. This induced convergence is an analytical continuation in the sense of enlarging the convergence radius of the original sequence $\{A_n\}$ [28, 72, 75]. The original convergence radius may be even equal to zero [88] or such that the considered sequence diverges exponentially [237]. It then follows from the preceding that whenever we are given a geometric sequence $\{A_n\}$ of the type (16.1) from the onset, the Shanks transform (16.2) will yield the exact expression for the base B of the sequence through the complete filtering of all the transients $\{a_k q_k^n\}$ ($1 \leq k \leq K$).

Chapter 17

Quantification: harmonic inversion in the time domain

The Schmidt method [71] of solving a system of inhomogeneous linear equations uses physically plausible iterative relaxations to arrive at the exact solution in the analytical form (15.49) of a quotient of two Hankel determinants. To appreciate this very important result of direct relevance for solving the quantification problem via harmonic inversion in the time domain, it is instructive to recapitulate here the genesis under which the derivation has been made in chapter 15. There, the general rth component $x_{r,n} = x_r + t_{r,n}$ of the iterative solution $\mathbf{x}_n = \{x_{r,n}\}$ is obtained as a sum of the steady part x_r and the transient oscilations $t_{r,n}$ where n is the iteration number

$$x_{r,n} = x_r + t_{r,n} \qquad t_{r,n} \equiv \sum_{k=1}^{\nu} a_k^{(r)} q_k^n. \qquad (17.1)$$

However, at the same time, by the direct method, the quantity x_r is found to represent the non-iterative exact Cramer solution of the same system of linear equations (15.2). The assumption (15.4) guarantees the existence and uniqueness of the Cramer solution x_r from (15.3) for the system (15.2). Therefore, in order to have $\lim_{n\to\infty} x_{r,n} = x_r$, the transitory part $t_{r,n}$ in $x_{r,n}$ must die out, i.e. decay to zero $\lim_{n\to\infty} t_{r,n} = 0$, which will happen only if $q_k < 1 (1 \leq k \leq \nu)$. In such a case, the Schmidt iterative solution $x_{r,n}$ would coincide with the exact result x_r. It is fascinating that the exact solution x_r via the iterations $x_{r,n}$ proceeds via an intermediate stage with the emergence of transients $t_{r,n}$ in the form of a geometric sequence. This is due to the fact that the iterative solution is achieved simply by converting the original system of inhomogeneous linear equations (15.2) into the corresponding system of difference equations (15.11) with constant coefficients. The exact solution of this latter system is known to be a geometric sequence (15.26).

This conversion of (15.2) to the difference equations (15.11) is accomplished in two steps. First, in the original system of equations, all the unknown components x_r of the vector $\mathbf{x} = \{x_r\}$ above the main diagonal on the lhs of (15.2) are replaced by the initial approximations $\{x_{r,0}\}$ to arrive at the new system (15.10). This latter system is uncoupled, as opposed to (15.2). The uncoupled equations (15.10) are solved exactly, one at a time, with the solutions $\{x_{r,1}\}$ that should represent an improvement over the corresponding initial guesses $\{x_{r,0}\}$. Next, the solutions $\{x_{r,1}\}$ are substituted back into the original system (15.2) as the new intial approximations to $\{x_r\}$ and the procedure of relaxations is repeated leading to the further improvements $\{x_{r,2}\}$. These successive substitutions are continued some n times leading to the nth order improvement $\{x_{r,n}\}$ called the nth iterate. After n such iterations, the original ansatz (15.2) is mapped onto the system of n simultaneous inhomogenous difference equations (15.11).

The system (15.11) has two kinds of iterates, $\{x_{r,n}\}$ and $\{x_{r,n+1}\}$, that are located symmetrically above and below the main diagonal, respectively. A solution of (15.11) would be possible if the iterating index could be homogenized, so that only the quantities $\{x_{r,n}\}$ appear explicitly in (15.11). The index $n+1$ in $\{x_{r,n+1}\}$ is scaled to n with the help of the shift-operator $x_{r,n+1} = \hat{E}x_{r,n}$, as per definition (15.12). This leads to the system (15.15) which contains only $x_{r,n}$ and, as such, the Cramer rule becomes applicable with the result (15.16). However, as a consequence of homogenizing the iteration indices n and $n+1$ in (15.16), the scalar elements $a_{i,j}$ of the matrix $\mathbf{A} = \{a_{i,j}\}$ from (15.1) now became operator-valued quantities $\{a_{i,j}\hat{E}\}$ that also appear in both determinants $D(\hat{E})$ and $D_r(\hat{E})$ from the solution (15.16) of (15.15).

Up to this step, the solution (15.16) is only formal, since it is given via the shifting operator (15.12). The way in which these determinantal operators act on $x_{r,n}$ is defined by the Cauchy expansion yielding an operator-valued polynomial in $\hat{E}$ as in (15.20). On the lhs of (15.20) the operator polynomial contains the powers of the operator $\hat{E}$ acting on $x_{r,n}$. There, each term $\hat{E}^m x_{r,n}$ becomes $x_{r,n+m}$ by virtue of the shifting (15.12). This gives the single inhomogeneous difference equation (15.20) whose unique solution is the geometric sequence $x_{r,n} = x_r + t_{r,n} = x_r + \sum_{k=1}^{\nu} a_k^{(r)} q_k^n$ as in (15.26). This is how the geometric sequence emerged as an interim solution of the system (15.2) of linear equations. We see that, in the search for limiting value x_r, the only role played by the transients $\{t_{r,n}\}$ is to make themselves vanish identically ($\lim_{n\to\infty} t_{r,n} = 0$). Of course, this is the only way in which the term $x_{r,n}$ from $x_{r,n} = x_r + t_{r,n}$ could coincide with the exact solution x_r of (15.2) via $\lim_{n\to\infty} x_{r,n} = x_r$ as per (15.29).

The first immediate advantage of the Schmidt iterative relaxations is provided already in the very process of filtering out the transient $t_{r,n}$ from the nth iterate $x_{r,n}$. For this to happen *exactly* through $\lim_{n\to\infty} t_{r,n} = 0$, it is sufficient to perform a simple algebraic transformation of the original system of equations for $\{x_{r,\nu}\}$ in such a way that the knowns (minors) and the unknowns (iterates) exchange roles. In this way, a modified system of linear equations (15.44) is obtained with the solution x_r given by (15.49) in the form of the determinantal quotient $x_r = \mathrm{H}_{\nu+1}(x_{r,0})/\mathrm{H}_\nu(\Delta^2 x_{r,0})$. Hence, this latter ratio is the exact filter for all the transients and, as such, it is recognized as the Shanks transform $e_\nu(x_{r,0})$ for the iterates $\{x_{r,n}\}$ of the form of a geometric sequence $x_{r,n} = x_r + \sum_{k=1}^{\nu} a_k^{(r)} q_k^n$. In other words, the demand for the exact filtering of the transients $t_{r,n}$ from $x_{r,n}$ resulted beneficially in the rigorous *derivation* of the Shanks transform $e_\nu(x_{r,0})$. This illustrates how the Shanks transform is naturally ingrained in the Schmidt exact iterative solution of a system of inhomogeneous linear equations.

Finally, from the Schmidt iterative relaxations, one more advantage of utmost importance becomes evident in the course of assessing the domain of validity of the mapping $e_\nu(x_{r,0})$ from (15.49) within the realm of the *exact* transforms. To find this domain, i.e. to determine for which sequences the Shanks transform becomes exact, it is sufficient to return to the modified system of linear equations (15.44) to eliminate both the minors and the base x_r. The result of these elimations is the equation $\mathrm{H}_{\nu+1}(\Delta x_{r,0}) = 0$, as in (15.58). This equation tells us that $e_\nu(x_{r,0})$ will be exact if the sequence $\{x_{r,n}\}$ is such that it maintains a constant difference between its successive terms. This is the defining feature of the geometric sequence. Hence the Shanks transform is exact for a geometric sequence. Moreover, if the information that the sequence $\{x_{r,n}\}$ is geometric with ν transients is not available to us before the application of the Shanks transform, then the eventual subsequent finding $\mathrm{H}_{\nu+1}(\Delta x_{r,0}) = 0$, for a given numeric sequence $\{x_{r,n}\}$ sampled at various n, will assure that the set $\{x_{r,n}\}$ is indeed a geometric sequence $\sum_{k=1}^{\nu} a_k^{(r)} q_k^n$ with precisely ν transients. In other words, the number ν of transients is *derivable* exactly from the given sequence $\{x_{r,n}\}$ and one of the easiest ways to accomplish this is to check whether the relation $\mathrm{H}_{\nu+1}(\Delta x_{r,0}) = 0$ holds true.

Having expounded the virtues of the Schmidt iterative relaxations, it becomes obvious that the transients $\{a_k^{(r)} q_k^n\}$ in the iterates $\{x_{r,n}\}$ are of precisely the same type as those encountered in the time signals from data matrices and their determinants in signal processing. In this way, the mathematical model (5.11) for the time signal $\{c_n\}$ in the form of the geometric sequence $c_n = \sum_{k=1}^{K} d_k u_k^n$ can, in fact, be conceived as derivable from the mechanism of successive relaxations, rather than being imposed on the spectral analysis in an *ad hoc* manner, as customarily done in various fitting recipes encountered in spectral analysis of experimental data. Moreover, the same general methodological framework from chapter 15 can also be employed to solve exactly the spectral problem in signal processing. When the quantities $\{q_k\}$ are the damped complex exponentials as those encountered in the time signal (7.10), the notation $e_K(A_n)$ is indicative, since the symbol e refers in a transparent way to the exponential decay of transients. The symbol $e_K(A_n)$ for the Shanks transform (15.49) has originally been introduced in [72]. In signal processing, the sequence $\{A_n\}$ is a given set of time signal points $\{c_n\}(0 \leq n \leq N-1)$ that can be quantitatively described by the mathematical model $c_n = \sum_{k=1}^{K} d_k \exp(-in\omega_k \tau)$ where $\mathrm{Im}(\omega_k) < 0$, as in (7.10) or by a slightly more general signal (7.59). In other words, for the given signal $\{c_n\}$ we construct an auxiliary sequence $\{\bar{c}_n\}$ with the elements $\bar{c}_n = \bar{c}_\infty + \sum_{k=1}^{K} d_k \exp(-in\omega_k\tau)$ as in (7.59). Then the extended harmonic inversion problem (EHIP) consists of extracting from $\{\bar{c}_n\}$ the quartet of spectral parameters such as the order K, the base constant $\bar{c}_\infty$ as well as the complex-valued amplitudes $\{d_k\}$ and frequencies $\{\omega_k\} = \{(i/\tau)\ln(u_k)\}$.

The first step in the EHIP is to find the order or rank K by computing the sequence of the Hankel determinant $\{H_m(\Delta c_0)\}$ $(m = 1, 2, \ldots)$. Then using $\Delta c_n = c_{n+1} - c_n$ instead of $\Delta x_{r,n}$ in (15.58), it follows that the first value $m = M$ for which we find that

$$H_{M+1}(\Delta c_0) \equiv \begin{vmatrix} \Delta c_0 & \Delta c_1 & \Delta c_2 & \cdots & \Delta c_M \\ \Delta c_1 & \Delta c_2 & \Delta c_3 & \cdots & \Delta c_{M+1} \\ \Delta c_2 & \Delta c_3 & \Delta c_4 & \cdots & \Delta c_{M+2} \\ \vdots & \vdots & \vdots & \ddots & \vdots \\ \Delta c_M & \Delta c_{M+1} & \Delta c_{M+2} & \cdots & \Delta c_{2M} \end{vmatrix} = 0 \quad (17.2)$$

will give straight the sought number of transients $K = M$. Of course, according to chapter 68, the Hankel determinants $\{H_m(\Delta c_0)\}$ $(m = 1, 2, \ldots)$ are not computed directly from their definition, but rather via the recursive product-difference algorithm of Gordon [170]. This procedure based upon (17.2) determines unequivocally the number K of the transients in the signal $\{\bar{c}_n\}$ from (7.59).

Once the order K has become available, we proceed to the next step in the EHIP and that is finding the base constant $\bar{c}_\infty$. This is readily accomplished via the Kth order Shanks transform $e_K(\bar{c}_n)$ applied to the sequence $\{\bar{c}_n\}$ with the result

$$e_K(\bar{c}_n) = \bar{c}_\infty \quad (17.3)$$

as announced in (7.63). Of course, for (7.10) we have $\bar{c}_n = c_n$, since $\bar{c}_\infty = 0$ and this maps (17.3) to $e_K(c_n) = 0$ which is the signature for the information of exactly K transients $\{d_k u_k\}$ in the usual time signal $\{c_n\}$ from (5.11). Moreover, applying (7.11) and (7.12) to (16.2) we obtain $B \equiv \bar{c}_\infty = 0$, which also would follow in this case from (17.3) on account of (7.60) and (7.61).

The remaining part of the EHIP is to extract the transient frequencies $\omega_k = (i/\tau)\ln(u_k)$ and the corresponding amplitudes $\{d_k\}$, i.e. the spectral pairs $\{u_k, d_k\}$ $(1 \leq k \leq K)$. The first subset $\{u_k\}$ of this latter couple is obtained from the following relationship [246]

$$u_k = \lim_{s \to \infty} \frac{\varepsilon_{2k-2}^{(s+1)}}{\varepsilon_{2k-2}^{(s)}} \quad (1 \leq k \leq K) \quad (17.4)$$

where $\{\varepsilon_n^{(m)}\}$ is the Wynn epsilon sequence generated recursively via (7.54). In this way, the ratios $\{\varepsilon_{2k-2}^{(s+1)}/\varepsilon_{2k-2}^{(s)}\}$ are computed by increasing the index s and watching for convergence. Whenever this latter sequence converges, its limiting value will represent the frequency exponential $u_k = \exp(-i\omega_k \tau)$. In principle, this permits the extraction of all the nodal frequences $\{\omega_k\}$ ($1 \le k \le K$) directly from the time domain data $\{c_n\}$ ($1 \le n \le N-1$) without using any classical root searching routine. However, in practice, we are usually given time signals or auto-correlation functions with a finite length N. Clearly, in such a case, the limit $s \longrightarrow \infty$ in (17.4) cannot be taken, so that the approximately converged u_k might be of insufficient accuracy. However, such u_k could still be reasonable and, moreover, they might be further refined if taken as the initial values in the modified Newton–Raphson algorithm from chapter 28. With this refinement, we have customarily obtained in practice at least eight significant decimal places in u_k with a remarkably small iteration number of the order of ten or less.

Once all the K quantities u_k are obtained in this way, the final step of the EHIP is to generate the amplitude d_k which corresponds to the given u_k and, of course, this should be performed for every k across the whole interval $1 \le k \le K$. Inserting u_k into the model (7.59) for the $\bar{c}_n$, we obtain a system of K linear equations for d_k according to $\bar{c}_n = \bar{c}_\infty + \sum_{k=1}^{K} d_k u_k^n$ ($0 \le n \le K-1$). This latter system can be solved analytically with the result [57, 247]

$$d_k = \frac{c_\downarrow \prod_{r=1, r \ne k}^{K} (u - u_r)}{\prod_{r=1, r \ne k}^{K} (u_k - u_r)} \qquad (17.5)$$

$$c_\downarrow \prod_{r=1, s \ne k}^{K} (u - u_r) = \sum_{r=1}^{K} b_r^{(k)} c_r. \qquad (17.6)$$

Here, the label $c_\downarrow$ is a symbolic operation which produces the image c_m when acting upon the power function u^m according to the following prescription which has been originally introduced by Prony [57]

$$c_\downarrow u^m = c_m. \qquad (17.7)$$

The quantity $b_r^{(k)}$ in (17.6) is the coefficient multiplying the rth power function u^r in the polynomial $\prod_{r=1, r \ne k}^{K} (u - u_r)$

$$\prod_{r=1, r \ne k}^{K} (u - u_r) = (u - u_1)(u - u_2) \cdots (u - u_{k-1})(u - u_{k+1}) \cdots (u - u_K)$$

$$\equiv \sum_{r=1, r \ne k}^{K} b_r^{(k)} u^r. \qquad (17.8)$$

The set $b_r^{(k)}$ ($r \neq k$) can be extracted from the lhs of (17.8) in a recursive manner [248]

$$\mathcal{B}_{k,r} = \mathcal{B}_{k-1,r-1} - u_k \mathcal{B}_{k-1,r} \qquad 2 \leq k \leq K \qquad 1 \leq r \leq k$$
$$\mathcal{B}_{k,0} = -u_k \mathcal{B}_{k-1,0} \qquad \mathcal{B}_{k,k} = \mathcal{B}_{k-1,k-1}$$
$$\mathcal{B}_{1,0} = -u_1 \qquad \mathcal{B}_{1,1} = 1$$
$$\therefore \quad b_r^{(k)} = \mathcal{B}_{K,r} \qquad (r \neq k). \tag{17.9}$$

This recursion terminates at $k = K$ yielding the final result $b_r^{(k)}$ for a fixed k via $b_r^{(k)} = \mathcal{B}_{K,r}$. The coefficient $b_r^{(k)}$ depends parametrically upon K like $b_r^{(k)} \equiv b_r^{(k)}(K)$, but this was unnecessary to show explicitly in (17.8) and (17.9). The algorithm (17.9) is an efficient recursion, since it has a significantly reduced number of multiplications relative to a classical relationship between symmetric functions, such as the polynomial on the lhs of (17.8) [248]. Obviously, the same algorithm can be used to extract the expansion coefficients b_s from a polynomial with no missing term $(u - u_k)$ and that is the characteristic or secular polynomial $\prod_{r=1}^{K}(u - u_r) = (u - u_1) \cdots (u - u_{k-1})(u - u_k)(u - u_{k+1}) \cdots (u - u_K)$.

We emphasize that in this chapter the presently defined extended harmonic inversion problem, the EHIP, a part of which is the usual harmonic inversion, the HIP [108, 109, 115] or the quantification problem from MRS [19] is solved fully while working exlusively in the time domain where the signal has been experimentally encoded. This accomplishment is achieved without the need for solving any eigenvalue problem, in contrast to the state space estimators like FD [108, 109, 115], DSD [113], HLSVD [60], etc. Moreover, the presented time domain spectral analysis does not necessitate rooting the characteristic or secular polynomial as opposed to DLP or DPA [115]. Since here we do not use explicitly these latter polynomials at all, there is no need to generate their coefficients either. As mentioned earlier, such polynomial coefficients are also the predicting coefficients in the LP or DLP estimators and, moreover, they are equal to the coefficients of the denominator polynomial in the PA or DPA. However, if these coefficients are desired, they can be readily obtained by means of the recursive algorithm (17.9). Once this is done, the coefficents of the numerator polynomial in the PA or DPA could also be at once generated from the explicit, analytical formula (11.11). Of course, to obtain the frequency spectrum for the given time signal $\{c_n\}$ from (5.11), it is not necessary to construct the numerator and denominator polynomials in the PA. This is because such a spectrum is automatically available via the closed formula $\sum_{k=1}^{K} d_k/(u - u_k)$ from (5.23), requiring only the spectral parameters $\{u_k, d_k\}$ which we obtain here exclusively via the time domain spectral analysis. The formula (17.5) for the kth amplitude d_k is obtained by solving *analytically* the system $c_n = \sum_{k=1}^{K} d_k u_k^n$ ($0 \leq n \leq N-1$). Obviously, this is advantageous relative to the corresponding *numerical* solution of the same system as customarily done in the LP, DLP, HLSVD, etc. However, it should be emphasized that both the analytical expression (17.5) and its numerical counterpart obtain every amplitude d_k for each fixed k as a function of *all* the computed u_k from the set $\{u_k\}$ ($1 \leq k \leq K$). In principle, this is not a problem if every u_k is computed exactly. In practice, the accuracy of d_k could be undermined if some spurious u_k have creeped in along with the genuine ones. This is unlikely to happen in (17.5) which would not converge for extraneous roots. In contrast to this, however, the corresponding polynomial root search in the LP or DLP and solving the eigenvalue problem of the data matrix in the HLSVD are prone to produce spurious frequencies that could deteriorate the accuracy of the amplitudes $\{d_k\}$.

Chapter 18

Enhanced convergence of sequences by nonlinear transforms

The Aitken extrapolation [70], the Shanks transform [71, 72] and the Wynn algorithm [73] can all be inductively derived from a single geometric sequence $\{A_n\}$ of the following simple type [235]

$$A_{n+1} - B = \gamma(A_n - B). \tag{18.1}$$

Here, B is the limiting value of $\{A_n\}$ as $n \longrightarrow \infty$ and γ is a constant independent of n. Since there are two unknowns (γ and B), we need one more equation in addition to (18.1). To this end, we specify another form of (18.1) through the replacement of the index n by $n+1$

$$A_{n+2} - B = \gamma(A_{n+1} - B). \tag{18.2}$$

Along with the sequence $\{A_n\}$ we shall also consider the sequence of the first-order forward differences $\{\Delta A_n\}$ with the general element ΔA_n

$$\Delta A_n = A_{n+1} - A_n. \tag{18.3}$$

Subtracting (18.2) from (18.1), it follows $A_{n+2} - A_{n+1} = \gamma(A_{n+1} - A_n)$ and, therefore

$$\frac{\Delta A_{n+1}}{\Delta A_n} = \gamma. \tag{18.4}$$

This shows that the set $\{A_n\}$ from (18.1) is indeed a geometric sequence, since the ratios of the successive differences are all equal to the constant γ. The parameter γ can be eliminated altogether from the model through the division of (18.1) by (18.2), yielding $(B - A_{n+1})^2 = (B - A_n)(B - A_{n+2})$. From here, the limit B to which the sequence $\{A_n\}$ converges as $n \longrightarrow \infty$ is found as

$$\mathcal{S}(A_n) \equiv B = \frac{A_n A_{n+2} - A_{n+1}^2}{A_n - 2A_{n+1} + A_{n+2}}. \tag{18.5}$$

This is the Aitken extrapolation formula [70] which transforms the old sequence $\{A_n\}$ into a new sequence $\{S(A_n)\}$. The quotient from (18.5) is not the only variant of the Aitken formula. Namely, as pointed out in [72] and [176], one is allowed in (18.5) to replace $n+1$ and $n-1$ by $n+r$ and $n-r$, respectively, for any integer number r [70]. For example, the Aitken interpolation formula can equivalently be written via $B = (A_{n-1}A_{n+1} - A_n^2)/(A_{n-1} - 2A_n + A_{n+1})$ as we shall do later on. If the denominator in (18.5) vanishes, i.e. $A_n - 2A_{n+1} + A_{n+2} = 0$ then we shall have $A_n = B$ for all the values of n. The denominator $A_n - 2A_{n+1} + A_{n+2}$ from (18.5) can be equivalently expressed as the second-order forward difference $\Delta^2 A_n$. By employing $\Delta^2 A_n$ as $\Delta^2 A_n = \Delta(A_{n+1} - A_n) = \Delta A_{n+1} - \Delta A_n = (A_{n+2} - A_{n+1}) - (A_{n+1} - A_n) = A_{n+2} - 2A_{n+1} - A_n$ and $\Delta^2 A_n = \Delta A_{n+1} - \Delta A_n$, $A_n A_{n+2} - A_{n+1}^2 = A_n \Delta A_{n+1} - A_{n+1} \Delta A_n$, we obtain the 2×2 determinants for both terms in (18.5)

$$A_n A_{n+2} - A_{n+1}^2 = \begin{vmatrix} A_n & A_{n+1} \\ \Delta A_n & \Delta A_{n+1} \end{vmatrix} \tag{18.6}$$

$$A_n - 2A_{n+1} + A_{n+2} = \begin{vmatrix} 1 & 1 \\ \Delta A_n & \Delta A_{n+1} \end{vmatrix}. \tag{18.7}$$

With these formulae, the Aitken extrapolation B from (18.5) becomes

$$B = \frac{A_n A_{n+2} - A_{n+1}^2}{A_n - 2A_{n+1} + A_{n+2}} = \frac{\begin{vmatrix} A_n & A_{n+1} \\ \Delta A_n & \Delta A_{n+1} \end{vmatrix}}{\begin{vmatrix} 1 & 1 \\ \Delta A_n & \Delta A_{n+1} \end{vmatrix}}. \tag{18.8}$$

We can write the quotient of the two 2×2 determinants from (18.8) as

$$\frac{\begin{vmatrix} A_n & A_{n+1} \\ \Delta A_n & \Delta A_{n+1} \end{vmatrix}}{\begin{vmatrix} 1 & 1 \\ \Delta A_n & \Delta A_{n+1} \end{vmatrix}} = \frac{A_n \Delta A_{n+1} - A_{n+1} \Delta A_n}{\Delta A_{n+1} - \Delta A_n}$$

$$= \frac{[A_{n+1} - (A_{n+1} - A_n)]\Delta A_{n+1} - A_{n+1} \Delta A_n}{\Delta A_{n+1} - \Delta A_n}$$

$$= \frac{(A_{n+1} - \Delta A_n)\Delta A_{n+1} - A_{n+1} \Delta A_n}{\Delta A_{n+1} - \Delta A_n}$$

$$= \frac{A_{n+1}(\Delta A_{n+1} - \Delta A_n) - \Delta A_n \Delta A_{n+1}}{\Delta A_{n+1} - \Delta A_n}$$

$$= A_{n+1} + \frac{\Delta A_n \Delta A_{n+1}}{\Delta A_n - \Delta A_{n+1}}$$

$$= A_{n+1} + \frac{1}{\dfrac{1}{\Delta A_{n+1}} - \dfrac{1}{\Delta A_n}}.$$

This result can be equivalently stated as

$$\varepsilon^{(2)}(A_n) = \varepsilon^{(0)}(A_{n+1}) + \frac{1}{\varepsilon^{(1)}(A_{n+1}) - \varepsilon^{(1)}(A_n)} \tag{18.9}$$

with

$$\varepsilon^{(0)}(A_n) = A_n \qquad \varepsilon^{(1)}(A_n) = \frac{1}{\Delta A_n} \tag{18.10}$$

$$\varepsilon^{(2)}(A_n) \equiv \frac{\begin{vmatrix} A_n & A_{n+1} \\ \Delta A_n & \Delta A_{n+1} \end{vmatrix}}{\begin{vmatrix} 1 & 1 \\ \Delta A_n & \Delta A_{n+1} \end{vmatrix}}. \tag{18.11}$$

The calculation carried out along these lines also for 3×3 matrices yields the result $\varepsilon^{(3)}(A_n) = \varepsilon^{(1)}(A_{n+1}) + [\varepsilon^{(2)}(A_{n+1}) - \varepsilon^{(2)}(A_n)]^{-1}$. This inductive procedure gives the following general relation

$$\varepsilon^{(2k)}(A_n) = e_k(A_n) \qquad \varepsilon^{(2k+1)}(A_n) = \frac{1}{e_k(\Delta A_n)} \tag{18.12}$$

$$e_k(A_n) = \frac{\begin{vmatrix} A_n & A_{n+1} & \cdots & A_{n+k} \\ \Delta A_n & \Delta A_{n+1} & \cdots & \Delta A_{n+k} \\ \vdots & \vdots & \ddots & \vdots \\ \Delta A_{n+k-1} & \Delta A_{n+k} & \cdots & \Delta A_{n+2k-1} \end{vmatrix}}{\begin{vmatrix} 1 & 1 & \cdots & 1 \\ \Delta A_n & \Delta A_{n+1} & \cdots & \Delta A_{n+k} \\ \vdots & \vdots & \ddots & \vdots \\ \Delta A_{n+k-1} & \Delta A_{n+k} & \cdots & \Delta A_{n+2k-1} \end{vmatrix}} \tag{18.13}$$

where

$$\varepsilon^{(k+1)}(A_n) = \varepsilon^{(k-1)}(A_{n+1}) + \frac{1}{\varepsilon^{(k)}(A_{n+1}) - \varepsilon^{(k)}(A_n)} \tag{18.14}$$

$$\varepsilon^{(-1)}(A_n) = 0 \qquad \varepsilon^{(0)}(A_n) = A_n. \tag{18.15}$$

For a given sequence $\{A_n\}$ the mappings from $\{A_n\}$ to (18.5), (18.13) and (18.14) are called the Aitken extrapolation [70], the Shanks transform [71, 72] and the Wynn algorithm [73], respectively. In the above derivation, the Shanks transform appears as a direct extension of the Aitken extrapolation to an arbitrary order k. Of course, these two sequence-to-sequence nonlinear transformations coincide with each other in the first-order $k = 1$

$$e_1(A_n) = \{B\}_{n+1 \to n} = \frac{A_{n+1}A_{n-1} - A_n^2}{A_{n+1} - 2A_n + A_{n-1}}. \tag{18.16}$$

In this context, the Wynn algorithm for generating the array $\varepsilon^{(k)}(A_n)$ is not an independent transform on its own, since it represents a numerical technique for avoiding inefficient computations of the determinants for higher ranks that are encountered in applications of the Shanks transform $e_k(A_n)$. Nevertheless, the significance of [73] is considerable, since without the ε-algorithm the Shanks transform (18.13) would not be useful in any practical computations for large values of the order k.

Chapter 19

The Shanks transform as the exact filter for harmonic inversion

We saw that the Shanks transform $e_k(A_n)$ deals simultaneously with $2k + 1$ elements of the given sequence $\{A_n\}$. In chapter 18, this transform was introduced by searching for the limit of a simple model sequence given by the geometric progression (18.1). In the present chapter, we shall tie the Shanks transform (18.13) to the spectral harmonic analysis by referring to the representations (5.9) and (5.11) for the discrete auto-correlation function C_n and time signal c_n, respectively. With this goal, let us consider another model sequence, which is more general than (18.1). This sequence $\{A_n\}$ will hereafter be viewed as the input data to the so-called reconstruction/retrieval/decomposition problem or the generalized harmonic inversion problem (GHIP) as announced before in (7.59)

$$A_n = B + p_n \qquad (0 \le n \le 2k) \tag{19.1}$$

$$p_n = \sum_{r=1}^{k} a_r q_r^n \qquad (n = 0, 1, 2, \ldots, 2k - 1) \qquad q_r \ne 0 \qquad q_r \ne 1 \tag{19.2}$$

$$q_r = e^{-\tau \alpha_r} \qquad \operatorname{Re}(\alpha_r) > 0. \tag{19.3}$$

Setting $k = \infty$ and $B = 0$ in (19.1) yields the well-known Dirichlet series. The parameters $\{q_r, a_r\}$ are called the quotients and the amplitudes, whereas B in (19.1) is named the base. Here B is the limiting value of A_n obtained at $n \longrightarrow \infty$ provided that $|q_r| < 1$ which is fulfilled for $\operatorname{Re}(\alpha_r) > 0$ for each r from the interval $1 \le r \le k$. The sum p_n is of the type of a geometric progression and it represents the discrete counterpart of the continuous function $p(t)$ of the time dependent variable

$$p(t) = \sum_{r=1}^{k} a_r e^{-\alpha_r t} \qquad \operatorname{Re}(\alpha_r) > 0. \tag{19.4}$$

As before, the time discretization is done through $t = t_n = n\tau$ where τ is the constant time increment. The reconstruction problem is a general spectral or inverse problem in the sense that a given input sequence $\{A_n\}$ is described by the mathematical model (19.1) which seeks to extract the order k and the spectrum $\{q_r, a_r\}_{r=1}^k$ from the data $\{A_n\}$. If the parameter α_r is replaced by $i\omega_r\tau$ in (19.2), then the term p_n is reduced to the auto-correlation function C_n from (5.9), or equivalently, to a time signal c_n from (5.11) with $\text{Im}(\omega_r) < 0$. Using (18.3), we have from (19.1)

$$\Delta A_n = \sum_{r=1}^{k} a_r q_r^n (q_r - 1) \qquad (0 \le n \le 2k-1). \tag{19.5}$$

The amplitudes $\{a_r\}$ and the nonlinear parameters $\{\alpha_r\}$ are unknown and they can be complex numbers. There are two general themes (i) and (ii) in studying the reconstruction problem for the given/known sequence $\{A_n\}$.

(i) If in (19.1) we formally set the baseline constant to be equal to zero

$$B = 0 \tag{19.6}$$

from the onset, then the model A_n becomes a pure linear combination of k exponentials that are the solutions of the ODE of the kth order from chapter 7. Moreover, when setting $\alpha_r = i\omega_r\tau$ the model sequence A_n is found to coincide with the auto-correlation function C_n from (5.9) or a time signal c_n from (5.11)

$$A_n = C_n = c_n \tag{19.7}$$

in which case the generalized harmonic inversion problem reduces to the harmonic inversion for determining the unknown parameters $\{u_r, d_r\}$ and the order k from the given set $\{c_n\} = \{C_n\} = \{\sum_{r=1}^{k} d_r \exp(-in\tau\omega_r)\}$, where we set $a_r = d_r$ to cohere with (5.9) and (5.11).

(ii) By filtering out the sum of exponentials $\sum_{r=1}^{k} a_r q_r^n$ from (19.1) where the set $\{q_r, a_r\}$ is known, one aims at searching for the limit B of the convergent sequence $\{A_n\}$ and this reduces the model sequence to its baseline constant. If the sequence $\{A_n\}$ is divergent, the concept of the so-called anti-limit is invoked through analytic continuation [28, 72, 75].

We shall illustrate both themes (i) and (ii) in the first few cases with a rank k which is fixed in advance as, e.g. $k \le 2$. We begin with $k = 1$. In (i) we are given the input data $\{A_n\}$ with at least three equidistant time elements (p_0, p_1 and p_2), where $p_n \equiv p(t_n) = p(n\tau)$

$$p_n = \sum_{r=1}^{1} a_r q_r^n = a_1 q_1^n \qquad (n = 0 \text{ and } 1). \tag{19.8}$$

In 1949 Shanks [72] re-emphasized that it is the equidistant sampling, $p_n \equiv p(t_n) = p(n\tau)$ which is the key point in converting the original transcendental reconstruction problem (19.1) into a much easier algebraic problem (19.2) for determining the unknown quotients and amplitudes $\{q_r, a_r\}$. However, Shanks [72] was apparently unaware of the seminal work of Prony [57] who was the first to demonstrate the power of this critical sampling which he used in the ensuing exact representation of given data $\{p_n\}$ in terms of a linear combination of exponentials. The characteristic equation related to (19.8) is

$$\begin{vmatrix} p_0 & 1 \\ p_1 & q \end{vmatrix} = 0 = \begin{vmatrix} 1 & q \\ p_0 & p_1 \end{vmatrix} \implies p_0 q - p_1 = 0 \tag{19.9}$$

which gives the root $q \equiv q_1 = p_1/p_0$

$$q_1 = \frac{p_1}{p_0}. \tag{19.10}$$

The corresponding amplitude is obtained by substituting the root q_1 into (19.8) and solving for a_1 with the result $p_n = a_1(p_1/p_0)^n$ which for $n = 0$ becomes

$$a_1 = p_0. \tag{19.11}$$

The correctness of this procedure is established if, with the obtained spectral parameters $\{q_1, a_1\}$, we could reconstruct *exactly* the two input data p_0 and p_1. Indeed, inserting (19.10) and (19.11) into (19.8) yields for $n = 0$

$$p_0 = a_1 = p_0 \quad \text{(QED)} \tag{19.12}$$

and for $n = 1$

$$p_1 = a_1 q_1 = p_0 \left(\frac{p_1}{p_0}\right) = p_1 \quad \text{(QED)}. \tag{19.13}$$

In (19.8) the base B is zero and to get the spectral parameters $\{q_1, a_1\}$ we need only two input values p_0 and p_1. However, if B is desired, then we need p_0, p_1 and p_2, such that (19.1) should be used instead of (19.8) [57]

$$A_n = B + p_n = B + a_1 q_1^n \quad (0 \leq n \leq 2). \tag{19.14}$$

Solving the system of equations (19.14) gives the base

$$B = \frac{\begin{vmatrix} A_0 & A_1 \\ 1 & q_1 \end{vmatrix}}{\begin{vmatrix} 1 & 1 \\ 1 & q_1 \end{vmatrix}}. \tag{19.15}$$

The correctness of (19.15) is verified if the substitution of (19.14) into the rhs of (19.15) leads to an identity. This is indeed the case

$$B = \frac{\begin{vmatrix} A_0 & A_1 \\ 1 & q_1 \end{vmatrix}}{\begin{vmatrix} 1 & 1 \\ 1 & q_1 \end{vmatrix}} = \frac{(B + a_1)q_1 - (B + a_1 q_1)}{q_1 - 1} = \frac{B(q_1 - 1)}{q_1 - 1} = B \quad \text{(QED)}.$$

It is seen from (19.10) that the root q_1 depends only on the input data p_0 and p_1. Therefore, such a root can be eliminated from (19.15). This can be done in a systematic way, which will pave the road for a generalization beyond $k = 1$. Using (19.5) for $k = 1$ and (19.14), we find

$$\begin{vmatrix} A_0 & A_1 \\ \Delta A_0 & \Delta A_1 \end{vmatrix} = Ba_1(q_1-1)^2$$

$$\begin{vmatrix} 1 & 1 \\ \Delta A_0 & \Delta A_1 \end{vmatrix} = a_1(q_1-1)^2. \tag{19.16}$$

Thus the ratio between the two determinants from (19.16) is equal to B

$$B = \frac{\begin{vmatrix} A_0 & A_1 \\ \Delta A_0 & \Delta A_1 \end{vmatrix}}{\begin{vmatrix} 1 & 1 \\ \Delta A_0 & \Delta A_1 \end{vmatrix}} = e_1(A_0) \tag{19.17}$$

where (18.16) is utilized. Therefore, as an alternative to the previous expression (19.15), the base B can be written in the form (19.17), which is independent of the root q_1 (QED).

Next we consider the case $k = 2$. Here, while working within the theme (i), we are given the input data $\{A_n\}$ with at least four equidistantly spaced time elements (p_0, p_1, p_2 and p_3)

$$p_n = \sum_{r=1}^{2} a_r q_r^n \quad (0 \leq n \leq 3). \tag{19.18}$$

The system of equations from (19.18) will have non-trivial solutions if

$$\begin{vmatrix} 1 & 1 & p_0 \\ q_1 & q_2 & p_1 \\ q_1^2 & q_2^2 & p_2 \end{vmatrix} = 0 \quad \begin{vmatrix} 1 & 1 & p_1 \\ q_1 & q_2 & p_2 \\ q_1^2 & q_2^2 & p_3 \end{vmatrix} = 0. \tag{19.19}$$

The characteristic polynomial of the system (19.18) is defined by

$$\begin{vmatrix} p_0 & p_1 & 1 \\ p_1 & p_2 & q \\ p_2 & p_3 & q^2 \end{vmatrix} = 0 = \begin{vmatrix} 1 & q & q^2 \\ p_0 & p_1 & p_2 \\ p_1 & p_2 & p_3 \end{vmatrix}. \tag{19.20}$$

Carrying out the computation in (19.20) via the Cauchy expansion of the 3×3 determinant in terms of its 2×2 minors $\{m_{2,0}, m_{2,1}, m_{2,2}\}$, we have

$$m_{2,2} q^2 - m_{2,1} q + m_{2,0} = 0 \tag{19.21}$$

$$m_{2,0} = \begin{vmatrix} p_1 & p_2 \\ p_2 & p_3 \end{vmatrix} \quad m_{2,1} = \begin{vmatrix} p_0 & p_2 \\ p_1 & p_3 \end{vmatrix} \quad m_{2,2} = \begin{vmatrix} p_0 & p_1 \\ p_1 & p_2 \end{vmatrix}.$$

$$\tag{19.22}$$

The first index 2 in the minors $\{m_{2,1}, m_{2,2}, m_{2,3}\}$ denotes the rank $k = 2$. The roots of the quadratic equation (19.21) yield one half of the sought pair of the unknown parameters

$$q_{1/2} = \frac{m_{2,1} \pm \sqrt{m_{2,1}^2 - 4m_{2,0}m_{2,2}}}{2m_{2,2}}. \tag{19.23}$$

The other pair of the unknown constants $\{a_1, a_2\}$ is obtained via substitution of the roots (19.22) into the system (19.18) so that

$$a_1 = \frac{p_1 - p_0 q_2}{q_1 - q_2} \qquad a_2 = \frac{p_1 - p_0 q_1}{q_2 - q_1}. \tag{19.24}$$

Since we provisionally fixed the value of k to be equal to 2 prior to spectral analysis, we must have the following condition fulfilled according to (7.12)

$$\begin{vmatrix} p_0 & p_1 & p_2 \\ p_1 & p_2 & p_3 \\ p_2 & p_3 & p_4 \end{vmatrix} = 0 \tag{19.25}$$

$$\therefore \qquad m_{2,2} p_4 = m_{2,1} p_3 - m_{2,0} p_2. \tag{19.26}$$

Hence, if $k = 2$ is chosen in advance, then (19.26) represents the extrapolation from the given set $\{p_0, p_1, p_2, p_3\}$ to the new signal point p_4. Clearly, there is no limitation to perform further extrapolations to predict p_n for $n \geq 5$ and this can be done by generalizing (19.26) to

$$m_{2,2} p_n = m_{2,1} p_{n-1} - m_{2,0} p_{n-2}. \tag{19.27}$$

The validity of all of the above-outlined steps rests upon the assumption that the integer k is preassigned as $k = 2$. However, the order k is customarily unknown in advance. Then adding the number k to the list of the unknown parameters, as required by the definition of the harmonic inversion, the condition (19.25) *will* become the criterion for verifying whether indeed we have $k = 2$ as the condition (7.12) would demand. Therefore, if (19.25) is satisfied for all sets of five consecutive data points $p_n (0 \leq n \leq 4)$, this would mean that we have $k = 2$. If on the other hand, the determinant in (19.25) is non-zero, then it follows that $k > 2$. It could also happen that all three minors $\{m_{2,0}, m_{2,1}, m_{2,2}\}$ are zero, in which case we have $k < 2$, that is $k = 1$. In order to solve a system of linear equations, we must fix the number of such equations in advance. In the harmonic inversion, the number k of these linear equations is unknown prior to computation and, as such, k is surmised as done in, e.g. the LP. If in the considered example, we make the guess that $k = 2$, then the criterion (19.25) should be checked to see whether our guess is in the right direction for the given set of all the available data points. If the base B is required, then the restriction (19.6) must be lifted. Taking again the case $k = 2$ for an illustration, we need an additional signal point p_4, since we now have five unknown parameters $\{B, q_1, q_2, a_1, a_2\}$

$$A_n = B + \sum_{r=1}^{2} a_r q_r^n \qquad (0 \le n \le 4). \tag{19.28}$$

The first three of this system of five linear equations will give the constant B in the form

$$B = \frac{\begin{vmatrix} A_0 & A_1 & A_2 \\ 1 & q_1 & q_1^2 \\ 1 & q_2 & q_2^2 \end{vmatrix}}{\begin{vmatrix} 1 & 1 & 1 \\ 1 & q_1 & q_1^2 \\ 1 & q_2 & q_2^2 \end{vmatrix}}. \tag{19.29}$$

In the general case, the limiting value B of the form (19.29) is impractical, since it requires knowledge of the roots. However, according to (19.22) and (19.23) the roots $q_{1/2}$ are themselves functions of p_n only and this is true for any positive integer k. This implies that the presence of q_1 and q_2 in B from (19.29) is superfluous and, as such, could be eliminated. To this end, we set $k = 2$ in (19.5) and compare the resulting four equations, $\Delta A_n = \sum_{r=1}^{2} a_r q_r^n (q_r - 1)$ $(0 \le n \le 3)$ with their counterparts in (19.18). This yields

$$\begin{vmatrix} 1 & q & q^2 \\ 1 & q_1 & q_1^2 \\ 1 & q_2 & q_2^2 \end{vmatrix} = 0 = \begin{vmatrix} 1 & q & q^2 \\ \Delta A_0 & \Delta A_1 & \Delta A_2 \\ \Delta A_1 & \Delta A_2 & \Delta A_3 \end{vmatrix}. \tag{19.30}$$

Here, the ratios of the minors of either of the two determinants are symmetric functions of the roots q_1 and q_2. Hence, these ratios of such minors must be equal to each other. This is verified by using (19.5) for $k = 2$ with the results

$$\begin{vmatrix} 1 & q_1^2 \\ 1 & q_2^2 \end{vmatrix} = q_2^2 - q_1^2 \qquad \begin{vmatrix} 1 & q_1 \\ 1 & q_2 \end{vmatrix} = q_2 - q_1 \tag{19.31}$$

$$\begin{vmatrix} \Delta A_0 & \Delta A_1 \\ \Delta A_1 & \Delta A_2 \end{vmatrix} = a_1 a_2 (q_1 - 1)(q_2 - 1)(q_1 - q_2)^2 \tag{19.32}$$

$$\begin{vmatrix} \Delta A_0 & \Delta A_2 \\ \Delta A_1 & \Delta A_3 \end{vmatrix} = a_1 a_2 (q_1 - 1)(q_2 - 1)(q_1 + q_2)(q_1 - q_2)^2 \tag{19.33}$$

$$\frac{\begin{vmatrix} 1 & q_1^2 \\ 1 & q_2^2 \end{vmatrix}}{\begin{vmatrix} 1 & q_1 \\ 1 & q_2 \end{vmatrix}} = q_1 + q_2 = \frac{\begin{vmatrix} \Delta A_0 & \Delta A_2 \\ \Delta A_1 & \Delta A_3 \end{vmatrix}}{\begin{vmatrix} \Delta A_0 & \Delta A_1 \\ \Delta A_1 & \Delta A_2 \end{vmatrix}} \tag{19.34}$$

and so on. Hence (19.29) can be cast into an equivalent form as

$$B = \frac{\begin{vmatrix} A_0 & A_1 & A_2 \\ \Delta A_0 & \Delta A_1 & \Delta A_2 \\ \Delta A_1 & \Delta A_2 & \Delta A_3 \end{vmatrix}}{\begin{vmatrix} 1 & 1 & 1 \\ \Delta A_0 & \Delta A_1 & \Delta A_2 \\ \Delta A_1 & \Delta A_2 & \Delta A_3 \end{vmatrix}} = e_2(A_0) \qquad (19.35)$$

where (18.13) is used for $k = 2$. This is a clear extension to $k = 2$ of the result (19.17) for $k = 1$.

The above analysis for the first two simplest cases ($k = 1$ and $k = 2$) is presently carried out in a way which can be easily extended to $k > 2$. Thus proceeding inductively, we find that the following general expression is valid for an arbitrary integer $k \geq 1$

$$B = \frac{\begin{vmatrix} A_0 & A_1 & \cdots & A_k \\ \Delta A_0 & \Delta A_1 & \cdots & \Delta A_k \\ \vdots & \vdots & \ddots & \vdots \\ \Delta A_{k-1} & \Delta A_k & \cdots & \Delta A_{2k-1} \end{vmatrix}}{\begin{vmatrix} 1 & 1 & \cdots & 1 \\ \Delta A_0 & \Delta A_1 & \cdots & \Delta A_k \\ \vdots & \vdots & \ddots & \vdots \\ \Delta A_{k-1} & \Delta A_k & \cdots & \Delta A_{2k-1} \end{vmatrix}} = e_k(A_0) \qquad (19.36)$$

where (18.13) is employed. An explicit calculation shows that in the case of the 2×2 determinants from (18.6) and (18.7), we have the following identities

$$\begin{vmatrix} A_n & A_{n+1} \\ \Delta A_n & \Delta A_{n+1} \end{vmatrix} = A_n A_{n+2} - A_{n+1}^2 = \begin{vmatrix} A_n & A_{n+1} \\ A_{n+1} & A_{n+2} \end{vmatrix} = H_2(A_n) \qquad (19.37)$$

$$\begin{vmatrix} 1 & 1 \\ \Delta A_n & \Delta A_{n+1} \end{vmatrix} = A_n - 2A_{n+1} + A_{n+2} = \Delta^2 A_n = H_1(\Delta^2 A_n) \qquad (19.38)$$

where $H_m(A_n)$ is the Hankel determinant (5.15) with $H_1(A_n) = A_n$. Thus (19.36) now reads

$$B = \frac{\begin{vmatrix} A_0 & A_1 & \cdots & A_k \\ A_1 & A_2 & \cdots & A_{k+1} \\ \vdots & \vdots & \ddots & \vdots \\ A_k & A_{k+1} & \cdots & A_{2k} \end{vmatrix}}{\begin{vmatrix} \Delta^2 A_0 & \Delta^2 A_1 & \cdots & \Delta^2 A_{k-1} \\ \Delta^2 A_1 & \Delta^2 A_2 & \cdots & \Delta^2 A_k \\ \vdots & \vdots & \ddots & \vdots \\ \Delta^2 A_{k-1} & \Delta^2 A_k & \cdots & \Delta^2 A_{2k-2} \end{vmatrix}} \qquad (19.39)$$

which then becomes the quotient of the two Hankel determinants that are taken from (5.15)

$$B = \frac{H_{k+1}(A_0)}{H_k(\Delta^2 A_0)}. \qquad (19.40)$$

This is the Shanks transform [71, 72] from (16.2). A more general formulation of the transform can be given by considering $2k+1$ elements of the sequence $\{A_r\}$ centred around A_n, i.e. $n-k \leq r \leq n+k$ where $n \geq k$. Such a sequence is then represented by the kth order form of the reconstruction problem which determines $2k+1$ parameters $\{B_{k,n}, q_{r,n}, a_{r,n}\}_{r=1}^{k}$ $(n = k, k+1, k+2, \ldots)$

$$\left. \begin{array}{l} A_m = B_{k,n} + \sum_{r=1}^{k} a_{n,r}(q_{n,r})^m \qquad q_{n,r} \neq 0 \qquad q_{n,r} \neq 1 \\ n-k \leq m \leq n+k \qquad n \geq k \end{array} \right\}. \qquad (19.41)$$

Of course, the parameters $\{a_{n,r}, q_{n,r}\}$ depend upon k and this could be explicitly emphasized by writing, e.g. $\{a_{n,r}^{(k)}, q_{n,r}^{(k)}\}$, but this is unnecessary. The scalar quantity $B_{k,n}$ is called the local kth order base of the sequence A_n which represents the input data. The explicit expression for $B_{k,n} \equiv B_{kn}$ is obtained from (19.36) through increasing all the indices by $n-k$

$$B_{k,n} = \frac{\begin{vmatrix} A_{n-k} & A_{n-k+1} & \cdots & A_n \\ \Delta A_{n-k} & \Delta A_{n-k+1} & \cdots & \Delta A_n \\ \vdots & \vdots & \ddots & \vdots \\ \Delta A_{n-1} & \Delta A_n & \cdots & \Delta A_{n+k-1} \end{vmatrix}}{\begin{vmatrix} 1 & 1 & \cdots & 1 \\ \Delta A_{n-k} & \Delta A_{n-k+1} & \cdots & \Delta A_n \\ \vdots & \vdots & \ddots & \vdots \\ \Delta A_{n-1} & \Delta A_n & \cdots & \Delta A_{n+k-1} \end{vmatrix}} = e_k(A_n) \qquad (19.42)$$

$$B_{0,n} = A_n \qquad (n = 0, 1, 2, 3, \ldots). \qquad (19.43)$$

When the order k is held fixed, the local base $B_{k,n}$ as well as the remaining $2k$ parameters $\{q_{n,r}, a_{n,r}\}$ are generally dependent upon the integers n and k. However, when $\{A_n\}$ is itself nearly of the order k, then the base $B_{k,n}$ is expected to change only slightly relative to the variations of A_n. Therefore, it is instructive to study the sequence of the local bases $\{B_{k,n}\}$ for $n \geq k$. Such an investigation could be motivated by the modality of generation of the sequence $\{B_{k,n}\}$ indicating that if $\{A_n\}$ converges to a constant, then the sequence $\{B_{k,n}\}$ could converge faster to the same constant. More importantly, if $\{A_n\}$ diverges, the sequence $\{B_{k,n}\}$ may still converge in the sense of analytical continuation [28, 72, 75]. The application of the Cauchy expansion for determinants to both the numerator and the denominator of the Shanks quotient (19.42) leads to the relation

$$e_k(A_n) = \frac{v_{n-k} A_{n-k} + v_{n-k+1} A_{n-k+1} + \cdots + v_n A_n}{v_{n-k} + v_{n-k+1} + \cdots + v_n}. \qquad (19.44)$$

Here, the elements v_m of the set $\{v_m\}$ represent the cofactors of the determinants in (19.42) [228]. The result (19.44) shows that the transform $B_{k,n}$ is a weighted combination of the input sequence $\{A_m\}$. The weight factors $\{v_m\}$ are cofactors that themselves depend upon the functions A_m. This implies that $B_{k,n}$ is a *nonlinear* transform of the given sequence $\{A_n\}$. It is well-known [68] that when the representation (19.2) is valid, the ratios $\{q_r\}$ are given by the roots of the polynomial

$$\begin{vmatrix} 1 & q & q^2 & \cdots & q^k \\ p_0 & p_1 & p_2 & \cdots & p_k \\ p_1 & p_2 & p_3 & \cdots & p_{k+1} \\ \vdots & \vdots & \vdots & \ddots & \vdots \\ p_{k-1} & p_k & p_{k+1} & \cdots & p_{2k-1} \end{vmatrix} = 0. \tag{19.45}$$

With the help of (18.3), (19.2) and (19.41) the elements p_s of the set $\{p_s\}$ can be expressed as

$$\Delta A_{n-k+s} = A_{n-k+s+1} - A_{n-k+s}$$

$$= \left[B_{k,n} + \sum_{r=1}^{k} a_{n,r}(q_{n,r})^{n-k+s+1} \right] - \left[B_{k,n} + \sum_{r=1}^{k} a_{n,r}(q_{n,r})^{n-k+s} \right]$$

$$= \sum_{r=1}^{k} a_{n,r}[(q_{n,r})^{n-k+1} - (q_{n,r})^{n-k}](q_{n,r})^s \equiv \sum_{r=1}^{k} a_r q_r^s$$

$$\Delta A_{n-k+s} = \sum_{r=1}^{k} a_r q_r^s = p_s \tag{19.46}$$

where we put $a_r \equiv a_{n,r}[(q_{n,r})^{n-k+1} - (q_{n,r})^{n-k}]$ and $q_r \equiv q_{n,r}$. Replacing the set $\{p_s\}$ by $\{\Delta A_{n-k+s}\}$ in (19.45), we obtain the characteristic equation

$$\begin{vmatrix} 1 & q & q^2 & \cdots & q^k \\ \Delta A_{n-k} & \Delta A_{n-k+1} & \Delta A_{n-k+2} & \cdots & \Delta A_n \\ \Delta A_{n-k+1} & \Delta A_{n-k+2} & \Delta A_{n-k+3} & \cdots & \Delta A_{n+1} \\ \vdots & \vdots & \vdots & \ddots & \vdots \\ \Delta A_{n-1} & \Delta A_n & \Delta A_{n+1} & \cdots & \Delta A_{n+k-1} \end{vmatrix} = 0. \tag{19.47}$$

The Cauchy expansion of the determinant (19.47) in terms of the same cofactors $\{v_m\}$ as those encountered in (19.44) leads to

$$v_n q^k + v_{n-1} q^{k-1} + \cdots + v_{n-k-1} q + v_{n-k} = 0. \tag{19.48}$$

Substituting (19.41) into (19.44) and using (19.48), it follows

$$e_k(A_n)[v_{n-k} + v_{n-k+1} + \cdots + v_n]$$
$$= v_{n-k}\left[B_{k,n} + \sum_{r=1}^{k} a_{n,r}(q_{n,r})^{n-k}\right] + v_{n-k+1}\left[B_{k,n} + \sum_{r=1}^{k} a_{n,r}(q_{n,r})^{n-k+1}\right]$$
$$+ \cdots + v_n\left[B_{k,n} + \sum_{r=1}^{k} a_{n,r}(q_{n,r})^{n}\right]$$
$$= B_{k,n}[v_{n-k} + v_{n-k+1} + \cdots + v_n]$$
$$+ \sum_{r=1}^{k} a_{n,r}(q_{n,r})^{n-k}[v_{n-k} + v_{n-k+1}q_{n,r} + \cdots + v_n(q_{n,r})^k]$$
$$= B_{k,n}[v_{n-k} + v_{n-k+1} + \cdots + v_n] \qquad (19.49)$$

with $q_{n,r}$ being the rth root of (19.48) where $q_{n,r} \neq 0$ and $q_{n,r} \neq 1$. Thus after cancelling the common factor $(v_{n-k}+v_{n-k+1}+\cdots+v_n) \neq 0$ in the sequel (19.49), it follows

$$B_{k,n} = e_k(A_n) \qquad \text{(QED)} \qquad (19.50)$$

as in (19.42). Consequently, for the given input data (19.2), the transform $B_{k,n}$ coincides precisely with the Shanks transform $e_k(A_n)$ from (18.13). As mentioned earlier, the introduction of the symbol e_k has the intention of pointing at k exponentials of the mathematical model (19.41). Once the roots $\{q_r\}_{r=1}^{k}$ of the characteristic equation (19.48) are obtained, the amplitudes $\{a_r\}_{r=1}^{k}$ follow readily in the analytical form of the type (11.13) or (11.14) by using the Cauchy residue theorem. In this way the Shanks transform $e_k(A_n)$ finds *exactly* all the unknown parameters $\{q_r, a_r\}_{r=1}^{k}$ and, hence, completes the final task of spectral analysis. The order k is also obtained in the course of the application of the Shanks transform through checking the conditions (7.11) and (7.12). Namely, if the input sequence has k geometrically decaying transients of the type (19.2), then the Shanks transform $e_m(A_n)$ is the exact parameter estimator for $m \geq k$.

It is often advantageous to iterate the Shanks transform to obtain higher-order accelerators [72, 88]. For example, in addition to the first transformation from $\{A_n\}$ to $\{B_{k,n}\}$ leading to the Shanks transform $B_{k,n} = e_k(A_n)$ for $n \geq k$, we could subject the newly created sequence $\{B_{k,n}\}$ to the Shanks acceleration and so forth

$$\begin{aligned} B_{k,n} &= e_k(A_n) & (n \geq k) \\ C_{k,n} &= e_k(B_{k,n}) = e_k^2(A_n) & (n \geq 2k) \\ D_{k,n} &= e_k(C_{k,n}) = e_k^3(A_n) & (n \geq 3k) \quad \text{etc.} \end{aligned} \qquad (19.51)$$

The transforms $B_{k,n}$, $C_{k,n}$ and $D_{k,n}$ are connected to their sequences A_n, B_n and C_n, respectively. This can be written according to the rules

$$A_n = B_{k,n} + \sum_{r=1}^{k} a_{n,r} e^{-n\alpha_{n,r}}$$

$$B_n = C_{k,n} + \sum_{r=1}^{k} b_{n,r} e^{-n\beta_{n,r}}$$

$$C_n = D_{k,n} + \sum_{r=1}^{k} c_{n,r} e^{-n\gamma_{n,r}} \quad \text{etc} \tag{19.52}$$

where $\{\alpha_{n,r}, a_{n,r}\}$, $\{\beta_{n,r}, b_{n,r}\}$, $\{\gamma_{n,r}, c_{n,r}\}$ are the corresponding iterated spectral parameters, respectively. The string of iterated Shanks transforms

$$A_n \longrightarrow B_{k,n} \longrightarrow C_{k,n} \longrightarrow D_{k,n} \longrightarrow \cdots \tag{19.53}$$

will be designated as the kth order transform $\tilde{e}_k(A_n)$ of A_n. The simplest example of these hybrid sequence-to-sequence nonlinear transformations is the iterated Aitken transform $\tilde{e}_1(A_n)$

$$A_n \longrightarrow B_{1,n} \longrightarrow C_{1,n} \longrightarrow D_{1,n} \longrightarrow \cdots \tag{19.54}$$

$$B_{1,n} = \frac{A_{n+1} A_{n-1} - A_n^2}{A_{n+1} + A_{n-1} - 2A_n}$$

$$C_{1,n} = \frac{B_{1,n+1} B_{1,n-1} - B_{1,n}^2}{B_{1,n+1} + B_{1,n-1} - 2B_{1,n}}$$

$$D_{1,n} = \frac{C_{1,n+1} C_{1,n-1} - C_{1,n}^2}{C_{1,n+1} + C_{1,n-1} - 2C_{1,n}} \quad \text{etc.} \tag{19.55}$$

Nonlinearity of $e_k(A_n)$ implies that the following inequality holds true

$$e_k(X_n + Y_n) \neq e_k(X_n) + e_k(Y_n) \tag{19.56}$$

where X_n and Y_n are any two given sequences. However, if here $Y_n = 0$ and X_n are replaced by ξX_n, where ξ is a constant independent of n, or if X_n is unchanged, but Y_n is independent of n, e.g. $Y_n = \xi$, then the two *linear* relationships are valid

$$e_k(\xi X_n) = \xi e_k(X_n) \qquad e_k(\xi + X_n) = \xi + e_k(X_n). \tag{19.57}$$

The following expression is also a useful rule in applications for $0 \leq \ell \leq k$

$$e_k(A_n) = \frac{v_{n-k} A_{n-k+\ell} + v_{n-k+1} A_{n-k+1+\ell} + \cdots + v_n A_{n+\ell}}{v_{n-k} + v_{n-k+1} + \cdots + v_n} \tag{19.58}$$

where v_m are the associated cofactors as before. The formula (19.58) can be derived from (19.41) and (19.42). To this end, one replaces the first row in the numerator in (19.42) by the sum of the 1st + 2nd + 3rd + $\cdots$ + $(\ell + 1)$st rows of the numerator. Then one applies the Cauchy expansion by the first new row of the numerator to obtain (19.58) which is an extension of (19.44).

Another special case is also important when the Shanks transform is applied to a sequence $\{A_n\}$ which is comprised of partial sums

$$A_n \equiv g(z) = \sum_{r=0}^{n} c_r z^r. \tag{19.59}$$

Using (18.3) for the forward difference $\Delta A_n = A_{n+1} - A_n$ in the partial sums $\{A_n\}$ from (19.59) it is immediately seen that subtraction of two partial sums leads to a single term as follows

$$\Delta A_n = \sum_{r=0}^{n+1} c_r z^r - \sum_{r=0}^{n} c_r z^r = c_{n+1} z^{n+1}. \tag{19.60}$$

Thus, for this problem, the Shanks transform e_k from (19.42) of the sequence $\{A_n\}$ of partial sums from (19.59) is given by the following quotient of the determinants

$$e_k(A_{n-k}) = \frac{\begin{vmatrix} \sum_{r=0}^{n-k} c_r z^r & \sum_{r=0}^{n-k+1} c_r z^r & \cdots & \sum_{r=0}^{n} c_r z^r \\ c_{n-k+1} z^{n-k+1} & c_{n-k+2} z^{n-k+2} & \cdots & c_{n+1} z^{n+1} \\ \vdots & \vdots & \ddots & \vdots \\ c_n z^n & c_{n+1} z^{n+1} & \cdots & c_{n+k} z^{n+k} \end{vmatrix}}{\begin{vmatrix} 1 & 1 & \cdots & 1 \\ c_{n-k+1} z^{n-k+1} & c_{n-k+2} z^{n-k+2} & \cdots & c_{n+1} z^{n+1} \\ \vdots & \vdots & \ddots & \vdots \\ c_n z^n & c_{n+1} z^{n+1} & \cdots & c_{n+k} z^{n+k} \end{vmatrix}}. \tag{19.61}$$

To illustrate the connection between the Shanks transform and the Padé approximant in the case of sequences of partial sums, we consider the simplest case with $k = 1$ and $n \neq k$. In the 2 × 2 version of (19.61), we multiply the first columns of both the numerator and denominator by $z \neq 0$. In the obtained result, we subsequently divide the second row by z^n. This factors out the common term z from the numerator and denominator. Finally, cancelling this joint term $z \neq 0$

yields the sought relationship

$$\frac{\begin{vmatrix} \sum_{r=0}^{n-1} c_r z^r & \sum_{r=0}^{n} c_r z^r \\ c_n z^n & c_{n+1} z^{n+1} \end{vmatrix}}{\begin{vmatrix} 1 & 1 \\ c_n z^n & c_{n+1} z^{n+1} \end{vmatrix}} = \frac{\begin{vmatrix} z\left(\sum_{r=0}^{n-1} c_r z^r\right) & \sum_{r=0}^{n} c_r z^r \\ z(c_n z^n) & c_{n+1} z^{n+1} \end{vmatrix}}{\begin{vmatrix} z & 1 \\ z(c_n z^n) & c_{n+1} z^{n+1} \end{vmatrix}}$$

$$= \frac{\begin{vmatrix} z\left(\sum_{r=0}^{n-1} c_r z^r\right) & \sum_{r=0}^{n} c_r z^r \\ z^{-n}\{z(c_n z^n)\} & z^{-n}\{c_{n+1} z^{n+1}\} \end{vmatrix}}{\begin{vmatrix} z & 1 \\ z^{-n}\{z(c_n z^n)\} & z^{-n}\{c_{n+1} z^{n+1}\} \end{vmatrix}}$$

$$= \frac{\begin{vmatrix} z\left(\sum_{r=0}^{n-1} c_r z^r\right) & \sum_{r=0}^{n} c_r z^r \\ c_n z & c_{n+1} z \end{vmatrix}}{\begin{vmatrix} z & 1 \\ c_n z & c_{n+1} z \end{vmatrix}}$$

$$= \frac{z \begin{vmatrix} z\left(\sum_{r=0}^{n-1} c_r z^r\right) & \sum_{r=0}^{n} c_r z^r \\ c_n & c_{n+1} \end{vmatrix}}{z \begin{vmatrix} z & 1 \\ c_n & c_{n+1} \end{vmatrix}}$$

$$= \frac{\begin{vmatrix} z\left(\sum_{r=0}^{n-1} c_r z^r\right) & \sum_{r=0}^{n} c_r z^r \\ c_n & c_{n+1} \end{vmatrix}}{\begin{vmatrix} z & 1 \\ c_n & c_{n+1} \end{vmatrix}}$$

$$= [n/1]_g(z)$$

$$\therefore \quad \frac{\begin{vmatrix} \sum_{r=0}^{n-1} c_r z^r & \sum_{r=0}^{n} c_r z^r \\ c_n z^n & c_{n+1} z^{n+1} \end{vmatrix}}{\begin{vmatrix} 1 & 1 \\ c_n z^n & c_{n+1} z^{n+1} \end{vmatrix}} = [n/1]_g(z). \tag{19.62}$$

Using (19.60), the determinantal quotient from (19.62) can be written as

$$\frac{\begin{vmatrix} \sum_{r=0}^{n-1} c_r z^r & \sum_{r=0}^{n} c_r z^r \\ c_n z^n & c_{n+1} z^{n+1} \end{vmatrix}}{\begin{vmatrix} 1 & 1 \\ c_n z^n & c_{n+1} z^{n+1} \end{vmatrix}} = \frac{\begin{vmatrix} A_{n-1} & A_n \\ A_n - A_{n-1} & A_{n+1} - A_n \end{vmatrix}}{\begin{vmatrix} 1 & 1 \\ A_n - A_{n-1} & A_{n+1} - A_n \end{vmatrix}}$$

$$= \frac{A_{n-1}(A_{n+1} - A_n) - A_n(A_n - A_{n-1})}{(A_{n+1} - A_n) - (A_n - A_{n-1})}$$

$$= \frac{A_{n-1}(A_{n+1} - 2A_n + A_{n-1}) - (A_n - A_{n-1})^2}{A_{n+1} - 2A_n + A_{n-1}}$$

$$= A_{n-1} - \frac{(\Delta A_{n-1})^2}{\Delta^2 A_{n-1}} = e_1(A_{n-1}) \tag{19.63}$$

$$\therefore \quad e_1(A_n) = [(n+1)/1]_g(z). \tag{19.64}$$

In the general case of the numerator and denominator determinants of (19.61), we multiply the pth column ($p = 1, 2, \ldots, k + 1$) by z^{k+1-p} and divide the qth row ($q = 2, 3, \ldots, k + 1$) by z^{n+q-1}. Then in (19.61) the numerator and denominator will contain the common power function z^{n+k}, which can be cancelled out to give the result

$$e_k(A_{n-k}) = \frac{\begin{vmatrix} z^k \sum_{r=0}^{n-k} c_r z^r & z^{k-1} \sum_{r=0}^{n-k+1} c_r z^r & \cdots & \sum_{r=0}^{n} c_r z^r \\ c_{n-k+1} & c_{n-k+2} & \cdots & c_{n+1} \\ \vdots & \vdots & \ddots & \vdots \\ c_n & c_{n+1} & \cdots & c_{n+k} \end{vmatrix}}{\begin{vmatrix} z^k & z^{k-1} & \cdots & 1 \\ c_{n-k+1} & c_{n-k+2} & \cdots & c_{n+1} \\ \vdots & \vdots & \ddots & \vdots \\ c_n & c_{n+1} & \cdots & c_{n+k} \end{vmatrix}}.$$
(19.65)

According to the definition (19.59), the sum $A_{n-k} = \sum_{r=0}^{n-k} c_r z^r$ is a polynomial of the $(n-k)$th degree in the variable z. Therefore, the product $z^k A_{n-k} = z^k \sum_{r=0}^{n-k} a_r z^r$ is a polynomial of the nth degree. In this way we see that the Shanks transform from (19.65) is a quotient of two polynomials such that the numerator and the denominator are the polynomials of the nth and kth degree, respectively. In other words, the transform $e_k(A_{n-k})$ for the partial sums A_n coincides with the Padé approximant $[n/k]_g(z)$ constructed for the rhs of (19.59). This generalizes (19.64) as follows

$$e_k(A_{n-k}) = [n/k]_g(z). \tag{19.66}$$

As a very useful application of this nonlinear method, we can quote the DFT which represents a sequence of partial sums for varying signal length [17]. When the harmonic inversion is formulated for this case, the Shanks transform, represents the *exact* solution of the problem.

As it stands, the original ε-algorithm of Wynn [73] is a non-parametric estimator, since (18.14) can provide only the shape of a spectrum. However, this method can also be a parametric estimator of spectra, due to the possibility of obtaining the natural damping factors and amplitudes $\{q_r, a_r\}$ as the constituent pair of the sequence $\{A_n\}$ from (19.1). Formally setting $B = 0$ in (19.1), we can retrieve the auto-correlation functions $\{C_n\}$ from (5.9) and time signal points $\{c_n\}$ from (5.11). Replacing z by $u = \exp(-i\omega\tau)$ and assuming that the $\varepsilon_k^{(m)}(u)$-vectors are meromorphic functions[1] of their variable u, we can obtain

---

[1] A given function $f(u)$ is meromorphic if its spectrum is comprised only of the poles $\{u_k\}$. In such a case, the same zeros $\{u_k\}$ of the function $f(u)$ are also the poles of the reciprocal function $1/f(u)$.

the fundamental frequencies $\{\omega_k\}$ from the zeros $\{u_k\}$ of the inverse function $\tilde{\varepsilon}_k^{(m)}(u)$ [249]

$$\tilde{\varepsilon}_k^{(m)}(u) = \frac{1}{\varepsilon_k^{(m)}(u)} \qquad (19.67)$$

and, therefore, the equation

$$\tilde{\varepsilon}_k^{(m)}(u_k) = 0 \qquad (19.68)$$

will yield the roots $\{u_k\}$. These roots provide the sought set $\{\omega_k\} = \{i\tau^{-1}\ln(u_k)\}$ of the natural frequencies $\{\omega_k\}$. Then the resulting complex spectrum is given as the partial fractions

$$\varepsilon(u) = \sum_{k=1}^{K} \frac{d_k}{u - u_k} \qquad d_k = \frac{1}{\tilde{\varepsilon}_{k,1}^{(m)}(u_k)} \qquad (19.69)$$

where $\tilde{\varepsilon}_{k,1}^{(m)}(u)$ is the first derivative of the inverse function (19.67), i.e. $\tilde{\varepsilon}_k^{(m)}(u)$

$$\tilde{\varepsilon}_{k,n}^{(m)}(u) = \left(\frac{d}{du}\right)^n \tilde{\varepsilon}_k^{(m)}(u) \qquad (n = 0, 1, 2, \ldots). \qquad (19.70)$$

One of the ways to find the zeros $\{u_k\}$ of the function $\tilde{\varepsilon}_k^{(m)}(u)$ is to use a modified Newton–Raphson algorithm from chapter 28 [176, 220]. Alternatively, the roots $\{u_k\}$ can also be obtained from a relation of the ratio of two Hankel determinants in the Shanks transform [246]

$$u_k = \lim_{m \to \infty} \frac{\varepsilon_{2k-2}^{(m+1)}(u)}{\varepsilon_{2k-2}^{(m)}(u)} \qquad (19.71)$$

as in (17.4). In other words, rather than solving (19.68), one computes the function $\varepsilon_k^{(m)}(u)$ at a sufficiently dense grid $\omega \in [\omega_{\min}, \omega_{\max}]$ to ensure that no eigenvalue $\omega_k = (i/\tau)\ln(u_k)$ is missed [237]. At each fixed variable u the vectors $\{\varepsilon_k^{(m)}(u)\}$ are computed and convergence of the ratios $\{\varepsilon_{2k-2}^{(m+1)}/\varepsilon_{2k-2}^{(m)}\}$ is monitored with respect to the increasing superscript m. If this latter sequence converges, then its limiting value *is* the sought root u_k. In such a way, we can find all the roots of the equation $\tilde{\varepsilon}_k^{(m)}(u) = 0$ without using any classical root search algorithm. Of course, in practice, the grid for u is never fine enough to yield u_k with high accuracy. Nevertheless, we have found in applications that such an obtained estimate for u_k is good enough as an initial value for the usual Newton–Raphson iteration [176, 220], which can efficiently yield a highly accurate approximation to the root u_k. An excellent refinement for u_k can readily be obtained to within at least eight significant decimals performing barely ten iterations of the modified Newton–Raphson algorithm from chapter 28.

As we have seen in chapter 18, the Shanks transform [71, 72] is a direct generalization of the Aitken extrapolation formula to an arbitrary order k and, as such, it represents the exact nonlinear mapping for sequences $\{A_n\} = \{B + \sum_{r=1}^{k} a_r q_r^n\}$ having k geometrically decaying transients $a_1 q_1^n + a_2 q_2^n + \cdots + a_k q_k^n$. Although the Aitken transform is exact for sequences with a single transient, this does not mean that (21.3) is limited only to such sequences. The formula (21.3) can also be applied to sequences with two or more transients giving certain approximate, but still instructive results [250].

In our recent applications [142], we have advantageously used both the direct and iterated Aitken transforms in spectral analysis. Especially important is the usage of these transforms for sequences of partial sums of FFTs, i.e. $\{F_k\} \equiv \{A_m(\tilde{u}_k)\}$ from (5.3), generated for different lengths 2^m of a given set of signal points $\{c_n\}$ ($0 \le n \le 2^m - 1$) [31]

$$A_m(\tilde{u}_k) = \frac{1}{2^m} \sum_{n=0}^{2^m-1} c_n \tilde{u}_k^n \qquad \tilde{u}_k = e^{-i\pi k/2^{m-1}}. \qquad (19.72)$$

Here, the iterated Aitken transforms can improve the usual FFTs that are simple trapezoidal quadrature rule [180] for the Fourier integral. Improvements over the fast Fourier transform are also possible by using some linear accelerators. One of them is the Euler transform of $A_m(\tilde{u}_k) \equiv A_m$ which can be conceived as sequential averaging [251]–[256]

$$\left.\begin{aligned} R_{1,m} &= \tfrac{1}{2}(A_m + A_{m+1}) \\ R_{2,m} &= \tfrac{1}{2}(R_{1,m} + R_{1,m+1}) \\ R_{3,m} &= \tfrac{1}{2}(R_{2,m} + R_{2,m+1}) \\ &\vdots \\ R_{n,m} &= \tfrac{1}{2}(R_{n-1,m} + R_{n-1,m+1}) \end{aligned}\right\}. \qquad (19.73)$$

Using this averaging technique, the following explicit sum rule is readily established by induction

$$R_{n,m} = \frac{1}{2^n} \sum_{k=0}^{n} \binom{n}{k} A_{m-k} \qquad \binom{n}{k} = \frac{n!}{k!(n-k)!}. \qquad (19.74)$$

Nevertheless, in practice, to avoid limitations of factorials with large numbers, the Euler sequential averaging (19.73) should be used instead of the formula (19.74). The linear acceleration technique (19.73) has been found to be very convenient in many studies, e.g. in [251]–[256], etc. The reason for the success of the transform (19.73) is in averaging the closest terms in the given sequence. These adjacent terms have more chance to be similar to each other than those residing far apart. In such a case, averaging similar members of a slowly convergent sequence is expected to naturally phase out the outliers and also increase the convergence rate^2. This is what often has been observed in practice when dealing with alternating slowly converging sequences, as well as with some diverging sequences [251]–[256].

---

^{2} Practitioners dealing with statistical data often encounter so-called outliers, i.e. data that single themselves out by being very different from the rest. Such data are often discarded or artificially manipulated when simple averaging is used. The Euler sequential averaging is expected to mitigate this problem with outliers in a more satisfactory and systematic way. In signal processing, the outliers are conceived as possible rare spectral features that cannot be reconstructed by using only the principal components.

Chapter 20

The base-transient concept in signal processing

Suppose we are given a sequence of partial sums, like those encountered in the Fourier based spectroscopy as time signals, or a generic sequence $\{A_n\}$ of any other numbers in the input data that do not need to be partial sums. In spectral analysis of such a sequence, and in the search for its *base* which is the limiting value B of A_n attained at $n \longrightarrow \infty$, the main attention is focused on improving the convergence rate of $\{A_n\}$ as n is infinitely increased. This is very important especially for the Fourier transform $F(\bar{\omega}_k)$ of time signal points $\{c_n\}$. In order to resolve closely spaced frequencies $\{\omega_k\}$ the FFT must have a sufficiently long epoch T implying that the points $\{c_n\}$ must be recorded at large values of the integer n which measures time ($t = t_n = n\tau$). At these large n the envelope of any time signal $\{c_n\}$ usually reaches its asymptotic tail where it becomes heavily corrupted with noise. Then the key question arises as to whether it would be possible to extract most of the essential information (and thus to achieve the required resolution on the frequency scale) from the earlier recorded signal points $\{c_n\}$ without necessarily entering the asymptotic time domain dominated mainly by random noise. This is a subject of general significance which encompasses not only the given slowly converging sequences, but also divergent sequences or series encountered in many research areas across interdisciplinary fields. For example, the three basic problems treated within the perturbation method of quantum mechanics (anharmonic oscillator, atoms in external electric and magnetic fields) possess their energy expansions that are divergent for most of the physically important values of the coupling constants. Even Fourier transforms of many realistic sequences are often divergent, as are, e.g. the periodic orbit sums in the theory of semi-classical quantization [114, 237], etc. Then one would like to have a simple and robust method which would be capable of both accelerating slowly converging sequences, as well as inducing convergence into diverging series through the concept of analytical continuation [28, 72, 75]. Shanks [72] has stated that no linear methods can achieve these two goals. However, the

The base-transient concept in signal processing 151

well-known linear accelerator of Euler can [251]–[256]. The Euler method is a powerful linear transformation of slowly converging *alternating* series or sequences. This linear transform can also analytically continue some diverging series and, as such, it can be used for resummation of asymptotic series [142]. The main idea of introducing nonlinear transformations is that they can both improve convergence of sequences that converge slowly and force divergent sequences to converge. This is the consequence of nonlinearity which manifests itself in an adaptive sum of the elements A_n of the original sequence $\{A_n\}$. A nonlinear transformation is a weighted average of the original members from the set $\{A_n\}$ with the weight functions that themselves depend upon A_n as in (19.44). In this way a nonlinear transform becomes able to weigh heavily the earlier A_n with smaller values of n if the input sequence $\{A_n\}$ is divergent. If a given sequence converges, but achieves this slowly, the same nonlinear transform would weigh heavily the later A_n. In either case (convergence or divergence), nonlinear transforms exhibit distinct advantages over linear transforms. Linear transforms, e.g. the FFT or the Euler transform are also certain averages of the elements $\{A_n\}$, i.e. $[\sum_{n=0}^{N} w_n A_n]/[\sum_{n=0}^{N} w_n]$, but here the pre-assigned weights w_n are independent of A_n.

After these introductory remarks, we shall focus our attention on geometric sequences encountered in signal processing. For this special purpose, it will be more convenient to employ the more familiar symbol c_n instead of A_n. Likewise, to cohere with the notation in the other chapters on signal processing, the transients $\{a_r q_r^n\}_{r=1}^{k}$ from A_n will be denoted by $\{d_k u_k^n\}_{k=1}^{K}$ in c_n where $u_k = \exp(-i\omega_k \tau)$ as in (5.8). Moreover, to make a link to chapter 19, we shall consider a slightly more general time signal $\bar{c}_n$ representing the usual c_n appended by a baseline constant $\bar{c}_\infty$ which plays the role of the limit B in A_n. Of course, $\bar{c}_\infty$ extrapolates $\bar{c}_n$ to $n = \infty$. The first idea which comes to mind while attempting to improve the convergence rate of the sequence of 'generalized' signal points $\{\bar{c}_n\} \equiv \{\bar{c}_\infty + c_n\}$ is to eliminate the terms with the most pronounced transient behaviour in the limit $n \longrightarrow \infty$. Recall that the standard time signal c_n is modelled by the geometric sequence (5.11)

$$\bar{c}_n = \bar{c}_\infty + c_n \qquad c_n = \sum_{k=1}^{K} d_k u_k^n \qquad u_k = e^{-i\omega_k \tau} \qquad (20.1)$$

where $\mathrm{Im}(\omega_k) < 0$. In the next chapter we shall demonstrate explicitly how the Shanks transform filters out all the transients $\{d_k u_k^n\}$ $(1 \leq k \leq K)$ from the sequence $\{\bar{c}_n\}$. The calculations will be done explicitly for $K \leq 3$ and the obtained results in these particular cases will clearly show the pattern for derivation of the general formula for any positive integer K.

Chapter 21

Extraction of the exact number of transients from time signals

In signal processing, it is of utmost importance to be able to find the exact number K of the constituent transients/harmonics of the investigated time signal $\{c_n\}$. These transients $\{d_k u_k^n\}$, that represent metabolites in MRS, lead to peaks/resonances in the corresponding frequency spectrum that need to be quantified. The first step in any reliable spectral quantification is an accurate determination of the number of transients. The absence of a method for an unequivocal extraction of K led to a common practice consisting of guessing the number, as customarily done in, e.g. the LP, HLSVD, LCModel [11], etc. In this way, the answer which one seeks is simply put by hand into the analysis. Inevitably, any guess K' would either underestimate or overestimate the true number K. This would yield some $K' - K$ spurious (non-physical) or $K - K'$ undetected genuine (physical) transients. Both of these drawbacks become severe and, as such, particularly unacceptable in medical diagnostics based upon MRS or MRSI. However, as we are going to show, surmising K is unnecessary, since the number of transients can be determined exactly through the Shanks transform $e_K(c_n)$ for both non-degenerate (Lorentzian) and degenerate (non-Lorentzian) spectra. The proof is in using $e_K(c_n)$ to filter out all the transients from the given time signal expressed as a sum of damped complex exponentials with constant or time-dependent amplitudes.

21.1 Explicit filtering of one transient $c_n = d_1 u_1^n$

As a prototype of a physical transient which decays after a sufficiently long time has elapsed, we shall first consider the simplest one-harmonic signal c_n described by a single mathematical *transient* $d_1 u_1^n$ from (20.1)

$$\bar{c}_n = \bar{c}_\infty + d_1 u_1^n \equiv \bar{c}_\infty + c_n \tag{21.1}$$

Explicit filtering of one transient $c_n = d_1 u_1^n$ 153

where $d_1 \neq 0$ and $0 < |u_1| < 1$. This is a mathematical transient of the corresponding physical transient, because the remainder $d_1 u_1^n$ of the sequence $\{\bar{c}_n\}$ vanishes as $n \longrightarrow \infty$ due to the feature $|u_1| < 1$. For nonlinear transforms of $\{\bar{c}_n\}$, the condition $|u_1| < 1$ is only a matter of convenience, but not necessity. The nonlinear transforms such as those of Aitken, Shanks, Padé and the like are meaningful also for $|u_1| \geq 1$. In this latter case, the sequence $\{\bar{c}_n\}$ is manifestly divergent, but nonlinear transforms can provide the anti-limit $\bar{c}_\infty$ in the sense of analytic continuation [28, 72, 75], which extends the radius of convergence of $\{\bar{c}_n\}$ even when such a radius is originally zero [88]. In the simplest case (21.1) with one transient, every member $\bar{c}_n$ of the sequence $\{\bar{c}_n\}$ depends only on the three parameters d_1, u_1 and $\bar{c}_\infty$. Therefore, the algebraic condition for determining the limiting value $\bar{c}_\infty$ of the sequence (21.1) is given by supplying merely three consecutive terms $\bar{c}_{n-1}, \bar{c}_n$ and $\bar{c}_{n+1}$ from the set $\{\bar{c}_n\}$

$$\bar{c}_{n-1} = \bar{c}_\infty + d_1 u_1^{n-1} \qquad \bar{c}_n = \bar{c}_\infty + d_1 u_1^n \qquad \bar{c}_{n+1} = \bar{c}_\infty + d_1 u_1^{n+1}. \quad (21.2)$$

Solving these equations for the constant $\bar{c}_\infty$ we obtain the result

$$\bar{c}_\infty \equiv \mathcal{S}(\bar{c}_n) = e_1(\bar{c}_n) = \frac{\bar{c}_{n+1} \bar{c}_{n-1} - \bar{c}_n^2}{\bar{c}_{n+1} - 2\bar{c}_n + \bar{c}_{n-1}}. \quad (21.3)$$

The transform $\mathcal{S}(\bar{c}_n)$ of $\bar{c}_n$ is the Aitken [70] extrapolation formula which was encountered previously in (18.16) as the Shanks transform $e_1(\bar{c}_n)$ of the first order. This can alternatively be denoted by $\bar{c}_\infty = \bar{c}_\infty^{\text{extr}}$ where the symbol $\bar{c}_\infty^{\text{extr}} = \mathcal{S}(\bar{c}_n)$ points at the extrapolation nature of the Aitken transform. Adding and subtracting the term $\bar{c}_{n-1}(\bar{c}_{n-1} - 2\bar{c}_n)$ in the numerator of (21.3), as in (19.63), the following equivalent form of the Aitken transform emerges

$$\mathcal{S}(\bar{c}_n) = \bar{c}_{n-1} - \frac{(\Delta \bar{c}_{n-1})^2}{\Delta^2 \bar{c}_{n-1}} \quad (21.4)$$

where $\Delta \bar{c}_n$ is the first-order forward difference taken from (18.3). In (21.4), the term $\Delta^2 \bar{c}_n$ represents the second-order forward difference $\Delta^2 \bar{c}_{n-1} = \Delta(\bar{c}_n - \bar{c}_{n-1}) = \bar{c}_{n+1} + \bar{c}_{n-1} - 2\bar{c}_n$. The expression (21.4) shows that the Aitken extrapolation *has no first-order error* and, therefore, it is frequently called the Aitken Δ^2-process. The Aitken transform is exact only if the sequence $\{\bar{c}_n\}$ possesses just one transient, as in (21.1). In particular, the formula (21.4) of the Δ^2-process is illustrative, since it clearly demonstrates the capability of the Aitken interpolation to filter out, i.e. to eliminate the transient $d_1 u_1^n$, thus arriving straight at the limit $\bar{c}_\infty$ of the sequence $\{\bar{c}_n\}$. This can immediately be seen by carrying out the computations of the first- and the second-order finite differences as follows

$$\Delta \bar{c}_n = d_1 u_1^n (u_1 - 1) \qquad \Delta^2 \bar{c}_n = d_1 u_1^n (u_1 - 1)^2 \quad (21.5)$$

$$\mathcal{S}(\bar{c}_n) = e_1(\bar{c}_n) = \bar{c}_{n-1} - \frac{(\Delta \bar{c}_{n-1})^2}{\Delta^2 \bar{c}_{n-1}} = \bar{c}_{n-1} - d_1 u_1^{n-1}$$

$$= (\bar{c}_\infty + d_1 u_1^{n-1}) - d_1 u_1^{n-1} = \bar{c}_\infty \qquad \text{(QED)}. \quad (21.6)$$

The same result for $e_1(\bar{c}_n)$ as in (21.6) can also be obtained in another way which is more convenient for an extension to the most general case $e_k(\bar{c}_n)$. This is a direct way from the determinantal quotient

$$e_1(\bar{c}_n) = \frac{\begin{vmatrix} \bar{c}_n & \bar{c}_{n+1} \\ \Delta\bar{c}_n & \Delta\bar{c}_{n+1} \end{vmatrix}}{\begin{vmatrix} 1 & 1 \\ \Delta\bar{c}_n & \Delta\bar{c}_{n+1} \end{vmatrix}}$$

$$= \frac{v_1 \bar{c}_n - v_2 \bar{c}_{n+1}}{v_1 - v_2} \equiv \frac{\mathcal{N}_1}{\mathcal{D}_1} \quad (21.7)$$

$$\Delta\bar{c}_n = d_1 u_1^n (u_1 - 1) \qquad v_1 = \Delta\bar{c}_{n+1} \qquad v_2 = \Delta\bar{c}_n. \quad (21.8)$$

The determinants $\mathcal{D}_1$ and $\mathcal{N}_1$ can be reduced to

$$\mathcal{D}_1 = d_1 u_1^n (u_1 - 1)^2 \quad (21.9)$$

$$\mathcal{N}_1 = (\bar{c}_\infty + d_1 u_1^n)[d_1 u_1^{n+1}(u_1 - 1)] - (\bar{c}_\infty + d_1 u_1^{n+1})[d_1 u_1^n (u_1 - 1)]$$

$$= d_1 u_1^n (u_1 - 1)[u_1(\bar{c}_\infty + d_1 u_1^n) - (\bar{c}_\infty + d_1 u_1^{n+1})]$$

$$= d_1 u_1^n (u_1 - 1)\{\bar{c}_\infty(u_1 - 1) + [(1-1)d_1]u_1^{n+1}\}$$

$$= d_1 u_1^n (u_1 - 1)\{\bar{c}_\infty(u_1 - 1) + (0 \cdot d_1)u_1^{n+1}\}$$

$$= d_1 u_1^n (u_1 - 1)^2 \bar{c}_\infty = \mathcal{D}_1 \bar{c}_\infty$$

$$\mathcal{N}_1 = \bar{c}_\infty \mathcal{D}_1 \quad (21.10)$$

$$e_1(\bar{c}_n) = \frac{\mathcal{N}_1}{\mathcal{D}_1} = \bar{c}_\infty \quad \text{(QED)}. \quad (21.11)$$

The introduction of the base-transient concept is critical to spectral analysis of sequences and series. As mentioned before, the base $\bar{c}_\infty$ is the limit (the antilimit) of the given convergent (divergent) sequence $\{\bar{c}_n\}$. The intrinsic structure of the sequence $\{\bar{c}_n\}$ is concentrated in its transient part, which represents an obstacle in the approach to $\bar{c}_\infty$ as $n \longrightarrow \infty$. Therefore, the goal is to filter out the obstructing transient part of $\{\bar{c}_n\}$ while simultaneously preserving the base $\bar{c}_\infty$. This goal is achieved *exactly* by the Aitken transform when $\{\bar{c}_n\}$ has only one transient as in (21.1).

21.2 Explicit filtering of two transients $c_n = d_1 u_1^n + d_2 u_2^n$

The simple derivation (21.6) or (21.11) shows the efficiency of the Aitken transform $\mathcal{S}(\bar{c}_n) = e_1(\bar{c}_n)$ to reach the exact result for a sequence with only one transient $d_1 u_1^n$. Similarly, we can demonstrate that the Shanks second-order transform $e_2(\bar{c}_n)$ represents the exact solution for a sequence $\{\bar{c}_n\}$ with two geometrically decaying transients $d_1 u_1^n$ and $d_2 u_2^n$

$$\bar{c}_n = \bar{c}_\infty + (d_1 u_1^n + d_2 u_2^n) \equiv \bar{c}_\infty + c_n \quad (21.12)$$

Explicit filtering of two transients $c_n = d_1 u_1^n + d_2 u_2^n$ 155

where $d_{1,2} \neq 0$ and $0 < |u_{1,2}| < 1$. Thus we have

$$e_2(\bar{c}_n) = \frac{\begin{vmatrix} \bar{c}_n & \bar{c}_{n+1} & \bar{c}_{n+2} \\ \Delta\bar{c}_n & \Delta\bar{c}_{n+1} & \Delta\bar{c}_{n+2} \\ \Delta\bar{c}_{n+1} & \Delta\bar{c}_{n+2} & \Delta\bar{c}_{n+3} \end{vmatrix}}{\begin{vmatrix} 1 & 1 & 1 \\ \Delta\bar{c}_n & \Delta\bar{c}_{n+1} & \Delta\bar{c}_{n+2} \\ \Delta\bar{c}_{n+1} & \Delta\bar{c}_{n+2} & \Delta\bar{c}_{n+3} \end{vmatrix}}$$

$$= \frac{v_1 \bar{c}_n - v_2 \bar{c}_{n+1} + v_3 \bar{c}_{n+2}}{v_1 - v_2 + v_3} \equiv \frac{\mathcal{N}_2}{\mathcal{D}_2} \quad (21.13)$$

$$\left. \begin{array}{l} v_1 = \Delta\bar{c}_{n+1}\Delta\bar{c}_{n+3} - (\Delta\bar{c}_{n+2})^2 \\ v_2 = \Delta\bar{c}_n\Delta\bar{c}_{n+3} - \Delta\bar{c}_{n+1}\Delta\bar{c}_{n+2} \\ v_3 = \Delta\bar{c}_n\Delta\bar{c}_{n+2} - (\Delta\bar{c}_{n+1})^2 \end{array} \right\}. \quad (21.14)$$

In principle, when dealing with the kth order transform $e_k(\bar{c}_n)$ of $\{\bar{c}_n\}$ we should use the labels $\bar{c}_\infty^{(K)}, c_n^{(K)}$ and $v_k^{(K)}$ ($1 \leq k \leq K$) for the limiting value, the partial sum $\sum_{k=1}^K d_k u_k^n$ and the cofactors, respectively. However, no confusion should arise when, for simplicity, we use the same labels $\bar{c}_\infty, c_n$ and v_k ($1 \leq k \leq K$) for any order K throughout. The first-order finite difference $\Delta\bar{c}_n$ for the sequence (21.12) takes the form

$$\Delta\bar{c}_n = d_1 u_1^n (u_1 - 1) + d_2 u_2^n (u_2 - 1) \quad (21.15)$$

$$\left. \begin{array}{l} v_1 = d_1 d_2 (u_1 - 1)(u_2 - 1)(u_2 - u_1)^2 u_1^{n+1} u_2^{n+1} \\ v_2 = d_1 d_2 (u_1 - 1)(u_2 - 1)(u_2 - u_1)^2 (u_1 + u_2) u_1^n u_2^n \\ v_3 = d_1 d_2 (u_1 - 1)(u_2 - 1)(u_2 - u_1)^2 u_1^n u_2^n \end{array} \right\}. \quad (21.16)$$

Inserting the results (21.16) into (21.13), we obtain

$$\mathcal{D}_2 = d_1 d_2 (u_1 - 1)^2 (u_2 - 1)^2 (u_2 - u_1)^2 u_1^n u_2^n \quad (21.17)$$

$$\mathcal{N}_2 = v_1 (\bar{c}_\infty + d_1 u_1^n + d_2 u_2^n) - v_2 (\bar{c}_\infty + d_1 u_1^{n+1} + d_2 u_2^{n+1})$$
$$+ v_3 (\bar{c}_\infty + d_1 u_1^{n+2} + d_2 u_2^{n+2})$$
$$= d_1 d_2 (u_1 - 1)(u_2 - 1)(u_2 - u_1)^2 u_1^n u_2^n$$
$$\times [(\bar{c}_\infty + d_1 u_1^n + d_2 u_2^n) u_1 u_2 - (\bar{c}_\infty + d_1 u_1^{n+1} + d_2 u_2^{n+1})(u_1 + u_2)$$
$$+ (\bar{c}_\infty + d_1 u_1^{n+2} + d_2 u_2^{n+2})]$$
$$= d_1 d_2 (u_1 - 1)(u_2 - 1)(u_2 - u_1)^2 u_1^n u_2^n \{\bar{c}_\infty (u_1 u_2 - u_1 - u_2 + 1)$$
$$+ [(u_1 + u_2 - u_1 - u_2) d_1] u_1^{n+1} + [(u_1 + u_2 - u_1 - u_2) d_2] u_2^{n+1}\}$$
$$= d_1 d_2 (u_1 - 1)(u_2 - 1)(u_2 - u_1)^2 u_1^n u_2^n$$
$$\times [\bar{c}_\infty (u_1 - 1)(u_2 - 1) + (0 \cdot d_1) u_1^{n+1} + (0 \cdot d_2) u_2^{n+1}]$$

$$= d_1 d_2 (u_1-1)^2 (u_2-1)^2 (u_2-u_1)^2 u_1^n u_2^n \bar{c}_\infty = \mathcal{D}_2 \bar{c}_\infty$$
$$\mathcal{N}_2 = \bar{c}_\infty \mathcal{D}_2 \qquad (21.18)$$
$$e_2(\bar{c}_n) = \frac{\mathcal{N}_2}{\mathcal{D}_2} = \bar{c}_\infty \quad \text{(QED)}. \qquad (21.19)$$

As seen in the ninth line of the string of equations preceding (21.18), the second-order Shanks transform $e_2(\bar{c}_n)$ explicitly eliminates both transients $d_1 u_1^n$ and $d_2 u_2^n$ from the sequence $\{\bar{c}_n\}$ supplied by (21.12) and thus arrives straight at the exact limiting value $\bar{c}_\infty$.

21.3 Explicit filtering of three transients $c_n = d_1 u_1^n + d_2 u_2^n + d_3 u_3^n$

Next we shall apply the Shanks transform of the third order, i.e. $e_3(\bar{c}_n)$ to the geometric sequence $\{\bar{c}_n\}$ whose general element $\bar{c}_n$ contains three transients $d_k u_k^n (1 \le k \le 3)$

$$\bar{c}_n = \bar{c}_\infty + (d_1 u_1^n + d_2 u_2^n + d_3 u_3^n) \equiv \bar{c}_\infty + c_n \qquad (21.20)$$

where $d_{1,2,3} \neq 0$ and $0 < |u_{1,2,3}| < 1$. In this case we shall have

$$e_3(\bar{c}_n) = \frac{\begin{vmatrix} \bar{c}_n & \bar{c}_{n+1} & \bar{c}_{n+2} & \bar{c}_{n+3} \\ \Delta\bar{c}_n & \Delta\bar{c}_{n+1} & \Delta\bar{c}_{n+2} & \Delta\bar{c}_{n+3} \\ \Delta\bar{c}_{n+1} & \Delta\bar{c}_{n+2} & \Delta\bar{c}_{n+3} & \Delta\bar{c}_{n+4} \\ \Delta\bar{c}_{n+2} & \Delta\bar{c}_{n+3} & \Delta\bar{c}_{n+4} & \Delta\bar{c}_{n+5} \end{vmatrix}}{\begin{vmatrix} 1 & 1 & 1 & 1 \\ \Delta\bar{c}_n & \Delta\bar{c}_{n+1} & \Delta\bar{c}_{n+2} & \Delta\bar{c}_{n+3} \\ \Delta\bar{c}_{n+1} & \Delta\bar{c}_{n+2} & \Delta\bar{c}_{n+3} & \Delta\bar{c}_{n+4} \\ \Delta\bar{c}_{n+2} & \Delta\bar{c}_{n+3} & \Delta\bar{c}_{n+4} & \Delta\bar{c}_{n+5} \end{vmatrix}}$$

$$= \frac{v_1 \bar{c}_n - v_2 \bar{c}_{n+1} + v_3 \bar{c}_{n+2} - v_4 \bar{c}_{n+3}}{v_1 - v_2 + v_3 - v_4} \equiv \frac{\mathcal{N}_3}{\mathcal{D}_3} \qquad (21.21)$$

$$\Delta\bar{c}_n = d_1 u_1^n (u_1-1) + d_2 u_2^n (u_2-1) + d_3 u_3^n (u_3-1) \qquad (21.22)$$

$$v_1 = \Delta\bar{c}_{n+1}[\Delta\bar{c}_{n+3}\Delta\bar{c}_{n+5} - (\Delta\bar{c}_{n+4})^2]$$
$$- \Delta\bar{c}_{n+2}(\Delta\bar{c}_{n+2}\Delta\bar{c}_{n+5} - \Delta\bar{c}_{n+3}\Delta\bar{c}_{n+4})$$
$$+ \Delta\bar{c}_{n+3}[\Delta\bar{c}_{n+2}\Delta\bar{c}_{n+4} - (\Delta\bar{c}_{n+3})^2] \qquad (21.23)$$

$$v_2 = \Delta\bar{c}_n[\Delta\bar{c}_{n+3}\Delta\bar{c}_{n+5} - (\Delta\bar{c}_{n+4})^2]$$
$$- \Delta\bar{c}_{n+2}(\Delta\bar{c}_{n+1}\Delta\bar{c}_{n+5} - \Delta\bar{c}_{n+2}\Delta\bar{c}_{n+4})$$
$$+ \Delta\bar{c}_{n+3}(\Delta\bar{c}_{n+1}\Delta\bar{c}_{n+4} - \Delta\bar{c}_{n+2}\Delta\bar{c}_{n+3}) \qquad (21.24)$$

$$v_3 = \Delta\bar{c}_n(\Delta\bar{c}_{n+2}\Delta\bar{c}_{n+5} - \Delta\bar{c}_{n+3}\Delta\bar{c}_{n+4})$$
$$- \Delta\bar{c}_{n+1}(\Delta\bar{c}_{n+1}\Delta\bar{c}_{n+5} - \Delta\bar{c}_{n+2}\Delta\bar{c}_{n+4})$$
$$+ \Delta\bar{c}_{n+3}[\Delta\bar{c}_{n+1}\Delta\bar{c}_{n+3} - (\Delta\bar{c}_{n+2})^2] \qquad (21.25)$$

Explicit filtering of three transients $c_n = d_1 u_1^n + d_2 u_2^n + d_3 u_3^n$ 157

$$v_4 = \Delta \bar{c}_n [\Delta \bar{c}_{n+2} \Delta \bar{c}_{n+4} - (\Delta \bar{c}_{n+3})^2]$$
$$- \Delta \bar{c}_{n+1} (\Delta \bar{c}_{n+1} \Delta \bar{c}_{n+4} - \Delta \bar{c}_{n+2} \Delta \bar{c}_{n+3})$$
$$+ \Delta \bar{c}_{n+2} [\Delta \bar{c}_{n+1} \Delta \bar{c}_{n+3} - (\Delta \bar{c}_{n+2})^2]. \tag{21.26}$$

Using (21.22) we obtain the results

$$v_1 = d_1 d_2 d_3 u_1^{n+1} u_2^{n+1} u_3^{n+1} (u_1 - 1)(u_2 - 1)(u_3 - 1)$$
$$\times (u_2 - u_1)^2 (u_3 - u_1)^2 (u_3 - u_2)^2 \tag{21.27}$$

$$v_2 = d_1 d_2 d_3 u_1^n u_2^n u_3^n (u_1 - 1)(u_2 - 1)(u_3 - 1)$$
$$\times (u_2 - u_1)^2 (u_3 - u_1)^2 (u_3 - u_2)^2 (u_1 u_2 + u_1 u_3 + u_2 u_3) \tag{21.28}$$

$$v_3 = d_1 d_2 d_3 u_1^n u_2^n u_3^n (u_1 - 1)(u_2 - 1)(u_3 - 1)$$
$$\times (u_2 - u_1)^2 (u_3 - u_1)^2 (u_3 - u_2)^2 (u_1 + u_2 + u_3) \tag{21.29}$$

$$v_4 = d_1 d_2 d_3 u_1^n u_2^n u_3^n (u_1 - 1)(u_2 - 1)(u_3 - 1)$$
$$\times (u_2 - u_1)^2 (u_3 - u_1)^2 (u_3 - u_2)^2. \tag{21.30}$$

With this, the determinants $\mathcal{D}_3$ and $\mathcal{N}_3$ from (21.21) now become

$$\mathcal{D}_3 = d_1 d_2 d_3 u_1^n u_2^n u_3^n (u_1 - 1)^2 (u_2 - 1)^2 (u_3 - 1)^2$$
$$\times (u_2 - u_1)^2 (u_3 - u_1)^2 (u_3 - u_2)^2 \tag{21.31}$$

$$\mathcal{N}_3 = d_1 d_2 d_3 u_1^n u_2^n u_3^n (u_1 - 1)(u_2 - 1)(u_3 - 1)$$
$$\times (u_2 - u_1)^2 (u_3 - u_1)^2 (u_3 - u_2)^2$$
$$\times \{\bar{c}_\infty [u_1 u_2 u_3 - (u_1 u_2 + u_1 u_3 + u_2 u_3) + (u_1 + u_2 + u_3) - 1]$$
$$+ [u_2 u_3 - (u_1 u_2 + u_1 u_3 + u_2 u_3) + (u_1^2 + u_1 u_2 + u_1 u_3) - u_1^2] d_1 u_1^{n+1}$$
$$+ [u_1 u_3 - (u_1 u_2 + u_1 u_3 + u_2 u_3) + (u_1 u_2 + u_2^2 + u_2 u_3) - u_2^2] d_2 u_2^{n+1}$$
$$+ [u_1 u_2 - (u_1 u_2 + u_1 u_3 + u_2 u_3) + (u_1 u_3 + u_2 u_3 + u_3^2) - u_3^2] d_3 u_3^{n+1}\}$$
$$= d_1 d_2 d_3 u_1^n u_2^n u_3^n (u_1 - 1)(u_2 - 1)(u_3 - 1)$$
$$\times (u_2 - u_1)^2 (u_3 - u_1)^2 (u_3 - u_2)^2$$
$$\times \{\bar{c}_\infty [u_1 u_2 u_3 - (u_1 u_2 + u_1 u_3 + u_2 u_3) + (u_1 + u_2 + u_3) - 1]$$
$$+ [(u_1 + u_2)(u_1 + u_3) - (u_1 + u_2)(u_1 + u_3)] d_1 u_1^{n+1}$$
$$+ [(u_1 + u_2)(u_2 + u_3) - (u_1 + u_2)(u_2 + u_3)] d_2 u_2^{n+1}$$
$$+ [(u_1 + u_3)(u_2 + u_3) - (u_1 + u_3)(u_2 + u_3)] d_3 u_3^{n+1}\}$$
$$= d_1 d_2 d_3 u_1^n u_2^n u_3^n (u_1 - 1)(u_2 - 1)(u_3 - 1)$$
$$\times (u_2 - u_1)^2 (u_3 - u_1)^2 (u_3 - u_2)^2 \{\bar{c}_\infty (u_1 - 1)(u_2 - 1)(u_3 - 1)$$
$$+ [(0 \cdot d_1) u_1^{n+1} + (0 \cdot d_2) u_2^{n+1} + (0 \cdot d_3) u_3^{n+1}]\}$$
$$= d_1 d_2 d_3 u_1^n u_2^n u_3^n (u_1 - 1)^2 (u_2 - 1)^2 (u_3 - 1)^2$$
$$\times (u_2 - u_1)^2 (u_3 - u_1)^2 (u_3 - u_2)^2 \bar{c}_\infty = \mathcal{D}_3 \bar{c}_\infty$$

$$\mathcal{N}_3 = \bar{c}_\infty \mathcal{D}_3. \tag{21.32}$$

Then, the Shanks transform $e_3(\bar{c}_n)$ from (21.21) finally yields

$$e_3(\bar{c}_n) = \frac{\mathcal{N}_3}{\mathcal{D}_3} = \bar{c}_\infty \quad \text{(QED)}. \tag{21.33}$$

21.4 Explicit filtering of K transients
$$c_n = d_1 u_1^n + d_2 u_2^n + d_3 u_3^n + \cdots + d_K u_K^n$$

These explicit derivations for $1 \leq K \leq 3$ can now be readily extended to the Kth order Shanks transform $e_K(\bar{c}_n)$ given in (19.42) for a general sequence $\{\bar{c}_n\}$ from (19.1). The task is to show, along the same lines as for the preceding three particular cases, that the application of the general transform $e_K(\bar{c}_n)$ to the sequence

$$\bar{c}_n = \bar{c}_\infty + (d_1 u_1^n + d_2 u_2^n + d_3 u_3^n + \cdots + d_K u_K^n)$$
$$= \bar{c}_\infty + \sum_{k=1}^{K} d_k u_k^n \equiv \bar{c}_\infty + c_n \tag{21.34}$$

will filter out completely all the K geometrically decaying transients $\{d_k u_k^n\}$ ($1 \leq k \leq K$) and, hence, reach the exact limiting value $\bar{c}_\infty$. To this end, it suffices to inspect our calculations for $e_1(\bar{c}_n)$, $e_2(\bar{c}_n)$ and $e_3(\bar{c}_n)$ which would yield the quotient $\mathcal{N}_K/\mathcal{D}_K$ for the general Kth transform $e_K(\bar{c}_n)$

$$\mathcal{D}_K = \mathbf{V}_K^2\{\bar{d}\bar{u}^n\}\mathbf{W}_K^2 \quad \bar{d} = \prod_{k=1}^{K} d_k \neq 0 \quad \bar{u} = \prod_{k=1}^{K} u_k \neq 0 \tag{21.35}$$

$$\mathcal{N}_K = \mathbf{V}_K^2\{\bar{d}\bar{u}^n\}\mathbf{W}_K(\bar{c}_\infty \mathbf{W}_K + c_n^\dagger) = \mathcal{D}_K \bar{c}_\infty \quad \mathbf{W}_K = \prod_{k=1}^{K}(u_k - 1) \tag{21.36}$$

$$c^\dagger \equiv 0 \cdot c_n = 0 \quad c_n = \sum_{k=1}^{K} d_k u_k^n \tag{21.37}$$

$$\therefore \quad e_K(\bar{c}_n) = \frac{\mathcal{N}_K}{\mathcal{D}_K} = \bar{c}_\infty \quad \text{(QED)} \tag{21.38}$$

where $\mathbf{V}_K$ is the Vandermonde determinant (7.14). As mentioned in chapter 7, $\bar{d}\bar{u}^n$ can be viewed as a single 'effective' transient. It should be emphasized that (21.38) is consistent with the scaling (19.57) which gives $e_K(\bar{c}_n) = e_K(\bar{c}_\infty + c_n) = \bar{c}_\infty + e_K(c_n) = \bar{c}_\infty$ provided that $e_K(c_n) = 0$. The equation $e_K(c_n) = 0$ indeed holds true, since the general result (21.38) for the auxiliary sequence $\{\bar{c}_n\}$ is also applicable to the time signal $\{c_n\}$ for which $\bar{c}_\infty = 0$ according to (21.34). In such a case (21.38) is reduced to $e_K(c_n) = 0$ as in (7.53)

$$e_K(c_n) = 0: \text{ the signature for exactly } K \text{ transients in signal } c_n. \tag{21.39}$$

Explicit filtering of K transients $c_n = d_1 u_1^n + d_2 u_2^n + d_3 u_3^n + \cdots + d_K u_K^n$ 159

The quantity $c_n^\dagger$ from (21.36) and (21.37) is the 'image' of the action of the Kth order Shanks transform onto c_n according to

$$e_K(c_n) = c_n^\dagger \qquad c_n^\dagger = \sum_{k=1}^{K} d_k^\dagger u_k^n = 0 \qquad d_k^\dagger = 0 \cdot d_k = 0. \qquad (21.40)$$

Thus, $e_K(c_n)$ is seen to act onto c_n as an annihilation operator. For this reason, in the prescription (21.40), the notation $c_n^\dagger$ is employed to symbolize 'annihilation' of all the K transients $\{d_k u_k^n\}_{k=1}^{K}$. Hence, the end result of subjecting the time sequence $\{c_n\} = \{\sum_{k=1}^{K} d_k u_k^n\}$ to the Kth order Shanks transform $e_K(c_n)$ is the complete annihilation of the input c_n. The concrete mechanism by which this annihilation is achieved is found here to be *the exact zeroing out of each of the K amplitudes* d_k $(1 \leq k \leq K)$ of the constituent transients $\{d_k u_k^n\}$ (see also chapter 57).

The interpretation of the conclusion (21.39) is the following: if the application of the Kth order Shanks transform $e_K(c_n)$ to a given tabulated time signal c_n (assumed to be comprised of exponentially damped sinusoidals) yields precisely zero, i.e. $e_K(c_n) = 0$ then we are sure that c_n has *exactly* K transients $\{d_k u_k^n\}$ $(1 \leq k \leq K)$. In such a case, the input data $\{c_n\}$ are said to strictly obey the mathematical model $c_n = \sum_{k=1}^{K} d_k u_k^n$. Of course, while verifying whether or not the condition $e_K(c_n) = 0$ is fulfilled, we have only the set $\{c_n\}$ available, and *not* the spectral parameters $\{u_k, d_k\}$. Nevertheless, the eventual finding $e_K(c_n) = 0$ would firmly establish the validity of the representation $c_n = \sum_{k=1}^{K} d_k u_k^n$ for the investigated time signal. This would, in turn, justify the subsequent spectral analysis, i.e. extraction of the spectral pairs $\{u_k, d_k\}_{k=1}^{K}$ from the given signal $\{c_n\}$. Clearly, we do not know the rank K in advance, but the signal c_n contains this information. Such information must be accessible and we can tease it out by computing a string of the Shanks transforms $\{e_m(c_n)\}$ for gradually increasing orders m. Then the first value of m (say m') for which we find that $e_{m'}(c_n) = 0$ will give the sought order $K = m'$. In this way, the Padé approximant through its time domain algorithm, the Shanks transform, emerges as a signal processor which is capable of determining the exact number K of transients in a given time signal c_n. Strictly speaking, this applies with certainty only to a time signal given by a linear combination of complex exponentials with constant complex amplitudes $c_n = \sum_{k=1}^{K} d_k \exp(-in\omega_k \tau)$, which leads to a sum of pure Lorentzians in the frequency domain, i.e. a non-degenerate spectrum. The same conclusion also extends to a degenerate spectrum for which one or more frequency dependent exponentials $u_k = \exp(-i\omega_k \tau)$ coincide with each other. This can be accomplished by, e.g. the limiting procedure $u_k \longrightarrow u_{k'}$ to the already performed anaysis for the non-degenerate case. The Padé degenerate (non-Lorentzian) spectrum corresponds to a non-stationary time signal. This latter signal is also a sum of damped exponentials $c_n = \sum_{k=1}^{J} d_{k,n} \exp(-in\omega_k \tau)$, but with the time-dependent amplitudes $d_{k,n} = \sum_{j=1}^{m_k} \tilde{d}_{k,j}[(n\tau)^{j-1}]$ where $m_k \geq 1$ $(m_1 + m_2 + \cdots + m_J = K)$ is the multiplicity of the kth quotient u_k and $\tilde{d}_{k,j_k}$ are some constants.

Chapter 22

The Lanczos recursive algorithm for state vectors $|\psi_n\rangle$

The Lanczos algorithm [93]–[102] is one of the most widely used eigenvalue solvers for large matrices. This recursive method is usually formulated with the dynamic operator $\hat{\Omega}$, but we shall presently use the evolution operator $\hat{U}(\tau)$. We shall study a general system and focus our attention on the resolvent operator $\hat{R}(u)$. This is the Green resolvent from (5.21) which is associated with the evolution operator $\hat{U}(\tau)$ from (5.6). The corresponding Green function $\mathcal{R}(u)$ is given in (5.22). We shall now perform a parallel study of the spectra of the operators $\hat{U}(\tau)$ and $\hat{R}(u)$. One of the possibilities for obtaining the spectrum of the evolution operator $\hat{U}(\tau)$ is to diagonalize directly the original eigenvalue problem (5.8) in a conveniently chosen complete *orthonormal* basis. Such a basis can be constructed from, e.g. the Lanczos states $\{|\psi_n\rangle\}$ that are associated with matrix $\mathbf{U} = \{U_{n,m}\}$. The Lanczos states $\{|\psi_n\rangle\}$ are presently generated from the recursion [17, 93, 97]

$$\beta_{n+1}|\psi_{n+1}\rangle = \{\hat{U}(\tau) - \alpha_n \hat{1}\}|\psi_n\rangle - \beta_n|\psi_{n-1}\rangle \qquad (n > 0) \qquad |\psi_0\rangle = |\Phi_0\rangle \tag{22.1}$$

$$\alpha_n = \frac{\langle\psi_n|\hat{U}(\tau)|\psi_n\rangle}{\langle\psi_n|\psi_n\rangle} \qquad \beta_n = \frac{\langle\psi_{n-1}|\hat{U}(\tau)|\psi_n\rangle}{\langle\psi_{n-1}|\psi_{n-1}\rangle} \qquad \beta_0 = 0. \tag{22.2}$$

The sum $\alpha_n|\psi_n\rangle + \beta_n|\psi_{n-1}\rangle$ appearing on the rhs of (22.1) represents the two orthogonalizing terms that are subtracted from $\hat{U}(\tau)|\psi_n\rangle$ to discard certain pathways from the intermediate state $|\psi_n\rangle$ such as the return to $|\Phi_0\rangle$. In applications, the Lanczos algorithm is customarily introduced with the matrix $\mathbf{U}(\tau)$ instead of the operator $\hat{U}(\tau)$ with the appropriate replacement of the wavefunction by the corresponding column state vector. In such a case the matrix-vector multiplication is carried out at each iteration and this is the most computer time consuming part of the matrix version of the Lanczos algorithm. Of course, there is a flexibility in the Lanczos algorithm, which permits the recursion

The Lanczos recursive algorithm for state vectors $|\psi_n)$ 161

to be carried out with tensors, matrices, operators or scalars [101, 102]. In chapter 34 we shall show that the operator version (22.1) can be used to establish the *analytical Lanczos algorithm* with an arbitrary state $|\psi_n)$ in an explicit form. The recurrence (22.1) is symmetric in the sense that the component $|\psi_{n+1})$ in $\hat{U}(\tau)|\psi_n)$ is the same as the constituent $|\psi_n)$ in $\hat{U}(\tau)|\psi_{n+1})$. Here, the general element $U_{n,m}$ of the evolution matrix $\mathbf{U}(\tau)$ is a complex symmetric matrix with the elements

$$U_{n,m}^{(s)} = (\psi_m|\hat{U}^s(\tau)\psi_n) = (\psi_n|\hat{U}^s(\tau)|\psi_m) = U_{m,n}^{(s)}. \qquad (22.3)$$

Of course, the matrix elements in the basis set of the Lanczos $\{|\psi_n)\}$ and Schrödinger $\{|\Phi_n)\}$ states are different but, for simplicity, we shall use the same formal label $U_{n,m}^{(s)}$ in both (5.13) and (22.3). Multiplying (22.1) throughout by the term $\beta_1\beta_2\cdots\beta_n$ leads to the results

$$(\beta_1\beta_2\cdots\beta_{n-1}\beta_n)\beta_{n+1}|\psi_{n+1})$$
$$= [\hat{U}(\tau) - \alpha_n](\beta_1\beta_2\cdots\beta_{n-1}\beta_n)|\psi_n) - (\beta_1\beta_2\cdots\beta_{n-1}\beta_n)\beta_n|\psi_{n-1})$$
$$[(\beta_1\beta_2\cdots\beta_{n-1}\beta_n\beta_{n+1})|\psi_{n+1})]$$
$$= [\hat{U}(\tau) - \alpha_n][(\beta_1\beta_2\cdots\beta_{n-1}\beta_n)|\psi_n)] - \beta_n^2[(\beta_1\beta_2\cdots\beta_{n-1})|\psi_{n-1})]$$
$$|\tilde{\psi}_{n+1}) = \{\hat{U}(\tau) - \alpha_n\}|\tilde{\psi}_n) - \beta_n^2|\tilde{\psi}_{n-1})$$
$$|\tilde{\psi}_n) \equiv \bar{\beta}_n|\psi_n) \qquad \bar{\beta}_n = \prod_{m=1}^n \beta_m \neq 0 \qquad n \geq 1 \qquad |\tilde{\psi}_0) = |\Phi_0). \qquad (22.4)$$

The alternative recursion (22.4) involves the monic Lanczos states $\{|\tilde{\psi}_n)\}$. Here, the term 'monic' serves to indicate that for any given integer n, the highest state $|\Phi_n)$ in the finite sum which defines the vector $|\tilde{\psi}_n)$ always has an overall multiplying coefficient equal to unity similarly to a monic polynomial [218]. The Lanczos states $\{|\tilde{\psi}_n)\}$ are orthogonal, but unnormalized, as opposed to orthonormalized Lanczos states $\{|\psi_n)\}$. By construction, both sequences $\{|\psi_n)\}$ and $\{|\tilde{\psi}_n)\}$ lead to certain linear combinations of powers of the operator $\hat{U}(\tau)$ acting on the initial state $|\Phi_0)$. Therefore, due to the relation $|\Phi_n) = \hat{U}^n(\tau)|\Phi_0)$ from (5.6), the vectors $\{|\psi_n)\}$ and $\{|\tilde{\psi}_n)\}$ are certain sums of Schrödinger states $\{|\Phi_n)\}$. For example, in addition to $|\tilde{\psi}_0) = |\Phi_0)$, we obtain

$$|\tilde{\psi}_1) = |\Phi_1) - \alpha_0|\Phi_0)$$
$$|\tilde{\psi}_2) = |\Phi_2) - (\alpha_0 + \alpha_1)|\Phi_1) + (\alpha_0\alpha_1 - \beta_1^2)|\Phi_0)$$
$$|\tilde{\psi}_3) = |\Phi_3) - (\alpha_0 + \alpha_1 + \alpha_2)|\Phi_2) + \{(\alpha_0\alpha_1 - \beta_1^2) + \alpha_2(\alpha_0 + \alpha_1) - \beta_2^2\}|\Phi_1)$$
$$+ \{\alpha_0\beta_2^2 - \alpha_2(\alpha_0\alpha_1 - \beta_1^2)\}|\Phi_0). \qquad (22.5)$$

Using (22.2), we computed analytically several coupling constants $\{\alpha_n, \beta_n\}$ as the functions of the signal points $\{c_n\}$ with the results

$$\alpha_0 = \frac{c_1}{c_0} \qquad \alpha_1 = \frac{c_0^2 c_3 - 2 c_0 c_1 c_2 + c_1^3}{c_0 (c_0 c_2 - c_1^2)}$$

$$\alpha_2 = [(2 c_1 c_2 c_3 + c_0 c_2 c_4 - c_0 c_3^2 - c_1^2 c_4 - c_2^3)(c_0 c_2 - c_1^2)]^{-1}$$
$$\times (c_0^2 c_2^2 c_5 - 2 c_0 c_1^2 c_2 c_5 + c_1^4 c_5 + 2 c_0 c_1 c_2^2 c_4 + 3 c_1^2 c_2^2 c_3 - 2 c_0 c_1 c_2 c_3^2$$
$$- 2 c_0^2 c_2 c_3 c_4 - 2 c_1^3 c_2 c_4 + 2 c_0 c_1^2 c_3 c_4 - c_1^2 c_3^2 - c_1 c_2^4 + c_0^2 c_3^3)$$

$$\beta_0 = 0 \qquad \beta_1^2 = \frac{c_0 c_2 - c_1^2}{c_0^2}$$

$$\beta_2^2 = c_0 \frac{2 c_1 c_2 c_3 + c_0 c_2 c_4 - c_0 c_3^2 - c_1^2 c_4 - c_2^3}{(c_0 c_2 - c_1^2)^2} \tag{22.6}$$

$$c_0^3 \beta_1^4 \beta_2^2 = 2 c_1 c_2 c_3 + c_0 c_2 c_4 - c_0 c_3^2 - c_1^2 c_4 - c_2^3. \tag{22.7}$$

Once the set $\{\alpha_n, \beta_n\}$ is generated, the matrix $\mathbf{U} = \{U_{n,m}\}$ becomes automatically available. This is seen by projecting (22.1) onto the state $(\psi_m|$ and using (22.3) for $s = 1$ to obtain $U_{n,m} = c_0(\alpha_n \delta_{n,m} + \beta_{n+1} \delta_{n+1,m} + \beta_n \delta_{n-1,m})$. Or by writing $\delta_{n-1,m}$ equivalently as $\delta_{n,m+1}$, we have [17, 97]

$$U_{n,m} = c_0(\alpha_n \delta_{n,m} + \beta_m \delta_{n+1,m} + \beta_n \delta_{n,m+1}) \equiv c_0 J_{n,m}. \tag{22.8}$$

Here, the elements $U_{n,m}$ are alternatively denoted by $c_0 J_{n,m}$ to indicate that the matrix $\mathbf{U}$ in the basis $\{|\psi_n\rangle\}$ is a tridiagonal matrix, which is equivalently called the Jacobi matrix or J-matrix and denoted by $\mathbf{J} = \{J_{n,m}\}$. Hence, in the Lanczos basis, the evolution matrix $\mathbf{U}$ automatically acquires its tridiagonal (codiagonal) form of the type of a J-matrix in a finite dimension, say $M \times M$, such that $\mathbf{U}_M = \text{trid}_M[\beta, \alpha, \beta] \equiv c_0 \mathbf{J}_M$. This is due to the definition of the basis $\{|\psi_n\rangle\}$ in which the above matrix element $U_{n,m}$ is equal to zero for $|m - n| > 1$ and otherwise $(\psi_m|\psi_m) = (\psi_0|\psi_0) = (\Phi_0|\Phi_0) = c_0 \neq 0$ for any non-negative integer m. Thus, in general, the orthogonality relation for the basis $\{|\psi_n\rangle\}$ is given by [17, 97]

$$(\psi_m|\psi_n) = c_0 \delta_{n,m} \tag{22.9}$$

as the explicit calculations of the first few overlaps of Lanczos states (22.1) indeed confirm. According to (22.9), both the denominators $(\psi_n|\psi_n)$ and $(\psi_{n-1}|\psi_{n-1})$ in the expression (22.2) for $\{\alpha_n, \beta_n\}$ are equal to $c_0 \neq 0$. It then follows from the above statements that the matrix $\mathbf{U}_M$ is zero everywhere except on the diagonal and the two codiagonals that are the subdiagonal and the superdiagonal whose elements reside immediately at the two opposite sides of the main diagonal [176]

$$\mathbf{U}_M = c_0 \mathbf{J}_M \qquad \mathbf{J}_M = \begin{pmatrix} \alpha_0 & \beta_1 & 0 & 0 & \cdots & \cdots & 0 \\ \beta_1 & \alpha_1 & \beta_2 & 0 & \cdots & \cdots & 0 \\ 0 & \beta_2 & \alpha_2 & \beta_3 & \cdots & \cdots & 0 \\ 0 & 0 & \beta_3 & \alpha_3 & \cdots & \cdots & 0 \\ \vdots & \vdots & \vdots & \ddots & \ddots & \ddots & \vdots \\ 0 & 0 & 0 & \cdots & \beta_{M-2} & \alpha_{M-2} & \beta_{M-1} \\ 0 & 0 & 0 & \cdots & 0 & \beta_{M-1} & \alpha_{M-1} \end{pmatrix}.$$
(22.10)

This is a band-structured matrix or a band-matrix which is in this particular tridiagonal form also called the Jacobi matrix. Projecting both sides of (22.1) onto $\beta_{n+1}\langle\psi_{n+1}|$ and using (22.9) we find

$$\beta_{n+1}^2 = \frac{\|\{\hat{U}(\tau) - \alpha_n \hat{1}\}|\psi_n\rangle - \beta_n|\psi_{n-1}\rangle\|^2}{c_0} \neq 0. \qquad (22.11)$$

This is an alternative definition of the Lanczos parameter β_n which is equivalent to (22.2). The expression (22.11) shows that β_{n+1}^2 is, in fact, proportional to the squared norm of the state $\{\hat{U}(\tau) - \alpha_n \hat{1}\}|\psi_n\rangle - \beta_n|\psi_{n-1}\rangle$. Returning to (22.10) it should be noted that there are many good programs available in the literature for finding eigenvalues and eigenvectors of the Jacobi matrix $\mathbf{J}$. Among these, we recommend the routines COMQR and F02AMF from EISPACK [177] and NAG [178] libraries, respectively. Of course, the matrix (22.10) in the studied model is an approximation to the exact evolution matrix $\mathbf{U}_N$. In principle, we should use a different label, say $\mathbf{U}_M^a = c_0 \mathbf{J}_M$, in (22.10) to distinguish this approximation from the exact matrix $\mathbf{U}_N$. Instead, to avoid clutter, it will be understood hereafter that the notation $\mathbf{U}_M = \{U_{n,m}\}_{n,m=0}^{M-1}$ refers to the approximation $\mathbf{U}_M^a$ which is $c_0 \mathbf{J}_M = \{c_0 J_{n,m}\}_{n,m=0}^{M-1}$ from (22.10). Moreover, the important thing to observe is that the approximation (22.10) to the exact evolution operator covers only a part of the generally infinite-dimensional vector space $\mathcal{H}$ for the general system under study. The relevant Lanczos subspace $\mathcal{L}_M = \text{span}\{|\psi_0\rangle, |\psi_1\rangle, |\psi_2\rangle, \ldots, |\psi_{M-1}\rangle\} \subset \mathcal{H}$ is spanned by the above-constructed set $\{|\psi_n\rangle\}$ ($1 \leq n \leq M$). Suppose that we wish to obtain the K eigenvalues $\{u_k\}_{k=1}^K$ of the evolution matrix $\mathbf{U}_N$. Despite the inherent limitations of the above model, by choosing M to be sufficiently large, the $\mathbf{J}$-matrix (22.10) will still be able to extract good approximate eigenvalues $\{u_k^M\}$ to the exact set $\{u_k\}$. This is the so-called *Lanczos phenomenon* which states that the exact $\mathbf{U}$-matrix and its approximation $c_0\mathbf{J}$ from (22.10) share a common set of eigenvalues provided that M is large enough [99]. In practice, M is usually chosen to be greater than K to assure that all the relevant eigenvalues are indeed extracted from the selected interval. The relation $M > K$ implies a local over-completeness of the basis $\{|\psi_n\rangle\}_{n=0}^{M-1}$ and this leads to the appearance of the so-called singular eigenvalues. Over-completeness means that the number of equations is larger than the number of unknown quantities ($M > K$). This mathematical ill-conditioning of the problem is usually tackled by several standard procedures, e.g. the Cholesky or the Householder decomposition or the SVD or the so-called QZ algorithm [176]–[179]. This is reminiscent of a situation encountered in solving the harmonic inversion problem through nonlinear fitting, where the lack of knowledge of K is handled by guessing an integer M which could lead to under-fitting ($M < K$) or over-fitting ($M > K$). Under-fitting leaves some $K - M$ physical (genuine) resonances undetected. Over-fitting yields some $M - K$ unphysical (spurious) resonances. Both situations are unacceptable from theoretical and practical standpoints. This highlights the necessity for having a method of known validity to determine the *exact* number K of resonances directly from the raw data that are available as the experimentally measured time signals $\{c_n\}$. In the FPT, however, the true number K of resonances is determined without any ambiguity from the uniqueness of the Padé polynomial quotient for a given series of the input signal points $\{c_n\}$ [17, 142].

In the Lanczos algorithm, a sufficiently large number M is not necessary only for obtaining accurate eigenvalues, but also for arriving at good eigenvectors if they are needed in the analysis. For a large enough M, the basis $\{|\psi_n\rangle\}_{n=0}^{M-1}$ will be capable of approximately spanning the vector space of the exact wavefunctions $\{|\Upsilon_k\rangle\}$ in the chosen frequency range. This circumstance permits arriving at an accurate approximation $|\Upsilon_k^M\rangle$ to the exact state vector $|\Upsilon_k\rangle$ by forming a linear combination of M functions $\{|\psi_n\rangle\}$ [17, 97]

$$|\Upsilon_k^M\rangle = \sum_{n=0}^{M-1} Q_{n,k} |\psi_n\rangle. \tag{22.12}$$

The unknown expansion coefficients $\{Q_{n,k}\}$ from (22.12) are the elements of the column matrix $\mathbf{Q}_k = \{Q_{n,k}\}$ with $Q_{n,k} \equiv Q_n(u_k)$. We insert (22.12) into the eigenvalue problem (5.8), then project the result onto the state vector $(\psi_m|$ and finally use the orthogonality condition (22.9) to write

$$\sum_{n=0}^{M-1} Q_{n,k} (\psi_m|\hat{U}(\tau)|\psi_n) = u_k \sum_{n=0}^{M-1} Q_{n,k} (\psi_m|\psi_n)$$

$$= c_0 u_k \sum_{n=0}^{M-1} Q_{n,k} \delta_{n,m} = c_0 u_k Q_{m,k}.$$

From here, with the help of (22.10), the eigenvalue problem for the **J**-matrix is deduced in the form of a finite system of linear equations

$$\sum_{n=0}^{M-1} J_{n,m} Q_{n,k} = u_k Q_{m,k} \qquad \mathbf{J}_M \mathbf{Q}_k = u_k \mathbf{Q}_k. \tag{22.13}$$

The eigenvector $\mathbf{Q}_k$ corresponds to the eigenvalue u_k in a finite chain model of the length M. As opposed to the Schrödinger states $\{|\Phi_n\rangle\}$, the Lanczos basis $\{|\psi_n\rangle\}$

produces the matrix elements $\{U_{n,m}\}$ as a *nonlinear* combination of c_n due to the presence of the normalization factor β_{n+1} of the state ψ_{n+1} in the recursion (22.1). The constant coefficients α_n and β_n are the coupling constants or the recurrence (or chain) parameters of the Lanczos states $\{|\psi_n\rangle\}$. The Lanczos algorithm (22.1) generates a wave packet propagation leading directly to tridiagonalization of the evolution matrix $\mathbf{U}(\tau)$. This method is such that $|\psi_{n-1}\rangle$ is automatically normalized to $c_0 \neq 0$ and orthogonalized to $|\psi_n\rangle$ and $|\psi_{n-1}\rangle$. However, by construction of the chain, it follows that $|\psi_{n+1}\rangle$ is also orthogonal to all the remaining previous elements $|\psi_{n-2}\rangle, \ldots, |\psi_0\rangle$ so that

$$\hat{1} = \sum_{k=1}^{K} |\psi_k\rangle(\psi_k|. \qquad (22.14)$$

The Lanczos algorithm (22.1) is a low-storage method, as opposed to the corresponding Gram–Schmidt orthogonalization (GSO) which uses all states at each stage of the computation. Otherwise, the final explicit results are rigorously the same in the GSO and the Lanczos orthogonalizations (see chapter 60). Physically, the state $|\psi_n\rangle$ is essentially the nth environment of $|\psi_0\rangle$. *But the coupling of $|\psi_n\rangle$ with its surroundings is assumed as significant only with the two nearest neighbours (environments) or 'orbitals' $|\psi_{n+1}\rangle$ and $|\psi_{n-1}\rangle$*. A useful consequence of this in practice is that the $(n+1)$st iteration in (22.1) yielding the state $\beta_{n+1}|\psi_{n+1}\rangle$ needs to store *only the two* preceding states $|\psi_{n-1}\rangle$ and $|\psi_n\rangle$, since all other vectors can safely be overwritten. This extreme storage economy is the key to the success of the scheme (22.1) relative to, e.g. the GSO, which requires a copy of the surrounding orbitals for each new generated state vector[1]. Hence, the recursion (22.1) which is also known as the nearest neighbour approximation, or the chain model, or the tight binding model, or localized orbital model is one of the ways to create a *local* representation of the 'Hamiltonian' $\hat{\Omega}$, which can be Hermitean or complex symmetric operator or the like. This model fulfils the plausible physical demand that the local density of states is predominantly determined by the local orbital itself, whereas the progressively more remote orbitals play a less important role. This is because the stronger effect on the state of interest is naturally expected to come from the nearest readily accessible orbitals without traversing many intermediate states that could act as barriers/obstacles to the passage. Thus the chain model with its Lanczos state recursion is a mathematical prescription for the physical concept of a local environment. Clearly, each successive orbital in the chain of orbitals will progressively cluster in the periphery of the environment of the initial orbital $|\psi_0\rangle = |\Phi_0\rangle$). The measure of the effect of this environment onto the investigated state of the system is given by the coupling parameters $\{\alpha_n, \beta_n\}$ of the recurrence (22.1). *It is then clear that the chain model, in fact, describes the*

---

[1] The majority of the elements of a large **J**-matrix are zero. For example, some 99.9% zeros can be found in a $10^4 \times 10^4$ matrix **J** from (22.10) of a typical size encountered in many problems in solid state physics to which the chain model has been extensively applied in the past [97].

overall evolution of the system from the given initial state $|\Phi_0\rangle$. The Lanczos algorithm converts the original, completely filled evolution matrix $\mathbf{U}_N$ into its Jacobi counterpart, a sparse matrix $c_0 \mathbf{J}_M$ of a considerably reduced dimension $M \ll N$. Strictly speaking, the respective dimensions of the original evolution matrix and its Jacobi counterpart are the same. In other words, tridiagonalization itself does not literally reduce the dimension of the original large problem, but rather it makes the underlying Jacobi matrix sparse. *However, the coupling parameters $\{\alpha_n, \beta_n\}$ that determine the elements of the obtained $\mathbf{J}$-matrix are of decreasing importance with increased dimension M.* This is, in turn, equivalent to an effective dimensionality reduction $M \ll N$ of the original problem of a large size N. As a result of the Lanczos conversion $\mathbf{U}_N \longrightarrow \mathbf{U}_M$, the transformed evolution matrix $\mathbf{U}_M = c_0 \mathbf{J}_M$ is capable of governing the motion of the studied system among its states in the vector space $\mathcal{L}_M \subset \mathcal{H}$ of a significantly reduced dimension relative to that for the original matrix $\mathbf{U}_N$. We may say that the Lanczos state localization is economical because it introduces only those intermediate states of $\hat{\Omega}$ or $\hat{U}(\tau)$ that are coupled to the initial state $|\Phi_0\rangle$. By construction, the chain $\{|\psi_n\rangle\}$ contains progressively less important environments of the initial state $|\Phi_0\rangle$ and this is adequate, since the essential physics is often ingrained in only the first few neighbouring states of $|\Phi_0\rangle$. This is also economical, as it actually implies that the full transformation leading to a long chain $\{|\psi_n\rangle\}$ is not necessary. Such a circumstance is reminiscent of the equivalent situation encountered in the time domain, where the envelope of the signal $c(t)$ decays exponentially as t increases. Moreover, not only that the signal $\{c_n\}$ possesses larger intensities $|d_k|$ at earlier recordings, but it also exhibits more rapid changes in oscillations. At large times, the intensities $\{d_k\}$ eventually decrease to the level where the signal becomes heavily corrupted with background noise. Thus, it is advantageous to complete the experimental encoding of the time signal reasonably quickly by trying to avoid the excessively long tail of $c(t)$, since with a large acquisition time T one will inevitably measure mainly noise in the asymptotic region of the variable t [76]. This is at variance with the request of the Fourier based spectroscopy, which demands a large T, since the resolution of the FFT is pre-assigned as $2\pi/T$.

The quest for the locality of operators $\hat{\Omega}$ or $\hat{U}$ is natural in view of the fact that, in practice, one rarely needs the whole informational content of the Green resolvent $\hat{R}(u)$, but rather only a pre-assigned matrix element [17]. An important example of the need for such a local information is the Green function $\mathcal{R}(u)$ from (5.22). Physically, $\mathcal{R}(u)$ describes the effect of the rest of the system onto its one selected part. Therefore, it is plausible that the examined local orbital itself should play the major role and that the successively more distant neighbours are expected to exhibit lesser effects. The Lanczos algorithm of nearest neighbours achieves this hierarchy of environments whose relative influence to the local density of states is explicitly weighted and displayed. Each element of the set $\{|\psi_n\rangle\}$ has the symmetry of $|\psi_0\rangle$ as a result of the repeated action of $\hat{\Omega}$ or $\hat{U}$ onto the initial state $|\psi_0\rangle = |\Phi_0\rangle$. If the set $\{|\psi_n\rangle\}$ is required to contain functions of different symmetry, it will be necessary to consider different initial orbitals. The chain (22.1) does not contain those orbitals $|\psi_n\rangle$ that are uncoupled to $|\psi_0\rangle$ indicating the zero survival probability $(\psi_n|\psi_0) = 0$ of the state vector $|\psi_n\rangle$ for $n \neq 0$ as in (22.9). Any matrix can be transformed into its corresponding Jacobi matrix. It is then clear that the chain model is equivalent to expressing the matrix **U** as the corresponding Jacobi matrix or tridiagonal matrix $c_0 \mathbf{J}$ [17, 97]. An original problem under study might be of a high dimension N, i.e. of a large number of degrees of freedom that could be strongly coupled to each other leading to serious storage and computer time consuming problems. In such a case, the standard diagonalization of the associated dynamic matrix $\mathbf{U}_N$ would require N^2 registers. This constraint can be dramatically relaxed to the amount of stores of only $2N$ if $\mathbf{U}_N$ possesses a local representation stored in a compact form (22.10), which emerges from the recursion (22.1).

Suppose that $\hat{\Omega} = \hat{H}$, where $\hat{H}$ is a Hermitean Hamiltonian of the studied system. Let the set $\{|\psi_n\rangle\}$ be generated from (22.1) with $\hat{U}(\tau)$ replaced by $\hat{H}$. In such a case, the coupling parameters $\{\alpha_n, \beta_n\}$ would have a direct physical meaning and interpretation. An example of a physical system, which is itself a realization of the chain model, is the one-dimensional ion crystallization of multiple charged ions in a storage ring with electron cooling. Such a special crystallization of ions in storage rings has recently been achieved experimentally [257]. A multi-electron ionic system of this type, which represents a one-dimensional chain of clustered ions, could be treated within the frozen-core approximation, in which only one electron is active, whereas the other electrons have a passive role consisting of screening the nuclear charge. The Hamiltonian of this one-electron model in the case of a semi-infinite strip $(-\infty, n)$ could be readily set up and spectrally analysed in the chain model. In the simplifying circumstance of a one-dimensional chain of hydrogen-like ions, the chain model would allow the electron to hop from one orbital (e.g. an s-orbital with zero value of the angular momentum) to another orbital of a neighbouring ion. Here, all the recurrence parameters $\{\alpha_n\}$ are the same constant, equal to a given s-orbital energy, whereas the couplings $\{\beta_n\}$ are the 'hopping' elements from one ion to the next. According to (22.11), the parameter β_{n+1} is proportional to the norm of the state $|\psi_{n+1}\rangle = [\{\hat{H} - \alpha_n \hat{1}\}|\psi_n\rangle - \beta_n |\psi_{n-1}\rangle]/\beta_{n+1}$ for $\hat{\Omega} = \hat{H}$. Therefore, the Lanczos coupling constants β_{n+1} and β_{n-1} can be considered as a measure of the strength of the interaction between the states $|\psi_n\rangle$ and $|\psi_{n\pm 1}\rangle$. Likewise, α_n is the self-energy of the orbital $|\psi_n\rangle$. The state vector $|\psi_n\rangle$ can be viewed as a linear combination of the wavefunctions $\hat{H}^n|\psi_0\rangle$ and this itself might be considered as the sum of all the pathways consisting of n steps, starting from the initial configuration $|\psi_0\rangle = |\Phi_0\rangle$. Here, the term 'step' means a hop from one orbital to another, connected by the non-zero hopping matrix element of $\hat{H}$ with the assumption that the steps between the nearest neighbours are the most important.

Chapter 23

The Lanczos orthogonal polynomials $P_n(u)$ and $Q_n(u)$

The Lanczos algorithm of tridiagonalization is the key step for converting the original (presumably large) matrix $\mathbf{U}_N$ into its corresponding sparse $\mathbf{J}$-matrix $c_0 \mathbf{J}_M$ as given in (22.10) [97]. The kth eigenstate of the $\mathbf{J}$-matrix is given by the vector $\mathbf{Q}_k$ which corresponds to the eigenvalue u_k. By definition, the quantities $\{\mathbf{Q}_k\}$ are the column vectors $\mathbf{Q}_k = \{Q_{n,k}\}$ whose elements $Q_{n,k}$ are the expansion coefficients of the unnormalized total eigenstate $|\Upsilon_k^\infty)$ developed in the complete Lanczos set $\{|\psi_n)\}$ of the most general infinite chain model

$$|\Upsilon_k^\infty) = \sum_{n=0}^{\infty} Q_{n,k} |\psi_n) \qquad Q_n(u_k) \equiv Q_{n,k} \qquad (23.1)$$

which is an extension of (22.12) to the infinite chain ($M = \infty$). This is an unnormalized state vector. Its norm is obtained by using the orthogonalization from (3.6) yielding

$$\|\Upsilon_k^\infty\|^2 = (\Upsilon_k^\infty | \Upsilon_k^\infty) = c_0 \sum_{n=0}^{\infty} Q_{n,k}^2. \qquad (23.2)$$

The normalized total wavefunction $|\Upsilon_k^+)$ is determined by

$$|\Upsilon_k^+) = \|\Upsilon_k^\infty\|^{-1} |\Upsilon_k^\infty) = \left[c_0 \sum_{n=0}^{\infty} Q_{n,k}^2 \right]^{-1/2} |\Upsilon_k^\infty) \qquad (23.3)$$

where $|\Upsilon_k^\infty)$ is taken from (23.1). Substituting the state vector (23.1) into (5.8), which is afterwards multiplied by $(\psi_m|$ from the left, an infinite system of linear equations follows upon using (22.9)

$$\sum_{n=0}^{\infty} J_{n,m} Q_{n,k} = u_k Q_{m,k}$$

$$\mathbf{J}_\infty \mathbf{Q}_k = u_k \mathbf{Q}_k. \qquad (23.4)$$

This is an ordinary eigenvalue problem in which the tridiagonal Jacobi matrix $\mathbf{J}_\infty$ is given in (22.10) with $M = \infty$. The residues $\{d_k\}$ are defined by (3.10) where $|\Upsilon_k\rangle$ is the exact complete state vector normalized to $c_0 \neq 0$ according to (3.8). The same type of definition is valid for an approximation such as (23.1) provided that normalization is properly included according to (23.3)

$$d_k = (\Phi_0|\Upsilon_k^+)^2. \tag{23.5}$$

Inserting (23.1) into (23.3) and substituting afterwards the obtained result into (23.5) yields

$$d_k^\infty = \frac{c_0 Q_{0,k}^2}{\sum_{n=0}^\infty Q_{n,k}^2} \tag{23.6}$$

where the superscript ∞ in the residue d_k^∞ is introduced to point at the infinite chain ($M = \infty$). Comparing (23.2) and (23.6), it follows

$$d_k^\infty = \frac{c_0^2 Q_{0,k}^2}{\|\Upsilon_k^\infty\|^2}. \tag{23.7}$$

It should be emphasized that we do not need to solve (23.4) to obtain the wavefunction $|\Upsilon_k^\infty\rangle$ from (23.1). This is because the whole sequence $\{Q_{n,k}\}$ can also be generated recursively using the already obtained sets $\{\alpha_n, \beta_n\}$ and $\{u_k\}$. To this end, we substitute (23.1) into the time-independent Schrödinger equation (5.8) and exploit the linearity of $\hat{U}(\tau)$ to write

$$\hat{U}(\tau) \sum_{n=0}^\infty Q_{n,k}|\psi_n\rangle = u_k \sum_{n=0}^\infty Q_{n,k}|\psi_n\rangle. \tag{23.8}$$

Eliminating the term $\hat{U}(\tau)|\psi_n\rangle$ from the lhs of (23.8) by means of the recursion (22.1), we get

$$\sum_{n=0}^\infty Q_{n,k}\{\alpha_n|\psi_n\rangle + \beta_{n+1}|\psi_{n+1}\rangle + \beta_n|\psi_{n-1}\rangle\} = u_k \sum_{n=0}^\infty Q_{n,k}|\psi_n\rangle. \tag{23.9}$$

From here, the sought recursion for $\{Q_{n,k}\}$ can immediately be extracted if the whole inner part of the sum on the lhs of (23.9) could also be reduced to the nth state $|\psi_n\rangle$ just as in $u_k \sum_{n=0}^\infty Q_{n,k}|\psi_n\rangle$ on the rhs of the same equation. Then a comparison of the multipliers of the same vector $|\psi_n\rangle$ would give the sought result. This is indeed possible by the index changes $n \pm 1 \longrightarrow n'$ in the two terms containing $|\psi_{n\pm 1}\rangle$ on the lhs of (23.9), so that

$$\sum_{n=0}^\infty v_n^{(k)}|\psi_n\rangle \equiv \sum_{n=0}^\infty \{(\alpha_n - u_k)Q_n(u_k)$$
$$+ \beta_{n+1}Q_{n+1}(u_k) + \beta_n Q_{n-1}(u_k)\}|\psi_n\rangle = |0\rangle \tag{23.10}$$

where the relation $\beta_0 = 0$ from (22.2) is used along with the defining initialization $Q_{-1}(u_k) \equiv 0$. In (23.10), the symbol $|0\rangle$ denotes the zero state vector as before. Since the set $\{|\psi_n\rangle\}$ is a basis, its elements $|\psi_n\rangle$ are linearly independent. This implies that (23.10) can be satisfied only if all the auxiliary coefficients $\{v_n^{(k)}\}$ multiplying the nth term $|\psi_n\rangle$ are equal to zero. This yields the Lanczos recursion for the state vectors $\mathbf{Q}_k = \{Q_n(u_k)\}$ corresponding to the eigenvalues $\{u_k\}$

$$\beta_{n+1} Q_{n+1}(u_k) = (u_k - \alpha_n) Q_n(u_k) - \beta_n Q_{n-1}(u_k)$$
$$Q_{-1}(u_k) = 0 \qquad Q_0(u_k) = 1. \qquad (23.11)$$

This recurrence relation and the available triple set $\{\alpha_n, \beta_n; u_k\}$ are sufficient to determine completely the state vector $\mathbf{Q}_k$ without any diagonalization of the associated Jacobi matrix $\mathbf{U} = c_0 \mathbf{J}$ which is given in (22.10). Of course, diagonalization of $\mathbf{J}$ might be used to obtain the eigenvalues $\{u_k\}$, but this is not the only approach at our disposal. Alternatively, the same set $\{u_k\}$ is also obtainable by rooting the characteristic polynomial or eigenpolynomial from a secular equation[1], which is entirely equivalent to the eigenvalue problem (5.8) [17]. Moreover, this characteristic polynomial is, in fact, proportional to $Q_K(u)$. Hence, the eigenvalue set $\{u_k\}_{k=1}^K$ coincides with the roots of the following characteristic equation

$$Q_K(u_k) = 0 \qquad 1 \le k \le K. \qquad (23.12)$$

Returning to (23.2) and (23.6), we combine $Q_0(u_k) = 1$ from (23.11) to obtain the following expression for the residue d_k

$$d_k^\infty = \frac{c_0}{\sum_{n=0}^\infty Q_{n,k}^2} \implies d_k^\infty = \frac{c_0^2}{\|\Upsilon_k^\infty\|^2}. \qquad (23.13)$$

The Green operator $\hat{G}(\omega)$ from (3.9) can provide information about the studied physical system at any frequency ω and not just at ω_k that belong to $|\Upsilon_k\rangle$, as opposed to the Schrödinger eigenvalue problem (3.5). This is an apparent paradox, since the same operator $\hat{\Omega}$ constitutes the only source of information in both the Green and the Schrödinger formalisms. Indeed the Green function is known to be a powerful tool which can analytically continue the Schrödinger spectrum $\{\omega_k\}$ to encompass any value of ω. But the same ought to be also true for the stationary Schrödinger state vector, which one should be able to compute numerically at any ω, in addition to the eigenvalues ω_k as permitted by the argument of analytical continuation [28, 72, 75]. This follows from the fact that, once the Green function is available at a fixed frequency ω, one can compute the

---

[1] As noted before, the optimal method for finding all the roots of a given polynomial is the use of the corresponding Hessenberg matrix to solve its well-conditioned eigenvalue problem [17, 176]. By contrast, any direct rooting of polynomials is known to be an ill-conditioned procedure as discussed in chapter 28.

corresponding state vector at the same ω by means of certain well-known integral transforms [145, 258]. Alternatively, we can show that (23.11) is valid at an arbitrary value of $u = \exp(-i\omega\tau)$ and not just at $u_k = \exp(-i\omega_k\tau)$, if we have a straightforward definition of the stationary Schrödinger state $|\Upsilon^\infty(u))$ for a given u [97]. Fixing $u = u_k$ here would retrieve the eigenstate (23.1) corresponding to the eigenvalue u_k. Similar to the pure eigenvector (23.1), the state $|\Upsilon^\infty(u))$ at any u can also be conceived as a linear combination of the Lanczos basis states

$$|\Upsilon^\infty(u)) = \sum_{n=0}^{\infty} Q_n(u)|\psi_n). \tag{23.14}$$

The quantity $|\Upsilon^\infty(u))$ will describe an extended spectrum of $\hat{U}(\tau)$, beyond the Schrödinger eigenvalue set $\{u_k\}$, if the image $\hat{U}(\tau)|\Upsilon^\infty(u))$ of the operator $\hat{U}(\tau)$ is proportional to the same vector $|\Upsilon^\infty(u))$, for example

$$\hat{U}(\tau)|\Upsilon^\infty(u)) = u|\Upsilon^\infty(u)) \tag{23.15}$$

which happens to be true. This is nothing but rephrasing the stationary Schrödinger equation at any u, and the Schrödinger eigenvalue problem (5.8) follows at once by the specification to the ridge $u = u_k$ as previously pointed out in [17, 97]. Likewise, evaluating $|\Upsilon^\infty(u))$ at the eigenvalue $u = u_k$ leads to $|\Upsilon_k^\infty) \equiv |\Upsilon^\infty(u_k))$, where $|\Upsilon_k^\infty)$ is the eigenvector as in (23.1). Adopting the infinite chain model and proceeding exactly along the lines (23.8) and (23.10) by using u in lieu of u_k we obtain the Lanczos recursive algorithm for $Q_n(u)$

$$\beta_{n+1} Q_{n+1}(u) = (u - \alpha_n) Q_n(u) - \beta_n Q_{n-1}(u)$$
$$Q_{-1}(u) = 0 \qquad Q_0(u) = 1. \tag{23.16}$$

We see that the expansion coefficients in (23.14) are polynomials of degree n in the complex variable u. The same polynomials $Q_n(u_k) = Q_{n,k}$ evaluated at $u = u_k$ are also present in (23.1). Having obtained the whole set $\{Q_n(u)\}$ from the recursion (23.16), the usage of (23.14) would give the sought state vector $|\Upsilon^\infty(u))$ at any u. This is a simple computational tool, which resolves the above paradox, making the stationary Schrödinger equation (23.15) also formally usable at any u, as is the case with the corresponding Green function from (5.22) [17, 97].

It is possible to interconnect the matrices $\mathbf{J}$ and $\mathbf{Q}_k$ through a compact matrix equation [97]. This follows from the tridiagonal structure of the matrix in (22.10), which allows one to deduce (23.11) directly from the product $\mathbf{JQ}_k$ at the set of the eigenvalues $\{u_k\}$. By the same reasoning, this conclusion can also be extended to encompass (23.16) at values of u other than u_k. Thus the polynomials $\{Q_v(u)\}$

satisfy the matrix equation [259]

$$\begin{pmatrix} u-\alpha_0 & -\beta_1 & 0 & 0 & \cdots & 0 \\ -\beta_1 & u-\alpha_1 & -\beta_2 & 0 & \cdots & 0 \\ 0 & -\beta_2 & u-\alpha_2 & -\beta_3 & \cdots & 0 \\ \vdots & \vdots & \ddots & \ddots & \ddots & \vdots \\ 0 & 0 & 0 & -\beta_{n-1} & u-\alpha_{n-1} & -\beta_n \\ 0 & 0 & 0 & \cdots & -\beta_n & u-\alpha_n \end{pmatrix} \begin{pmatrix} Q_0(u) \\ Q_1(u) \\ Q_2(u) \\ \vdots \\ Q_{n-1}(u) \\ Q_n(u) \end{pmatrix}$$

$$= \begin{pmatrix} 0 \\ 0 \\ 0 \\ \vdots \\ 0 \\ \beta_{n+1} Q_{n+1}(u) \end{pmatrix} \qquad (23.17)$$

where a finite sequence of $n+1$ polynomials $\{Q_\nu\}_{\nu=0}^n$ is considered. Obtaining the results of the indicated multiplication on the lhs of (23.17) and comparing them with the elements of the column matrix from the rhs of the same equation, we readily arrive at (23.16).

The recursion (23.16) can be extended to operators and matrices. This is done by using the Cayley–Hamilton theorem [228], which states that for a given analytic scalar function $f(u)$, the expression for its operator counterpart $f(\hat{U})$ is obtained via replacement of u by $\hat{U}$ as in (3.5). In this way, we can introduce the Lanczos operator and matrix polynomials defined by the recursions [260]

$$\beta_{n+1} Q_{n+1}(\hat{U}) = (\hat{U} - \alpha_n \hat{1}) Q_n(\hat{U}) - \beta_n Q_{n-1}(\hat{U})$$
$$Q_{-1}(\hat{U}) = 0 \qquad Q_0(\hat{U}) = \hat{1}$$
$$\beta_{n+1} Q_{n+1}(\mathbf{U}) = (\mathbf{U} - \alpha_n \mathbf{1}) Q_n(\mathbf{U}) - \beta_n Q_{n-1}(\mathbf{U})$$
$$Q_{-1}(\mathbf{U}) = 0 \qquad Q_0(\mathbf{U}) = \mathbf{1} \qquad (23.18)$$

where $\mathbf{0}$ and $\mathbf{1}$ are the zero and unit matrices of the same dimension as $\mathbf{U}$. The eigenstates $\{Q_n(u_k)\}$ of the **J**-matrix from (22.10) are the regular solutions of the three-term contiguous recurrence (23.11). This recurrence is a difference equation which is a discrete counterpart of the corresponding second-order differential equation. Given the recursions (22.1) for $|\psi_n\rangle$ and (23.18) for $Q_n(\hat{U})$, we can observe inductively that the following relationship is valid

$$|\psi_n\rangle = Q_n(\hat{U})|\psi_0\rangle \qquad (23.19)$$

where $\hat{U} \equiv \hat{U}(\tau)$ and $|\psi_0\rangle = |\Phi_0\rangle$. This equation defines the Lanczos polynomial propagation of the initial wave packet $|\Phi_0\rangle$. Within second-order differential equations satisfied by the classical orthogonal polynomials [180], we always introduce a pair of solutions, one regular and the other irregular, called the polynomials of the first and the second kind, respectively. Using the analogy between differential and difference equations, we can identify the solution $Q_n(u)$ of the difference equation (23.16) as the Lanczos polynomial of the first kind. Likewise, the same equation has another solution which is the Lanczos polynomial of the second kind $P_n(u)$. In other words, $P_n(u)$ satisfies the same recurrence (23.16), so that [17, 97]

$$\beta_{n+1} P_{n+1}(u) = (u - \alpha_n) P_n(u) - \beta_n P_{n-1}(u)$$
$$P_0(u) = 0 \qquad P_1(u) = 1. \qquad (23.20)$$

Notice that the initial conditions for $P_n(u)$ are different from those for $Q_n(u)$ as seen in (23.16) and (23.20). In particular, the initialization to the recurrence (23.20) scales the degree of the polynomial $P_n(u)$ downward by 1. This implies that $P_n(u)$ is a polynomial of degree $n - 1$, as opposed to $Q_n(u)$, which is the nth degree polynomial. Thus the polynomial $P_n(u)$ has $n - 1$ zeros and that is by one fewer than in the case of $Q_n(u)$.

Chapter 24

Recursions for derivatives of the Lanczos polynomials

Once the polynomial set $\{Q_n(u)\}$ becomes available from (23.16), the same type of recursion can automatically be derived for general derivatives $\{(d/du)^m Q_n(u)\}$. This is greatly simplified by the following relation

$$\left(\frac{d}{du}\right)^m [u Q_n(u)] = u Q_{n,m}(u) + m Q_{n,m-1}(u) \qquad (24.1)$$

$$Q_{n,m}(u) = \left(\frac{d}{du}\right)^m Q_n(u) \qquad Q_{n,0}(u) = Q_n(u) \qquad (24.2)$$

$$Q_{n,m}(0) = \left[\left(\frac{d}{du}\right)^m Q_n(u)\right]_{u=0}. \qquad (24.3)$$

The expression (24.1) can easily be established in an inductive manner as pointed out in [261]. Hereafter, we use the notation $Q_{n,s}(u)$ for the sth derivative of $Q_n(u)$ with respect to u instead of the more standard notation $Q_n^{(s)}(u)$. The reason for this is that the notation $Q_n^{(s)}(u)$ is customarily employed in the theory and applications of the so-called delayed orthogonal polynomials constructed from power moments or auto-correlation functions or signal points $\{\mu_{n+s}\} = \{C_{n+s}\} = \{c_{n+s}\}$ where the first $s < N$ points $\{\mu_r\} = \{C_r\} = \{c_r\}$ ($0 \le r \le s-1$) are either skipped or simply missing from the full set of N elements. We shall thoroughly study the delayed time signals $\{c_{n+s}\}$ in chapters 65–72. If the expression (23.16) is differentiated m times and (24.1) is used, the following recursion is obtained inductively for the derivatives $\{Q_{n,m}(u)\}$ [17]

$$\beta_{n+1} Q_{n+1,m}(u) = (u - \alpha_n) Q_{n,m}(u) - \beta_n Q_{n-1,m}(u) + m Q_{n,m-1}(u) \quad (24.4)$$

$$Q_{-1,m}(u) = 0 \qquad Q_{0,0}(u) = Q_0(u) = 1. \qquad (24.5)$$

This is a fast and accurate way of computing derivatives $\{Q_{n,m}(u)\}$ of any order $m > 0$. Likewise, the availability of the polynomials $\{P_n(u)\}$ from (23.20) enables the derivation of a recursive way of computing derivatives $\{P_{n,m}(u)\}$ via the analogy with (24.4)

$$\beta_{n+1} P_{n+1,m}(u) = (u - \alpha_n) P_{n,m}(u) - \beta_n P_{n-1,m}(u) + m P_{n,m-1}(u) \quad (24.6)$$

$$P_{0,m}(u) = 0 \qquad P_{1,0}(u) = P_1(u) = 1 \quad (24.7)$$

$$P_{n,m}(u) = \left(\frac{d}{du}\right)^m P_n(u) \qquad P_{n,0}(u) = P_n(u) \quad (24.8)$$

$$P_{n,m}(0) = \left[\left(\frac{d}{du}\right)^m P_n(u)\right]_{u=0}. \quad (24.9)$$

Chapter 25

The secular equation and the characteristic polynomial

The eigenequation (23.4) can formally be rewritten as a problem of finding zeros of the matrix equation $(u\mathbf{1}-\mathbf{U})\mathbf{Q}_k = \mathbf{0}$. This matrix equation can operationally be solved by conversion to its corresponding scalar counterpart, i.e. a determinantal equation known as the secular equation [212]

$$\det[u\mathbf{1} - \mathbf{U}] = 0. \tag{25.1}$$

The roots $\{u_k\}_{k=1}^K$ of (25.1) coincide with the eigenvalues of matrix $\mathbf{U}$. On the other hand, the function $\det[u\mathbf{1} - \mathbf{U}]$ is proportional to its Kth degree characteristic polynomial, which is equal to $Q_K(u)$ up to an overall multiplicative constant of the normalization type

$$Q_K(u) = \frac{\det[u\mathbf{1} - \mathbf{U}]}{\bar{\beta}_K} \tag{25.2}$$

where $\bar{\beta}_K$ is defined by (22.4). The K zeros $\{u_k\}_{k=1}^K$ of the characteristic polynomial $Q_K(u)$ coincide with the K eigenvalues of the U-matrix (22.10) for $M = K$. Let us now link the secular equation with the Lanczos recursion. To this end, we insert (23.14) into (23.15) and then multiply the results from the left by $(\psi_m|$ to obtain the following system of linear equations

$$\sum_{n=0}^{\infty} U_{n,m} Q_n(u) = u Q_m(u). \tag{25.3}$$

The tridiagonal structure (22.8) of the elements $U_{n,m}$ implies that the matrix form of (25.3) coincides with (23.17) if the matrix of the infinite dimension in (25.3) is appropriately truncated. Then, let the determinant of the tridiagonal matrix

The secular equation and the characteristic polynomial 177

$\text{trid}_{n+1}[\beta, \alpha - u, \beta]$ for a finite chain be denoted by $\tilde{Q}_{n+1}(u)$

$$\tilde{Q}_{n+1}(u) = \begin{vmatrix} u-\alpha_0 & -\beta_1 & 0 & 0 & \cdots & 0 \\ -\beta_1 & u-\alpha_1 & -\beta_2 & 0 & \cdots & 0 \\ 0 & -\beta_2 & u-\alpha_2 & -\beta_3 & \cdots & 0 \\ \vdots & \vdots & \vdots & \vdots & \ddots & \vdots \\ 0 & 0 & 0 & -\beta_{n-1} & u-\alpha_{n-1} & -\beta_n \\ 0 & 0 & 0 & \cdots & -\beta_n & u-\alpha_n \end{vmatrix}.$$

(25.4)

As usual, one solves (25.3) by first finding the zero of the determinant

$$\tilde{Q}_K(u_k) = 0 \Longrightarrow \omega_k = i\tau^{-1}\ln(u_k) \tag{25.5}$$

and then one searches for the eigenvectors [97]. Let us assume that the determinants $\tilde{Q}_{n-1}(u)$ and $\tilde{Q}_n(u)$ are known and we wish to obtain $\tilde{Q}_{n+1}(u)$. For this purpose, we use the fact that the determinant in (25.4) is sparse and has only two non-zero elements in the last $(n + 1)$st row such as $-\beta_n$ and $u - \alpha_n$. This means that the Cauchy expansion theorem for determinants employed to develop $\tilde{Q}_{n+1}$ in the elements of the $(n + 1)$st row will yield the sum of only two determinants

$\tilde{Q}_{n+1}(u) = (u - \alpha_n)$

$$\times \begin{vmatrix} u-\alpha_0 & -\beta_1 & 0 & 0 & \cdots & 0 \\ -\beta_1 & u-\alpha_1 & -\beta_2 & 0 & \cdots & 0 \\ 0 & -\beta_2 & u-\alpha_2 & -\beta_3 & \cdots & 0 \\ \vdots & \vdots & \vdots & \vdots & \ddots & \vdots \\ 0 & 0 & 0 & -\beta_{n-2} & u-\alpha_{n-2} & -\beta_{n-1} \\ 0 & 0 & 0 & \cdots & -\beta_{n-1} & u-\alpha_{n-1} \end{vmatrix}$$

$$+ \beta_n \begin{vmatrix} u-\alpha_0 & -\beta_1 & 0 & 0 & \cdots & 0 \\ -\beta_1 & u-\alpha_1 & -\beta_2 & 0 & \cdots & 0 \\ 0 & -\beta_2 & u-\alpha_2 & -\beta_3 & \cdots & 0 \\ \vdots & \vdots & \vdots & \vdots & \ddots & \vdots \\ 0 & 0 & 0 & -\beta_{n-2} & u-\alpha_{n-2} & 0 \\ 0 & 0 & 0 & \cdots & -\beta_{n-1} & -\beta_n \end{vmatrix}$$

$= (u - \alpha_n)\tilde{Q}_n(u)$

$$- \beta_n^2 \begin{vmatrix} u-\alpha_0 & -\beta_1 & 0 & 0 & \cdots & 0 \\ -\beta_1 & u-\alpha_1 & -\beta_2 & 0 & \cdots & 0 \\ 0 & -\beta_2 & u-\alpha_2 & -\beta_3 & \cdots & 0 \\ \vdots & \vdots & \vdots & \vdots & \ddots & \vdots \\ 0 & 0 & 0 & -\beta_{n-3} & u-\alpha_{n-3} & -\beta_{n-2} \\ 0 & 0 & 0 & \cdots & -\beta_{n-2} & u-\alpha_{n-2} \end{vmatrix}.$$

(25.6)

In the first line of (25.6), the determinant multiplying the term $u - \alpha_n$ is equal to $\tilde{Q}_n(u)$ according to (25.4). In the second line of (25.6), the determinant multiplying β_n is expanded in the elements of its end row from which two additional determinants emerge, one with the multiplying coefficient $-\beta_n$, and the other having the overall multiplicative factor β_{n-1}. However, only the former determinant is seen in the third line of (25.6) thus giving $-\beta_n^2 \tilde{Q}_{n-1}(u)$, whereas the determinant multiplying the term β_{n-1} is zero, since its last column contains only zero elements. Therefore

$$\tilde{Q}_{n+1}(u) = (u - \alpha_n)\tilde{Q}_n(u) - \beta_n^2 \tilde{Q}_{n-1}(u)$$
$$\tilde{Q}_{-1}(u) = 0 \qquad \tilde{Q}_0(u) = 1 \qquad (25.7)$$

and this coincides precisely with the definition of the Askey–Wilson polynomials [262]

$$X_{n+1}(z) - (z - a_n)X_n(z) + b_n^2 X_{n-1}(z) = 0. \qquad (25.8)$$

Multiplying both sides of (23.16) by the term $\beta_1 \beta_2 \cdots \beta_n$ and proceeding the same way as in deriving the recursion (22.4) for the state vectors $\{|\psi_n\rangle\}$, it follows

$$[(\beta_1 \beta_2 \cdots \beta_{n-1} \beta_n \beta_{n+1}) Q_{n+1}(u)] = (u - \alpha_n)[(\beta_1 \beta_2 \cdots \beta_{n-1} \beta_n) Q_n(u)]$$
$$- \beta_n^2 [(\beta_1 \beta_2 \cdots \beta_{n-1}) Q_{n-1}(u)]$$

and this will be reduced to (25.7) if we define

$$\tilde{Q}_n(u) \equiv \bar{\beta}_n Q_n(u) \qquad (25.9)$$

where $\bar{\beta}_n = \beta_1 \beta_2 \cdots \beta_n \neq 0$ as in (22.4). Therefore, the determinant $\tilde{Q}_n(u)$ is also a polynomial of degree n in variable u. Recalling (23.1) we see that every total eigenstate $|\Upsilon_k^\infty\rangle$ is proportional to the sum of the products of the determinant $Q_{n,k} = Q_n(u_k)$ and the Lanczos vector $|\psi_n\rangle$. Taking into account (25.5) and (25.9), it follows that the K eigenvalues $\{u_k\}$ of the chain of K states can equivalently be conceived as the roots of the polynomial $Q_K(u)$ of the Kth degree

$$Q_K(u_k) = 0 \qquad Q_K(u) = \frac{1}{\beta_K}[(u - \alpha_{K-1})Q_{K-1}(u) - \beta_{K-1}Q_{K-2}(u)]$$
$$(25.10)$$

as in (23.12). This is also clear from (23.17) which links the matrix $\mathbf{J}_n$ with the polynomials $\{Q_n(u)\}$. The eigenvalue problem (23.4) will coincide exactly with (23.17) at $u = u_k$ and $n = K - 1$ if $Q_K(u_k) = 0$ in which case the rhs of (23.17) is the column zero vector. Hence, the eigenvalues $\{u_k\}$ can be found either by diagonalizing the $\mathbf{J}$-matrix from (22.10), through solving the ordinary eigenvalue problem (23.4) with a truncation of n, or equivalently, by searching for the zeros of the polynomial $Q_K(u)$, i.e. finding the roots in (25.10). If (23.4) is used to obtain u_k, one would not know in advance whether the $\mathbf{J}$-matrix (22.10) possesses degenerate eigenvalues $u_{k'} = u_k$ ($k' \neq k$). But, such information can be deduced from the fact that all the eigenvalues $\{u_k\}$ from (23.4) are also the roots of the characteristic polynomial $Q_K(u)$. To show this, let us assume that there is, e.g. one such degenerate eigenvalue. This means that the rank of the J-matrix trid$_K[\beta, \alpha, \beta]$ of the type (22.10) for the specified dimension $M = K$ will automatically be reduced from K to $K - 1$. Consequently every Kth minor of $\det[u\mathbf{1} - \mathbf{U}]$, where $\mathbf{U} = c_0 \mathbf{J}_K$ according to (22.10), will be zero and, therefore, $\tilde{Q}_{K-1}(u_k) = 0$. This also implies $Q_{K-1}(u_k) = 0$ due to (25.9), since $\bar{\beta}_{K-1} \neq 0$ according to (22.4). However, if $u = u_k$ is a root, then the secular equation would imply that $Q_K(u_k) = 0$. Therefore, using (23.11) we also have $Q_{K+1}(u_k) = 0$. In such a case, the recurrence (23.11) would mean that every element $Q_{K-m}(u)$ of the set $\{Q_{K-m}\}_{m=0}^{K}$ is zero and, therefore, $Q_0(u) = 0$ [97, 259]. This contradicts the initialization $Q_0(u) = 1$ from (23.11). Hence, all the eigenvalues $\{u_k\}$ of the J-matrix (22.10) are *non-degenerate* ($u_{k'} \neq u_k$ for $k' \neq k$) and, by implication

$$Q_{K-1}(u_k) \neq 0. \tag{25.11}$$

Another line of reasoning proving (25.11) runs as follows. Suppose that all the roots $\{u_k\}_{k=1}^{K}$ are non-degenerate $u_{k'} \neq u_k (k' \neq k)$. This assumption is equivalent to saying that the polynomial $Q_K(u)$ has no multiple roots. In such a case, the following factorization is possible $Q_K(u) = q_{K,0}(u - u_1)(u - u_2) \cdots (u - u_k) \cdots (u - u_K)$, where $q_{K,0}$ is the coefficient multiplying the highest power u^K in the power series representation of $Q_K(u)$. If the opposite were true, i.e. if $Q_{K-1}(u_k) = 0$, then the polynomials $Q_{K-1}(u)$ and $Q_K(u)$ would have a common factor, say $(u - u_{k'})$. However, this means that the same common divisor $(u - u_{k'})$ will also be contained in $Q_{K-2}(u)$, as well as in $Q_{K-3}(u)$, etc, all the way down to $Q_0(u)$, since K is any non-negative integer. Hence, $Q_0(u) = q_{K,0}(u - u_{k'})$, which is a contradiction with $Q_0(u) = 1$ from (23.16). Therefore, non-degeneracy of the roots $\{u_k\}_{k=1}^{K}$ of the polynomial $Q_K(u)$ implies $Q_{K-1}(u_k) \neq 0$ as stated in (25.11). Conversely, the structure of the recursion (23.16), with its initial conditions, implies that all the roots $\{u_k\}_{k=1}^{K}$ originating from $Q_K(u_k) = 0$ are non-degenerate [17, 97, 259].

Chapter 26

Power series representations for two Lanczos polynomials

For many purposes, it is important to have the explicit representations of both polynomials $P_n(u)$ and $Q_n(u)$ as the finite linear combinations of the powers of u, such as

$$P_n(u) = \sum_{r=0}^{n-1} p_{n,n-r} u^r \qquad Q_n(u) = \sum_{r=0}^{n} q_{n,n-r} u^r \qquad (26.1)$$

where the constants $\{p_{n,n-r}, q_{n,n-r}\}$ are the expansion coefficients. We recall that the polynomials $P_n(u)$ and $Q_n(u)$ are of degree $n-1$ and n, respectively. Obviously, with a change of the dummy index $r \longrightarrow n-r$ in (26.1) we could equivalently write $Q_n(u) = \sum_{r=0}^{n} q_{n,r} u^{n-r}$. Inserting (26.1) into (23.20) and equating afterwards the multipliers of like powers of u, we obtain the recursion [263]

$$\beta_{n+1} p_{n+1,n+1-r} = p_{n,n+1-r} - \alpha_n p_{n,n-r} - \beta_n p_{n-1,n-1-r}$$
$$p_{n,-1} = 0 \qquad p_{n,m} = 0 \quad (m > n) \qquad p_{0,0} = 0 \qquad p_{1,1} = 1. \quad (26.2)$$

Repeating the same procedure with (23.16) and (26.1) will produce the recursive relation [263]

$$\beta_{n+1} q_{n+1,n+1-r} = q_{n,n+1-r} - \alpha_n q_{n,n-r} - \beta_n q_{n-1,n-1-r}$$
$$q_{n,-1} = 0 \qquad q_{n,m} = 0 \quad (m > n) \qquad q_{0,0} = 1. \quad (26.3)$$

It then follows that the Lanczos expansion coefficients form a lower triangular matrix $\mathbf{q} = \{q_{i,j}\}$ with zero elements above the main diagonal [176]

$$\mathbf{q} = \begin{pmatrix} q_{0,0} & 0 & 0 & 0 & 0 & \cdots & 0 \\ q_{1,0} & q_{1,1} & 0 & 0 & 0 & \cdots & 0 \\ q_{2,0} & q_{2,1} & q_{2,2} & 0 & 0 & \cdots & 0 \\ q_{3,0} & q_{3,1} & q_{3,2} & q_{3,3} & 0 & \cdots & 0 \\ \vdots & \vdots & \vdots & \vdots & \vdots & \ddots & \vdots \\ q_{n,0} & q_{n,1} & q_{n,2} & q_{n,3} & q_{n,4} & \cdots & q_{n,n} \end{pmatrix}. \quad (26.4)$$

The two leading coefficients $p_{K,1}$ and $q_{K,0}$ in the polynomials $P_K(u)$ and $Q_K(u)$ are, respectively

$$p_{K,1} = \frac{\beta_1}{\bar{\beta}_K} = \frac{1}{\prod_{m=2}^{K} \beta_m} \qquad q_{K,0} = \frac{1}{\bar{\beta}_K} = \frac{1}{\prod_{m=1}^{K} \beta_m} \qquad (26.5)$$

where $\bar{\beta}_K$ is from (22.4). The expansion coefficients $\{q_{n,n-r}\}$ are the prediction coefficients of the LP method [61, 64]. The first several of these coefficients from the set $\{p_{n,n-r}\}$ and $\{q_{n,n-r}\}$ are

$$p_{0,0} = 0 \qquad p_{1,1} = 1 \qquad p_{2,1} = \frac{1}{\beta_2} \qquad p_{2,2} = -\frac{\alpha_1}{\beta_2}$$

$$p_{3,1} = \frac{1}{\beta_2 \beta_3} \qquad p_{3,2} = -\frac{\alpha_1 + \alpha_2}{\beta_2 \beta_3} \qquad p_{3,3} = \frac{\alpha_1 \alpha_2 - \beta_2^2}{\beta_2 \beta_3}$$

$$q_{0,0} = 1 \qquad q_{1,0} = \frac{1}{\beta_1} \qquad q_{1,1} = -\frac{\alpha_0}{\beta_1}$$

$$q_{2,0} = \frac{1}{\beta_1 \beta_2} \qquad q_{2,1} = -\frac{\alpha_0 + \alpha_1}{\beta_1 \beta_2} \qquad q_{2,2} = \frac{\alpha_0 \alpha_1 - \beta_1^2}{\beta_1 \beta_2}$$

$$q_{3,0} = \frac{1}{\beta_1 \beta_2 \beta_3} \qquad q_{3,1} = -\frac{\alpha_0 + \alpha_1 + \alpha_2}{\beta_1 \beta_2 \beta_3}$$

$$q_{3,2} = \frac{\alpha_0 \alpha_1 + (\alpha_0 + \alpha_1)\alpha_2 - \beta_1^2 - \beta_2^2}{\beta_1 \beta_2 \beta_3}$$

$$q_{3,3} = \frac{\alpha_0 \beta_2^2 - \alpha_2(\alpha_0 \alpha_1 - \beta_1^2)}{\beta_1 \beta_2 \beta_3}. \qquad (26.6)$$

The resulting expressions for $\{P_n(u), Q_n(u)\}$ with $n \leq 4$ are given by

$$P_0(u) = 0 \qquad P_1(u) = 1 \qquad \beta_2 P_2(u) = u - \alpha_1$$

$$\beta_2 \beta_3 P_3(u) = (\alpha_1 \alpha_2 - \beta_2^2) - (\alpha_1 + \alpha_2)u + u^2$$

$$\beta_2 \beta_3 \beta_4 P_4(u) = [\alpha_1 \beta_3^2 - \alpha_3(\alpha_1 \alpha_2 - \beta_2^2)]$$
$$+ [(\alpha_1 \alpha_2 - \beta_2^2) + \alpha_3(\alpha_1 + \alpha_2) - \beta_3^2]u$$
$$- (\alpha_1 + \alpha_2 + \alpha_3)u^2 + u^3. \qquad (26.7)$$

$$Q_0(u) = 1 \qquad \beta_1 Q_1(u) = u - \alpha_0$$

$$\beta_1 \beta_2 Q_2(u) = (\alpha_0 \alpha_1 - \beta_1^2) - (\alpha_0 + \alpha_1)u + u^2$$

$$\beta_1 \beta_2 \beta_3 Q_3(u) = [\alpha_0 \beta_2^2 - \alpha_2(\alpha_0 \alpha_1 - \beta_1^2)]$$
$$+ [(\alpha_0 \alpha_1 - \beta_1^2) + \alpha_2(\alpha_0 + \alpha_1) - \beta_2^2]u$$
$$- (\alpha_0 + \alpha_1 + \alpha_2)u^2 + u^3. \qquad (26.8)$$

The ratio of the coefficients $p_{K,1}$ and $q_{K,0}$ from (26.5) is reduced to the parameter β_1 according to

$$\frac{p_{K,1}}{q_{K,0}} = \beta_1. \qquad (26.9)$$

An alternative way of computing the coefficients $\{p_{n,n-r}\}$ and $\{q_{n,n-r}\}$ comes from observing that the two power series representations in (26.1) are the truncated Maclaurin expansions. Therefore

$$p_{n,n-r} = \frac{1}{r!} P_{n,r}(0) \qquad q_{n,n-r} = \frac{1}{r!} Q_{n,r}(0) \qquad (26.10)$$

where the derivatives $Q_{n,r}(u)$ and $P_{n,r}(u)$ are defined in (24.2) and (24.8), respectively. The two polynomials from (26.1) can now be written as

$$P_n(u) = \sum_{r=0}^{n-1} \frac{1}{r!} P_{n,r}(0) u^r \qquad Q_n(u) = \sum_{r=0}^{n} \frac{1}{r!} Q_{n,r}(0) u^r \qquad (26.11)$$

where $P_0(u) \equiv 0$ as in (23.20). As a check, we computed the coefficients $\{p_{n,n-r}, q_{n,n-r}\}$ from (26.10) using the derivative recursions (24.4) as well as (24.6) and retrieved (26.6). In order to express the polynomials $P_n(u)/p_{n,1}$ and $Q_n(u)/q_{n,0}$ in terms of the signal points $\{c_n\}$, as well as through the variable u, we use (22.6) and carry out an extensive algebraic calculation with the following results

$$\alpha_0 + \alpha_1 = \frac{c_0 c_3 - c_1 c_2}{c_0 c_2 - c_1^2} \qquad \alpha_0 \alpha_1 - \beta_1^2 = \frac{c_1 c_3 - c_2^2}{c_0 c_2 - c_1^2}$$

$$\alpha_1 + \alpha_2 = -\frac{p_1}{p_0} \qquad \alpha_0 + \alpha_1 + \alpha_2 = -\frac{q_1}{q_0} \qquad \alpha_1 \alpha_2 - \beta_2^2 = \frac{p_2}{p_0}$$

$$\alpha_0 \alpha_1 + \alpha_2 (\alpha_0 + \alpha_1) - \beta_1^2 - \beta_2^2 = \frac{q_2}{q_0}$$

$$\alpha_0 \beta_2^2 - \alpha_2 (\alpha_0 \alpha_1 - \beta_1^2) = \frac{q_3}{q_0} \qquad (26.12)$$

$$p_0 = c_0 (2 c_1 c_2 c_3 + c_0 c_2 c_4 - c_0 c_3^2 - c_1^2 c_4 - c_2^3) = c_0 q_0$$
$$p_1 = 2 c_1^2 c_2 c_3 - 2 c_0 c_1 c_3^2 + c_0^2 c_3 c_4 - c_0^2 c_2 c_5 + c_0 c_1^2 c_5 + c_0 c_2^2 c_3 - c_1^3 c_4 - c_1 c_2^3$$
$$p_2 = 2 c_1 c_2^2 c_3 + 2 c_0 c_2^2 c_4 - 2 c_1^2 c_2 c_4 - 2 c_0 c_2 c_3^2 + 2 c_0 c_1 c_3 c_4 - 2 c_0 c_1 c_2 c_5$$
$$\quad + c_1 c_2^2 c_3 + c_0^2 c_3 c_5 - c_0^2 c_4^2 - c_1^2 c_3^2 + c_1^3 c_5 - c_2^4 \qquad (26.13)$$
$$q_0 = 2 c_1 c_2 c_3 + c_0 c_2 c_4 - c_0 c_3^2 - c_1^2 c_4 - c_2^3$$
$$q_1 = c_0 c_3 c_4 - c_0 c_2 c_5 - c_1 c_2 c_4 - c_1 c_3^2 + c_1^2 c_5 + c_2^2 c_3$$
$$q_2 = c_0 c_3 c_5 + c_1 c_3 c_4 - c_1 c_2 c_5 - c_0 c_4^2 - c_2 c_3^2 + c_2^2 c_4$$
$$q_3 = c_3^3 - c_1 c_3 c_5 - 2 c_2 c_3 c_4 + c_1 c_4^2 + c_2^2 c_5. \qquad (26.14)$$

In this derivation, we used the formula $c_0^3 \beta_1^4 \beta_2^2 = q_0$ which also appears in (22.7). According to (5.18), the constant q_0 is the 3×3 Hankel determinant $H_3(c_0) = \det S_3 = q_0$. With these findings, the polynomials $P_n(u)$ and $Q_n(u)$

become

$$P_1(u) = 1 \qquad \beta_2 P_2(u) = \frac{(2c_0c_1c_2 - c_0^2 c_3 - c_1^3) + c_0(c_0c_2 - c_1^2)u}{c_0(c_0c_2 - c_1^2)}$$

$$\beta_2\beta_3 P_3(u) = \frac{p_0 u^2 + p_1 u + p_2}{p_0} \qquad (26.15)$$

$$\beta_1 Q_1(u) = u - \frac{c_1}{c_0}$$

$$\beta_1\beta_2 Q_2(u) = \frac{(c_1c_3 - c_2^2) - (c_0c_3 - c_1c_2)u + (c_0c_2 - c_1^2)u^2}{c_0c_2 - c_1^2}$$

$$\beta_1\beta_2\beta_3 Q_3(u) = \frac{q_0 u^3 + q_1 u^2 + q_2 u + q_3}{q_0}. \qquad (26.16)$$

These are the sought formulae that exhibit an explicit dependence of the polynomials $P_n(u)$ and $Q_n(u)$ upon the variable u and the given signal points $\{c_n\}$ for $0 \leq n \leq 3$.

Chapter 27

The Wronskian for the Lanczos polynomials

The Wronskian of a regular and irregular solution of a second-order differential equation is a constant. In the present context of difference equations, one can introduce the following Wronskian [17, 97]

$$W[P_n(u), Q_n(u)] \equiv \beta_{n+1}[P_{n+1}(u)Q_n(u) - Q_{n+1}(u)P_n(u)]. \quad (27.1)$$

If we multiply (23.16) by $P_n(u)$ and (23.20) by $Q_n(u)$ and subtract the obtained results, it follows

$$\beta_{n+1}[P_{n+1}(u)Q_n(u) - Q_{n+1}(u)P_n(u)] = \beta_n[P_n(u)Q_{n-1}(u) - Q_n(u)P_{n-1}(u)]$$
$$W[P_n(u), Q_n(u)] = [\beta_n P_n(u)]Q_{n-1}(u) - [\beta_n Q_n(u)]P_{n-1}(u). \quad (27.2)$$

Eliminating $\beta_n Q_n(u)$ and $\beta_n P_n(u)$ from the rhs of (27.2) using the recursions (23.16) and (23.20), respectively, we have $W[P_n(u), Q_n(u)] = [\beta_{n-1}P_{n-1}(u)]Q_{n-2}(u) - [\beta_{n-1}Q_{n-1}(u)]P_{n-2}(u)$. Here, the same procedure can be repeated by eliminating the term $\beta_{n-1}Q_{n-1}(u)$, as well as $\beta_{n-1}P_{n-1}(u)$ and continuing this chaining some $n-1$ times downward, we obtain

$$W[P_n(u), Q_n(u)] = [\beta_n P_n(u)]Q_{n-1}(u) - [\beta_n Q_n(u)]P_{n-1}(u)$$
$$= [\beta_{n-1}P_{n-1}(u)]Q_{n-2}(u) - [\beta_{n-1}Q_{n-1}(u)]P_{n-2}(u)$$
$$= [\beta_{n-2}P_{n-2}(u)]Q_{n-3}(u) - [\beta_{n-2}Q_{n-2}(u)]P_{n-3}(u)$$
$$= \cdots = \beta_1 P_1(u)Q_0(u) - \beta_1 Q_1(u)P_0(u) = \beta_1$$
$$W[P_n(u), Q_n(u)] = \beta_1 \quad (27.3)$$

where the relations $P_0(u) = 0$, $P_1(u) = 1$ and $Q_0(u) = 1$ from (23.16) and (23.20) are used. Hence, the Wronskian $W[P_n(u), Q_n(u)]$ is equal to the constant β_1 for any n, as it should be, due to the analogy between difference and differential equations [17, 97].

The eigenstates of the **J**-matrix from (22.10) are the elements of the set $\{Q_n(u_k)\}$. In the above relationships between the Lanczos polynomials of the first and second kind $Q_n(u)$ and $P_n(u)$, respectively, the variable u does not need to be an eigenvalue. However, if u is one of the elements from the eigenroot set $\{u_k\}$ of the characteristic polynomial $Q_K(u)$, then certain important expressions can be derived for the eigenstates. For instance, in a finite chain of length K we can set $n = K$ and $u = u_k$ in (27.2). In such a case, a combination of the characteristic equation (25.10) and the Wronskian (27.2) with (27.3) leads to

$$\beta_K P_K(u_k) Q_{K-1}(u_k) = \beta_1 \implies P_K(u_k) = \frac{\beta_1}{\beta_K Q_{K-1}(u_k)}. \tag{27.4}$$

This result shows that the Lanczos polynomial of the second kind $P_K(u_k)$ is proportional to the reciprocal of the eigenstate $Q_{K-1}(u_k)$. Obviously (27.4) is meaningful only if $Q_{K-1}(u_k) \neq 0$ and this is true according to (25.11). Both β_1 and β_K from (27.4) are non-zero on account of being the norms of states $|\psi_1\rangle$ and $|\psi_K\rangle$, respectively, according to (22.4) and (22.11).

Chapter 28

Finding accurate zeros of high-degree polynomials

The Lanczos polynomial of the first kind $Q_K(u)$ with its K zeros $\{u_k\}_{k=1}^{K}$ can be expressed as

$$Q_K(u) = \frac{1}{\bar{\beta}_K} \prod_{k=1}^{K}(u - u_k) \tag{28.1}$$

where $1/\bar{\beta}_K = q_{K,0}$ from (22.4) is the leading coefficient in the power series representation (26.1) of the polynomial $Q_K(u)$. One of the ways to find the zeros $\{u_k\}$ of the polynomial $Q_K(u)$ is to use the Newton–Raphson algorithm [176, 220]. This is a nonlinear iterative method which, as such, needs the starting values $\{u_{k,0}\}$ that can be chosen, e.g. randomly on the unit circle, but never on the real axis [125, 218]. In other words, the initial guess should be a complex number $\text{Im}(u_{k,0}) \neq 0$. Then, the successively better approximations $\{u_{k,n+1}\}(n = 0, 1, 2, \ldots)$ for the kth root of the equation $Q_K(u_k) = 0$ are constructed from the iteration formula

$$u_{k,n+1} = u_{k,n} - v_k \qquad v_k = \frac{Q_K(u_{k,n})}{Q'_K(u_{k,n})} \qquad (n = 0, 1, 2, \ldots) \tag{28.2}$$

where the first derivative $Q'_K(u)$ is computed exactly from (24.4) for $m = 1$ at $u = u_{k,n}$

$$\bar{\beta}_K Q'_K(u) = (u - \alpha_{K-1})Q'_{K-1}(u) - \beta_{K-1}Q'_{K-2}(u) + Q_{K-1}(u)$$
$$Q'_m(u) = 0 \qquad Q_0(u) = 1 \tag{28.3}$$

where $m = -1$ and $m = 0$. Given a prescribed accuracy threshold $\xi > 0$, the value $u_{k,n}$, as the approximation of the kth root at the nth iteration, will be satisfactory if either $Q_K(u_{k,n}) \approx 0$ or $|u_{k,n+1} - u_{k,n}| < \xi$. If neither of the two latter conditions is fulfilled after, e.g. some 100 iterations or so, the process is considered as diverging from the exact root u_k. In such a case, a new initial value for $u_{k,0}$ is selected and the procedure is repeated [125, 218]. When convergence is reached at a given iteration $n = m$, the sought root u_k is set to be $u_k \equiv u_{k,m}$. The Newton–Raphson method has a quadratic convergence rate, as can be seen from the development of the rhs of (28.2) in a power series in terms of $u_k^{(n)} - u_k$ [264]

$$u_k^{(n+1)} - u_k^{(n)} = a_{2,k}[u_k^{(n)} - u_k]^2 + a_{3,k}[u_k^{(n)} - u_k]^3 + \cdots$$
$$+ a_{m,k}[u_k^{(n)} - u_k]^m + \cdots \qquad (28.4)$$

where the expansion coefficients $\{a_{m,k}\}$ depend only upon the root u_k. Due to the absence of the linear term $a_{1,k}[u_k^{(n)} - u_k]$ on the rhs of (28.4), it is clear that from a certain point on, the error $u_k^{(n+1)} - u_k^{(n)}$ will be of the order of the magnitude of the square of the preceding error $u_k^{(n)} - u_k^{(n-1)}$. Once this step has been reached, every further iteration will approximately double the number of correct decimal places relative to the previously achieved approximation. Hence the name, 'quadratic convergence' [264]. To mimic the higher-order derivatives that are absent from (28.2), we found in practice that the following modification of the Newton–Raphson algorithm is very useful for obtaining complex roots of any function, including our characteristic equation $Q_K(u_k) = 0$ [142]

$$u_{k,n+1} = u_{k,n} - v_k \left\{ \frac{Q_K(u_{k,n} - v_k) - Q_K(u_{k,n})}{2Q_K(u_{k,n} - v_k) - Q_K(u_{k,n})} \right\} \qquad (28.5)$$

which differs from (28.2) by the presence of the rational function in the curly brackets multiplying the Newton quotient $v_k = Q_K(u_{k,n})/Q'_K(u_{k,n})$. In practice, we observed that the method in (28.5) converges quickly and consistently to at least eight decimal places in less than ten iterations [142].

In general, rooting a higher-degree polynomial, say $K \geq 100$, is known to be an ill-conditioned nonlinear problem. Therefore, for high degrees K it is better to solve (23.12) by more robust methods of linear algebra. One such method is diagonalization of the corresponding $K \times K$ square Hessenberg matrix (also known as the companion matrix) denoted by $\tilde{\mathbf{H}}$ which is defined by [176]

$$\tilde{\mathbf{H}} = \begin{pmatrix} -\frac{q_{K,1}}{q_{K,0}} & -\frac{q_{K,2}}{q_{K,0}} & \cdots & -\frac{q_{K,K-1}}{q_{K,0}} & -\frac{q_{K,K}}{q_{K,0}} \\ 1 & 0 & \cdots & 0 & 0 \\ 0 & 1 & \cdots & 0 & 0 \\ \vdots & \vdots & \vdots & \vdots & \vdots \\ 0 & 0 & \cdots & 1 & 0 \end{pmatrix} \qquad (28.6)$$

$$Q_K(u) = \frac{\det[u\mathbf{1} - \tilde{\mathbf{H}}]}{\bar{\beta}_K}. \qquad (28.7)$$

Notice that (28.7) is equivalent to (25.2) on account of (26.5). When all the roots $\{u_k\}_{k=1}^{K}$ are found, a simple analytical expression for $Q'_K(u_k)$ can also be obtained. Using (28.1), it follows

$$Q'_K(u) = \frac{1}{\bar{\beta}_K} \sum_{k''=1}^{K} \left\{ \prod_{k'=1(k'\neq k'')}^{K} (u - u_{k'}) \right\}. \tag{28.8}$$

Here, exactly at $u = u_k$ the whole product in the curly brackets vanishes due to the presence of the factor $(u - u_k)|_{u \to u_k}$, so that [265]

$$Q'_K(u_k) = \frac{1}{\bar{\beta}_K} \prod_{k'=1(k'\neq k)}^{K} (u_k - u_{k'})$$

$$Q'_K(u_k) \neq 0 \qquad u_{k'} \neq u_k \qquad (k' \neq k). \tag{28.9}$$

Due to its remarkable sparseness, the matrix $\tilde{\mathbf{H}}$ from (28.6) is easy to diagonalize even for higher dimensions. In this way, we have explicitly obtained very accurate eigenvalues $\{u_k\}_{k=1}^{K}$ for $K \sim 1000$ [114, 115] via the NAG library [178]. The same library, however, fails when we used its corresponding routine in an attempt to root directly polynomials of much lower degrees $K \geq 120$.

Chapter 29

Recursions for sums involving Lanczos polynomials

It is frequently required to compute certain linear combinations of orthogonal polynomials. An example is the finite sum of weighted characteristic polynomials $\{Q_n(u)\}$

$$Y_N = \sum_{n=0}^{N-1} a_n Q_n(u) \qquad a_{-n} \equiv 0 \qquad (N = 1, 2, 3, \ldots). \qquad (29.1)$$

The coefficients $\{a_n\}$ from (29.1) are some given general constants that are real or complex numbers independent of variable u. The set of orthogonal characteristic polynomials $\{Q_n(u)\}$ satisfies the three contiguous recursive relation (23.16) with the appropriate initial conditions $Q_{-1}(u) = 0$ and $Q_0(u) = 1$. The quantity Y_N is itself a polynomial of order N and it will hereafter be called a generalized Lanczos polynomial for the given set of coefficients $\{a_n\}_{n=0}^{N-1}$. The straightforward way to evaluate the sum Y_N from (29.1) is to first generate the family $\{Q_n(u)\}$ from (23.16) and subsequently add the individual products $a_0 Q_0(u) + a_1 Q_1(u) + \cdots + a_{N-1} Q_{N-1}(u)$. Obviously, it would be advantageous to fuse these two steps together and obtain the sum Y_N in the course of the recursion without having to store all the vectors $\{Q_n\}$. This would require a modification of recursion (23.16). One of the simplest adjustments consists of introducing an auxiliary self-generating set of polynomials $\{S_n\}$ defined by

$$S_n(u) = a_n + \frac{u - \alpha_n}{\beta_{n+1}} S_{n+1}(u) - \frac{\beta_{n+1}}{\beta_{n+2}} S_{n+2}(u) \qquad (29.2)$$

with the initial conditions

$$S_{N-1}(u) = a_n \qquad S_N(u) = 0 \qquad S_{N+1}(u) = 0. \qquad (29.3)$$

This relation is of the type of (23.16), except that the elements from the sequence $\{S_n\}$ are generated by a backward rather than by a forward recurrence. By recurring all the way downward to $n = 0$, we see that (29.2) will provide us with $S_0(u)$ which is the sought sum Y_N

$$S_0(u) = Y_N(u). \tag{29.4}$$

Solving (29.2) formally for the a_n we write

$$a_n = S_n(u) - \frac{u - \alpha_n}{\beta_{n+1}} S_{n+1}(u) + \frac{\beta_{n+1}}{\beta_{n+2}} S_{n+2}(u) \tag{29.5}$$

which upon insertion into the defining relation (29.1) gives

$$Y_N(u) = \sum_{n=0}^{N-1} \left[S_n(u) - \frac{u - \alpha_n}{\beta_{n+1}} S_{n+1}(u) + \frac{\beta_{n+1}}{\beta_{n+2}} S_{n+2}(u) \right] Q_n(u). \tag{29.6}$$

At first glance, this may seem as a vicious circle. However, this is not the case, as can be illustrated by selecting N to be any finite fixed number, say $N = 5$, and calculating from (29.6)

$$Y_5(u) = \left[S_5(u) - \frac{u - \alpha_5}{\beta_6} S_6(u) + \frac{\beta_6}{\beta_7} S_7(u) \right] Q_5(u)$$

$$+ \left[S_4(u) - \frac{u - \alpha_4}{\beta_5} S_5(u) + \frac{\beta_5}{\beta_6} S_6(u) \right] Q_4(u)$$

$$+ \left[S_3(u) - \frac{u - \alpha_3}{\beta_4} S_4(u) + \frac{\beta_4}{\beta_5} S_5(u) \right] Q_3(u)$$

$$+ \left[S_2(u) - \frac{u - \alpha_2}{\beta_3} S_3(u) + \frac{\beta_3}{\beta_4} S_4(u) \right] Q_2(u)$$

$$+ \left[S_1(u) - \frac{u - \alpha_1}{\beta_2} S_2(u) + \frac{\beta_2}{\beta_3} S_3(u) \right] Q_1(u)$$

$$+ \left[a_0 - \frac{\beta_1}{\beta_2} S_2(u) + \frac{\beta_1}{\beta_2} S_2(u) \right] Q_0(u) \tag{29.7}$$

where in the last line the term $(\beta_1/\beta_2) S_2(u)$ is added and subtracted for convenience. Using (29.3) with $S_7(u) = S_6(u) = 0$ and collecting the like terms $S_m(u)$ for $m = 2 - 5$ on the rhs of (29.7), we obtain

$$Y_5(u) = \frac{1}{\beta_5} S_5(u)[\beta_5 Q_5(u) - (u - \alpha_4) Q_4(u) + \beta_4 Q_3(u)]$$

$$+ \frac{1}{\beta_4} S_4(u)[\beta_4 Q_4(u) - (u - \alpha_3) Q_3(u) + \beta_3 Q_2(u)]$$

$$+ \frac{1}{\beta_3} S_3(u)[\beta_3 Q_3(u) - (u - \alpha_2) Q_2(u) + \beta_2 Q_1(u)]$$

$$+ \frac{1}{\beta_2} S_2(u)[\beta_2 Q_2(u) - (u - \alpha_1) Q_1(u) + \beta_1 Q_0(u)]$$

$$+ \left[a_0 Q_0(u) + S_1(u) Q_1(u) - \frac{\beta_1}{\beta_2} S_2(u) Q_0(u) \right]. \tag{29.8}$$

Now by virtue of the recursion (23.16) for the polynomials $\{Q_n(u)\}$ each of the square brackets in (29.8) multiplying the terms $\beta_m^{-1} S_m(u)$ for $m = 2 - 5$ is identically equal to zero so that, in the end, only the quantities in the curly brackets survive

$$Y_5(u) = a_0 Q_0(u) + S_1(u) Q_1(u) - \frac{\beta_1}{\beta_2} S_2(u) Q_0(u). \tag{29.9}$$

The same type of conclusion would also hold true if we started from any other value of positive integer N higher or lower than $N = 5$. Therefore, in general the whole sum in (29.1) is reduced to only three terms as follows

$$Y_N(u) = a_0 + \frac{u - \alpha_0}{\beta_1} S_1(u) - \frac{\beta_1}{\beta_2} S_2(u) \tag{29.10}$$

where (29.4) is used together with the condition $Q_0(u) = 1$ from (23.16). A similar procedure can be found in [176]. The results (29.10) constitute the Clenshaw sum rule [266] for calculation of the quantity Y_N which is defined by (29.1). This rule speeds up the computation since it avoids a direct addition of the N terms $a_n Q_n(u)$ from $n = 0$ to $n = N - 1$ in (29.1), which necessitates precomputation of the polynomial set $\{Q_n(u)\}$. Instead, only the backward recursion (29.2) of the auxiliary quantity $S_n(u)$ should be run together with (29.3) from $n = N$ to $n = 0$ yielding $S_0(u) = a_0 + S_1(u) Q_1(u) - (\beta_1/\beta_2) S_2(u)$ which is equal to $Y_N(u)$ according to (29.4). This latter quantity $S_0(u)$ *is* the sought result Y_N according to (29.4). Such a finding is stated in (29.10).

As an illustration of the outlined procedure, we choose $N = 1 - 4$ and apply the sum rule to arrive at the following explicit results

$$Y_1(u) = a_0 \qquad Y_2(u) = \left(a_0 - a_1 \frac{\alpha_0}{\beta_1}\right) + \frac{a_1}{\beta_1} u \tag{29.11}$$

$$Y_3(u) = \left(a_0 - a_1 \frac{\alpha_0}{\beta_1} + a_2 \frac{\alpha_0 \alpha_1 - \beta_1^2}{\beta_1 \beta_2}\right) + \left(\frac{a_1}{\beta_1} - a_2 \frac{\alpha_0 + \alpha_1}{\beta_1 \beta_2}\right) u + \frac{a_2}{\beta_1 \beta_2} u^2 \tag{29.12}$$

$$Y_4(u) = \left[a_0 - a_1 \frac{\alpha_0}{\beta_1} + a_2 \frac{\alpha_0 \alpha_1 - \beta_1^2}{\beta_1 \beta_2} - a_3 \frac{(\alpha_0 \alpha_1 - \beta_1^2)\alpha_2 - \alpha_0 \beta_2^2}{\beta_1 \beta_2 \beta_3}\right]$$
$$+ \left[\frac{a_1}{\beta_1} - a_2 \frac{\alpha_0 + \alpha_1}{\beta_1 \beta_2} + a_3 \frac{(\alpha_0 + \alpha_1)\alpha_2 + \alpha_0 \alpha_1 - \beta_1^2 - \beta_2^2}{\beta_1 \beta_2 \beta_3}\right] u$$
$$+ \left(\frac{a_2}{\beta_1 \beta_2} - a_3 \frac{\alpha_0 + \alpha_1 + \alpha_2}{\beta_1 \beta_2 \beta_3}\right) u^2 + \frac{a_3}{\beta_1 \beta_2 \beta_3} u^3. \tag{29.13}$$

For a check, the same results from (29.13) are obtained by substituting the expressions from (26.8) for $Q_1(u)$, $Q_2(u)$ and $Q_3(u)$ into (29.1).

On many occasions, especially when computing power spectra, it is useful to have an expression for the sum Y_N in a form of the explicit expansion in powers of u. This can be done by inserting the development (26.1) into (29.1) as follows

$$Y_N(u) = \sum_{n=0}^{N-1} a_n \sum_{r=0}^{n} q_{n,n-r} u^r = \sum_{n=0}^{N-1} a_n \sum_{r=0}^{n} q_{n,n-r} \theta(n-r) u^r$$

$$= \sum_{r=0}^{N-1} \left\{ \sum_{n=0}^{N-1} a_n q_{n,n-r} \theta(n-r) \right\} u^r = \sum_{r=0}^{N-1} \left\{ \sum_{n=r}^{N-1} a_n q_{n,n-r} \right\} u^r$$

$$= \sum_{r=0}^{N-1} \left\{ \sum_{n=0}^{N-r-1} a_{n+r} q_{n+r,n} \right\} u^r$$

$$Y_N(u) = \sum_{r=0}^{N-1} \left\{ \sum_{n=0}^{N-r-1} a_{n+r} q_{n+r,n} \right\} u^r \tag{29.14}$$

where θ is the Heaviside [132] step function from (9.20). The θ-function is used in (29.14) to make the derivation more explicit. However, the same result follows without (9.20) by relying upon the definition (26.3), where $q_{n,n-r} = 0$ for $r > n$ which is equivalent to multiplication of $q_{n,n-r}$ by $\theta(n-r)$, as done in (29.14). All told, the sought development of $Y_N(u)$ in powers of u is

$$Y_N(u) = \sum_{n=0}^{N-1} \gamma_n^{(N)} u^n \tag{29.15}$$

where the new expansion coefficients $\{\gamma_n^{(N)}\}$ are

$$\gamma_n^{(N)} = \sum_{r=0}^{N-n-1} a_{n+r} q_{n+r,r} = \sum_{r=n}^{N-1} a_r q_{r,r-n} \qquad \gamma_{N-1}^{(N)} = \frac{a_{N-1}}{\bar{\beta}_{N-1}}. \tag{29.16}$$

Here the set $\{q_{n+r,n}\}$ is generated from the recursion (26.3) and $\bar{\beta}_n$ is defined in (22.4). As a verification of the result (29.15), we carried out an explicit calculation of the first few coefficients $\{\gamma_n^{(N)}\}$ for $N = 1 - 4$ using (29.16) and obtained the expressions

$$\gamma_0^{(1)} = a_0 \qquad \gamma_0^{(2)} = a_0 - a_1 \frac{\alpha_0}{\beta_1} \qquad \gamma_1^{(2)} = \frac{a_1}{\beta_1}$$

$$\gamma_0^{(3)} = a_0 - a_1 \frac{\alpha_0}{\beta_1} + a_2 \frac{\alpha_0 \alpha_1 - \beta_1^2}{\beta_1 \beta_2}$$

$$\gamma_1^{(3)} = \frac{a_1}{\beta_1} - a_2 \frac{\alpha_0 + \alpha_1}{\beta_1 \beta_2} \qquad \gamma_2^{(3)} = \frac{a_2}{\beta_1 \beta_2}$$

$$\gamma_0^{(4)} = a_0 - a_1 \frac{\alpha_0}{\beta_1} + a_2 \frac{\alpha_0 \alpha_1 - \beta_1^2}{\beta_1 \beta_2} - a_3 \frac{(\alpha_0 \alpha_1 - \beta_1^2) \alpha_2 - \alpha_0 \beta_2^2}{\beta_1 \beta_2 \beta_3}$$

$$\gamma_1^{(4)} = \frac{a_1}{\beta_1} - a_2 \frac{\alpha_0 + \alpha_1}{\beta_1 \beta_2} + a_3 \frac{(\alpha_0 + \alpha_1) \alpha_2 + \alpha_0 \alpha_1 - \beta_1^2 - \beta_2^2}{\beta_1 \beta_2 \beta_3}$$

$$\gamma_2^{(4)} = \frac{a_2}{\beta_1 \beta_2} - a_3 \frac{\alpha_0 + \alpha_1 + \alpha_2}{\beta_1 \beta_2 \beta_3} \qquad \gamma_3^{(4)} = \frac{a_3}{\beta_1 \beta_2 \beta_3}. \qquad (29.17)$$

When these results are inserted in (29.15), we obtained the formulae for $\{Y_N\}_{N=1}^4$ that coincide with the previous findings (29.13).

Chapter 30

The Lanczos finite-dimensional linear vector space $\mathcal{L}_M$

In chapter 22 we have introduced the finite-dimensional Lanczos subspace $\mathcal{L}_M \subset \mathcal{H}$ spanned by the Lanczos state vectors $\{|\psi_n)\}_{n=0}^{M-1}$ via

$$\mathcal{L}_M = \text{span}\{|\psi_0), |\psi_1), |\psi_2), \ldots, |\psi_{M-1})\}. \tag{30.1}$$

The Lanczos vector space $\mathcal{L}_M$ can be defined through its basis and the appropriate scalar product. A finite sequence of the Lanczos orthogonal polynomials of the first kind is complete, as will be shown in chapter 31 and, therefore, the set $\{Q_n(u)\}$ with K elements represents a basis. Thus, the polynomial set $\{Q_n(u)\}_{n=0}^{K}$ will be our fixed choice for the basis in $\mathcal{L}_K$. Of particular importance is the set $\mathcal{K}$ of the zeros $\{u_k\}_{k=1}^{K}$ of the Kth degree characteristic polynomial $Q_K(u)$

$$\mathcal{K} \equiv \{u_k\}_{k=1}^{K} \qquad Q_K(u_k) = 0. \tag{30.2}$$

As seen earlier, these zeros $\{u_k\}$ of $Q_K(u)$ coincide with the eigenvalues of both the evolution matrix $\mathbf{U}$ and the corresponding Hessenberg matrix $\tilde{\mathbf{H}}$ from (5.17) and (28.6), respectively. That is why the zeros of $Q_K(u)$ are called eigenzeros. The structure of $\mathcal{L}_M$ is determined by its scalar product for analytic functions of complex variable z or u. For any two regular functions $f(u)$ and $g(u)$ from $\mathcal{L}_M$, the scalar product in $\mathcal{L}_M$ is defined by means of the generalized Stieltjes integral

$$(f(u)|g(u)) = (g(u)|f(u)) = \frac{1}{2\pi i} \oint_C f(z)g(z)d\sigma_0(z) \qquad f, g \in \mathcal{L}_M \tag{30.3}$$

which is symmetric, i.e. without conjugation of the soft 'bra' vector $(f(u)|$. Here, $d\sigma_0(z)$ is a general, *complex valued* Lebesgue measure and the closed contour C encircles counterclockwise the eigenvalue set $\mathcal{K}$. Since the polynomials $\{Q_n(u)\}$ have no multiple zeros, all the spectral representations of functions in $\mathcal{L}_M$ will exhibit only simple poles. Hereafter, the generalized Lebesgue measure $\sigma_0(z)$

from (30.3) is specified as

$$\sigma_0(z) = \sum_{k=1}^{K} d_k \vartheta(z - u_k) \qquad \vartheta(z - u_k) = \begin{cases} 1 & z \in \mathcal{K} \\ 0 & z \notin \mathcal{K} \end{cases} \qquad (30.4)$$

where d_k is the residue (3.10) and $\vartheta(z - u_k)$ is the generalized real valued Heaviside function of the complex argument $z - u_k$, which is a generalization of the real case (9.20). Using the formula for the derivative $d\vartheta(z)/dz = \delta(z)$ where $\delta(z)$ is the complex valued Dirac function, we have the following expression for the measure in $\mathcal{L}_M$

$$d\sigma_0(z) = \rho_0(z)dz \qquad \rho_0(z) = \sum_{k=1}^{K} d_k \delta(z - u_k). \qquad (30.5)$$

The generalized Dirac function of a complex variable from (30.5) belongs to the class of the so-called ultra distributions [267, 268]. In the present context $\delta(z - u_k)$ has the same operational property as the usual Dirac function with a real argument, except that the contour integrals are involved viz

$$\oint_C dz\, f(z)\delta(z - u_k) = \sum_{k=1}^{K} f(u_k) \qquad (30.6)$$

where the function $f(z)$ of complex variable z is analytic throughout the same contour C as in (30.3). More generally, if $f(z)$ is regular within and on the contour C and $g(z)$ has K simple zeros $\{v_k\}_{k=1}^{K}$ within C, we can apply the Cauchy residue theorem to write

$$\oint_C dz f(z)\delta(g(z)) = \sum_{k=1}^{K} \frac{f(v_k)}{g'(v_k)} \qquad (30.7)$$

where $g'(z) = (d/dz)g(z)$ and $g(v_k) = 0$ $(1 \le k \le K)$. The weight function $\rho_0(z)$ from (30.5) is reminiscent of the so-called 'complex impulse train function' in signal processing [61, 64]. If one formally sets the eigenvalues $\{\omega_k\}$ of the operator $\hat{\Omega}$ to be equal to the Fourier grid points $\{\tilde{\omega}_k\}$, one would equate the residues $\{d_k\}$ with the complex Fourier amplitude $\{F_k\}$. Both $\{\tilde{\omega}_k\}$ and $\{F_k\}$ are given in (5.3). In such a case, the special impulse train function $\sum_k F_k \delta(\omega - \tilde{\omega}_k)$ would represent the Fourier stick spectrum with jumps or heights $|F_k|$ at the grid points $\omega = \tilde{\omega}_k$ and would otherwise be zero elsewhere. In (30.5), the quantities $\{|d_k|\}$ also have the meaning of heights or jumps in the spectrum (5.23) constructed from the peak parameters $\{u_k, d_k\}$.

With the help of the measure (30.5), the generalized Stieltjes integral in (30.3) is reduced to the sum of the usual contour Cauchy integrals, where d$\sigma_0(z)$ is equal to $\sigma_0(z)dz$

$$(f(u)|g(u)) = \frac{1}{2\pi i}\oint_C f(z)g(z)d\sigma_0(z) = \frac{1}{2\pi i}\oint_C f(z)g(z)\rho_0(z)dz$$

$$= \sum_{k=1}^{K} \frac{d_k}{2\pi i}\oint_C f(z)g(z)\delta(z-u_k)dz = \sum_{k=1}^{K} d_k f(u_k)g(u_k).$$

The discrete counterpart of the symmetric inner product (30.3) for the pair $\{f(u), g(u)\} \in \mathcal{L}_M$ is given by

$$(f(u)|g(u)) = (g(u)|f(u)) = \sum_{k=1}^{K} d_k f(u_k)g(u_k) \qquad (30.8)$$

where the residue $d_k = (\Phi_0|\Upsilon_k)^2$ from the definition (3.10) is a complex valued weight function.

Chapter 31

Completeness proof for the Lanczos polynomials

In practice, the complete state vector $|\Upsilon_k^\infty\rangle$ is computed from the infinite sum over n in (23.1) by monitoring convergence of the corresponding sequence of partial sums $\sum_{n=0}^{M-1} Q_{n,k}|\psi_n\rangle$ with the progressively increasing integer M. For a fixed large integer M the total state vector (23.14) is approximated by its finite chain counterpart $|\Upsilon^\infty(u)\rangle \approx |\Upsilon^M(u)\rangle$

$$|\Upsilon^M(u)\rangle = \sum_{n=0}^{M-1} Q_n(u)|\psi_n\rangle. \quad (31.1)$$

At the eigenvalues $u = u_k$ we have the wavefunction $|\Upsilon_k^M\rangle$ whose corresponding residue d_k^M is an approximation to d_k^∞ from (23.13), such that $d_k^\infty \approx d_k^M$ for large M, where

$$d_k^M = \frac{c_0}{\sum_{n=0}^M Q_{n,k}^2}. \quad (31.2)$$

The numerator of the quotient in this equation is the norm of the state vector $|\Upsilon_k^M\rangle$ and, therefore

$$\|\Upsilon_k^M\|^2 = \langle\Upsilon_k^M|\Upsilon_k^M\rangle = c_0 \sum_{n=0}^{M-1} Q_{n,k}^2 \implies d_k^M = \frac{c_0^2}{\|\Upsilon_k^M\|^2}. \quad (31.3)$$

The finite sum in the denominator of (31.2) can be evaluated analytically for any finite integer M. Moreover, we shall show that a more general expression of the type $\sum_{n=0}^{M-1} P_n(u) Q_n(u')$ can be derived algebraically in a closed form. In this regard, we first write the two recursions (23.16) and (23.20) at two arbitrary values of the variables u and u' as

$$\beta_n P_n(u) = (u - \alpha_{n-1}) P_{n-1}(u) - \beta_{n-1} P_{n-2}(u)$$
$$P_0(u) = 0 \qquad P_1(u) = 1 \qquad (31.4)$$
$$\beta_n Q_n(u') = (u' - \alpha_{n-1}) Q_{n-1}(u') - \beta_{n-1} Q_{n-2}(u')$$
$$Q_{-1}(u') = 0 \qquad Q_0(u') = 1. \qquad (31.5)$$

Multiplying (31.4) and (31.5) by $Q_{n-1}(u')$ and $P_{n-1}(u)$, respectively, and subtracting the results from each other, we obtain

$$P_{n-1}(u) Q_{n-1}(u') = \beta_n \frac{P_n(u) Q_{n-1}(u') - P_{n-1}(u) Q_n(u')}{u - u'}$$
$$- \beta_{n-1} \frac{P_{n-1}(u) Q_{n-2}(u') - P_{n-2}(u) Q_{n-1}(u')}{u - u'}. \qquad (31.6)$$

Writing (31.6) explicitly for, e.g. $1 \leq n \leq 3$, it will follow

$$P_0(u) Q_0(u') = \left[\beta_1 \frac{P_1(u) Q_0(u') - P_0(u) Q_1(u')}{u - u'} \right]_1$$
$$- \left\{ \beta_0 \frac{P_0(u) Q_{-1}(u') - P_{-1}(u) Q_0(u')}{u - u'} \right\}$$

$$P_1(u) Q_1(u') = \left[\beta_2 \frac{P_2(u) Q_1(u') - P_1(u) Q_2(u')}{u - u'} \right]_2$$
$$- \left[\beta_1 \frac{P_1(u) Q_0(u') - P_0(u) Q_1(u')}{u - u'} \right]_1$$

$$P_2(u) Q_2(u') = \left[\beta_3 \frac{P_3(u) Q_2(u') - P_2(u) Q_3(u')}{u - u'} \right]_3$$
$$- \left[\beta_2 \frac{P_2(u) Q_1(u') - P_1(u) Q_2(u')}{u - u'} \right]_2. \qquad (31.7)$$

If we add together these three equations, the square brackets labelled symbolically as $[\cdots]_j$ and $[\cdots]_{j'}$ with the same indices $j' = j$ will cancel each other, whereas the curly bracket is equal to zero, since $Q_{-1}(u') = 0$ and $P_0(u) = 0 = P_{-1}(u)$. Thus the final result for the sum of the three equations in (31.7) is given by $\sum_{n=0}^{2} P_n(u) Q_n(u') = \beta_3 [P_3(u) Q_2(u') - P_2(u) Q_3(u')]/(u - u')$. This process in (31.6) can be extended inductively to higher values of n up to $n = M - 1$, where M is an arbitrarily large and finite positive integer. When such M terms are added together, it follows

$$\sum_{n=0}^{M-1} P_n(u) Q_n(u') = \beta_M \frac{P_M(u) Q_{M-1}(u') - P_{M-1}(u) Q_M(u')}{u - u'} \qquad (31.8)$$

which we call the extended Christoffel–Darboux formula for the sum of the products of the Lanczos polynomials of the first and second kind $Q_n(u)$ and

$P_n(u)$, respectively. Of course, the sum in (31.8) could start from $n=1$ due to $P_0(u) = 0$, but this is not done to allow the possibility to replace $P_n(u)$ by $Q_n(u)$ in the sequel. When P_n is used instead of Q_n, the result (31.8) is reduced to

$$\sum_{n=0}^{M-1} Q_n(u) Q_n(u') = \beta_M \frac{Q_M(u) Q_{M-1}(u') - Q_{M-1}(u) Q_M(u')}{u - u'}. \quad (31.9)$$

This is the original Christoffel–Darboux formula which includes only the polynomials of the first kind [17, 94, 97]. The formula (31.8) with mixed polynomials P_n and Q_m has no meaning at $u' = u$. However, this is not so in the case of (31.9) which is well defined in the limit $u \longrightarrow u'$. The indeterminate expression $0/0$ is regularized by means of the l'Hôpital rule which gives

$$\sum_{n=0}^{M-1} Q_n^2(u) = \beta_M [Q'_M(u) Q_{M-1}(u) - Q'_{M-1}(u) Q_M(u)] \quad (31.10)$$

where $Q'_n(u) = dQ_n(u)/du$. Setting here $u = u_k$ and inserting the resulting expression for the term $\sum_{n=0}^{M-1} Q_n^2(u_k)$ into (31.2) for the residue d_k^M in the case of the finite chain of any fixed length M, we have

$$d_k^M = \frac{c_0}{\beta_M} \frac{1}{Q'_M(u_k) Q_{M-1}(u_k) - Q'_{M-1}(u_k) Q_M(u_k)}. \quad (31.11)$$

Let us now set $M = K$, where K is the number of the roots $\{u_k\}_{k=1}^K$ of $Q_K(u)$. In such a case, taking u as the eigenvalue u_k and using $Q_K(u_k) = 0$ from (25.10), we simplify (31.10) as

$$\sum_{n=0}^{K-1} Q_n^2(u_k) = \beta_K Q'_K(u_k) Q_{K-1}(u_k). \quad (31.12)$$

This reduces (31.11) to the Christoffel formula for the Gauss numerical integration/quadrature [176]

$$d_k = \frac{c_0}{\beta_K} \frac{1}{Q'_K(u_k) Q_{K-1}(u_k)} \qquad d_k \equiv d_k^K \quad (31.13)$$

where $Q_{K-1}(u_k) \neq 0$, as per (25.11). If the norm of the state vector $|\Upsilon_k^K)$

$$\|\Upsilon_k\|^2 = (\Upsilon_k|\Upsilon_k) = c_0 \sum_{n=0}^{K-1} Q_{n,k}^2 \qquad |\Upsilon_k) \equiv |\Upsilon_k^K) \quad (31.14)$$

is substituted into (31.12) and (31.13), it follows

$$d_k = \frac{c_0^2}{\|\Upsilon_k\|^2}. \quad (31.15)$$

At $u = u_k$ we abbreviate $|\Upsilon_k) \equiv |\Upsilon_k^K)$ and write the expansion (31.1) for $M = K$ as

$$|\Upsilon_k) = \sum_{n=0}^{K-1} Q_{n,k} |\psi_n). \tag{31.16}$$

The result (31.13) should be contrasted to the usual applications of the Lanczos algorithm, in which the residue d_k is computed numerically from the defining relation $d_k = (\Phi_0|\Upsilon_k^+)^2$ as in (23.5) via squaring the overlap between the initial state $|\Phi_0)$ and an approximate total wavefunction $|\Upsilon_k^+)$. Taking u and u' as the eigenvalues $u = u_k$ and $u' = u_{k'}$ the results (31.9) and (31.10) become

$$\sum_{n=0}^{K-1} Q_n(u_k) Q_n(u_{k'}) = \beta_K \frac{Q_K(u_k) Q_{K-1}(u_{k'}) - Q_{K-1}(u_k) Q_K(u_{k'})}{u_k - u_{k'}} \tag{31.17}$$

$$\sum_{n=0}^{K-1} Q_n^2(u_k) = \beta_K [Q_K'(u_k) Q_{K-1}(u_k) - Q_{K-1}'(u_k) Q_K(u_k)]. \tag{31.18}$$

The use of the characteristic equation $Q_K(u_k) = 0 = Q_K(u_{k'})$ in (31.17) and (31.18) will give

$$\sum_{n=0}^{K-1} Q_n(u_k) Q_n(u_{k'}) = 0 \qquad k' \neq k$$

$$\sum_{n=0}^{K-1} Q_n^2(u_k) = \beta_K Q_K'(u_k) Q_{K-1}(u_k) \qquad k' = k. \tag{31.19}$$

With the help of the Kronecker δ-symbol $\delta_{k,k'}$ from (3.7), the two separate results from (31.19) for $k' \neq k$ and $k' = k$ can be combined into a single equation as

$$\sum_{n=0}^{K-1} Q_n(u_k) Q_n(u_{k'}) = c_0 \frac{\delta_{k,k'}}{d_k}. \tag{31.20}$$

In (31.19), the result (31.13) is used to identify the residue d_k. Note that the upper limit $K - 1$ in (31.20) could be replaced by K on the account of the characteristic equation $Q_K(u_k) = 0$. The expression (31.20) represents the *local completeness* relation or closure for the Lanczos polynomial $\{Q_n(u_k)\}$.

Next, we are interested in considering the weighted products $d_k Q_n(u_k) \times Q_m(u_k)$ summed over all the eigenvalues $\{u_k\}_{k=1}^K$ for two arbitrary degrees n and m

$$I_{n,m} \equiv \sum_{k=1}^{K} d_k Q_n(u_k) Q_m(u_k). \tag{31.21}$$

Inserting d_k from (3.10) into (31.21) and employing (3.6), (5.8), (23.19) and $|\Phi_0\rangle = |\psi_0\rangle$, we have

$$\begin{aligned}
I_{n,m} &= \sum_{k=1}^{K} d_k Q_n(u_k) Q_m(u_k) = \sum_{k=1}^{K} (\Phi_0|\Upsilon_k)^2 Q_n(u_k) Q_m(u_k) \\
&= \sum_{k=1}^{K} (\Phi_0|Q_m(u_k)|\Upsilon_k)(\Upsilon_k|Q_n(u_k)|\Phi_0) \\
&= \sum_{k=1}^{K} (\Phi_0|Q_m(\hat{U})|\Upsilon_k)(\Upsilon_k|Q_n(\hat{U})|\Phi_0) \\
&= (\Phi_0|Q_m(\hat{U})\left\{\sum_{k=1}^{K}|\Upsilon_k)(\Upsilon_k|\right\}Q_n(\hat{U})|\Phi_0) \\
&= (\Phi_0|Q_m(\hat{U})Q_n(\hat{U})|\Phi_0) = (\psi_m|\psi_n) = c_0 \delta_{n,m} \\
I_{n,m} &= c_0 \delta_{n,m}.
\end{aligned} \qquad (31.22)$$

Finally, using (30.8) we see that the integral $I_{n,m}$ coincides with the scalar product $(Q_m(u)|Q_n(u))$

$$(Q_m(u)|Q_n(u)) = \sum_{k=1}^{K} d_k Q_n(u_k) Q_m(u_k) = c_0 \delta_{n,m}. \qquad (31.23)$$

This is the orthogonality relation of the two Lanczos polynomials $Q_n(u)$ and $Q_m(u)$ with the weight function which is the residue d_k [97]. We recall that the sequence $\{Q_{n,k}\} \equiv \{Q_n(u_k)\}$ coincides with the set of eigenvectors of the Jacobi matrix (22.10).

Given the inner product (30.8), the two polynomials $Q_n(u)$ and $Q_m(u)$ from the set $\{Q_\nu(u)\}$ are said to be orthogonal to each other with respect to the complex weight d_k. Since the set $\{Q_n(u)\}$ is complete, every function $f(u) \in \mathcal{L}_M$ can be expanded in a series in terms of $\{Q_n(u)\}$

$$f(u) = \sum_{n=0}^{\infty} \gamma_n Q_n(u). \qquad (31.24)$$

The expansion coefficients $\{\gamma_n\}$ can be obtained by taking the scalar products of both sides of (31.24) with $Q_m(u)$ and using the orthogonality relation (31.20)

$$(Q_m(u)|f(u)) = \sum_{n=0}^{\infty} \gamma_n (Q_m(u)|Q_n(u)) = \sum_{n=0}^{\infty} \gamma_n c_0 \delta_{n,m} = c_0 \gamma_m$$

$$\gamma_n = \frac{(f(u)|Q_n(u))}{c_0}. \qquad (31.25)$$

Thus we have proven that the eigenvectors $\mathbf{Q}_k = \{Q_{n,k}\}$ of the **J**-matrix (22.10) form an orthonormal complete set of vectors and, hence, represent a basis set. We emphasize that orthonormality is a matter of convenience, but completeness is essential both in theory and in practice [157, 193, 226].

Chapter 32

Duality: the states $|\psi_n)$ and polynomials $Q_n(u)$

Besides the signal points $\{c_n\}$, the key ingredients of spectral analysis are the Lanczos coupling constants $\{\alpha_n, \beta_n\}$. They are defined in terms of the elements of the basis set $\{|\psi_n)\}$ of the Lanczos state vectors as stated in (22.2). The common matrix element from (22.2) is of the following type

$$L_{n,m}^{(s)} \equiv (\psi_m|\hat{U}^s|\psi_n). \tag{32.1}$$

We can eliminate $\hat{U}^s(\tau)$ from (32.1) by using the closure (3.6) for the normalized full state vectors $\{|\Upsilon_k)/\|\Upsilon_k\|\}$ and employing the eigenproblem $\hat{U}^s|\Upsilon_k) = u_k^s|\Upsilon_k)$ which comes from (5.8)

$$L_{n,m}^{(s)} = \sum_{k=1}^{K} \frac{(\psi_m|\hat{U}^s|\Upsilon_k)(\Upsilon_k|\psi_n)}{\|\Upsilon_k\|^2} = \sum_{k=1}^{K} u_k^s \frac{(\psi_m|\Upsilon_k)(\Upsilon_k|\psi_n)}{\|\Upsilon_k\|^2}$$

$$L_{n,m}^{(s)} = \frac{1}{c_0^2}\sum_{k=1}^{K} d_k u_k^s (\psi_m|\Upsilon_k)(\Upsilon_k|\psi_n) \tag{32.2}$$

where (31.15) is utilized. Inserting the expansion (31.16) into (32.2) and using (22.9) will give

$$L_{n,m}^{(s)} = \frac{1}{c_0^2}\sum_{k=1}^{K}\sum_{n'=0}^{K-1}\sum_{n''=0}^{K-1} d_k u_k^s Q_{n',k} Q_{n'',k} (\psi_m|\psi_{n'})(\psi_{n''}|\psi_n)$$

$$= \frac{1}{c_0^2}\sum_{k=1}^{K}\sum_{n'=0}^{K-1}\sum_{n''=0}^{K-1} d_k u_k^s Q_{n',k} Q_{n'',k} \{c_0^2 \delta_{n',m} \delta_{n,n''}\}$$

$$= \sum_{k=1}^{K} d_k u_k^s Q_{m,k} Q_{n,k} = (Q_m(u)|u^s|Q_n(u))$$

$$L_{n,m}^{(s)} \equiv (\psi_m|\hat{U}^s|\psi_n) = (Q_m(u)|u^s|Q_n(u)) \tag{32.3}$$

203

where the definition (30.8) of the scalar product in the Lanczos space $\mathcal{L}_K$ is employed. With the result (32.3) at hand, the couplings $\{\alpha_n, \beta_n\}$ from (22.2) can be equivalently written as

$$\alpha_n = \frac{(Q_n(u)|u|Q_n(u))}{(Q_n(u)|Q_n(u))} \qquad \beta_n = \frac{(Q_{n-1}(u)|u|Q_n(u))}{(Q_{n-1}(u)|Q_{n-1}(u))}. \qquad (32.4)$$

The significance of this finding is in establishing the Lanczos dual representation $\{|\psi_n), Q_n(u)\}$ which enables the following equivalent definitions

$$\alpha_n = \frac{(\psi_n|\hat{U}|\psi_n)}{(\psi_n|\psi_n)} = \frac{(Q_n(u)|u|Q_n(u))}{(Q_n(u)|Q_n(u))}$$

$$\beta_n = \frac{(\psi_{n-1}|\hat{U}|\psi_n)}{(\psi_{n-1}|\psi_{n-1})} = \frac{(Q_{n-1}(u)|u|Q_n(u))}{(Q_{n-1}(u)|Q_{n-1}(u))}. \qquad (32.5)$$

This duality enables switching from the work with the Lanczos state vectors $\{|\psi_n)\}$ to the analysis with the Lanczos polynomials $\{Q_n(u)\}$. A change from one representation to the other is readily accomplished along the lines indicated in this chapter together with the basic relations from chapters 30 and 31, in particular, the definition (30.8) of the inner product in the Lanczos space $\mathcal{L}_K$, the completeness (31.20) and orthogonality (31.23) of the polynomial basis $\{Q_{n,k}\}$.

Chapter 33

An analytical solution for the overlap determinant $\det S_n$

The results of the explicit calculation of the first few particular determinants $\{\det S_n\} = \{H_n(c_0)\}$ from (5.15) for $n = 1 - 4$ are given by (5.18). On the other hand, the calculation of the couplings $\{\beta_n^2\}$ from (22.6) for $n \leq 3$ encounters the following intermediate results

$$\beta_1^2 = \frac{c_0 c_2 - c_1^2}{c_0^2} \qquad \beta_2^2 = \frac{q_0}{c_0^3 \beta_1^4}$$

$$\beta_3^2 = \frac{q_0 c_6 + q_1 c_5 + q_2 c_4 + q_3 c_3}{c_0^4 \beta_1^6 \beta_2^4} \tag{33.1}$$

with $\{q_n\}_{n=0}^{3}$ given in (26.14). We found that all the numerators in each of the elements of the set $\{\beta_n^2\}$ from (33.1) are the determinants $H_n(c_0)$ from (5.18)

$$\beta_1^2 = \frac{H_2(c_0)}{c_0^2} \qquad \beta_2^2 = \frac{H_3(c_0)}{c_0^3 \beta_1^4} \qquad \beta_3^2 = \frac{H_4(c_0)}{c_0^4 \beta_1^6 \beta_2^4}. \tag{33.2}$$

Such a regular pattern shows that the coupling parameters $\{\beta_n\}$ satisfy the following recursion

$$\beta_{n-1}^2 = \frac{H_n(c_0)}{c_0^n \beta_1^{2n-2} \beta_2^{2n-4} \beta_3^{2n-6} \cdots \beta_{n-2}^4}. \tag{33.3}$$

This yields the general result for $H_n(c_0)$ in terms of c_0 and $\{\beta_\nu\}_{\nu=1}^{n-1}$. Moreover, it is possible to obtain an alternative expression for $H_n(c_0)$ by referring to (26.5) for the leading coefficient $q_{n,0} = 1/(\beta_1 \beta_2 \beta_3 \cdots \beta_n)$ in the power series representation (26.1) of the polynomial $Q_n(u)$. Thus, given the set $\{q_{n,0}\}$ ($n =$

205

0, 1, 2, 3, ...), we rewrite (5.18) as

$$H_1(c_0) = c_0 \qquad H_2(c_0) = \frac{c_0^2}{q_{1,0}^2}$$

$$H_3(c_0) = \frac{c_0^3}{q_{1,0}^2 q_{2,0}^2} \qquad H_4(c_0) = \frac{c_0^4}{q_{1,0}^2 q_{2,0}^2 q_{3,0}^2}$$

$$H_n(c_0) = \frac{c_0^n}{q_{1,0}^2 q_{2,0}^2 q_{3,0}^2 \cdots q_{n-1,0}^2} = \frac{c_0^n}{\prod_{m=1}^{n-1} q_{m,0}^2}. \qquad (33.4)$$

Therefore, the most general result for the determinant $H_n(c_0)$ for any n can be expressed in the following closed, i.e. analytical form

$$H_n(c_0) \equiv \det S_n = c_0^n \beta_1^{2n-2} \beta_2^{2n-4} \beta_3^{2n-6} \cdots \beta_{n-1}^2 = c_0^n \prod_{m=1}^{n-1} \beta_m^{2n-2m}$$

$$= \frac{c_0^n}{q_{1,0}^2 q_{2,0}^2 q_{3,0}^2 \cdots q_{n-1,0}^2} = \frac{c_0^n}{\prod_{m=1}^{n-1} q_{m,0}^2}. \qquad (33.5)$$

This finding, which was announced in chapter 5, exhibits a remarkably regular factorability of the general result for $H_n(c_0)$ as a direct consequence of a judicious combination of symmetry of the Hankel determinant (5.15) and the Lanczos orthogonal polynomials. Thus, given either the set $\{\beta_\nu\}$ or $\{q_{\nu,0}\}$, the Hankel determinant $H_n(c_0)$, or equivalently, the overlap determinant $\det S_n$ in the Schrödinger, i.e. Krylov basis $\{|\Phi_n\rangle\}$ can be constructed at once from (33.5).

Chapter 34

The explicit Lanczos algorithm

In this chapter we shall develop the explicit version of the Lanczos algorithm. With this goal, we first notice that the coefficients multiplying the powers of u in the numerators in (26.16) are the minors of the Hankel determinant $H_3(c_0) = q_0$. Thus (26.16) can be rewritten as

$$\beta_1 Q_1(u) = \frac{-1}{c_0} \begin{vmatrix} 1 & u \\ c_0 & c_1 \end{vmatrix}$$

$$\beta_1 \beta_2 Q_2(u) = \frac{(-1)^2}{c_0^2 \beta_1^2} \begin{vmatrix} 1 & u & u^2 \\ c_0 & c_1 & c_2 \\ c_1 & c_2 & c_3 \end{vmatrix}$$

$$\beta_1 \beta_2 \beta_3 Q_3(u) = \frac{(-1)^3}{c_0^3 \beta_1^4 \beta_2^3} \begin{vmatrix} 1 & u & u^2 & u^3 \\ c_0 & c_1 & c_2 & c_3 \\ c_1 & c_2 & c_3 & c_4 \\ c_2 & c_3 & c_4 & c_5 \end{vmatrix}. \qquad (34.1)$$

Continuing inductively, this leads to the general explicit expression for the Lanczos polynomial [161, 167]

$$\tilde{Q}_n(u) = \left\{ \frac{(-1)^n}{c_0^n \prod_{m=1}^n \beta_m^{2n-2m}} \right\} \begin{vmatrix} 1 & u & u^2 & \cdots & u^n \\ c_0 & c_1 & c_2 & \cdots & c_n \\ c_1 & c_2 & c_3 & \cdots & c_{n+1} \\ c_2 & c_3 & c_4 & \cdots & c_{n+2} \\ \vdots & \vdots & \vdots & \ddots & \vdots \\ c_{n-1} & c_n & c_{n+1} & \cdots & c_{2n-1} \end{vmatrix} \qquad (34.2)$$

$$\tilde{Q}_n(u) = \left\{ \frac{(-1)^n}{c_0^n} \prod_{m=1}^n q_{m,0}^2 \right\} \begin{vmatrix} 1 & u & u^2 & \cdots & u^n \\ c_0 & c_1 & c_2 & \cdots & c_n \\ c_1 & c_2 & c_3 & \cdots & c_{n+1} \\ c_2 & c_3 & c_4 & \cdots & c_{n+2} \\ \vdots & \vdots & \vdots & \ddots & \vdots \\ c_{n-1} & c_n & c_{n+1} & \cdots & c_{2n-1} \end{vmatrix} \qquad (34.3)$$

where $\tilde{Q}_n(u) = \bar{\beta}_n Q_n(u)$ as in (25.9). The result (34.2) or (34.3) represents the explicit orthonormalization of the polynomial sequence $\{Q_n(u)\}$. This is the explicit Lanczos algorithm as opposed to the recursive relation (23.16). The expressions (22.1), (23.16) and (23.20) will be called the implicit Lanczos algorithms, since the solution to the problem is not immediately available, but rather it is attainable through recursions. Also for $\{|\psi_n\rangle\}$, it is possible to design the explicit Lanczos algorithm by deriving the expression which holds the whole result with no recourse to recurrence relations at all. Extending the Cayley–Hamilton theorem to encompass determinants with a mixed structure of scalars and operators/matrices as elements, we can generalize (34.2) to

$$\tilde{Q}_n(\hat{U}) = \left\{\frac{(-1)^n}{c_0^n \prod_{m=1}^n \beta_m^{2n-2m}}\right\} \begin{vmatrix} \hat{1} & \hat{U} & \hat{U}^2 & \cdots & \hat{U}^n \\ c_0 & c_1 & c_2 & \cdots & c_n \\ c_1 & c_2 & c_3 & \cdots & c_{n+1} \\ c_2 & c_3 & c_4 & \cdots & c_{n+2} \\ \vdots & \vdots & \vdots & \ddots & \vdots \\ c_{n-1} & c_n & c_{n+1} & \cdots & c_{2n-1} \end{vmatrix} \quad (34.4)$$

$$\tilde{Q}_n(U) = \left\{\frac{(-1)^n}{c_0^n \prod_{m=1}^n \beta_m^{2n-2m}}\right\} \begin{vmatrix} 1 & U & U^2 & \cdots & U^n \\ c_0 & c_1 & c_2 & \cdots & c_n \\ c_1 & c_2 & c_3 & \cdots & c_{n+1} \\ c_2 & c_3 & c_4 & \cdots & c_{n+2} \\ \vdots & \vdots & \vdots & \ddots & \vdots \\ c_{n-1} & c_n & c_{n+1} & \cdots & c_{2n-1} \end{vmatrix}. \quad (34.5)$$

Here, the premultiplying factor could equivalently be written in a form containing expansion coefficients of the characteristic polynomial using the relationship $1/\prod_{m=1}^n \beta_m^{2n-2m} = \prod_{m=1}^n q_{m,0}^2$ as in (34.2) and (34.3). The explicit Lanczos polynomial operator $Q_n(\hat{U})$ from (34.4) applied to $|\psi_0\rangle = |\Phi_0\rangle$ generates the wave packet $|\psi_n\rangle$ according to $|\psi_n\rangle = Q_n(\hat{U})|\psi_0\rangle$ as in (23.19). Therefore, the final result is the following expression for the explicit Lanczos states $\{|\psi_n\rangle\}$

$$|\tilde{\psi}_n\rangle = \left\{\frac{(-1)^n}{c_0^n \prod_{m=1}^n \beta_m^{2n-2m}}\right\} \begin{vmatrix} |\Phi_0\rangle & |\Phi_1\rangle & |\Phi_2\rangle & \cdots & |\Phi_n\rangle \\ c_0 & c_1 & c_2 & \cdots & c_n \\ c_1 & c_2 & c_3 & \cdots & c_{n+1} \\ c_2 & c_3 & c_4 & \cdots & c_{n+2} \\ \vdots & \vdots & \vdots & \ddots & \vdots \\ c_{n-1} & c_n & c_{n+1} & \cdots & c_{2n-1} \end{vmatrix} \quad (34.6)$$

where $|\tilde{\psi}_n\rangle = \bar{\beta}_n |\psi_n\rangle$ as in (22.4). This expression for the explicit Lanczos orthogonalization of the Schrödinger state vectors $\{|\Phi_n\rangle\}$ coincides with the result obtained by the Gram–Schmidt orthogonalization [142]. By projecting both sides of (34.6) onto $\langle\Phi_0|$, we will end up with a determinant with the identical first two rows $\{c_0, c_1, c_2, \ldots, c_n\}$ and, as such, equal to zero

$$(\Phi_0|\psi_n) = \left\{\frac{(-1)^n}{c_0^n \prod_{m=1}^n \beta_m^{2n-2m+1}}\right\} \begin{vmatrix} c_0 & c_1 & c_2 & \cdots & c_n \\ c_0 & c_1 & c_2 & \cdots & c_n \\ c_1 & c_2 & c_3 & \cdots & c_{n+1} \\ c_2 & c_3 & c_4 & \cdots & c_{n+2} \\ \vdots & \vdots & \vdots & \ddots & \vdots \\ c_{n-1} & c_n & c_{n+1} & \cdots & c_{2n-1} \end{vmatrix}$$

$$\therefore \quad (\Phi_0|\psi_n) = 0 \tag{34.7}$$

where $n \geq 1$, in agreement with the orthogonality condition (22.9) recalling that $|\Phi_0) = |\psi_0)$.

Chapter 35

The explicit Lanczos wave packet propagation

Given the initial state $|\Phi_0\rangle$ of a system at time $t = 0$, the Schrödinger state $|\Phi_n\rangle$ at a later time $n\tau$ is obtained by propagation via the evolution operator $\hat{U}(\tau)$ such that $|\Phi_n\rangle = \hat{U}^n(\tau)|\Phi_0\rangle$ as in (5.6). To get the spectrum of $\hat{U}(\tau)$, diagonalization techniques could be used with the Schrödinger basis set $\{|\Phi_n\rangle\}$. Such a basis is delocalized, causing matrices to be full, and this enhances the ill-conditioning of the problem. One of the ways to counteract this is to switch to a localized basis set given by the Fourier sum of $\{|\Phi_n\rangle\}$ as done, e.g. in the FD method [108, 109]. The result of this change of the basis is a modification of the Schrödinger matrix $\mathbf{U}$ which then becomes diagonally dominated and, hence, of significantly reduced ill-conditioning [17, 109]. Another way to obtain sparse diagonalizing matrices is provided by switching from $\{|\Phi_n\rangle\}$ to the Lanczos basis set $\{|\psi_n\rangle\}$ where the vectors $\{|\psi_n\rangle\}$ are implicitly generated via the recursion (22.1). The resulting evolution matrix $\mathbf{U}(\tau)$ is tridiagonal as in (22.10). We see from the definition (22.1) that each state vector $|\psi_n\rangle$ is a result of repeated applications of $\hat{U}(\tau)$ onto the initial wave packet $|\psi_0\rangle = |\Phi_0\rangle$. This is expected to yield an expression in which the general vector $|\psi_n\rangle$ is given by a linear combination of Schrödinger states $|\Phi_n\rangle = \hat{U}^n(\tau)|\Phi_0\rangle$. Such a feature can be well analysed within, e.g. the concept of a *polynomial* Lanczos propagation of wave packets $\{|\psi_n\rangle\}$ from the initial state vector $|\psi_0\rangle = |\Phi_0\rangle$, i.e. $|\psi_n\rangle = Q_n(\hat{U})|\psi_0\rangle$ as in (23.19). This equation defines the Lanczos polynomial propagation of the initial wave packet $|\Phi_0\rangle$. Here, $Q_n(\hat{U})$ is the Lanczos operator polynomial defined by the recursion (23.18) which results from applying the Cayley–Hamilton theorem to (23.16). Of course, the same theorem can also be applied to power series representation (26.1) giving

$$Q_n(\hat{U}) = \sum_{r=0}^{n} q_{n,n-r} \hat{U}^r \qquad (35.1)$$

where the coefficients $\{q_{n,n-r}\}$ are given recursively in (26.3). Inserting the finite sum (35.1) into (23.19) and using (5.6), we derive immediately the following result

$$|\psi_n\rangle = \sum_{r=0}^{n} q_{n,n-r}|\Phi_r\rangle. \qquad (35.2)$$

Hence, as expected, the general Lanczos state $|\psi_n\rangle$ is indeed a linear combination of the Schrödinger states $\{|\Phi_r\rangle\}$. Taking the few first coefficients $\{q_{n,n-r}\}$ from (26.6) gives

$$\beta_1|\psi_1\rangle = |\Phi_1\rangle - \alpha_0|\Phi_0\rangle$$
$$\beta_1\beta_2|\psi_2\rangle = |\Phi_2\rangle - (\alpha_0 + \alpha_1)|\Phi_1\rangle + (\alpha_0\alpha_1 - \beta_1^2)|\Phi_0\rangle$$
$$\beta_1\beta_2\beta_3|\psi_3\rangle = |\Phi_3\rangle - (\alpha_0 + \alpha_1 + \alpha_2)|\Phi_2\rangle$$
$$+ \{(\alpha_0\alpha_1 - \beta_1^2) + \alpha_2(\alpha_0 + \alpha_1) - \beta_2^2\}|\Phi_1\rangle$$
$$+ \{\alpha_0\beta_2^2 - \alpha_2(\alpha_0\alpha_1 - \beta_1^2)\}|\Phi_0\rangle. \qquad (35.3)$$

These results agree with our earlier findings from (22.5) that were obtained using the recursion (22.4) with the help of (5.6) and $|\tilde{\psi}_n\rangle = \bar{\beta}_n|\psi_n\rangle$. The availability of the explicit formula (35.2) for any state vector $|\psi_n\rangle$ also permits deriving the explicit expressions for the coupling constants $\{\alpha_n, \beta_n\}$. Thus, inserting (35.2) into (22.2), we have

$$\alpha_n = \frac{1}{c_0} \sum_{r=0}^{n} \sum_{r'=0}^{n} q_{n,n-r} q_{n,n-r'} c_{r+r'+1}$$

$$\beta_n = \frac{1}{c_0} \sum_{r=0}^{n} \sum_{r'=0}^{n-1} q_{n,n-r} q_{n-1,n-r'-1} c_{r+r'+1}. \qquad (35.4)$$

The double sums over r and r' will now be simplified by introducing the auxiliary expression $Y_{n,n'}^{(s)}$ via

$$Y_{n,n'}^{(s)} = \frac{1}{c_0} \sum_{r=0}^{n} \sum_{r'=0}^{n'} q_{n,n-r} q_{n',n'-r'} \tilde{c}_{r+s,r'} \qquad (35.5)$$

such that

$$\alpha_n = Y_{n,n}^{(1)} \qquad \beta_n = Y_{n,n-1}^{(1)} \qquad \tilde{c}_{r+s,r'} \equiv c_{r+r'+s}. \qquad (35.6)$$

Introducing an additional sum over m with the help of the Kronecker δ-symbol (3.7), we rewrite (35.5) as

$$Y_{n,n'}^{(s)} = \frac{1}{c_0} \sum_{m=0}^{n+n'} \sum_{r=0}^{n} \sum_{r'=0}^{n'} q_{n,n-r} q_{n',n'-r'} \tilde{c}_{r+s,r'} \delta_{r+r',m}. \qquad (35.7)$$

This yields

$$Y^{(s)}_{n,n'} = \frac{1}{c_0} \sum_{m=0}^{n+n'} \sum_{\substack{r+r'=0 \ (r,r'\geq 0)}}^{r+r'=m} q_{n,n-r} q_{n',n'-r'} \tilde{c}_{r+s,r'}$$

$$= \frac{1}{c_0} \sum_{m=0}^{n+n'} \sum_{r=r_1}^{r_2} q_{n,n-r} q_{n',n'-m+r} \tilde{c}_{r+s,m-r}$$

$$Y^{(s)}_{n,n'} = \frac{1}{c_0} \sum_{m=0}^{n+n'} \sum_{r=r_1}^{r_2} q_{n,n-r} q_{n',n'-m+r} \tilde{c}_{r+s,m-r} \tag{35.8}$$

$$r_1 = \max\{0, m-n\} \qquad r_2 = \min\{n, m\}. \tag{35.9}$$

Replacing the coefficient $\tilde{c}_{j,j'}$ in (35.8) by $\tilde{c}_{j,j'} = c_{j+j'}$, we can finally write down the results for the parameters $\{\alpha_n, \beta_n\}$ from (35.4) as

$$\alpha_n = \frac{1}{c_0} \sum_{m=0}^{2n} a_{n,m} c_{m+1} \qquad \beta_n = \frac{1}{c_0} \sum_{m=0}^{2n-1} b_{n,m} c_{m+1}$$

$$a_{n,m} = \sum_{r=r_1^a}^{r_2} q_{n,n-r} q_{n,n-m+r} \qquad b_{n,m} = \sum_{r=r_1^b}^{r_2} q_{n,n-r} q_{n-1,n-1-m+r}$$

$$r_1^a = \max\{0, m-n\} \qquad r_1^b = \max\{0, m-n+1\} \qquad r_2 = \min\{n, m\}. \tag{35.10}$$

An explicit computation by means of (35.10) leads to

$$c_0 \alpha_0 = c_1 \qquad c_0 \alpha_1 = c_1 q_{1,1}^2 + 2 c_2 q_{1,0} q_{1,1} + c_3 q_{1,0}^2$$
$$c_0 \alpha_2 = c_1 q_{2,2}^2 + 2 c_2 q_{2,1} q_{2,2} + c_3 (q_{2,1}^2 + 2 q_{2,0} q_{2,2}) + 2 c_4 q_{2,0} q_{2,1} + c_5 q_{2,0}^2$$
$$\beta_0 = 0 \qquad c_0 \beta_1 = q_{0,0}(c_1 q_{1,1} + c_2 q_{1,0})$$
$$c_0 \beta_2 = c_1 q_{1,1} q_{2,2} + c_2 (q_{1,0} q_{2,2} + q_{1,1} q_{2,1})$$
$$\qquad + c_3 (q_{1,0} q_{2,1} + q_{2,0} q_{1,1}) + c_4 q_{1,0} q_{2,0}. \tag{35.11}$$

For checking purposes, we insert the particular values of the coefficients $\{q_{n,m}\}$ from (26.6) into (35.11) and obtain again the results (22.6) for the couplings $\{\alpha_n, \beta_n\}$ with $0 \leq n \leq 2$.

Chapter 36

The Padé–Lanczos approximant (PLA)

Given the eigenvalue set $\mathcal{K}$ from (30.2), the polynomial $Q_K(u)$ can be expressed through a product of the type $Q_K(u) = q_{K,0}(u - u_1)(u - u_2) \cdots (u - u_K)$ as in (28.1) where $q_{K,0} = 1/\bar{\beta}_K$ is the leading coefficient (26.5) in the power series representation (26.1) of the polynomial $Q_K(u)$. On the other hand, the Lanczos orthogonal polynomial of the second kind $P_n(u)$ from (26.1) has one zero less than the corresponding polynomial of the first kind $Q_n(u)$, since the degrees of the polynomials $P_n(u)$ and $Q_n(u)$ are $n-1$ and n, respectively. This suggests that there ought to be a linear transformation, say $\mathcal{T}$, which is capable of deducing the polynomial $P_K(u)$ from a function containing the quotient $\propto Q_K(z)/(u-z)$. It appears that the contour integral in (30.3) is well suited for such a task of the operator $\mathcal{T}$. Clearly, this is contingent upon a judicious modification of the function $Q_K(z)/(u-z)$ to meet the requirement of a regular behaviour throughout the complex z-plane bounded by the contour C which encompasses the eigenset $\mathcal{K}$. Such an adjustment is possible through a replacement of the ansatz $Q_K(z)/(u-z)$ by the following first-order finite difference

$$g(u,z) = \frac{Q_K(u) - Q_K(z)}{u - z} \qquad P_K(u) = \mathcal{T} g(u, z) \qquad (36.1)$$

so that the polynomial $P_K(u)$ should be deducible from the linear mapping $P_K(u) = \mathcal{T} g(u, z)$. This is indeed possible by selecting the operator $\mathcal{T}$ to be proportional to the contour integral from (30.3). Such a choice leads naturally to the following integral representation for the Lanczos polynomial of the second kind $P_n(u)$ [168]

$$P_n(u) = \frac{\beta_1}{c_0} \frac{1}{2\pi i} \oint_C d\sigma_0(z) \frac{Q_n(u) - Q_n(z)}{u - z} \qquad (36.2)$$

where C is the contour from (30.3). In the special case $n = K$ in (36.2), we use the definition of the inner product (30.8), as well as the characteristic equation

$Q_K(u_k) = 0$ from (25.10), to obtain

$$P_K(u) = \frac{\beta_1}{c_0} \frac{1}{2\pi i} \oint_C d\sigma_0(z) \frac{Q_K(u) - Q_K(z)}{u - z}$$

$$= \frac{\beta_1}{c_0} \sum_{k=1}^{K} d_k \frac{Q_K(u) - Q_K(u_k)}{u - u_k} = \frac{\beta_1}{c_0} Q_K(u) \sum_{k=1}^{K} \frac{d_k}{u - u_k} \quad (36.3)$$

where the residue $d_k = (\Phi_0|\Upsilon_k)^2$ is taken from the definition (3.10). Since $\beta_1/c_0 \neq 0$ according to (4.5) and (22.11), we can divide (36.3) by the term $(\beta_1/c_0) Q_K(u) \neq 0$, so that

$$\frac{c_0}{\beta_1} \frac{P_K(u)}{Q_K(u)} = \sum_{k=1}^{K} \frac{d_k}{u - u_k}. \quad (36.4)$$

The rhs of (36.4) is equal to the spectral representation of the Kth rank approximation to the total Green function $\mathcal{R}(u)$ from (5.23). The representation $(c_0/\beta_1) P_K(u)/Q_K(u)$ from (36.4) for the Green function $\mathcal{R}(u)$ will be called the Padé–Lanczos approximant (PLA) of the order $(K-1, K)$ and, as such, will be denoted by $\mathcal{R}^{\text{PLA}}_{K,K}(u) \equiv \mathcal{R}^{\text{PLA}}_K(u)$ [17, 97, 120]

$$\mathcal{R}(u) \approx \mathcal{R}^{\text{PLA}}_K(u) \qquad \mathcal{R}^{\text{PLA}}_K(u) \equiv \frac{c_0}{\beta_1} \frac{P_K(u)}{Q_K(u)}. \quad (36.5)$$

The Lanczos algorithm and the Padé approximant have also been combined in other research fields, e.g. in computational and applied mathematics [85], as well as in engineering via circuit theory [120]. The authors of [120] from 1995, being apparently unaware of the earlier work of [97] from 1972, rederived through a different procedure the Padé–Lanczos approximant and called it the Padé via Lanczos (PVL) method.

The numerator $P_K(u)$ and denominator $Q_K(u)$ polynomials from (36.5) are of degree $K - 1$ and K, respectively, so that the ensuing PLA belongs to the 'subdiagonal' case of the general order Padé approximant [81]. The PLA can also be introduced by starting from inversion of the matrix $(\Phi_0|\{u\mathbf{1}-\mathbf{U}\}^{-1}|\Phi_0)$, as will be analysed in chapter 37. In either case, the name Padé–Lanczos approximant and the associated acronym PLA is used to emphasize that the PA, i.e. the representation $(c_0/\beta_1) P_K(u)/Q_K(u)$ for the Green function $\mathcal{R}(u)$ from (5.22), is generated by means of the Lanczos algorithms for the numerator and denominator polynomials $P_K(u)$ and $Q_K(u)$, respectively. In addition to the subdiagonal PLA, we can also introduce the general PLA as a ratio of two Lanczos polynomials of degrees $L - 1$ and K as

$$\mathcal{R}^{\text{PLA}}_{L,K}(u) \equiv \frac{c_0}{\beta_1} \frac{P_L(u)}{Q_K(u)} \quad (36.6)$$

where $K \geq L$. If $L = K$, we have the subdiagonal PLA, whereas the case $L = K + m$, where m is a positive integer ($m \geq 2$), represents the mth 'paradiagonal' PLA. The pure diagonal PLA corresponds to $L - 1 = K$. We saw in chapter 23 that the polynomials $P_K(u)$ and $Q_K(u)$ are generated individually in the course of producing the Lanczos states $\{|\psi_n\rangle\}$ in K iterations. Such a computation rests upon the availability of the coupling parameters $\{\alpha_n, \beta_n\}$ that are generated within the recursion for the state vectors $\{|\psi_n\rangle\}$. However, this is not how the Lanczos polynomials should be constructed in practice. It would be advantageous to dissociate completely the recursions for $\{P_n(u), Q_n(u)\}$ from the Lanczos algorithm for $\{|\psi_n\rangle\}$. To this end, one needs an autonomous strategy for computations of the constants $\{\alpha_n, \beta_n\}$. Such are the powerful Chebyshev, Gragg and Wheeler algorithms linked to the nearest neighbour method in [17]. *Generation of the constants $\{\alpha_n, \beta_n\}$ is of paramount importance, since these parameters contain the whole information about the studied system.* This is because the Green function, density of states, evolution matrix, auto-correlation function, etc, can all be expressed solely in terms of the couplings $\{\alpha_n, \beta_n\}$. In this way, one can bypass altogether any reliance of the polynomials $\{P_n(u), Q_n(u)\}$ upon the Lanczos recursion for the state vectors $\{|\psi_n\rangle\}$.

The spectral representation of $\mathcal{R}_K^{\text{PLA}}(u)$ from (36.5) is given by the Heaviside partial fractions [132]

$$\mathcal{R}_K^{\text{PLA}}(u) = \frac{c_0}{\beta_1} \frac{P_K(u)}{Q_K(u)} = \sum_{k=1}^{K} \frac{d_k}{u - u_k}. \tag{36.7}$$

This expression for $\mathcal{R}_K^{\text{PLA}}(u)$ represents a meromorphic function, since its poles are the only singularities encountered. Therefore, the zeros of $P_K(u)$ and $Q_K(u)$ are the zeros and poles of $\mathcal{R}_K^{\text{PLA}}(u)$, respectively. There are K poles $\{u_k\}_{k=1}^{K}$ of $\mathcal{R}_K^{\text{PLA}}(u)$, since $Q_K(u)$ is a polynomial of the Kth degree. The Heaviside partial fraction (36.7) includes only the first-order poles $\{u_k\}$ since the polynomials $P_K(u)$ and $Q_K(u)$ have no multiple zeros as shown in chapter 25. The definition (36.5) of the PLA obeys the main property of the general PA, according to which the Maclaurin expansions of the Green functions $\mathcal{R}(u)$ and $\mathcal{R}_K^{\text{PLA}}(u)$ in powers of the variable u agree with each other, exactly term-by-term, including the power $2K - 1$

$$\mathcal{R}(u) - \frac{c_0}{\beta_1} \frac{P_K(u)}{Q_K(u)} = \mathcal{O}(u^{2K}). \tag{36.8}$$

Here, as usual, the remainder symbol $\mathcal{O}(u^{2K})$ represents a Maclaurin series in powers of u, such that the starting term is u^{2K}.

The above derivation shows that the definition of the residue $d_k = (\Phi_0|\Upsilon_k)^2$ from the general expression (3.10) is transferred as intact to the spectral representation (36.7) in the PLA. Therefore, such an introduction of the PLA falls precisely within the so-called state expansion methods in signal processing [17, 64]. This establishes the PLA as a complete eigenvalue solver, since the computed Lanczos pair $\{|\psi_n\rangle, Q_n(u)\}$, as well as the zeros of the characteristic polynomial

$Q_K(u)$, also provide the whole eigenset $\{u_k, |\Upsilon_k\rangle\}$ according to (23.1). However, while the residues $\{d_k\}$ in other state expansion methods, e.g. FD [108, 109] or DSD [115] necessitate explicit computations to obtain some approximations of the full exact eigenstate vector $|\Upsilon_k\rangle$, the PLA possesses an alternative procedure. This procedure does not require any knowledge about the state vector $|\Upsilon_k\rangle$. It is an explicit expression for d_k as the residue of the quotient $P_K(u)/Q_K(u)$

$$d_k = \frac{c_0}{\beta_1} \lim_{u \to u_k} (u - u_k) \frac{P_K(u)}{Q_K(u)}. \tag{36.9}$$

The simplest way to calculate d_k from (36.9) is given by the following direct prescription

$$\frac{\beta_1}{c_0} d_k = \lim_{u \to u_k} (u - u_k) \frac{P_K(u)}{Q_K(u)}$$

$$= P_K(u_k) \left\{ \lim_{u \to u_k} \frac{Q_K(u) - Q_K(u_k)}{u - u_k} \right\}^{-1} = \frac{P_K(u_k)}{\{(d/du)Q_K(u)\}_{u=u_k}}$$

where we 'extended' the denominator $Q_K(u)$ by subtracting the zero term $Q_K(u_k) = 0$. Moreover, we used the definition of the first derivative $[(d/du)Q_K(u)]_{u=u_k} = \lim_{u \to u_k}[Q_K(u) - Q_K(u_k)]/(u - u_k)$. Thus, the PLA yields the result for the residue d_k in the closed analytical form as

$$d_k = \frac{c_0}{\beta_1} \frac{P_K(u_k)}{Q'_K(u_k)} \qquad Q'_K(u) = \frac{dQ_K(u)}{du}. \tag{36.10}$$

Therefore (36.7) becomes

$$\mathcal{R}_K^{\text{PLA}}(u) = \sum_{k=1}^{K} \left\{ \frac{c_0}{\beta_1} \frac{P_K(u_k)}{Q'_K(u_k)} \right\} \frac{1}{u - u_k}. \tag{36.11}$$

We computed the first three PLAs from (36.5) and the results are as follows

$$\mathcal{R}_1^{\text{PLA}}(u) = \frac{c_0}{u - \alpha_0}$$

$$\mathcal{R}_2^{\text{PLA}}(u) = c_0 \frac{u - \alpha_1}{u^2 - (\alpha_0 + \alpha_1)u + (\alpha_0\alpha_1 - \beta_1^2)}$$

$$\mathcal{R}_3^{\text{PLA}}(u) = c_0 \frac{u^2 - (\alpha_1 + \alpha_2)u + (\alpha_1\alpha_2 - \beta_2^2)}{u^3 - (\alpha_0 + \alpha_1 + \alpha_2)u^2 + \alpha_\beta u + \beta_\alpha} \tag{36.12}$$

$$\alpha_\beta = \alpha_0\alpha_1 + \alpha_2(\alpha_0 + \alpha_1) - \beta_1^2 - \beta_2^2$$

$$\beta_\alpha = \alpha_0\beta_2^2 - \alpha_2(\alpha_0\alpha_1 - \beta_1^2). \tag{36.13}$$

Using the respective expressions (26.12) and (26.14) for the quantities p_k and $q_k(0 \leq m \leq 3)$, the results from (36.12) can be put in the following forms that contain only the signal points $\{c_n\}$ and the variable u

$$\mathcal{R}_1^{\text{PLA}}(u) = \frac{c_0^2}{c_0 u - c_1}$$

$$\mathcal{R}_2^{\text{PLA}}(u) = \frac{c_0(c_1^2 - c_0 c_2)u + (c_0^2 c_3 - 2 c_0 c_1 c_2 + c_1^3)}{(c_1^2 - c_0 c_2)u^2 + (c_0 c_3 - c_1 c_2)u + (c_2^2 - c_1 c_3)}$$

$$\mathcal{R}_3^{\text{PLA}}(u) = \frac{p_0 u^2 + p_1 u + p_2}{q_0 u^3 + q_1 u^2 + q_2 u + q_3}. \tag{36.14}$$

Chapter 37

Inversion of the Schrödinger matrix $u\mathbf{1} - \mathbf{U}$ by the PLA

Once the Lanczos coupling constants $\{\alpha_n, \beta_n\}$ have been computed, we could construct the Green function $\mathcal{R}(u) = c_0(\Phi_0|\{u\mathbf{1} - \mathbf{U}\}^{-1}|\Phi_0)$ which is defined in (5.22). If in the formula (5.22) for $\mathcal{R}(u)$ we use the Lanczos representation for the matrix $u\mathbf{1} - \mathbf{U}$ based upon (22.10) within the infinite chain model, we shall have

$$\mathcal{R}(u) = c_0 \begin{pmatrix} u-\alpha_0 & -\beta_1 & 0 & \cdots & \cdots & 0 & \cdots \\ -\beta_1 & u-\alpha_1 & -\beta_2 & \cdots & \cdots & 0 & \cdots \\ 0 & -\beta_2 & u-\alpha_2 & \cdots & \cdots & 0 & \cdots \\ \vdots & \vdots & \vdots & \ddots & \ddots & \vdots & \vdots \\ 0 & 0 & 0 & -\beta_{n-1} & u-\alpha_{n-1} & -\beta_n & \cdots \\ 0 & 0 & 0 & 0 & -\beta_n & u-\alpha_n & \cdots \\ \vdots & \vdots & \vdots & \vdots & \vdots & \vdots & \vdots \end{pmatrix}^{-1}_{00}$$

(37.1)

where the symbol $(\cdots)_{00}^{-1}$ denotes $(\cdots)_{00}^{-1} \equiv (\Phi_0|(\cdots)^{-1}|\Phi_0)$ and $(\cdots)^{-1} \equiv (u\mathbf{1} - \mathbf{U})^{-1}$. However, by definition, the matrix element (37.1) of the inverse matrix $(u\mathbf{1} - \mathbf{U})^{-1}$ is given by the ratio of the corresponding determinant $\det[u\mathbf{1} - \mathbf{U}]$ and the associated main cofactor. Therefore

$$\mathcal{R}(u) = c_0 \frac{D_1(u)}{D_0(u)}. \tag{37.2}$$

Inversion of the Schrödinger matrix $u\mathbf{1} - \mathbf{U}$ by the PLA   219

Here, $D_0(u)$ is the determinant of the matrix $u\mathbf{1} - \mathbf{U}$ for the infinite chain model ($M = \infty$)

$$D_0(u) = \begin{vmatrix} u - \alpha_0 & -\beta_1 & 0 & \cdots & \cdots & 0 & \cdots \\ -\beta_1 & u - \alpha_1 & -\beta_2 & \cdots & \cdots & 0 & \cdots \\ 0 & -\beta_2 & u - \alpha_2 & \cdots & \cdots & 0 & \cdots \\ \vdots & \vdots & \vdots & \ddots & \ddots & \vdots & \vdots \\ 0 & 0 & 0 & -\beta_{n-1} & u - \alpha_{n-1} & -\beta_n & \cdots \\ 0 & 0 & 0 & 0 & -\beta_n & u - \alpha_n & \cdots \\ \vdots & \vdots & \vdots & \vdots & \vdots & \vdots & \vdots \end{vmatrix}$$

(37.3)

and $D_1(u)$ is the corresponding main cofactor of $D_0(u)$ at the position $u - \alpha_0$ (see Appendix A). In other words $D_1(u)$ is obtained by removing the first row and the first column from $D_0(u)$

$$D_1(u) \equiv \begin{vmatrix} u - \alpha_1 & -\beta_2 & 0 & \cdots & \cdots & 0 & \cdots \\ -\beta_2 & u - \alpha_2 & -\beta_3 & \cdots & \cdots & 0 & \cdots \\ 0 & -\beta_3 & u - \alpha_3 & \cdots & \cdots & 0 & \cdots \\ \vdots & \vdots & \vdots & \ddots & \ddots & \vdots & \vdots \\ 0 & 0 & \cdots & -\beta_{n-1} & u - \alpha_{n-1} & -\beta_n & \cdots \\ 0 & 0 & \cdots & 0 & -\beta_n & u - \alpha_n & \cdots \\ \vdots & \vdots & \vdots & \vdots & \vdots & \vdots & \vdots \end{vmatrix}.$$

(37.4)

The results of truncation of the infinite-order determinants $D_0(u)$ and $D_1(u)$ at $n = K$ will be denoted by $D_{0,K}(u)$ and $D_{1,K}(u)$, respectively, such that $D_{0,K}(u) = \tilde{Q}_K(u)$ according to (25.4)

$$D_{0,K}(u) \equiv \begin{vmatrix} u - \alpha_0 & -\beta_1 & 0 & \cdots & \cdots & 0 \\ -\beta_1 & u - \alpha_1 & -\beta_2 & \cdots & \cdots & 0 \\ 0 & -\beta_2 & u - \alpha_2 & \cdots & \cdots & 0 \\ \vdots & \vdots & \vdots & \ddots & \ddots & \vdots \\ 0 & 0 & 0 & -\beta_{K-2} & u - \alpha_{K-2} & -\beta_{K-1} \\ 0 & 0 & 0 & 0 & -\beta_{K-1} & u - \alpha_{K-1} \end{vmatrix}$$

(37.5)

$$D_{1,K}(u) \equiv \begin{vmatrix} u - \alpha_1 & -\beta_2 & 0 & \cdots & \cdots & 0 \\ -\beta_2 & u - \alpha_2 & -\beta_3 & \cdots & \cdots & 0 \\ 0 & -\beta_3 & u - \alpha_3 & \cdots & \cdots & 0 \\ \vdots & \vdots & \vdots & \ddots & \ddots & \vdots \\ 0 & 0 & \cdots & -\beta_{K-2} & u - \alpha_{K-2} & -\beta_{K-1} \\ 0 & 0 & \cdots & 0 & -\beta_{K-1} & u - \alpha_{K-1} \end{vmatrix}.$$

(37.6)

The ensuing Green function $c_0 D_{1,K}(u)/D_{0,K}(u)$ is an approximation to $\mathcal{R}(u)$ from (37.2)

$$\mathcal{R}(u) \approx \mathcal{R}_K(u) \qquad \mathcal{R}_K(u) \equiv c_0 \frac{D_{1,K}(u)}{D_{0,K}(u)}. \tag{37.7}$$

The determinant $D_0(u)$ can be computed iteratively by expressing it in terms of its subdeterminants $D_1(u)$ and $D_2(u)$ via the usage of the Cauchy expansion for determinants, as we have done earlier in (25.6). The determinant $D_2(u)$ is obtained by deleting the first two rows and the first two columns from $D_0(u)$. Thus, in a complete analogy with (25.7) we have

$$D_0(u) = (u - \alpha_0) D_1(u) - \beta_1^2 D_2(u). \tag{37.8}$$

In order to continue this iterative process further by descending to lower ranks, we denote by $D_n(u)$ the determinant which is obtained by deleting the first n rows and columns from $D_0(u)$. Then by working inductively with the help of the Cauchy expansion for determinants, we obtain the generalization of (37.8) as

$$D_{n-1}(u) = (u - \alpha_{n-1}) D_n(u) - \beta_n^2 D_{n+1}(u) \tag{37.9}$$

with the initializations $D_{-1}(u) = 0$ and $D_0(u) = 1$. This three-term contiguous relation for $D_n(u)$ falls precisely into the category of the Askey–Wilson polynomials (25.8) [262], just like the oppositely recurring counterpart of (25.7). When the Askey–Wilson polynomials are divided by the constant $\bar{\beta}_n$ from (22.4), the Lanczos polynomials $\{Q_n(u)\}$ are obtained as in (25.9). In particular

$$\begin{aligned} D_{0,K}(u) &= (\beta_1 \beta_2 \beta_2 \cdots \beta_K) Q_K(u) = \tilde{Q}_K(u) \\ D_{1,K}(u) &= (\beta_2 \beta_2 \cdots \beta_K) P_K(u) = \tilde{P}_K(u). \end{aligned} \tag{37.10}$$

Here, $\tilde{P}_K(u)$ is the minor of the determinant (25.4) at the position $u - \alpha_0$ for $n = K - 1$, obtained by deleting the first column and the first row from (25.4). Thus, $\tilde{P}_K(u) = D_{1,K}(u)$, as per (37.6)

$$c_0 \frac{D_{1,K}(u)}{D_{0,K}(u)} = c_0 \frac{\tilde{P}_K(u)}{\tilde{Q}_K(u)} = c_0 \frac{(\beta_2 \beta_3 \cdots \beta_K) P_K(u)}{(\beta_1 \beta_2 \beta_2 \cdots \beta_K) Q_K(u)} = \frac{c_0}{\beta_1} \frac{P_K(u)}{Q_K(u)} \tag{37.11}$$

where $Q_K(u)$ and $P_K(u)$ are the Lanczos polynomials (23.16) and (23.20) of degree K and $K - 1$, respectively. Therefore, truncation of the infinite-order determinants $D_0(u)$ and $D_1(u)$ at the finite order $n = K - 1$ leads to the following approximation for the quotient $\mathcal{R}(u)$ from (37.7)

$$\mathcal{R}_K(u) = \frac{c_0}{\beta_1} \frac{P_K(u)}{Q_K(u)} = \mathcal{R}_K^{\text{PLA}}(u) \tag{37.12}$$

in agreement with the previously established expression (36.5) for the resolvent $\mathcal{R}_K^{\text{PLA}}(u)$ in the PLA. The result $\mathcal{R}_K^{\text{PLA}}(u)$ from (37.12) is the PLA of the order

$(K-1, K)$ for the exact expression for $\mathcal{R}(u)$ from (37.2). The above derivation demonstrates that the explicit and exact inversion of scaled evolution matrix $u\mathbf{1}-\mathbf{U}$ is possible for any finite rank and the analytical result is precisely the Padé–Lanczos approximant. It should be recalled that in the literature the algorithm of the Lanczos continued fractions for iterative computation of the diagonal or off-diagonal Green functions is also known as *the recursion method* [97], or equivalently, the recursive residue generation method (RRGM) [100].

Chapter 38

The exact spectrum via the Green function series

The Green function $\mathcal{R}(u)$ is defined in (5.22) as a matrix element of the corresponding Green operator or resolvent $\hat{R}(u) = [\hat{1}u - \hat{U}(\tau)]^{-1}$ from (5.21) taken over the initial state $|\Phi_0\rangle$. In chapter 37 we wrote down the expression (37.2) for $\mathcal{R}(u)$ using the definition of a general inverse matrix. This leads to the PLA when the matrix $u\mathbf{1} - \mathbf{U}(\tau)$ to be inverted is tridiagonalized via the Lanczos algorithm. The PLA is a hybrid method with two steps consisting of: (i) the Lanczos tridiagonalization of the original large matrix and (ii) the subsequent inversion of the sparse Jacobi matrix $u\mathbf{1} - \mathbf{U}(\tau)$ by means of the Padé approximant. Given the function $\mathcal{R}(u)$, its PLA of a fixed order is unique and, hence, the same result must be obtained irrespective of the method used to generate the numerator and denominator polynomials. Therefore, the results for $\mathcal{R}_K^{\text{PLA}}(u)$ from (36.5) obtained within the PLA must coincide with any other technique of construction of the Padé rational polynomial for $\mathcal{R}(u)$. Let us, therefore, apply a different procedure to evaluate $\mathcal{R}(u)$ within the usual PA without the Lanczos algorithm and denote the obtained result with $\mathcal{R}_K^{\text{PA}}(u)$. To this end, we use the binomial series for the resolvent operator from (5.21)

$$\hat{R}(u) = [\hat{1}u - \hat{U}(\tau)]^{-1} = \sum_{n=0}^{\infty} \hat{U}^n(\tau) u^{-n-1}. \qquad (38.1)$$

Upon insertion of this expansion into the matrix element for the Green function $\mathcal{R}(u) = (\Phi_0|\hat{R}(u)|\Phi_0)$ from (5.22), we shall obtain a series in powers of u^{-1} with the coefficients $(\Phi_0|\hat{U}^n(\tau)|\Phi_0)$ that are the auto-correlation functions C_n according to (5.9). In the case of a Lorentzian spectrum, the expression (5.12) holds true, so that $C_n = c_n$ where $\{c_n\}$ is the corresponding time signal set and, therefore

$$\mathcal{R}(u) = \sum_{n=0}^{\infty} c_n u^{-n-1}. \qquad (38.2)$$

The exact spectrum via the Green function series

It is convenient to introduce an auxiliary function $\mathcal{G}(u^{-1})$ which is inherently contained in (38.2) as

$$\mathcal{G}(u^{-1}) = \sum_{n=0}^{\infty} c_n u^{-n} \qquad \mathcal{R}(u) = u^{-1}\mathcal{G}(u^{-1}). \qquad (38.3)$$

Then, the most general PAs, as symbolized by $[L/K]_{\mathcal{G}}(u^{-1})$ and $[L/K]_{\mathcal{R}}(u)$ for $\mathcal{G}(u^{-1})$ and $\mathcal{R}(u)$ for the series (38.3) and (38.2), respectively, are given by [17, 18]

$$\mathcal{G}(u^{-1}) \approx [L/K]_{\mathcal{G}}(u^{-1}) \equiv \frac{A_L(u^{-1})}{B_K(u^{-1})}$$

$$\mathcal{R}(u) \approx \mathcal{R}^{PA}_{L,K}(u) \equiv [L/K]_{\mathcal{R}}(u) = u^{-1}\frac{A_L(u^{-1})}{B_K(u^{-1})} \qquad (38.4)$$

where $A_L(u^{-1})$ and $B_K(u^{-1})$ are the unknown polynomials in variable u^{-1} of degrees L and K

$$A_L(u^{-1}) = \sum_{\ell=0}^{L} a_\ell u^{-\ell} \qquad B_K(u^{-1}) = \sum_{k=0}^{K} b_k u^{-k} \qquad b_0 \neq 0. \qquad (38.5)$$

We then say that the function $[L/K]_{\mathcal{G}}(u^{-1})$ from (38.4) is the PA of the order (L, K) for the ansatz $\mathcal{G}(u^{-1})$. The constant coefficients $\{a_\ell, b_k\}$ can be extracted uniquely from the given set $\{c_n\}_{n=0}^{K+L}$. The meaning of (38.4) is that the first $K+L$ coefficients of the Maclaurin expansions of the quotient $A_L(u^{-1})/B_K(u^{-1})$ agree exactly with the given subset $\{c_n\}$ for $0 \leq n \leq K+L$. Clearly (38.2) itself is the Maclaurin expansion of the function $\mathcal{R}(u)$. The above clarification of the significance of (38.4), in fact, leads to an alternative definition of the PA using $|u| \longrightarrow \infty$ via

$$\mathcal{G}(u^{-1}) - \frac{A_L(u^{-1})}{B_K(u^{-1})} = \mathcal{O}(u^{-L-K-1}),$$

$$\mathcal{G}(u^{-1})B_K(u^{-1}) - A_L(u^{-1}) = \mathcal{O}(u^{-L-1}). \qquad (38.6)$$

The remainder $\mathcal{O}(u^{-L-K-1})$ in (38.6) denotes a power series containing the terms proportional to u^{-L-K-m} ($m = 1, 2, 3, \ldots$). In other words, the term $\mathcal{O}(u^{-L-K-1})$ from (38.6) indicates that the Maclaurin expansions of $\mathcal{G}(u^{-1})$ and $[L/K]_{\mathcal{G}}(u^{-1})$ differ only in the coefficients multiplying the powers u^{-m} with $m \geq K+L+1$. Also, the relation (38.6) can alternatively be viewed as the extrapolation condition for $\mathcal{G}(u^{-1})$ at infinity ($u \longrightarrow \infty$). All poles and zeros in the PA are *free* in the sense that they are not given in advance, but instead they are determined from the extrapolation condition (38.6). In practice, one does not need to deal with the whole $L \times K$ two-dimensional Padé table of the order (L, K), but only with, e.g. the 'subdiagonal' PA $(L = K - 1)$ of the order $(K - 1, K)$

$$\mathcal{R}_K^{PA}(u) = u^{-1} \frac{A_{K-1}(u^{-1})}{B_K(u^{-1})}. \tag{38.7}$$

Here, $\mathcal{R}_K^{PA}(u) \equiv \mathcal{R}_{K,K}^{PA}(u)$ is the standard PA for $\mathcal{R}(u)$ of the order $(K - 1, K)$. This form of the usual subdiagonal PA helps a comparison with the PLA of the order $(K - 1, K)$, i.e. $\mathcal{R}_K^{PLA}(u)$ for $\mathcal{R}(u)$.

At this point, it is pertinent to return to the definition (38.6) of the standard PA. This is a practical definition, since it permits determination of all the coefficients $\{a_\ell\}$ and $\{b_k\}$. The lhs of the second equation in (38.6) is exact to the order u^{-m} with $m \leq L$, in which case $\mathcal{O}(u^{-L-1}) = 0$, so that $\mathcal{G}(u^{-1})B_K(u^{-1}) - A_L(u^{-1}) = 0$. From here, equating the like power of u^{-1} we obtain the system of linear equations (11.10) for b_k and the explicit formula (11.11) for a_ℓ. Of course, for small values of L and K, the analytical solutions of the system (11.10) are readily obtained by the standard Cramer rule. In this way, the first few results for the polynomials $A_L(u^{-1})$ and $B_K(u^{-1})$ with $L \leq 2$ and $K \leq 3$ read as

$$A_0(u^{-1}) = c_0 b_0$$
$$A_1(u^{-1}) = b_0 \frac{c_0(c_0 c_2 - c_1^2) + (2c_0 c_1 c_2 - c_0^2 c_3 - c_1^3)u^{-1}}{c_0 c_2 - c_1^2}$$
$$A_2(u^{-1}) = b_0 \frac{p_0 + p_1 u^{-1} + p_2 u^{-2}}{p_0} \tag{38.8}$$
$$B_1(u^{-1}) = b_0 \left(1 - \frac{c_1}{c_0} u^{-1}\right)$$
$$B_2(u^{-1}) = b_0 \frac{(c_0 c_2 - c_1^2) - (c_0 c_3 - c_1 c_2)u^{-1} + (c_1 c_3 - c_2^2)u^{-2}}{c_0 c_2 - c_1^2}$$
$$B_3(u^{-1}) = b_0 \frac{q_0 + q_1 u^{-1} + q_2 u^{-2} + q_3 u^{-3}}{q_0} \tag{38.9}$$

where the constants p_n and q_n are given by (26.13) and (26.14). Note that the common overall multiplying constant b_0 from the polynomials $A_{K-1}(u^{-1})$ and $B_K(u^{-1})$ is cancelled in their ratio $A_{K-1}(u^{-1})/B_K(u^{-1})$. Comparing (26.15) or (26.16) with (38.8) or (38.9), we find that the following relationship holds true for the general degrees $K - 1$ and K of the polynomial pairs $\{P_K(u), Q_K(u)\}$ and $\{A_{K-1}(u^{-1}), Q_K(u^{-1})\}$

$$u^{K-1} A_{K-1}(u^{-1}) = \frac{c_0 b_0}{p_{K,1}} P_K(u) = c_0 b_0 \tilde{P}_K(u)$$
$$u^K B_K(u^{-1}) = \frac{b_0}{q_{K,0}} Q_K(u) = b_0 \tilde{Q}_K(u) \tag{38.10}$$

where $\{p_{K,1}, q_{K,0}\}$ are the leading coefficients from (26.5) in the power series representations of the polynomials $\{P_K(u), Q_K(u)\}$, respectively[1]. The polynomial members of the set $\{Q_n(u)\}$ are mutually orthogonal for different degrees n ($n = 0, 1, 2, \ldots$) with respect to the weight function d_k which is the residue as stated in (31.23). This fact, together with the relationship (38.10), implies that the related elements from the sequence $\{u^n B_n(u^{-1})\}$ are also orthogonal polynomials. A similar conclusion is also valid for the polynomials $\{P_n(u)\}$ and $\{u^{n-1} A_{n-1}(u^{-1})\}$ with a different weight function $d'_k \neq d_k$. Substituting the results (38.8) and (38.9) into (38.7), we find

$$\mathcal{R}_1^{PA}(u) = \frac{c_0^2}{c_0 u - c_1}$$

$$\mathcal{R}_2^{PA}(u) = \frac{(c_0^2 c_3 - 2c_0 c_1 c_2 + c_1^3) + c_0(c_1^2 - c_0 c_2)u}{(c_0^2 - c_1 c_3) + (c_0 c_3 - c_1 c_2)u + (c_1^2 - c_0 c_2)u^2}$$

$$\mathcal{R}_3^{PA}(u) = \frac{p_2 + p_1 u + p_0 u^2}{q_3 + q_2 u + q_1 u^2 + q_0 u^3}. \tag{38.11}$$

This leads to the following equalities

$$\mathcal{R}_1^{PA}(u) = \mathcal{R}_1^{PLA}(u) \quad \mathcal{R}_2^{PA}(u) = \mathcal{R}_2^{PLA}(u) \quad \mathcal{R}_3^{PA}(u) = \mathcal{R}_3^{PLA}(u) \tag{38.12}$$

where $\mathcal{R}_1^{PLA}(u)$, $\mathcal{R}_2^{PLA}(u)$ and $\mathcal{R}_3^{PLA}(u)$ are the results from (36.14). Therefore, we can inductively arrive at the general relationship

$$\mathcal{R}_K^{PA}(u) = \mathcal{R}_K^{PLA}(u). \tag{38.13}$$

This result proves that the Maclaurin series (38.2) of the Green function $\mathcal{R}(u)$ in powers of u^{-1}, with the auto-correlation functions C_n or time signal points c_n as the expansion coefficients, is mathematically equivalent to the Padé–Lanczos approximant.

Of great importance in practice is the uniqueness of the PLA. This means that there is one and only one quotient of two extracted polynomials $A_L(u^{-1})/B_K(u^{-1})$ for the function $\mathcal{G}(u^{-1})$ from the series (38.3). If there were two such quotients $A_L(u^{-1})/B_K(u^{-1})$ and $\tilde{A}_L(u^{-1})/\tilde{B}_K(u^{-1})$, then they would both obey the definition of the PA. The polynomial ratio $A_L(u^{-1})/B_K(u^{-1})$ will comply with (38.6), whereas the corresponding quotient $\tilde{A}_L(u^{-1})/\tilde{B}_K(u^{-1})$ would satisfy the like condition

$$\mathcal{G}(u^{-1})\tilde{B}_K(u^{-1}) - \tilde{A}_L(u^{-1}) = \tilde{\mathcal{O}}(u^{-L-1}) \tag{38.14}$$

where $\tilde{\mathcal{O}}(u^{-L-1})$ is a power series which starts from u^{-L-1}. Multiplying (38.6) and (38.14) by $\tilde{B}_K(u^{-1})$ and $B_K(u^{-1})$, respectively, and subtracting the results

---

[1] Again we recall that the degree of the polynomial $P_K(u)$ is $K - 1$ and, hence, it is the same as the degree of the polynomial $A_{K-1}(u^{-1})$. On the other hand, the polynomials $Q_K(u)$ and $B_K(u^{-1})$ are both of degree K in their variables u and u^{-1}, respectively.

from each other, it follows

$$A_L(u^{-1})\tilde{B}_K(u^{-1}) - \tilde{A}_L(u^{-1})B_K(u^{-1})$$
$$= \mathcal{O}(u^{-L-1})\tilde{B}_K(u^{-1}) - \tilde{\mathcal{O}}(u^{-L-1})B_K(u^{-1}). \quad (38.15)$$

The lhs of (38.15) represents a polynomial whose degree is at most $L + K$ in variable u^{-1}, whereas the rhs of the same equation symbolically denotes a power series which begins with the term u^{-m-L-1} ($m \geq 0$). Hence, to the accuracy which is governed by the inclusion of all the terms $u^{-\nu}$ with $\nu \leq L + K$, as required by the PA of the general order (L, K), the rhs of (38.15) is exactly zero. Consequently, the two quotients $A_L(u^{-1})/B_K(u^{-1})$ and $\tilde{A}_L(u^{-1})/\tilde{B}_K(u^{-1})$ are the same

$$\frac{A_L(u^{-1})}{B_K(u^{-1})} = \frac{\tilde{A}_L(u^{-1})}{\tilde{B}_K(u^{-1})}. \quad (38.16)$$

Thus, there is only one ratio of two polynomials representing the Green function $u^{-1}\mathcal{G}(u^{-1}) \approx \mathcal{R}_K^{PA}(u)$ from (38.3), so that

$$\mathcal{R}_K^{PA}(u) = \frac{c_0}{\beta_1}\frac{P_K(u)}{Q_K(u)} = u^{-1}\frac{A_{K-1}(u^{-1})}{B_K(u^{-1})}. \quad (38.17)$$

The PLA also works directly in the time domain, in which case it can filter out exactly all the K transients $\{d_k u_k^n\}$ from the signal (5.11). Such a filter/transform exists only if the two conditions (7.11) and (7.12) are simultaneously satisfied, i.e. $H_K(c_n) \neq 0$ and $H_{K+1}(c_n) = 0$ [17]. This, in turn, determines unequivocally the true number K of resonances in the corresponding frequency spectrum.

Chapter 39

Uniqueness of the amplitudes $\{d_k\}$

Thus far, we have derived several spectral representations of the Green function with formally different expressions for the residues $\{d_k\}$. In particular, the residues $\{d_k\}$ in the defining equation (3.10) are given by the overlap $(\Phi_0|\Upsilon_k)^2$ as in FD [108, 109] and DSD [115]. On the other hand, the residues $\{d_k\}$ in the PLA are given by an apparently different formula $c_0 P_K(u_k)/[\beta_1 Q'_K(u_k)]$ in (36.10) [17]. However, the uniqueness of the PLA must imply that these two results for the residues are the same for the common set of the eigenvalues $\{u_k\}$. To prove this statement, we begin by assuming that, for the same number of available signal points $\{c_n\}$, one derives the following two, seemingly different, spectral representations for the PLA to the Green function $\mathcal{R}(u)$

$$\mathcal{R}_K^{\text{PLA}}(u) = \sum_{k=1}^{K} \left\{ \frac{c_0}{\beta_1} \frac{P_K(u_k)}{Q'_K(u_k)} \right\} \frac{1}{u - u_k}$$

$$\bar{\mathcal{R}}_K^{\text{PLA}}(u) = \sum_{k=1}^{K} \frac{(\Phi_0|\Upsilon_k)^2}{u - u_k} \qquad (39.1)$$

where the eigenvalues $\{u_k\}$ are common to both of the variants $\mathcal{R}_K^{\text{PLA}}(u)$ and $\bar{\mathcal{R}}_K^{\text{PLA}}(u)$. Given the simple poles $\{u_k\}$, the usual definitions of the residues d_k and $\bar{d}_k$ of the functions $\mathcal{R}_K^{\text{PLA}}(u)$ and $\bar{\mathcal{R}}_K^{\text{PLA}}(u)$ from (39.1) are

$$d_k = \lim_{u \to u_k} (u - u_k) \mathcal{R}_K^{\text{PLA}}(u) \qquad \bar{d}_k = \lim_{u \to u_k} (u - u_k) \bar{\mathcal{R}}_K^{\text{PLA}}(u)$$

$$d_k = \frac{c_0}{\beta_1} \frac{P_K(u_k)}{Q'_K(u_k)} \qquad \bar{d}_k = (\Phi_0|\Upsilon_k)^2. \qquad (39.2)$$

Every term $(\Phi_0|\Upsilon_k)^2/(u - u_k)$ in (39.1) is a meromorphic function, which is given simply by the quotient of a polynomial of the zeroth- and first-order d_k and $u - u_k$, respectively. This is the simplest PLA, i.e. [0/1] which is of the order (0, 1). This feature of having the numerator polynomial of the degree lower by one, relative to the denominator polynomial will, of course, persist if we keep adding together K such terms $(\Phi_0|\Upsilon_k)^2/(u-u_k)$, so that

$$\begin{aligned}\bar{\mathcal{R}}_K^{\text{PLA}}(u) &= \frac{(\Phi_0|\Upsilon_1)^2}{u-u_1} + \frac{(\Phi_0|\Upsilon_2)^2}{u-u_2} + \cdots + \frac{(\Phi_0|\Upsilon_K)^2}{u-u_K} \\ &= \frac{\sum_{k=1}^{K} (\Phi_0|\Upsilon_k)^2 \prod_{k'=1(k'\neq k)}^{K} (u-u_{k'})}{\prod_{k=1}^{K}(u-u_k)} \\ &= \left\{\frac{1}{\bar{\beta}_K Q_K(u)}\right\} \left\{\sum_{k=1}^{K} (\Phi_0|\Upsilon_k)^2 \prod_{k'=1(k'\neq k)}^{K} (u-u_{k'})\right\} \\ &\equiv \frac{c_0}{\beta_1} \frac{\bar{P}_K(u)}{Q_K(u)} \end{aligned} \qquad (39.3)$$

where $\bar{\beta}_K$ is from (22.4). We used (28.1) in (39.3) to identify the denominator polynomial $Q_K(u)$. The quantity $\bar{P}_K(u)$ from (39.3) is an auxiliary polynomial of degree $K-1$ defined by

$$\bar{P}_K(u) \equiv \frac{\beta_1}{c_0} \sum_{k''=1}^{K} (\Phi_0|\Upsilon_{k''})^2 \left\{\frac{1}{\bar{\beta}_K} \prod_{k'=1(k'\neq k'')}^{K} (u-u_{k'})\right\} \qquad (39.4)$$

$$\bar{\mathcal{R}}_K^{\text{PLA}}(u) = \frac{c_0}{\beta_1} \frac{\bar{P}_K(u)}{Q_K(u)}. \qquad (39.5)$$

At $u=u_k$ the term $u_k - u_k = 0$ always occurs in the product within the curly brackets in (39.4), causing the whole sum over k to collapse to a single term $(\Phi_0|\Upsilon_k)^2 \{\prod_{k'=1(k'\neq k)}^{K}(u_k-u_{k'})\}/\bar{\beta}_K$, which is equal to $(\Phi_0|\Upsilon_k)^2 Q'_K(u_k)$ according to (28.9) and, thus

$$\bar{P}_K(u_k) = \frac{\beta_1}{c_0}(\Phi_0|\Upsilon_k)^2 Q'_K(u_k). \qquad (39.6)$$

Since $(\beta_1/c_0)Q'_K(u_k) \neq 0$, we can divide both sides of (39.6) by the term $(\beta_1/c_0)Q'_K(u_k)$, so that

$$\frac{c_0}{\beta_1} \frac{\bar{P}_K(u_k)}{Q'_K(u_k)} = (\Phi_0|\Upsilon_k)^2. \qquad (39.7)$$

Both functions $(c_0/\beta_1)\bar{P}_K(u)/Q_K(u)$ and $(c_0/\beta_1)P_K(u)/Q_K(u)$ are the Padé approximants for the same Green function $\mathcal{R}(u)$. However, according to (38.16), uniqueness of the general PA also implies uniqueness of the PLA and, therefore

$$\frac{c_0}{\beta_1} \frac{P_K(u)}{Q_K(u)} = \frac{c_0}{\beta_1} \frac{\bar{P}_K(u)}{Q_K(u)} \Longrightarrow \bar{P}_K(u) = P_K(u). \qquad (39.8)$$

Uniqueness of the amplitudes $\{d_k\}$

Inserting (39.8) into (39.7) gives the sought result for the equality of the residues $\bar{d}_k$ and d_k

$$(\Phi_0|\Upsilon_k)^2 = \frac{c_0}{\beta_1} \frac{P_K(u_k)}{Q'_K(u_k)} \implies \bar{d}_k = d_k \quad \text{(QED)}. \tag{39.9}$$

Chapter 40

Partial fractions in the PLA only from the denominator polynomial

In our derivation of the residue d_k^∞ in the Lanczos formalism, we can start from the formula $d_k^\infty = (\Phi_0|\Upsilon_k^+)^2$ from the definition (23.5) via the normalized total state vector $|\Upsilon_k^+)$. We saw that by inserting $|\Upsilon_k^+)$ from (23.3) into (23.5), the analytical result (23.13) is obtained for the residue d_k^∞ which reduces to d_k from (31.13) for the finite chain of length K. On the other hand, as seen in chapter 36, the PLA offers yet another analytical expression (36.10) for the residue d_k with no recourse to the total eigenvector at all. Since the eigenvalues $\{u_k\}_{k=1}^K$ are the same in both the Lanczos formalism and in the PLA, the corresponding residues $\{d_k\}$ must also coincide with each other in these two methods, according to chapter 39. In other words, we ought to have

$$\{d_k\}_{\text{equation (31.13)}} = \{d_k\}_{\text{equation (36.10)}}$$

$$\frac{c_0}{\beta_K Q'_K(u_k) Q_{K-1}(u_k)} = \frac{c_0}{\beta_1} \frac{P_K(u_k)}{Q'_K(u_k)} \qquad (40.1)$$

$$\therefore \quad P_K(u_k) = \frac{\beta_1}{\beta_K Q_{K-1}(u_k)}. \qquad (40.2)$$

The result (40.2) is precisely the expression (27.4), which was established previously in an entirely different manner during the derivation of the Wronskian in chapter 27. Therefore, equation (40.2) is the correct expression and this, in turn, demonstrates the uniqueness of the residues computed by two different methods similarly to chapter 39. We re-emphasize that the special inverse proportionality between $P_K(u_k)$ and $Q_{K-1}^{-1}(u_k)$, as in (40.2), is valid only at the eigenvalues $u = u_k$. As a practical implication of the equivalence (40.1), we use (31.13) in (36.7) for the Green function $\mathcal{R}_K^{\text{PLA}}(u)$ and write

$$\mathcal{R}_K^{\text{PLA}}(u) = \sum_{k=1}^K \left\{ \frac{c_0}{\beta_K} \frac{1}{Q_{K-1}(u_k) Q'_K(u_k)} \right\} \frac{1}{u - u_k}. \qquad (40.3)$$

Partial fractions in the PLA only from the denominator polynomial

The expression (40.3) shows that one needs to deal only with the Lanczos polynomials of the first kind $Q_{K-1}(u_k)$ and $Q'_K(u_k)$. The presence of the Lanczos polynomial of the second kind $P_K(u_k)$ in the original PLA from (36.11) is eliminated by means of (40.2). Recall that the uniqueness of the residues was also established in chapter 39 within the PLA itself. Thus all the different ways for computing the residues by the Padé approximant must yield the same results for the same eigenvalue set $\{u_k\}$. The net outcome of the preceding chapter and of the present one is that the residues $\{d_k\}$ are unique, irrespective of the way they are computed, provided they belong to the same eigenvalue set $\{u_k\}$. The PLA computes the d_k as $(\Phi_0|\Upsilon_k)^2$ or $(c_0/\beta_1)P_K(u_k)/Q'_K(u_k)$ or $(c_0/\beta_K)/[Q_{K-1}(u_k)Q'_K(u_k)]$ and all the three results are shown *analytically* to be identical. The latter variant is the most efficient. The FD [108, 109] and DSD [115] obtain d_k exclusively from $(\Phi_0|\Upsilon_k)^2$ which, according to the proof from chapter 39, must be equal to the corresponding result of the PLA or DSD or DLP, if all the methods share the common eigenset $\{u_k\}$. All these parametric methods possess the joint eigenvalues $\{u_k\}$ because of the equivalence of the eigenvalue problem and the secular equation based upon the same evolution matrix **U** [228]. Of course, whenever a non-orthogonal basis set is used, as in DSD and FD, then the eigenvalue problem and the secular equation must be replaced by the generalized eigenvalue problem and the generalized secular equation, respectively [109, 115].

Chapter 41

The Lanczos continued fractions (LCF)

A general continued fraction (CF) [186] is another way of writing the PA as a staircase with descending quotients. There are several equivalent symbolic notations in use for a given CF and two of them are given by

$$\cfrac{A_1}{B_1 + \cfrac{A_2}{B_2 + \cfrac{A_3}{B_3 + \cdots}}} \equiv \frac{A_1}{B_1} \mathbin{\boldsymbol{+}} \frac{A_2}{B_2} \mathbin{\boldsymbol{+}} \frac{A_3}{B_3} \mathbin{\boldsymbol{+}} \cdots. \qquad (41.1)$$

The lhs of (41.1) is a natural way of writing the staircase-shaped continued fraction, but for frequent usage, the rhs of the same equation is more economical as it takes less space. It should be observed that the plus signs in bold-face on the rhs of (41.1) are *lowered* to remind us of a 'step-down' process in forming the CF. In other words, the rhs of (41.1) could also be equivalently written using the ordinary plus signs as $A_1/(B_1 + A_2/(B_2 + A_3/(B_3 + \cdots)))$ [17, 97]. In this chapter we shall connect the PLA to the CF by returning to the recursion (37.9) for the Lanczos determinants $\{D_n(u)\}$ [75, 97, 186]. Letting $n \longrightarrow n+1$ and dividing both sides of (37.9) by $D_n(u)$ will yield [97]

$$\frac{D_n(u)}{D_{n+1}(u)} = u - \alpha_n - \beta_{n+1}^2 \frac{D_{n+2}(u)}{D_{n+1}(u)}. \qquad (41.2)$$

Denoting the ratio $D_{n+1}(u)/D_n(u)$ by $\Gamma_n(u)$

$$\Gamma_n(u) = \frac{D_{n+1}(u)}{D_n(u)} \qquad (41.3)$$

we can rewrite (41.2) in the following form

$$\Gamma_n(u) = \frac{1}{u - \alpha_n - \beta_{n+1}^2 \Gamma_{n+1}(u)}. \qquad (41.4)$$

The Lanczos continued fractions (LCF)   233

In order to obtain the Nth order approximation to $\mathcal{R}(u)$, we use the backward recurrence (41.4) with the initialization $\Gamma_{N+1}(u) = 0$ and subsequently descend all the way down to $n = 0$. The final result is $\Gamma_0 = D_1(u)/D_0(u)$. Then using $\mathcal{R}(u) = c_0 D_1(u)/D_0(u)$ from (37.2), it follows

$$\mathcal{R}(u) = c_0 \Gamma_0(u) \tag{41.5}$$

so that with the help of (41.3) and (41.4), we arrive at

$$\mathcal{R}(u) = \mathcal{R}^{\text{LCF}}(u) \tag{41.6}$$

$$\mathcal{R}^{\text{LCF}}(u) = \frac{c_0}{u - \alpha_0} - \frac{\beta_1^2}{u - \alpha_1} - \frac{\beta_2^2}{u - \alpha_2} - \frac{\beta_3^2}{u - \alpha_3} - \cdots. \tag{41.7}$$

This procedure of obtaining the Green function $\mathcal{R}(u)$ is called hereafter the method of the Lanczos continued fractions (LCF) in the symbolic notation from the rhs of (41.1). Then, we say that the quantity $\mathcal{R}^{\text{LCF}}(u)$ is the infinite order LCF for the exact Green function $\mathcal{R}(u)$ from (5.22). A truncation at the nth term in (41.7) leads to the approximation

$$\mathcal{R}^{\text{LCF}}(u) \approx \mathcal{R}_n^{\text{LCF}}(u). \tag{41.8}$$

Here, $\mathcal{R}_n^{\text{LCF}}(u)$ is the LCF of the nth order

$$\mathcal{R}_n^{\text{LCF}}(u) = \{\mathcal{R}^{\text{LCF}}(u)\}_{\beta_n = 0} \quad (n = 1, 2, 3, \ldots) \tag{41.9}$$

$$\mathcal{R}_n^{\text{LCF}}(u) = \frac{c_0}{u - \alpha_0} - \frac{\beta_1^2}{u - \alpha_1} - \frac{\beta_2^2}{u - \alpha_2} - \frac{\beta_3^2}{u - \alpha_3} - \cdots - \frac{\beta_{n-1}^2}{u - \alpha_{n-1}}. \tag{41.10}$$

No matrix inversion is encountered in (41.7). Stated equivalently, the matrix $u\mathbf{1} - \mathbf{U}$ associated with the resolvent operator $\hat{R}(u)$ from (5.21) is inverted iteratively through its corresponding LCF. The LCF as a versatile convergence accelerator can yield the frequency spectrum (41.7) with a reduced number of terms $\{\alpha_n, \beta_n\}$. The meaning of the nearest neighbour approximation within the LCF is that the coefficients $\{\alpha_n, \beta_n\}$ become increasingly less significant for determination of the Green function $\mathcal{R}(u)$ as one progresses further down the continued fraction in (41.7). As a check, the three lowest orders LCFs are obtained from (41.7) in the forms

$$\mathcal{R}_1^{\text{LCF}}(u) = \frac{c_0}{u - \alpha_0}$$

$$\mathcal{R}_2^{\text{LCF}}(u) = \frac{c_0}{u - \alpha_0} - \frac{\beta_1^2}{u - \alpha_1}$$

$$\mathcal{R}_3^{\text{LCF}}(u) = \frac{c_0}{u - \alpha_0} - \frac{\beta_1^2}{u - \alpha_1} - \frac{\beta_2^2}{u - \alpha_2}. \tag{41.11}$$

These three results can be cast into their forms of simple quotients such as

$$\mathcal{R}_1^{LCF}(u) = \frac{c_0}{u - \alpha_0}$$

$$\mathcal{R}_2^{LCF}(u) = c_0 \frac{u - \alpha_1}{u^2 - (\alpha_0 + \alpha_1)u + (\alpha_0\alpha_1 - \beta_1^2)}$$

$$\mathcal{R}_3^{LCF}(u) = c_0 \frac{u^2 - (\alpha_1 + \alpha_2)u + (\alpha_1\alpha_2 - \beta_2^2)}{u^3 - (\alpha_0 + \alpha_1 + \alpha_2)u^2 + \alpha_\beta u + \beta_\alpha} \tag{41.12}$$

where α_β and β_α are given in (36.13). Comparison between these latter formulae and the corresponding findings (36.12) gives at once

$$\mathcal{R}_1^{LCF}(u) = \mathcal{R}_1^{PLA}(u) \qquad \mathcal{R}_2^{LCF}(u) = \mathcal{R}_2^{PLA}(u) \qquad \mathcal{R}_3^{LCF}(u) = \mathcal{R}_3^{PLA}(u). \tag{41.13}$$

Proceeding inductively in this direction for any non-negative integer n will lead to the general result

$$\mathcal{R}_n^{LCF}(u) = \mathcal{R}_n^{PLA}(u) \tag{41.14}$$

where the term $\mathcal{R}_n^{PLA}(u)$ is given in (36.5) for $K = n$ as the quotient of the two Lanczos polynomials $P_n(u)$ and $Q_n(u)$, i.e. $\mathcal{R}_n^{PLA}(u) = (c_0/\beta_1)P_n(u)/Q_n(u)$. Then according to (41.14), the nth order Lanczos continued fraction, which is derived from the expression (37.1) for the exact Green function $\mathcal{R}(u)$ from (5.22) is mathematically equivalent to the Padé–Lanczos approximant.

Chapter 42

Equations for eigenvalues u_k via continued fractions

If the set of the couplings $\{\alpha_n, \beta_n\}$ is precomputed, it is clear from the preceding chapter, that the LCF is technically more efficient than the PLA. This is because the PLA still needs to generate the Lanczos polynomials $\{P_K(u), Q_K(u)\}$ to arrive at (36.5), whereas LCF does not. The LCF is an accurate, robust and fast processor for computation of shape spectra with an easy way of programming implementations in practice. For parametric estimations of spectra, there are two options. We can search for the poles in (41.7) from the inherent polynomial equation, after the LCF is reduced to its polynomial quotient, which is precisely the PLA. In such a case, the efficiency of the LCF is the same as that of the PLA. However, the poles in (41.7) can be obtained without reducing $\mathcal{R}_n^{\text{LCF}}(u)$ to the polynomial quotient. Since $\bar{\beta}_K \neq 0$, as per (22.11), we can rewrite the characteristic equation (25.10) as $\tilde{Q}_K(u) = 0$. This can be stated as the tridiagonal secular equation

$$\tilde{Q}_K(u) \equiv \begin{vmatrix} u-\alpha_0 & -\beta_1 & 0 & \cdots & & \cdots & 0 \\ -\beta_1 & u-\alpha_1 & -\beta_2 & \cdots & & \cdots & 0 \\ 0 & -\beta_2 & u-\alpha_2 & \cdots & & \cdots & 0 \\ \vdots & \vdots & \vdots & \ddots & \ddots & & \vdots \\ 0 & 0 & 0 & -\beta_{K-2} & u-\alpha_{K-2} & -\beta_{K-1} \\ 0 & 0 & 0 & 0 & -\beta_{K-1} & u-\alpha_{K-1} \end{vmatrix} = 0. \tag{42.1}$$

Expanding the determinantal equation (42.1) by the Cauchy rule, we readily obtain the following continued fraction

$$u = \alpha_0 - \frac{\beta_1^2}{u-\alpha_1} - \frac{\beta_2^2}{u-\alpha_2} - \frac{\beta_3^2}{u-\alpha_3} - \cdots - \frac{\beta_{K-1}^2}{u-\alpha_{K-1}}. \tag{42.2}$$

The solutions $\{u_k\}_{k=1}^K$ of this equation are the eigenvalues of $\mathbf{U}(\tau)$. Clearly (42.2) is a transcendental equation. This is because the kth solution $u = u_k$ is given only implicitly, since u appears in the continued fraction from the rhs of (42.2) and the unknown u is also present on the lhs of the same equation. The modified Newton–Raphson iterative method used in (28.5) can be employed to solve $\tilde{F}(u) = 0$ for $u = u_k$, where $\tilde{F}(u)$ is the difference between the lhs and rhs of (42.2). Many quantum-mechanical eigenvalue problems with given potentials can be reduced to diagonalization of triangular matrices. Alternatively, the eigensolutions can be obtained using the corresponding secular determinantal equations similar to (42.1), but with the real variable ω instead of the complex exponential $u = \exp(-i\omega\tau)$. A good example for this is the extraction of extremely accurate eigenvalues (also called the 'characteristic numbers') of the Mathieu function [269].

Chapter 43

The exact analytical expression for the general Padé spectrum

In the preceding chapter, the continued fractions were introduced in an indirect way into the spectral analysis. Alternatively, they can be brought to the subject in a direct manner, as will be analysed in this chapter. By definition, the infinite-order CF to the time series (38.2) is given by

$$\mathcal{R}^{CF}(u) = \frac{a_1}{u} - \frac{a_2}{1} - \frac{a_3}{u} - \frac{a_4}{1} - \frac{a_5}{u} - \frac{a_6}{1} - \frac{a_7}{u} - \cdots. \quad (43.1)$$

The unknown set $\{a_n\}$ is determined from the condition of equality between the expansion coefficients of the series of the rhs of (43.1) developed in powers of u^{-1} and the signal points $\{c_n\}$ from (38.2), which we rewrite as the following ordinary summation

$$\mathcal{R}(u) = \frac{c_0}{u} + \frac{c_1}{u^2} + \frac{c_2}{u^3} + \frac{c_3}{u^4} + \frac{c_4}{u^5} + \frac{c_5}{u^6} + \frac{c_6}{u^7} + \cdots. \quad (43.2)$$

In practice, the rhs of (43.1) cannot be used to the infinite order and, therefore, a truncation is required. The infinite order CF, i.e. $\mathcal{R}^{CF}(u)$ which is truncated at the finite order n is called the nth order CF approximant $\mathcal{R}_n^{CF}(u)$, where n is a positive integer ($n = 1, 2, 3, \ldots$). This truncation is done by setting all the higher-order coefficients $\{a_{m \geq n+1}\}$ to zero from the onset

$$\mathcal{R}_n^{CF}(u) \equiv \{\mathcal{R}^{CF}(u)\}_{a_m=0, m \geq n+1} \quad (n = 1, 2, 3, \ldots). \quad (43.3)$$

The first few even-order CF approximants $\{\mathcal{R}_{2n}^{CF}(u)\}$ read explicitly as

$$\mathcal{R}_2^{CF}(u) = \frac{a_1}{u} - \frac{a_2}{1} \qquad \mathcal{R}_4^{CF}(u) = \frac{a_1}{u} - \frac{a_2}{1} - \frac{a_3}{u - a_4}$$

$$\mathcal{R}_6^{CF}(u) = \frac{a_1}{u} - \frac{a_2}{1} - \frac{a_3}{u} - \frac{a_4}{1} - \frac{a_5}{u - a_6}$$

$$\mathcal{R}_8^{CF}(u) = \frac{a_1}{u} - \frac{a_2}{1} - \frac{a_3}{u} - \frac{a_4}{1} - \frac{a_5}{u} - \frac{a_6}{1} - \frac{a_7}{u - a_8}. \quad (43.4)$$

These expressions can be written in the form of the Padé approximants. For instance, the first four CF approximants $\mathcal{R}_{2n}^{CF}(u)$ for $n = 2, 4, 6$ and $n = 8$ are

$$\mathcal{R}_2^{CF}(u) = \frac{a_1}{u - a_2} \qquad \mathcal{R}_4^{CF}(u) = a_1 \frac{u - (a_3 + a_4)}{u^2 - (a_2 + a_3 + a_4)u + a_2 a_4}$$

$$\mathcal{R}_6^{CF}(u) = a_1 \frac{u^2 - (a_3 + a_4 + a_5 + a_6)u + [a_3(a_5 + a_6) + a_4 a_6]}{u^3 - (a_2 + a_3 + a_4 + a_5 + a_6)u^2 + A'u - a_2 a_4 a_6}$$

$$\mathcal{R}_8^{CF}(u) = a_1 \frac{u^3 - (a_3 + a_4 + a_5 + a_6 + a_7 + a_8)u^2 + Au - B}{u^4 - B'u^3 + Cu^2 - Du + a_2 a_4 a_6 a_8} \qquad (43.5)$$

where

$$A = a_3(a_5 + a_6 + a_7 + a_8) + a_4(a_6 + a_7 + a_8) + a_5(a_7 + a_8) + a_6 a_8$$
$$B = a_3[a_5(a_7 + a_8) + a_6 a_8] + a_4 a_6 a_8$$
$$C = a_2(a_4 + a_5 + a_6 + a_7 + a_8) + a_3(a_5 + a_6 + a_7 + a_8)$$
$$\quad + a_4(a_6 + a_7 + a_8) + a_5(a_7 + a_8) + a_6 a_8$$
$$D = B + a_2[a_4(a_6 + a_7 + a_8) + a_5(a_7 + a_8) + a_6 a_8]$$
$$A' = a_2(a_4 + a_5 + a_6) + a_3(a_5 + a_6) + a_4 a_6$$
$$B' = a_2 + a_3 + a_4 + a_5 + a_6 + a_7 + a_8. \qquad (43.6)$$

From here, we can write the general expression for the even-ordered CF, i.e. $\mathcal{R}_{2n}^{CF}(u)$ in the form

$$\mathcal{R}_{2n}^{CF}(u) = a_1 \frac{\tilde{P}_n^{CF}(u)}{\tilde{Q}_n^{CF}(u)}. \qquad (43.7)$$

For $n \leq 4$ the closed expressions for $\tilde{P}_n^{CF}(u)$ and $\tilde{Q}_n^{CF}(u)$ are given in (43.5), but we want to derive the exact general analytical formulae that are valid for any positive integer n. The odd-order CF, i.e. $\mathcal{R}_{2n-1}^{CF}(u)$ are obtained directly from the associated even-order CF, $\mathcal{R}_{2n}^{CF}(u)$, by setting $a_{2n} \equiv 0$

$$\mathcal{R}_{2n-1}^{CF}(u) \equiv \{\mathcal{R}_{2n}^{CF}(u)\}_{a_{2n}=0} \qquad (n = 1, 2, 3, \ldots). \qquad (43.8)$$

For example

$$\mathcal{R}_3^{CF}(u) = \frac{a_1}{u} - \frac{a_2}{1} - \frac{a_3}{u}$$
$$\mathcal{R}_5^{CF}(u) = \frac{a_1}{u} - \frac{a_2}{1} - \frac{a_3}{u} - \frac{a_4}{1} - \frac{a_5}{u}$$
$$\mathcal{R}_7^{CF}(u) = \frac{a_1}{u} - \frac{a_2}{1} - \frac{a_3}{u} - \frac{a_4}{1} - \frac{a_5}{u} - \frac{a_6}{1} - \frac{a_7}{u}. \qquad (43.9)$$

Rewritten explicitly in the form of the polynomial quotients, these expressions now become

$$\mathcal{R}_3^{CF}(u) = a_1 \frac{u - a_3}{u^2 - (a_2 + a_3)u}$$

$$\mathcal{R}_5^{CF}(u) = a_1 \frac{u^2 - (a_3 + a_4 + a_5)u + a_3 a_5}{u^3 - (a_2 + a_3 + a_4 + a_5)u^2 + [a_2(a_4 + a_5) + a_3 a_5]u}$$

$$\mathcal{R}_7^{CF}(u) = a_1 \frac{u^3 - (a_3 + a_4 + a_5 + a_6 + a_7)u^2 + E'u - a_3 a_5 a_7}{u^4 - Eu^3 + F'u^2 - Fu} \quad (43.10)$$

$$E = a_2 + a_3 + a_4 + a_5 + a_6 + a_7$$
$$F = a_2[a_4(a_6 + a_7) + a_5 a_7] + a_3 a_5 a_7$$
$$E' = a_3(a_5 + a_6 + a_7) + a_4(a_6 + a_7) + a_5 a_7$$
$$F' = a_2(a_4 + a_5 + a_6 + a_7) + a_3(a_5 + a_6 + a_7) + a_4(a_6 + a_7) + a_5 a_7.$$
$$(43.11)$$

The polynomials $\tilde{P}_n^{CF}(u)$ and $\tilde{Q}_n^{CF}(u)$ from (43.7) can be written in their general power series representations as

$$\tilde{P}_n^{CF}(u) = \sum_{r=0}^{n-1} \tilde{p}_{n,n-r} u^r \qquad \tilde{Q}_n^{CF}(u) = \sum_{r=0}^{n} \tilde{q}_{n,n-r} u^r. \quad (43.12)$$

The expansion coefficients $\tilde{p}_{n,n-r}$ and $\tilde{q}_{n,n-r}$ of the polynomials $\tilde{P}_n^{CF}(u)$ and $\tilde{Q}_n^{CF}(u)$ can be identified from (43.5) and (43.10) in terms of the set $\{a_k\}$ alone. Then, the explicit expressions for the polynomials $\tilde{P}_n^{CF}(u)$ and $\tilde{Q}_n^{CF}(u)$ extracted from (43.5) with $1 \leq n \leq 4$ are

$$\tilde{P}_1^{CF}(u) = 1$$
$$\tilde{P}_2^{CF}(u) = u - (a_3 + a_4)$$
$$\tilde{P}_3^{CF}(u) = u^2 - (a_3 + a_4 + a_5 + a_6)u + [a_3(a_5 + a_6) + a_4 a_6]$$
$$\tilde{P}_4^{CF}(u) = u^3 - (a_3 + a_4 + a_5 + a_6 + a_7 + a_8)u^2$$
$$\qquad + [a_3(a_5 + a_6 + a_7 + a_8) + a_4(a_6 + a_7 + a_8)$$
$$\qquad + a_5(a_7 + a_8) + a_6 a_8]u$$
$$\qquad - \{a_3[a_5(a_7 + a_8) + a_6 a_8] + a_4 a_6 a_8\} \quad (43.13)$$
$$\tilde{Q}_1^{CF}(u) = u - a_2$$
$$\tilde{Q}_2^{CF}(u) = u^2 - (a_2 + a_3 + a_4)u + a_2 a_4$$
$$\tilde{Q}_3^{CF}(u) = u^3 - (a_2 + a_3 + a_4 + a_5 + a_6)u^2$$
$$\qquad + [a_2(a_4 + a_5 + a_6) + a_3(a_5 + a_6) + a_4 a_6]u - a_2 a_4 a_6$$
$$\tilde{Q}_4^{CF}(u) = u^4 - (a_2 + a_3 + a_4 + a_5 + a_6 + a_7 + a_8)u^3$$
$$\qquad + [a_2(a_4 + a_5 + a_6 + a_7 + a_8) + a_3(a_5 + a_6 + a_7 + a_8)$$

$$+ a_4(a_6 + a_7 + a_8) + a_5(a_7 + a_8) + a_6 a_8]u^2$$
$$- \{a_2[a_4(a_6 + a_7 + a_8) + a_5(a_7 + a_8) + a_6 a_8]$$
$$+ a_3[a_5(a_7 + a_8) + a_6 a_8] + a_4 a_6 a_8\} u + a_2 a_4 a_6 a_8. \quad (43.14)$$

Using (43.12) and (43.13), we can identify the coefficients $\tilde{p}_{n,n-r}$ as follows

$$\tilde{p}_{2,2} = -(a_3 + a_4)$$
$$\tilde{p}_{2,1} = 1$$
$$\tilde{p}_{3,3} = a_3(a_5 + a_6) + a_4 a_6$$
$$\tilde{p}_{3,2} = -(a_3 + a_4 + a_5 + a_6)$$
$$\tilde{p}_{3,1} = 1$$
$$\tilde{p}_{4,4} = -\{a_3[a_5(a_7 + a_8) + a_6 a_8] + a_4 a_6 a_8\}$$
$$\tilde{p}_{4,3} = a_3(a_5 + a_6 + a_7 + a_8) + a_4(a_6 + a_7 + a_8) + a_5(a_7 + a_8) + a_6 a_8$$
$$\tilde{p}_{4,2} = -(a_3 + a_4 + a_5 + a_6 + a_7 + a_8)$$
$$\tilde{p}_{4,1} = 1. \quad (43.15)$$

Similarly, employing (43.12) and (43.14), the coefficients $\tilde{q}_{n,n-r}$ are found to be in the forms

$$\tilde{q}_{1,1} = -a_2$$
$$\tilde{q}_{2,2} = a_2 a_4$$
$$\tilde{q}_{2,1} = -(a_2 + a_3 + a_4)$$
$$\tilde{q}_{2,0} = 1$$
$$\tilde{q}_{3,3} = -a_2 a_4 a_6$$
$$\tilde{q}_{3,2} = a_2(a_4 + a_5 + a_6) + a_3(a_5 + a_6) + a_4 a_6$$
$$\tilde{q}_{3,1} = -(a_2 + a_3 + a_4 + a_5 + a_6)$$
$$\tilde{q}_{3,0} = 1$$
$$\tilde{q}_{4,4} = a_2 a_4 a_6 a_8$$
$$\tilde{q}_{4,3} = -\{a_2[a_4(a_6 + a_7 + a_8) + a_5(a_7 + a_8) + a_6 a_8]$$
$$+ a_3[a_5(a_7 + a_8) + a_6 a_8] + a_4 a_6 a_8\}$$
$$\tilde{q}_{4,2} = a_2(a_4 + a_5 + a_6 + a_7 + a_8) + a_3(a_5 + a_6 + a_7 + a_8)$$
$$+ a_4(a_6 + a_7 + a_8) + a_5(a_7 + a_8) + a_6 a_8$$
$$\tilde{q}_{4,1} = -(a_2 + a_3 + a_4 + a_5 + a_6 + a_7 + a_8)$$
$$\tilde{q}_{4,0} = 1. \quad (43.16)$$

From these explicit formulae for particular values $0 \leq n \leq 4$, the general expressions follow as

$$\tilde{p}_{n,2} = -\sum_{r_1=3}^{2n} a_{r_1}$$

$$\tilde{p}_{n,3} = \sum_{r_1=3}^{2(n-1)} a_{r_1} \sum_{r_2=r_1+2}^{2n} a_{r_2}$$

$$\tilde{p}_{n,4} = -\sum_{r_1=3}^{2(n-2)} a_{r_1} \sum_{r_2=r_1+2}^{2(n-1)} a_{r_2} \sum_{r_3=r_2+2}^{2n} a_{r_3} \quad \text{etc} \quad (43.17)$$

$$\tilde{q}_{n,n} = (-1)^n \prod_{r_1=2}^{2n} a_{r_1}$$

$$\tilde{q}_{n,1} = -\sum_{r_1=2}^{2n} a_{r_1}$$

$$\tilde{q}_{n,2} = \sum_{r_1=2}^{2(n-1)} a_{r_1} \sum_{r_2=r_1+2}^{2n} a_{r_2}$$

$$\tilde{q}_{n,3} = -\sum_{r_1=2}^{2(n-2)} a_{r_1} \sum_{r_2=r_1+2}^{2(n-1)} a_{r_2} \sum_{r_3=r_2+2}^{2n} a_{r_3} \quad \text{etc.} \quad (43.18)$$

Therefore, the most general formulae for the polynomial expansion coefficients $\tilde{p}_{n,m}$ and $\tilde{q}_{n,m}$ are

$$\tilde{p}_{n,m} = (-1)^{m-1} \underbrace{\sum_{r_1=3}^{2(n-m+2)} a_{r_1} \sum_{r_2=r_1+2}^{2(n-m+3)} a_{r_2} \cdots \sum_{r_{m-1}=r_{m-2}+2}^{2n} a_{r_{m-1}}}_{m-1 \text{ summations}} \quad (43.19)$$

$$\tilde{q}_{n,m} = (-1)^m \underbrace{\sum_{r_1=2}^{2(n-m+1)} a_{r_1} \sum_{r_2=r_1+2}^{2(n-m+2)} a_{r_2} \sum_{r_3=r_2+2}^{2(n-m+3)} a_{r_3} \cdots \sum_{r_m=r_{m-1}+2}^{2n} a_{r_m}}_{m \text{ summations}}$$

$$(43.20)$$

where $n \geq m$. The explicit results (43.19) and (43.20) have been derived for the first time in [17]. They express both $\tilde{p}_{n,n-r}$ and $\tilde{q}_{n,n-r}$ as the compact analytical formulae for arbitrary values of non-negative integers $\{n, r\}$. Our closed analytical expressions (43.19) and (43.20) contain only the continued fraction coefficients $\{a_n\}$. Therefore, in order to design a purely algebraic analytical method, the remaining central problem is to obtain a general analytical formula for any CF coefficient a_n in the case of an arbitrary positive integer n. This problem will be tackled and solved in the next chapter. Notice that the forms of the general coefficients from (43.19) and (43.20) are similar to (7.44). Therefore, just like (7.44), the expansion coefficients from (43.19) and (43.20) could also be efficiently computed using a recursive algorithm of the type (17.9).

Chapter 44

The exact analytical result for any continued fraction coefficients

Here, the analysis is aimed at deriving the exact general analytical expression for all the expansion coefficients a_n in the CF from (43.1). To this end we shall use the theorem of uniqueness of two power series expansions for the same function. The rhs of (38.2) and (43.1) are two different representations of the same Green function (5.22). Therefore, if we develop the nth order CF, $\mathcal{R}_n^{\text{CF}}(u)$, from (43.1) in powers of u^{-1}, then the ensuing mth ($m \leq n$) expansion coefficient must be identical to the coefficients $\{c_n\}$ from the series (38.2). The series expansion of the rhs of (43.1) in terms of powers of u^{-1} will be obtained by first expressing $\mathcal{R}_n^{\text{CF}}(u)$ as the associated PA, i.e. $a_1 \tilde{P}_n^{\text{CF}}(u)/\tilde{Q}_n^{\text{CF}}(u)$ using (43.5) and (43.10). This will be followed by the inversion of the denominator polynomial $\tilde{Q}_n^{\text{CF}}(u)$ to obtain a series in u^{-1}. Such a series is afterwards multiplied by the numerator polynomial $\tilde{P}_n^{\text{CF}}(u)$ and the result is a new series. Finally, the first n terms of this latter series are equated to the first n time signal points from (38.2), according to the uniqueness expansion theorem for the function (5.22). This method provides the transformation between the general auto-correlation functions $\{C_n\} = \{c_n\}$ and the coefficients $\{a_n\}$ of the continued fractions (43.1). Such a uniquely defined transformation has been obtained in [17] in an explicit analytical form for any value of the non-negative integer n ($n = 0, 1, 2, 3, \ldots$). We have already derived the explicit expressions for the polynomials $\tilde{P}_n^{\text{CF}}(u)$ and $\tilde{Q}_n^{\text{CF}}(u)$ with the most general expansion coefficients (43.19) and (43.20). Therefore, the polynomial $\tilde{Q}_n^{\text{CF}}(u)$ can be explicitly inverted. Then it would be tempting to multiply $\tilde{P}_n^{\text{CF}}(u)$ by $[\tilde{Q}_n^{\text{CF}}(u)]^{-1}$ and arrive directly at the nth expansion coefficient of $\mathcal{R}_n^{\text{CF}}(u)$ developed as a series in powers of u^{-1}. However, it is less cumbersome to carry out separate calculations for each individual integer, as we did for $1 \leq n \leq 7$. The obtained results yield a clear pattern from which the general analytical expression for a_n can be deduced for any integer $n \geq 1$. The main outlines are given in this chapter using the following inversion formula for a given series [142, 240]

$$\frac{1}{\sum_{n=0}^{\infty} \gamma_n u^{-n}} = \sum_{n=0}^{\infty} \delta_n u^{-n} \qquad (44.1)$$

$$\delta_n = -\sum_{k=0}^{n-1} \delta_k \gamma_{n-k} \quad (n = 1, 2, \ldots) \qquad \delta_0 = 1 \qquad (44.2)$$

where, for the present application, it is appropriate to put $\gamma_0 = 1$. In the case of the coefficients $\{a_n\}$ with $1 \leq n \leq 7$, the following explicit expressions for δ_n ($1 \leq n \leq 6$) are found

$$\begin{aligned}
\delta_1 &= -\gamma_1 \\
\delta_2 &= -\gamma_2 + \gamma_1^2 \\
\delta_3 &= -\gamma_3 + 2\gamma_1\gamma_2 - \gamma_1^3 \\
\delta_4 &= -\gamma_4 + 2\gamma_1\gamma_3 + \gamma_2^2 - 3\gamma_1^2\gamma_2 + \gamma_1^4 \\
\delta_5 &= -\gamma_5 + 2\gamma_1\gamma_4 + 2\gamma_2\gamma_3 - 3\gamma_1^2\gamma_3 - 3\gamma_1\gamma_2^2 + 4\gamma_1^3\gamma_2 - \gamma_1^5 \\
\delta_6 &= -\gamma_6 + 2\gamma_1\gamma_5 + 2\gamma_2\gamma_4 - 3\gamma_1^2\gamma_4 + \gamma_3^2 - 6\gamma_1\gamma_2\gamma_3 + 4\gamma_1^3\gamma_3 - \gamma_2^3 \\
&\quad + 6\gamma_1^2\gamma_2^2 - 5\gamma_1^4\gamma_2 + \gamma_1^6.
\end{aligned} \qquad (44.3)$$

In the calculations that follow, we shall not invert a series, but rather a polynomial. However, if the denominator from (44.2) is replaced by a polynomial of degree m, then the same sum $\sum_{n=0}^{\infty} \delta_n u^{-n}$ will be formally preserved, provided that we redefine γ_n as $\gamma_n \equiv 0$ ($n \geq m+1$)

$$\frac{1}{\sum_{n=0}^{m} \gamma_n u^{-n}} = \sum_{n=0}^{\infty} \delta_n u^{-n} \qquad \gamma_n \equiv 0 \qquad n \geq m+1. \qquad (44.4)$$

An inspection of (43.4) and (43.9) would reveal that the first three coefficients a_1, a_2 and a_3 calculated from $\mathcal{R}_2^{CF}(u)$ and $\mathcal{R}_3^{CF}(u)$ can be obtained by using the simple binomial expansion, rather than the general inversion formula (44.4). Nevertheless, to establish the pattern of derivation of the general coefficient a_n, we shall use (44.4) even for inversion of a binomial.

44.1 Calculation of coefficients a_1 and a_2

The expressions for the coefficients a_1 and a_2 are derived by starting from the defining expression (43.4) for $\mathcal{R}_2^{CF}(u)$ rewritten as

$$\mathcal{R}_2^{CF}(u) = \frac{a_1}{u} \left(\sum_{n=0}^{1} \gamma_n u^{-n} \right)^{-1} \qquad (44.5)$$

$$\gamma_0 = 1 \qquad \gamma_1 = -a_2. \qquad (44.6)$$

The application of the inversion formula (44.4) requires the coefficients $\{\delta_0, \delta_1\}$ from (44.2) and (44.3). Therefore, (44.4) and (44.5) will yield

$$\mathcal{R}_2^{CF}(u) = a_1 \left(\frac{1}{u} - \frac{\gamma_1}{u^2} + \cdots \right). \tag{44.7}$$

Comparing (43.2) and (44.7) gives the expressions

$$c_0 = a_1$$
$$c_1 = -a_1 \gamma_1 \tag{44.8}$$

so that

$$c_0 = a_1$$
$$c_1 = a_1 a_2 \tag{44.9}$$

and this implies

$$a_1 = c_0$$
$$a_2 = \frac{c_1}{a_1}. \tag{44.10}$$

44.2 Calculation of coefficient a_3

The third coefficient a_3 is calculated from the third-order CF, i.e. $\mathcal{R}_3^{CF}(u)$ which is given in (43.9)

$$\mathcal{R}_3^{CF}(u) = a_1 \left(\frac{1}{u} - \frac{a_3}{u^2} \right) \left(\sum_{n=0}^{1} \gamma_n u^{-n} \right)^{-1} \tag{44.11}$$

$$\gamma_0 = 1 \qquad \gamma_1 = -(a_2 + a_3). \tag{44.12}$$

The inversion formula (44.4) should be used with $\gamma_n = 0$ for $n \geq 2$ and, therefore

$$\mathcal{R}_3^{CF}(u) = a_1 \left(\frac{1}{u} - \frac{a_3}{u^2} \right) \left(1 - \frac{\gamma_1}{u} + \frac{\gamma_1^2}{u^2} + \cdots \right) \tag{44.13}$$

$$\mathcal{R}_3^{CF}(u) = a_1 \left(\frac{1}{u} - \frac{\gamma_1 + a_3}{u^2} + \frac{\gamma_1^2 + a_3 \gamma_1}{u^3} + \cdots \right). \tag{44.14}$$

When this result is compared with (43.2), the following relationships are readily deduced

$$c_0 = a_1$$
$$c_1 = -a_1(\gamma_1 + a_3)$$
$$c_2 = a_1(\gamma_1^2 + a_3 \gamma_1) \tag{44.15}$$

and, hence

$$c_0 = a_1$$
$$c_1 = a_1 a_2$$
$$c_2 = a_1 a_2 a_3 + a_1 a_2^2. \qquad (44.16)$$

From here it follows

$$a_1 = c_0$$
$$a_2 = \frac{c_1}{a_1}$$
$$a_3 = \frac{c_2 - a_1 a_2^2}{a_1 a_2}. \qquad (44.17)$$

It is seen from the sequel (44.17) that the coefficient a_2 is expressed through c_1 and a_1. Likewise, we see from (44.17) that the coefficient a_3 is determined by the triple $\{c_2; a_1, a_2\}$. From these two simple, particular cases a_1 and a_2 alone, it is conjectured that the general coefficient a_n should be given in terms of $\{c_{n-1}; a_1, a_2, \ldots, a_{n-1}\}$. This is indeed the case as proven in the remainder of this chapter. In the upcoming demonstration, it will be apparent that the very dependence of a_n upon $\{c_{n-1}; a_1, a_2, \ldots, a_{n-1}\}$ for the first few values of n ($1 \leq n \leq 7$) will also be helpful in identifying the general form of the coefficient a_n.

44.3 Calculation of coefficient a_4

The fourth coefficient a_4 is calculated using the fourth-order CF, i.e. $\mathcal{R}_4^{CF}(u)$ from (43.5)

$$\mathcal{R}_4^{CF}(u) = \frac{a_1}{u}\left(1 - \frac{a_3 + a_4}{u}\right)\left(\sum_{n=0}^{2} \gamma_n u^{-n}\right)^{-1} \qquad (44.18)$$

$$\gamma_0 = 1 \qquad \gamma_1 = -(a_2 + a_3 + a_4) \qquad \gamma_2 = a_2 a_4. \qquad (44.19)$$

In (44.18), we need to invert a second-order polynomial through the use of (44.3). The required coefficients $\{\delta_1, \delta_2, \delta_3\}$ are given in (44.2) with $\gamma_n \equiv 0$ for $n \geq 3$, as prescribed by (44.4). Therefore, (44.18) will acquire the form

$$\mathcal{R}_4^{CF}(u) = \frac{a_1}{u}\left(1 - \frac{a_3 + a_4}{u}\right)\left(1 - \frac{\gamma_1}{u} + \frac{\gamma_1^2 - \gamma_2}{u^2} - \frac{\gamma_1^3 - 2\gamma_1\gamma_2}{u^3} + \cdots\right) \qquad (44.20)$$

or equivalently

$$\mathcal{R}_4^{CF}(u) = a_1\left[\frac{1}{u} - \frac{\gamma_1 + (a_3 + a_4)}{u^2} + \frac{(\gamma_1^2 - \gamma_2) + (a_3 + a_4)\gamma_1}{u^3} \right.$$
$$\left. - \frac{(\gamma_1^3 - 2\gamma_1\gamma_2) + (a_3 + a_4)(\gamma_1^2 - \gamma_2)}{u^4} + \cdots \right]. \qquad (44.21)$$

This series together with (43.2) will lead to

$$c_0 = a_1$$
$$c_1 = -a_1[\gamma_1 + (a_3 + a_4)]$$
$$c_2 = a_1[(\gamma_1^2 - \gamma_2) + (a_3 + a_4)\gamma_1]$$
$$c_3 = -a_1[(\gamma_1^3 - 2\gamma_1\gamma_2) + (a_3 + a_4)(\gamma_1^2 - \gamma_2)] \quad (44.22)$$

or, after simplifying

$$c_0 = a_1$$
$$c_1 = a_1 a_2$$
$$c_2 = a_1 a_2 a_3 + a_1 a_2^2$$
$$c_3 = a_1 a_2 a_3 a_4 + a_1 a_2 (a_2 + a_3)^2 \quad (44.23)$$

which implies

$$a_1 = c_0$$
$$a_2 = \frac{c_1}{a_1}$$
$$a_3 = \frac{c_2 - a_1 a_2^2}{a_1 a_2}$$
$$a_4 = \frac{c_3 - a_1 a_2 (a_2 + a_3)^2}{a_1 a_2 a_3}. \quad (44.24)$$

44.4 Calculation of coefficient a_5

The calculation of the fifth coefficient a_5 is performed by means of the fifth-order CF, i.e. $\mathcal{R}_5^{CF}(u)$ from (43.10)

$$\mathcal{R}_5^{CF}(u) = \frac{a_1}{u}\left(1 - \frac{a_3 + a_4 + a_5}{u} + \frac{a_3 a_5}{u^2}\right)\left(\sum_{n=0}^{2} \gamma_n u^{-n}\right)^{-1} \quad (44.25)$$

$$\gamma_0 = 1 \qquad \gamma_1 = -(a_2 + a_3 + a_4 + a_5)$$
$$\gamma_2 = a_2(a_4 + a_5) + a_3 a_5. \quad (44.26)$$

As in the case of $\mathcal{R}_4^{CF}(u)$ from (44.18), inversion of a second-order polynomial is needed in (44.25). Again we use (44.4) where this time we need the coefficients $\{\delta_1, \delta_2, \delta_3, \delta_4\}$. They are taken from (44.3) by letting $\gamma_n \equiv 0$ for $n \geq 3$, in accordance with (44.4). Thus, (44.25) can be cast into the form

$$\mathcal{R}_5^{CF}(u) = \frac{a_1}{u}\left(1 - \frac{a_3 + a_4 + a_5}{u} + \frac{a_3 a_5}{u^2}\right)$$
$$\times \left(1 - \frac{\gamma_1}{u} + \frac{\gamma_1^2 - \gamma_2}{u^2} - \frac{\gamma_1^3 - 2\gamma_1\gamma_2}{u^3} + \frac{\gamma_1^4 + \gamma_2^2 - 3\gamma_1^2\gamma_2}{u^4} + \cdots\right) \quad (44.27)$$

which can be reduced to

$$\mathcal{R}_5^{CF}(u) = a_1 \left[\frac{1}{u} - \frac{\gamma_1 + (a_3 + a_4 + a_5)}{u^2} \right.$$
$$+ \frac{(\gamma_1^2 - \gamma_2) + (a_3 + a_4 + a_5)\gamma_1 + a_3 a_5}{u^3}$$
$$- \frac{(\gamma_1^3 - 2\gamma_1\gamma_2) + (a_3 + a_4 + a_5)(\gamma_1^2 - \gamma_2) + a_3 a_5 \gamma_1}{u^4}$$
$$\left. + \frac{(\gamma_2^2 - 3\gamma_1^2\gamma_2 + \gamma_1^4) + (a_3 + a_4 + a_5)(\gamma_1^3 - 2\gamma_1\gamma_2) + a_3 a_5(\gamma_1^2 - \gamma_2)}{u^5} \right].$$
(44.28)

From a comparison of this series with (43.2), we obtain the following expressions

$$c_0 = a_1$$
$$c_1 = -a_1[\gamma_1 + (a_3 + a_4 + a_5)]$$
$$c_2 = a_1[(\gamma_1^2 - \gamma_2) + (a_3 + a_4 + a_5)\gamma_1 + a_3 a_5]$$
$$c_3 = -a_1[(\gamma_1^3 - 2\gamma_1\gamma_2) + (a_3 + a_4 + a_5)(\gamma_1^2 - \gamma_2) + a_3 a_5 \gamma_1]$$
$$c_4 = a_1[(\gamma_2^2 - 3\gamma_1^2\gamma_2 + \gamma_1^4) + (a_3 + a_4 + a_5)(\gamma_1^3 - 2\gamma_1\gamma_2) + a_3 a_5(\gamma_1^2 - \gamma_2)]$$
(44.29)

yielding

$$c_0 = a_1$$
$$c_1 = a_1 a_2$$
$$c_2 = a_1 a_2 a_3 + a_1 a_2^2$$
$$c_3 = a_1 a_2 a_3 a_4 + a_1 a_2 (a_2 + a_3)^2$$
$$c_4 = a_1 a_2 a_3 a_4 a_5 + a_1 a_2 a_3 (a_2 + a_3 + a_4)^2 + a_1 [a_2(a_2 + a_3)]^2 \quad (44.30)$$

and this leads to

$$a_1 = c_0$$
$$a_2 = \frac{c_1}{a_1}$$
$$a_3 = \frac{c_2 - a_1 a_2^2}{a_1 a_2}$$
$$a_4 = \frac{c_3 - a_1 a_2 (a_2 + a_3)^2}{a_1 a_2 a_3}$$
$$a_5 = \frac{c_4 - a_1 a_2 a_3 (a_2 + a_3 + a_4)^2 - a_1 [a_2(a_2 + a_3)]^2}{a_1 a_2 a_3 a_4}. \quad (44.31)$$

44.5 Calculation of coefficient a_6

In order to obtain the sixth coefficient a_6 we employ the sixth-order CF, i.e. $\mathcal{R}_6^{CF}(u)$ from (43.5)

$$\mathcal{R}_6^{CF}(u) = \frac{a_1}{u}\left[1 - \frac{a_3 + a_4 + a_5 + a_6}{u} + \frac{a_3(a_5 + a_6) + a_4 a_6}{u^2}\right]\left(\sum_{n=0}^{3}\gamma_n u^{-n}\right)^{-1}$$

(44.32)

$$\gamma_0 = 1 \qquad \gamma_1 = -(a_2 + a_3 + a_4 + a_5 + a_6)$$
$$\gamma_2 = a_2(a_4 + a_5 + a_6) + a_3(a_5 + a_6) + a_4 a_6 \qquad \gamma_3 = -a_2 a_4 a_6. \quad (44.33)$$

Inversion of the third-order polynomial from (44.32) is done via (44.4) which necessitates the coefficients $\{\delta_1, \delta_2, \delta_3, \delta_4, \delta_5\}$. These are taken from (44.3) by setting $\gamma_n \equiv 0$ for $n \geq 4$, following the general rule (44.4). In this way, (44.32) becomes

$$\mathcal{R}_6^{CF}(u) = \frac{a_1}{u}\left[1 - \frac{a_3 + a_4 + a_5 + a_6}{u} + \frac{a_3(a_5 + a_6) + a_4 a_6}{u^2}\right]$$

$$\times \left(1 - \frac{\gamma_1}{u} + \frac{\gamma_1^2 - \gamma_2}{u^2} - \frac{\gamma_3 - 2\gamma_1\gamma_2 + \gamma_1^3}{u^3}\right.$$

$$+ \frac{2\gamma_1\gamma_3 + \gamma_2^2 - 3\gamma_1^2\gamma_2 + \gamma_1^4}{u^4}$$

$$\left.+ \frac{2\gamma_2\gamma_3 - 3\gamma_1^2\gamma_3 - 3\gamma_1\gamma_2^2 + 4\gamma_1^3\gamma_2 - \gamma_1^5}{u^5} + \cdots\right) \quad (44.34)$$

so that

$$\mathcal{R}_6^{CF}(u) = \frac{a_1}{u} - [\gamma_1 + (a_3 + a_4 + a_5 + a_6)]\frac{a_1}{u^2}$$

$$+ \{(\gamma_1^2 - \gamma_2) + (a_3 + a_4 + a_5 + a_6)\gamma_1 + [a_3(a_5 + a_6) + a_4 a_6]\}\frac{a_1}{u^3}$$

$$- \{(\gamma_3 - 2\gamma_1\gamma_2 + \gamma_1^3) + (a_3 + a_4 + a_5 + a_6)(\gamma_1^2 - \gamma_2)$$

$$+ [a_3(a_5 + a_6) + a_4 a_6]\gamma_1\}\frac{a_1}{u^4} + \{(2\gamma_1\gamma_3 + \gamma_2^2 - 3\gamma_1^2\gamma_2 + \gamma_1^4)$$

$$+ (a_3 + a_4 + a_5 + a_6)(\gamma_3 - 2\gamma_1\gamma_2 + \gamma_1^3)$$

$$+ [a_3(a_5 + a_6) + a_4 a_6](\gamma_1^2 - \gamma_2)\}\frac{a_1}{u^5}$$

$$- \{(\gamma_1^5 - 2\gamma_2\gamma_3 + 3\gamma_1^2\gamma_3 + 3\gamma_1\gamma_2^2 - 4\gamma_1^3\gamma_2)$$

$$+ (a_3 + a_4 + a_5 + a_6)(2\gamma_1\gamma_3 + \gamma_2^2 - 3\gamma_1^2\gamma_2 + \gamma_1^4)$$

$$+ [a_3(a_5 + a_6) + a_4 a_6](\gamma_3 - 2\gamma_1\gamma_2 + \gamma_1^3)\}\frac{a_1}{u^6} + \cdots. \quad (44.35)$$

Relating this series to (43.2) will yield

$$c_0 = a_1$$
$$c_1 = -a_1[\gamma_1 + (a_3 + a_4 + a_5 + a_6)]$$
$$c_2 = a_1\{(\gamma_1^2 - \gamma_2) + (a_3 + a_4 + a_5 + a_6)\gamma_1 + [a_3(a_5 + a_6) + a_4 a_6]\}$$
$$c_3 = -a_1\{(\gamma_1^3 - 2\gamma_1\gamma_2 + \gamma_1^3) + (a_3 + a_4 + a_5 + a_6)(\gamma_1^2 - \gamma_2)$$
$$+ [a_3(a_5 + a_6) + a_4 a_6]\gamma_1\}$$
$$c_4 = a_1\{(2\gamma_1\gamma_3 + \gamma_2^2 - 3\gamma_1^2\gamma_2 + \gamma_1^4) + (a_3 + a_4 + a_5 + a_6)(\gamma_3 - 2\gamma_1\gamma_2 + \gamma_1^3)$$
$$+ [a_3(a_5 + a_6) + a_4 a_6](\gamma_1^2 - \gamma_2)\}$$
$$c_5 = -a_1\{(\gamma_1^5 - 2\gamma_2\gamma_3 + 3\gamma_1^2\gamma_3 + 3\gamma_1\gamma_2^2 - 4\gamma_1^3\gamma_2)$$
$$+ (a_3 + a_4 + a_5 + a_6)(2\gamma_1\gamma_3 + \gamma_2^2 - 3\gamma_1^2\gamma_2 + \gamma_1^4)$$
$$+ [a_3(a_5 + a_6) + a_4 a_6](\gamma_3 - 2\gamma_1\gamma_2 + \gamma_1^3)\}. \tag{44.36}$$

The results of the calculations indicated in (44.36) are

$$c_0 = a_1$$
$$c_1 = a_1 a_2$$
$$c_2 = a_1 a_2 a_3 + a_1 a_2^2$$
$$c_3 = a_1 a_2 a_3 a_4 + a_1 a_2 (a_2 + a_3)^2$$
$$c_4 = a_1 a_2 a_3 a_4 a_5 + a_1 a_2 a_3 (a_2 + a_3 + a_4)^2 + a_1[a_2(a_2 + a_3)]^2$$
$$c_5 = a_1 a_2 a_3 a_4 a_5 a_6 + a_1 a_2 a_3 a_4 (a_2 + a_3 + a_4 + a_5)^2$$
$$+ a_1 a_2 [a_2(a_2 + a_3) + a_3(a_2 + a_3 + a_4)]^2 \tag{44.37}$$

so that

$$a_1 = c_0$$
$$a_2 = \frac{c_1}{a_1}$$
$$a_3 = \frac{c_2 - a_1 a_2^2}{a_1 a_2}$$
$$a_4 = \frac{c_3 - a_1 a_2 (a_2 + a_3)^2}{a_1 a_2 a_3}$$
$$a_5 = \frac{c_4 - a_1 a_2 a_3 (a_2 + a_3 + a_4)^2 - a_1[a_2(a_2 + a_3)]^2}{a_1 a_2 a_3 a_4}$$
$$a_6 = \frac{c_5 - a_1 a_2 a_3 a_4 (a_2 + a_3 + a_4 + a_5)^2 - a_1 a_2 [a_2(a_2 + a_3) + a_3(a_2 + a_3 + a_4)]^2}{a_1 a_2 a_3 a_4 a_5}.$$
$$\tag{44.38}$$

44.6 Calculation of coefficient a_7

A derivation of the expression for the seventh coefficient a_7 necessitates the seventh-order CF, i.e. $\mathcal{R}_7^{CF}(u)$ from (43.10)

$$\mathcal{R}_7^{CF}(u) = a_1 \left[\frac{1}{u} - \frac{a_3 + a_4 + a_5 + a_6 + a_7}{u^2} \right.$$
$$+ \frac{a_3(a_5 + a_6 + a_7) + a_4(a_6 + a_7) + a_5 a_7}{u^3} - \frac{a_3 a_5 a_7}{u^4} \right]$$
$$\times \left(\sum_{n=0}^{3} \gamma_n u^{-n} \right)^{-1} \tag{44.39}$$

$$\gamma_0 = 1 \qquad \gamma_1 = -(a_2 + a_3 + a_4 + a_5 + a_6 + a_7)$$
$$\gamma_2 = a_2(a_4 + a_5 + a_6 + a_7) + a_3(a_5 + a_6 + a_7) + a_4(a_6 + a_7) + a_5 a_7$$
$$\gamma_3 = -\{a_2[a_4(a_6 + a_7) + a_5 a_7] + a_3 a_5 a_7\}. \tag{44.40}$$

Similarly to the preceding analysis of $\mathcal{R}_5^{CF}(u)$ from (44.25), inversion of a third-order polynomial is required in (44.39). To this end (44.4) is employed again, but in this case the coefficients $\{\delta_1, \delta_2, \delta_3, \delta_4, \delta_5, \delta_6\}$ are needed. They are given in (44.3), where $\gamma_n \equiv 0$ must be set for $n \geq 4$, as dictated by (44.4). Then (44.39) becomes

$$\mathcal{R}_7^{CF}(u) = a_1 \left[\frac{1}{u} - \frac{a_3 + a_4 + a_5 + a_6 + a_7}{u^2} \right.$$
$$+ \frac{a_3(a_5 + a_6 + a_7) + a_4(a_6 + a_7) + a_5 a_7}{u^3} - \frac{a_3 a_5 a_7}{u^4} \right]$$
$$\times \left[1 - \frac{\gamma_1}{u} + \frac{\gamma_1^2 - \gamma_2}{u^2} - \frac{\gamma_3 - 2\gamma_1 \gamma_2 + \gamma_1^3}{u^3} \right.$$
$$+ \frac{2\gamma_1 \gamma_3 + \gamma_2^2 - 3\gamma_1^2 \gamma_2 + \gamma_1^4}{u^4}$$
$$- \frac{\gamma_1^5 - 2\gamma_2 \gamma_3 + 3\gamma_1^2 \gamma_3 + 3\gamma_1 \gamma_2^2 - 4\gamma_1^3 \gamma_2}{u^5}$$
$$+ \frac{\gamma_3^2 - 6\gamma_1 \gamma_2 \gamma_3 + 4\gamma_1^3 \gamma_3 - \gamma_2^3 + 6\gamma_1^2 \gamma_2^2 - 5\gamma_1^4 \gamma_2 + \gamma_1^6}{u^6} + \cdots \right]$$
$$\tag{44.41}$$

or, when collecting the coefficients of the like powers of $1/u$

$$\mathcal{R}_7^{CF}(u) = \frac{a_1}{u} - [\gamma_1 + (a_3 + a_4 + a_5 + a_6 + a_7)]\frac{a_1}{u^2}$$
$$+ \{(\gamma_1^2 - \gamma_2) + (a_3 + a_4 + a_5 + a_6 + a_7)\gamma_1$$
$$+ [a_3(a_5 + a_6 + a_7) + a_4(a_6 + a_7) + a_5 a_7]\}\frac{a_1}{u^3}$$
$$- \{(\gamma_3 - 2\gamma_1\gamma_2 + \gamma_1^3) + (a_3 + a_4 + a_5 + a_6 + a_7)(\gamma_1^2 - \gamma_2)$$
$$+ [a_3(a_5 + a_6 + a_7) + a_4(a_6 + a_7) + a_5 a_7]\gamma_1 + a_3 a_5 a_7\}\frac{a_1}{u^4}$$
$$+ \{(2\gamma_1\gamma_3 + \gamma_2^2 - 3\gamma_1^2\gamma_2 + \gamma_1^4)$$
$$+ (a_3 + a_4 + a_5 + a_6 + a_7)(\gamma_3 - 2\gamma_1\gamma_2 + \gamma_1^3)$$
$$+ [a_3(a_5 + a_6 + a_7) + a_4(a_6 + a_7) + a_5 a_7](\gamma_1^2 - \gamma_2)$$
$$+ a_3 a_5 a_7 \gamma_1\}\frac{a_1}{u^5} - \{(\gamma_1^5 - 2\gamma_2\gamma_3 + 3\gamma_1^2\gamma_3 + 3\gamma_1\gamma_2^2 - 4\gamma_1^3\gamma_2)$$
$$+ (a_3 + a_4 + a_5 + a_6 + a_7)(2\gamma_1\gamma_3 + \gamma_2^2 - 3\gamma_1^2\gamma_2 + \gamma_1^4)$$
$$+ [a_3(a_5 + a_6 + a_7) + a_4(a_6 + a_7) + a_5 a_7](\gamma_3 - 2\gamma_1\gamma_2 + \gamma_1^3)$$
$$+ a_3 a_5 a_7(\gamma_1^2 - \gamma_2)\}\frac{a_1}{u^6}$$
$$+ \{(\gamma_3^2 - 6\gamma_1\gamma_2\gamma_3 + 4\gamma_1^3\gamma_3 - \gamma_2^3 + 6\gamma_1^2\gamma_2^2 - 5\gamma_1^4\gamma_2 + \gamma_1^6)$$
$$+ (a_3 + a_4 + a_5 + a_6 + a_7)$$
$$\times (\gamma_1^5 - 2\gamma_2\gamma_3 + 3\gamma_1^2\gamma_3 + 3\gamma_1\gamma_2^2 - 4\gamma_1^3\gamma_2)$$
$$+ [a_3(a_5 + a_6 + a_7) + a_4(a_6 + a_7) + a_5 a_7]$$
$$\times (2\gamma_1\gamma_3 + \gamma_2^2 - 3\gamma_1^2\gamma_2 + \gamma_1^4)$$
$$+ a_3 a_5 a_7(\gamma_3 - 2\gamma_1\gamma_2 + \gamma_1^3)\}\frac{a_1}{u^7}. \tag{44.42}$$

When comparing (43.2) with (44.42) we arrive at the expressions

$c_0 = a_1$
$c_1 = -[\gamma_1 + (a_3 + a_4 + a_5 + a_6 + a_7)]a_1$
$c_2 = \{\gamma_1^2 - \gamma_2 + (a_3 + a_4 + a_5 + a_6 + a_7)\gamma_1$
$\quad + [a_3(a_5 + a_6 + a_7) + a_4(a_6 + a_7) + a_5 a_7]\}a_1$
$c_3 = -\{(\gamma_3 - 2\gamma_1\gamma_2 + \gamma_1^3) + (a_3 + a_4 + a_5 + a_6 + a_7)(\gamma_1^2 - \gamma_2)$
$\quad + [a_3(a_5 + a_6 + a_7) + a_4(a_6 + a_7) + a_5 a_7]\gamma_1 + a_3 a_5 a_7\}a_1$
$c_4 = \{(2\gamma_1\gamma_3 + \gamma_2^2 - 3\gamma_1^2\gamma_2 + \gamma_1^4)$
$\quad + (a_3 + a_4 + a_5 + a_6 + a_7)(\gamma_3 - 2\gamma_1\gamma_2 + \gamma_1^3)$
$\quad + [a_3(a_5 + a_6 + a_7) + a_4(a_6 + a_7) + a_5 a_7](\gamma_1^2 - \gamma_2) + a_3 a_5 a_7 \gamma_1\}a_1$
$c_5 = -\{(\gamma_1^5 - 2\gamma_2\gamma_3 + 3\gamma_1^2\gamma_3 + 3\gamma_1\gamma_2^2 - 4\gamma_1^3\gamma_2)$

$$+ (a_3 + a_4 + a_5 + a_6 + a_7)(2\gamma_1\gamma_3 + \gamma_2^2 - 3\gamma_1^2\gamma_2 + \gamma_1^4)$$
$$+ [a_3(a_5 + a_6 + a_7) + a_4(a_6 + a_7) + a_5 a_7]$$
$$\times (\gamma_3 - 2\gamma_1\gamma_2 + \gamma_1^3) + a_3 a_5 a_7 (\gamma_1^2 - \gamma_2)\}a_1$$
$$c_6 = \{(\gamma_3^2 - 6\gamma_1\gamma_2\gamma_3 + 4\gamma_1^3\gamma_3 - \gamma_2^3 + 6\gamma_1^2\gamma_2^2 - 5\gamma_1^4\gamma_2 + \gamma_1^6)$$
$$+ (a_3 + a_4 + a_5 + a_6 + a_7)(\gamma_1^5 - 2\gamma_2\gamma_3 + 3\gamma_1^2\gamma_3 + 3\gamma_1\gamma_2^2 - 4\gamma_1^3\gamma_2)$$
$$+ [a_3(a_5 + a_6 + a_7) + a_4(a_6 + a_7) + a_5 a_7](2\gamma_1\gamma_3 + \gamma_2^2 - 3\gamma_1^2\gamma_2 + \gamma_1^4)$$
$$+ a_3 a_5 a_7 (\gamma_3 - 2\gamma_1\gamma_2 + \gamma_1^3)\}a_1. \tag{44.43}$$

The results of our protracted algebraic calculations are succinctly expressed as the following formulae

$$c_0 = a_1$$
$$c_1 = a_1 a_2$$
$$c_2 = a_1 a_2 a_3 + a_1 a_2^2$$
$$c_3 = a_1 a_2 a_3 a_4 + a_1 a_2 (a_2 + a_3)^2$$
$$c_4 = a_1 a_2 a_3 a_4 a_5 + a_1 a_2 a_3 (a_2 + a_3 + a_4)^2 + a_1 [a_2(a_2 + a_3)]^2$$
$$c_5 = a_1 a_2 a_3 a_4 a_5 a_6 + a_1 a_2 a_3 a_4 (a_2 + a_3 + a_4 + a_5)^2$$
$$\qquad + a_1 a_2 [a_2(a_2 + a_3) + a_3(a_2 + a_3 + a_4)]^2$$
$$c_6 = a_1 a_2 a_3 a_4 a_5 a_6 a_7 + a_1 a_2 a_3 a_4 a_5 (a_2 + a_3 + a_4 + a_5 + a_6)^2$$
$$\qquad + a_1 a_2 \{a_2 [a_2(a_2 + a_3) + a_3(a_2 + a_3 + a_4)]^2$$
$$\qquad + a_3 [a_2(a_2 + a_3) + a_3(a_2 + a_3 + a_4) + a_4(a_2 + a_3 + a_4 + a_5)]^2\}$$
$$\tag{44.44}$$

so that

$$a_1 = c_0$$
$$a_2 = \frac{c_1}{a_1}$$
$$a_3 = \frac{c_2 - a_1 a_2^2}{a_1 a_2}$$
$$a_4 = \frac{c_3 - a_1 a_2 (a_2 + a_3)^2}{a_1 a_2 a_3}$$
$$a_5 = \frac{c_4 - a_1 a_2 a_3 (a_2 + a_3 + a_4)^2 - a_1 [a_2(a_2 + a_3)]^2}{a_1 a_2 a_3 a_4}$$
$$a_6 = \frac{c_5 - a_1 a_2 a_3 a_4 (a_2 + a_3 + a_4 + a_5)^2 - a_1 a_2 [a_2(a_2 + a_3) + a_3(a_2 + a_3 + a_4)]^2}{a_1 a_2 a_3 a_4 a_5}$$
$$a_7 = \Big\{ c_6 - a_1 a_2 a_3 a_4 a_5 (a_2 + a_3 + a_4 + a_5 + a_6)^2$$
$$\qquad - a_1 a_2 \Big(a_2 [a_2(a_2 + a_3) + a_3(a_2 + a_3 + a_4)]^2$$

$$+ a_3[a_2(a_2 + a_3) + a_3(a_2 + a_3 + a_4)$$
$$+ a_4(a_2 + a_3 + a_4 + a_5)]^2 \bigg) \bigg\} \frac{1}{a_1 a_2 a_3 a_4 a_5 a_6}. \tag{44.45}$$

From here, our general results for any integer $n > 0$ can be readily inferred via

$$a_{n+1} = \frac{c_n - \left(\prod_{i=1}^{n-1} a_i\right)\left(\sum_{j=2}^{n} a_j\right)^2 - \left(\prod_{i=1}^{n-4} a_i\right) \sum_{j=\left[\frac{n-1}{2}\right]}^{n-3} a_j \left(\sum_{k=2}^{j+1} a_k \sum_{\ell=2}^{k+1} a_\ell\right)^2}{\prod_{i=1}^{n} a_i} \tag{44.46}$$

as well as

$$c_n = \prod_{i=1}^{n+1} a_i + \left(\prod_{i=1}^{n-1} a_i\right)\left(\sum_{j=2}^{n} a_j\right)^2 + \left(\prod_{i=1}^{n-4} a_i\right) \sum_{j=\left[\frac{n-1}{2}\right]}^{n-3} a_j \left(\sum_{k=2}^{j+1} a_k \sum_{\ell=2}^{k+1} a_\ell\right)^2 \tag{44.47}$$

where the symbol $[(n-1)/2]$ denotes the largest integer part of $(n-1)/2$ as in (10.1) and

$$\prod_{m=1}^{n}(\cdots) \equiv 0 \quad (n < m) \qquad \sum_{m=1}^{-|n|}(\cdots) \equiv 0. \tag{44.48}$$

The formulae (44.46) and (44.47) can be rewritten in their compact forms as

$$a_{n+1} = \frac{c_n - \sigma_n \pi_{n-1} - \lambda_n \pi_{n-4}}{\pi_n} \tag{44.49}$$

$$c_n = \pi_{n+1} + \sigma_n \pi_{n-1} + \lambda_n \pi_{n-4} \tag{44.50}$$

where

$$\pi_n = \prod_{i=1}^{n} a_i \qquad \sigma_n = \left(\sum_{j=2}^{n} a_j\right)^2$$

$$\lambda_n = \sum_{j=\left[\frac{n-1}{2}\right]}^{n-3} a_j \xi_j^2 \qquad \xi_j = \sum_{k=2}^{j+1} a_k \sum_{\ell=2}^{k+1} a_\ell. \tag{44.51}$$

In this notation, the two conventions from (44.48) become

$$\pi_n \equiv 0 \quad (n \leq 0) \qquad \sigma_n \equiv 0 \quad (n \leq 3). \tag{44.52}$$

If the time signal points $\{c_n\}$ or auto-correlation functions $\{C_n\}$ are given, the analytical formula (44.50) yields the whole set of the expansion coefficients

$\{a_n\}$ of the continued fractions $\mathcal{R}^{CF}(u)$ from (44.49) of the Green function $\mathcal{R}(u)$ from (43.2). Conversely, if the coefficients $\{a_n\}$ are given, the closed expression (44.49) supplies the entire set $\{c_n\}$. We see from (44.49) that each a_{n+1} is given in terms of one fixed c_n and the string $\{a_m\}$ ($1 \le m \le n$). These earlier coefficients $\{a_m\}$ ($1 \le m \le n$) can be eliminated altogether from any a_n. We show this inductively by writing down the first few explicit expressions for the products of a_n as

$$a_1 = c_0$$
$$a_1 a_2 = c_1$$
$$a_1 a_2 a_3 = \frac{c_0 c_2 - c_1^2}{c_0}$$
$$a_1 a_2 a_3 a_4 = \frac{c_1 c_3 - c_2^2}{c_1}$$
$$a_1 a_2 a_3 a_4 a_5 = \frac{c_0 c_2 c_4 - c_1^2 c_4 - c_2^3 - c_0 c_3^2 + 2 c_1 c_2 c_3}{c_0 c_2 - c_1^2}$$
$$a_1 a_2 a_3 a_4 a_5 a_6 = \frac{c_1 c_3 c_5 - c_2^2 c_5 - c_3^3 - c_1 c_4^2 + 2 c_2 c_3 c_4}{c_1 c_3 - c_2^2}. \qquad (44.53)$$

Thus, the products of the CF coefficients $\{a_n\}$ are rational functions that contain only the time signals $\{c_n\}$. Obviously this feature should also be preserved with each individual CF coefficient a_n. This is achieved if every two adjacent equations in (44.53) are divided by each other. In other words, dividing the above equation for $a_1 a_2$ by the one for a_1, then the equation for $a_1 a_2 a_3$ by that for $a_1 a_2$ and continuing in this way until the division of the equation for $a_1 a_2 a_3 a_4 a_5 a_6$ by the equation for $a_1 a_2 a_3 a_4 a_5$, we obtain the results

$$a_1 = c_0$$
$$a_2 = \frac{c_1}{c_0}$$
$$a_3 = \frac{c_0 c_2 - c_1^2}{c_0 c_1}$$
$$a_4 = \frac{c_0 (c_1 c_3 - c_2^2)}{c_1 (c_0 c_2 - c_1^2)}$$
$$a_5 = \frac{c_1 (c_0 c_2 c_4 - c_1^2 c_4 - c_2^3 - c_0 c_3^2 + 2 c_1 c_2 c_3)}{(c_0 c_2 - c_1^2)(c_1 c_3 - c_2^2)}$$
$$a_6 = \frac{(c_0 c_2 - c_1^2)(c_1 c_3 c_5 - c_2^2 c_5 - c_3^3 - c_1 c_4^2 + 2 c_2 c_3 c_4)}{(c_1 c_3 - c_2^2)(c_0 c_2 c_4 - c_1^2 c_4 - c_2^3 - c_0 c_3^2 + 2 c_1 c_2 c_3)}. \qquad (44.54)$$

As first glance, these expressions for $n \le 6$ do not seem to allow an obvious extension to the general CF coefficient a_n for any positive n. However, this is only apparent, since a comparison of (44.53) with the the Hankel determinants from (5.18) and (5.19) yields a systematic pattern

$$a_1 = \frac{c_0}{1} = \frac{H_1(c_0)}{H_0(c_0)} \tag{44.55}$$

$$a_1 a_2 = \frac{c_1}{1} = \frac{H_1(c_1)}{H_0(c_1)} \tag{44.56}$$

$$a_1 a_2 a_3 = \frac{\begin{vmatrix} c_0 & c_1 \\ c_1 & c_2 \end{vmatrix}}{c_0} = \frac{H_2(c_0)}{H_1(c_0)} \tag{44.57}$$

$$a_1 a_2 a_3 a_4 = \frac{\begin{vmatrix} c_1 & c_2 \\ c_2 & c_3 \end{vmatrix}}{c_1} = \frac{H_2(c_1)}{H_1(c_1)} \tag{44.58}$$

$$a_1 a_2 a_3 a_4 a_5 = \frac{\begin{vmatrix} c_0 & c_1 & c_2 \\ c_1 & c_2 & c_3 \\ c_2 & c_3 & c_4 \end{vmatrix}}{\begin{vmatrix} c_0 & c_1 \\ c_1 & c_2 \end{vmatrix}} = \frac{H_3(c_0)}{H_2(c_0)} \tag{44.59}$$

$$a_1 a_2 a_3 a_4 a_5 a_6 = \frac{\begin{vmatrix} c_1 & c_2 & c_3 \\ c_2 & c_3 & c_4 \\ c_3 & c_4 & c_5 \end{vmatrix}}{\begin{vmatrix} c_1 & c_2 \\ c_2 & c_3 \end{vmatrix}} = \frac{H_3(c_1)}{H_2(c_1)}. \tag{44.60}$$

Similarly, the expression (44.54) can be written as

$$a_1 = \frac{H_1(c_0)}{H_0(c_0)} \qquad a_2 = \frac{H_1(c_1) \, H_0(c_0)}{H_0(c_1) \, H_1(c_0)}$$

$$a_3 = \frac{H_2(c_0) \, H_0(c_1)}{H_1(c_0) \, H_1(c_1)} \qquad a_4 = \frac{H_2(c_1) \, H_1(c_0)}{H_1(c_1) \, H_2(c_0)}$$

$$a_5 = \frac{H_3(c_0) \, H_1(c_1)}{H_2(c_0) \, H_2(c_1)} \qquad a_6 = \frac{H_3(c_1) \, H_2(c_0)}{H_2(c_1) \, H_3(c_0)}. \tag{44.61}$$

These particular cases allow the extension to the general CF coefficients a_n

$$a_{2n} = \frac{H_n(c_1) \, H_{n-1}(c_0)}{H_{n-1}(c_1) \, H_n(c_0)} \tag{44.62}$$

$$a_{2n+1} = \frac{H_{n+1}(c_0) \, H_{n-1}(c_1)}{H_n(c_0) \, H_n(c_1)}. \tag{44.63}$$

Using (5.14), which states that the $n \times n$ Hankel matrix $\mathbf{H}_n(c_s)$ is equal to the matrix $\mathbf{U}_n^{(s)}$ of the evolution operator $\hat{U}$ raised to the power s, namely $\hat{U}^s$, we can write equivalently

$$a_{2n} = \frac{\det \mathbf{U}_n}{\det \mathbf{U}_{n-1}} \frac{\det \mathbf{S}_{n-1}}{\det \mathbf{S}_n}$$
$$a_{2n+1} = \frac{\det \mathbf{S}_{n+1}}{\det \mathbf{S}_n} \frac{\det \mathbf{U}_{n-1}}{\det \mathbf{U}_n}. \quad (44.64)$$

Here, according to (5.16) and (5.17), we used the relationships $\mathbf{U}^{(0)} \equiv \mathbf{S}$ and $\mathbf{U}^{(1)} \equiv \mathbf{U}$, where $\mathbf{U}$ is the evolution or relaxation matrix and $\mathbf{S}$ is the Schrödinger, i.e. the Krylov overlap matrix.

Chapter 45

Stieltjes' formula for any continued fraction coefficients

It is possible to derive an alternative form of the general coefficient a_n of continued fractions without using the method of the series inversion from chapter 44. This can be achieved simply if the first few explicit results (44.54) for a_n ($1 \leq n \leq 6$) are rewritten in a way which would permit an inductive derivation of a general expression for any a_n ($n \geq 1$). To this end, the sequel (44.54) is cast in the form

$$a_2 \equiv \frac{c_1}{c_0} = \frac{1}{\left(\dfrac{1}{c_0}\right)\left(\dfrac{c_0^2}{c_1}\right)} \equiv \frac{1}{b_{1,0}b_{2,0}}$$

$$a_3 \equiv \frac{c_0 c_2 - c_1^2}{c_0 c_1} = \frac{1}{\left(\dfrac{c_0^2}{c_1}\right)\left[\dfrac{c_1^2}{c_0(c_0c_2-c_1^2)}\right]} \equiv \frac{1}{b_{2,0}b_{3,0}}$$

$$a_4 \equiv \frac{c_0(c_1c_3 - c_2^2)}{c_1(c_0c_2 - c_1^2)} = \frac{1}{\left[\dfrac{c_1^2}{c_0(c_0c_2-c_1^2)}\right]\left[\dfrac{(c_0c_2-c_1^2)^2}{c_1(c_1c_3-c_2^2)}\right]} \equiv \frac{1}{b_{3,0}b_{4,0}}$$

$$a_5 \equiv \frac{c_1(c_0c_2c_4 - c_1^2c_4 - c_2^3 - c_0c_3^2 + 2c_1c_2c_3)}{(c_0c_2 - c_1^2)(c_1c_3 - c_2^2)}$$

$$= \frac{1}{\left[\dfrac{(c_0c_2-c_1^2)^2}{c_1(c_1c_3-c_2^2)}\right]\left[\dfrac{(c_1c_3-c_2^2)^2}{(c_0c_2-c_1^2)(c_0c_2c_4-c_1^2c_4-c_2^3-c_0c_3^2+2c_1c_2c_3)}\right]}$$

$$\equiv \frac{1}{b_{4,0}b_{5,0}}$$

258

$$a_6 \equiv \frac{(c_0c_2 - c_1^2)(c_1c_3c_5 - c_2^2c_5 - c_3^3 - c_1c_4^2 + 2c_2c_3c_4)}{(c_1c_3 - c_2^2)(c_0c_2c_4 - c_1^2c_4 - c_2^3 - c_0c_3^2 + 2c_1c_2c_3)}$$

$$= \frac{1}{\frac{(c_1c_3 - c_2^2)^2}{(c_0c_2 - c_1^2)(c_0c_2c_4 - c_1^2c_4 - c_2^3 - c_0c_3^2 + 2c_1c_2c_3)}}$$

$$\times \frac{1}{\frac{(c_0c_2c_4 - c_1^2c_4 - c_2^3 - c_0c_3^2 + 2c_1c_2c_3)^2}{(c_1c_3 - c_2^2)(c_1c_3c_5 - c_2^2c_5 - c_3^3 - c_1c_4^2 + 2c_2c_3c_4)}} \equiv \frac{1}{b_{5,0}} \times \frac{1}{b_{6,0}}$$

(45.1)

where

$$b_{1,n} = \frac{1}{c_n}$$

$$b_{2,n} = \frac{c_n^2}{c_{n+1}}$$

$$b_{3,n} = \frac{c_{n+1}^2}{c_n(c_nc_{n+2} - c_{n+1}^2)}$$

$$b_{4,n} = \frac{(c_nc_{n+2} - c_{n+1}^2)^2}{c_{n+1}(c_{n+1}c_{n+3} - c_{n+2}^2)}$$

$$b_{5,n} = \frac{(c_{n+1}c_{n+3} - c_{n+2}^2)^2}{\tilde{b}_{5,n}(c_nc_{n+2} - c_{n+1}^2)}$$

$$b_{6,n} = \frac{(c_nc_{n+2}c_{n+4} - c_{n+1}^2c_{n+4} - c_{n+2}^3 - c_nc_{n+3}^2 + 2c_{n+1}c_{n+2}c_{n+3})^2}{\tilde{b}_{6,n}(c_{n+1}c_{n+3} - c_{n+2}^2)}$$

(45.2)

$$\tilde{b}_{5,n} = c_nc_{n+2}c_{n+4} - c_{n+1}^2c_{n+4} - c_{n+2}^3 - c_nc_{n+3}^2 + 2c_{n+1}c_{n+2}c_{n+3}$$
$$\tilde{b}_{6,n} = c_{n+1}c_{n+3}c_{n+5} - c_{n+2}^2c_{n+5} - c_{n+3}^3 - c_{n+1}c_{n+4}^2 + 2c_{n+2}c_{n+3}c_{n+4}$$

(45.3)

or equivalently

$$b_{2,n} = b_{1,n+1}c_n^2$$
$$b_{4,n} = b_{3,n+1}(c_n - b_{2,n+1})^2$$
$$b_{6,n} = b_{5,n+1}(c_n - b_{2,n+1} - b_{4,n+1})^2$$
$$b_{8,n} = b_{6,n+1}(c_n - b_{2,n+1} - b_{4,n+1} - b_{6,n+1})^2 \quad (45.4)$$

and

$$b_{3,n} = \frac{b_{2,n+1}}{c_n(c_n - b_{2,n+1})}$$

$$b_{5,n} = \frac{b_{4,n+1}}{(c_n - b_{2,n+1})(c_n - b_{2,n+1} - b_{4,n+1})}$$

$$b_{7,n} = \frac{b_{6,n+1}}{(c_n - b_{2,n+1} - b_{4,n+1})(c_n - b_{2,n+1} - b_{4,n+1} - b_{6,n+1})}. \quad (45.5)$$

These particular expressions indicate clearly that the general result for the coefficient a_m in the case of any positive integer m is given by

$$a_m = \frac{1}{b_{m-1,0}b_{m,0}} \quad (45.6)$$

$$b_{2m+1,n-1} = \frac{b_{2m,n}}{d_{n,m}(d_{n,m} - b_{2m,n})}$$

$$b_{2m,n-1} = b_{2m-1,n}d_{n,m}^2$$

$$d_{n,m} = c_{n-1} - \sum_{k=1}^{m-1} b_{2k,n} \quad (m = 1, 2, 3\ldots). \quad (45.7)$$

The recursion (45.6) is the analytical formula of Stieltjes [163] derived in 1858 for any continued fraction coefficient a_m, in terms of the accompanying sequel (45.7) which depends only upon the signal points $\{c_n\}$. To compute the general coefficient a_m from $a_m = 1/(b_{m-1,0}b_{m,0})$ as dictated by (45.6), one needs the initial value $b_{1,n} = 1/c_n$, according to (45.2). The same initialization is also encountered in the quotient difference (QD) algorithm of Rutishauser [169]. Obviously, this initialization requires that for *all* the available discrete data we have $c_n \neq 0 \, (1 \leq n \leq N-1)$. Such a requirement is not invoked in our expression (44.49) for the general CF coefficient a_n.

Chapter 46

The exact analytical expressions for any Lanczos couplings $\{\alpha_n, \beta_n\}$

Written in the Padé form (43.7), the function $\mathcal{R}_{2n}^{CF}(u)$ ought to be identical to the corresponding result $\mathcal{R}_n^{PLA}(u)$ from the PLA, due to the uniqueness of the Padé approximant proven in chapter 38. More specifically, both expressions $\mathcal{R}_{2n}^{CF}(u)$ and $\mathcal{R}_n^{PLA}(u)$ must represent the same Padé approximants to the same series (38.2) for $\mathcal{R}(u)$. Therefore, we must have

$$\mathcal{R}_{2n}^{CF}(u) = \mathcal{R}_n^{PLA}(u) \tag{46.1}$$

where $\mathcal{R}_n^{PLA}(u)$ is given in (36.5) for $K = n$ as the quotient of the two Lanczos polynomials $P_n(u)$ and $Q_n(u)$, i.e. $\mathcal{R}_n^{PLA}(u) = (c_0/\beta_1) P_n(u)/Q_n(u)$. We use (36.5) to multiply and divide $\mathcal{R}_n^{PLA}(u)$ by $\{\beta_2\beta_3 \cdots \beta_n\}$ via $\mathcal{R}_n^{PLA}(u) = c_0[(\beta_2\beta_3 \cdots \beta_n)P_n(u)]/[(\beta_1\beta_2\beta_3 \cdots \beta_n)]Q_n(u)$, so that

$$\mathcal{R}_n^{PLA}(u) = c_0 \frac{\tilde{P}_n(u)}{\tilde{Q}_n(u)} \tag{46.2}$$

$$\tilde{P}_n(u) = \{\beta_2\beta_3 \cdots \beta_n\} P_n(u) = \tilde{P}_n^{CF}(u)$$
$$\tilde{Q}_n(u) = \{\beta_1\beta_2\beta_3 \cdots \beta_n\} Q_n(u) = \tilde{Q}_n^{CF}(u) \tag{46.3}$$

where $a_1 = c_0$ is from (44.43). Letting $n + 1 \longrightarrow n$ in (25.4), it follows that $\tilde{Q}_n(u)$ from the PLA becomes $\tilde{Q}_n(u) = D_{0,n}$ where $D_{0,n}$ is from (37.5) for $K = n$. In (46.3), the uniqueness of the PA yields the equalities, $\tilde{P}_n(u) = \tilde{P}_n^{CF}(u)$ and $\tilde{Q}_n(u) = \tilde{Q}_n^{CF}(u)$. Thus, using (46.3) and the expansions for $\{P_n(u), Q_n(u)\}$ and $\{\tilde{P}_n^{CF}(u), \tilde{Q}_n^{CF}(u)\}$ from (26.1) and (43.12), respectively, we have

$$\tilde{p}_{n,n-r} = \frac{\bar{\beta}_n}{\beta_1} p_{n,n-r} \qquad \tilde{q}_n = \bar{\beta}_n q_{n,n-r} \tag{46.4}$$

where $\bar{\beta}_n$ is from (22.4). The expansion coefficients $q_{n,n-r}$ are given in terms of the Lanczos coupling constants $\{\alpha_n, \beta_n\}$, whereas the parameters $\{a_n\}$ of the

262   The exact analytical expressions for any Lanczos couplings $\{\alpha_n, \beta_n\}$

continued fractions determine the coefficients $\tilde{q}_{n,n-r}$, as in (43.20). Therefore, the relationship (46.4) between $q_{n,n-r}$ and $\tilde{q}_{n,n-r}$ could be exploited to establish the relationship between the sets $\{\alpha_n, \beta_n\}$ and $\{a_n\}$. With this goal, we consider the first few coefficients $\{q_{n,n-r}\}$ and $\{\tilde{q}_{n,n-r}\}$ from (26.6) and (43.16)

$$\beta_1 q_{1,1} = -\alpha_0$$
$$\tilde{q}_{1,1} = -a_2 = -\alpha_0 + (\alpha_0 - a_2) = \beta_1 q_{1,1} + (\alpha_0 - a_2)$$
$$\therefore \quad \tilde{q}_{1,1} = \beta_1 q_{1,1} \iff \alpha_0 = a_2 \qquad (46.5)$$

$$\beta_1 \beta_2 q_{2,1} = -(\alpha_0 + \alpha_1)$$
$$\tilde{q}_{2,1} = -(a_2 + a_3 + a_4) = -[\alpha_0 + (a_3 + a_4)]$$
$$= -(\alpha_0 + \alpha_1) + [\alpha_1 - (a_3 + a_4)]$$
$$= \beta_1 \beta_2 q_{2,1} + [\alpha_1 - (a_3 + a_4)]$$
$$\therefore \quad \tilde{q}_{2,1} = \beta_1 \beta_2 q_{2,1} \iff \alpha_1 = a_3 + a_4 \qquad (46.6)$$

$$\beta_1 \beta_2 q_{2,2} = \alpha_0 \alpha_1 - \beta_1^2$$
$$\tilde{q}_{2,2} = a_2 a_4 = a_2(\alpha_1 - a_3) = \alpha_0 \alpha_1 - a_2 a_3$$
$$= (\alpha_0 \alpha_1 - \beta_1^2) + (\beta_1^2 - a_2 a_3)$$
$$= \beta_1 \beta_2 q_{2,2} + (\beta_1^2 - a_2 a_3)$$
$$\therefore \quad \tilde{q}_{2,2} = \beta_1 \beta_2 q_{2,2} \quad \therefore \quad \iff \beta_1^2 = a_2 a_3 \qquad (46.7)$$

$$\beta_1 \beta_2 \beta_3 q_{3,1} = -(\alpha_0 + \alpha_1 + \alpha_2)$$
$$\tilde{q}_{3,1} = -[a_2 + (a_3 + a_4) + (a_5 + a_6)] = -[\alpha_0 + \alpha_1 + (a_5 + a_6)]$$
$$= -(\alpha_0 + \alpha_1 + \alpha_2) + [\alpha_2 - (a_5 + a_6)]$$
$$= \beta_1 \beta_2 \beta_3 q_{3,1} + [\alpha_2 - (a_5 + a_6)]$$
$$\therefore \quad \tilde{q}_{3,1} = \beta_1 \beta_2 \beta_3 q_{3,1} \iff \alpha_2 = a_5 + a_6 \qquad (46.8)$$

$$\beta_1 \beta_2 \beta_3 q_{3,2} = (\alpha_0 + \alpha_1)\alpha_2 + \alpha_0 \alpha_1 - \beta_1^2 - \beta_2^2$$
$$\tilde{q}_{3,2} = a_2(a_4 + a_5 + a_6) + a_3(a_5 + a_6) + a_4 a_6$$
$$= (a_2 + a_3)(a_5 + a_6) + a_4(a_2 + a_6)$$
$$= (\alpha_0 + \alpha_1 - a_4)\alpha_2 + a_4(a_2 + a_6)$$
$$= (\alpha_0 + \alpha_1)\alpha_2 - a_4 \alpha_2 + a_4(a_2 + a_5 + a_6) - a_4 a_5$$
$$= (\alpha_0 + \alpha_1)\alpha_2 - a_4 \alpha_2 + a_4(a_2 + \alpha_2) - a_4 a_5$$
$$= (\alpha_0 + \alpha_1)\alpha_2 + a_2 a_4 - a_4 a_5$$
$$= (\alpha_0 + \alpha_1)\alpha_2 + a_2(\alpha_1 - a_3) - a_4 a_5$$
$$= (\alpha_0 + \alpha_1)\alpha_2 + a_2 \alpha_1 - a_2 a_3 - a_4 a_5$$
$$= (\alpha_0 + \alpha_1)\alpha_2 + \alpha_0 \alpha_1 - \beta_1^2 - a_4 a_5$$
$$= [(\alpha_0 + \alpha_1)\alpha_2 + \alpha_0 \alpha_1 - \beta_1^2 - \beta_2^2] + (\beta_2^2 - a_4 a_5)$$
$$= \beta_1 \beta_2 \beta_3 q_{3,2} + (\beta_2^2 - a_4 a_5)$$
$$\therefore \quad \tilde{q}_{3,2} = \beta_1 \beta_2 \beta_3 q_{3,2} \iff \beta_2^2 = a_4 a_5 \qquad (46.9)$$

$$\beta_1\beta_2\beta_3 q_{3,3} = \alpha_0\beta_2^2 - \alpha_0\alpha_1\alpha_2 + \alpha_2\beta_1^2$$

$$\tilde{q}_{3,3} = -a_2 a_4 a_6 = -\alpha_0 a_4 a_6 = -\alpha_0[a_4(a_5 + a_6) - a_4 a_5]$$
$$= -\alpha_0(a_4\alpha_2 - \beta_2^2) = \alpha_0\beta_2^2 - \alpha_0\alpha_2 a_4$$
$$= \alpha_0\beta_2^2 - \alpha_0\alpha_2(\alpha_1 - a_3) = \alpha_0\beta_2^2 - \alpha_0\alpha_1\alpha_2 + \alpha_2 a_2 a_3$$
$$= \alpha_0\beta_2^2 - \alpha_0\alpha_1\alpha_2 + \alpha_2\beta_1^2 = \beta_1\beta_2\beta_3 q_{3,3}$$
$$\therefore \quad \tilde{q}_{3,3} = \beta_1\beta_2\beta_3 q_{3,3}. \tag{46.10}$$

This is recapitulated as

$$\alpha_0 = a_2$$
$$\alpha_1 = a_3 + a_4 \qquad \beta_1^2 = a_2 a_3$$
$$\alpha_2 = a_5 + a_6 \qquad \beta_2^2 = a_4 a_5 \qquad \text{etc} \tag{46.11}$$

so that, in a general case of any positive integer n, we have

$$\alpha_n = a_{2n+1} + a_{2n+2} \qquad \beta_n^2 = a_{2n} a_{2n+1} \qquad (n \geq 1) \tag{46.12}$$

with $\alpha_0 = a_2$ and $\beta_0 = 0$. These are the sought general analytical expressions for any Lanczos coupling parameters $\{\alpha_n, \beta_n\}$ in terms of the continued fraction coefficients $\{a_n\}$, whose entire set is available from the single closed formula (44.49). For checking purposes, in the particular cases with $1 \leq n \leq 3$, we have used (44.49) to obtain the results for $a_{2n+1} + a_{2n+2}$ and $a_{2n} a_{2n+1}$ that are found to fully agree with α_n and β_n^2, respectively, from (22.6) which was calculated in a totally different way. Inserting the expressions for a_{2n} and a_{2n+1} from (44.63) into (46.12), we can write the following explicit analytical formulae for any Lanczos coupling parameters $\{\alpha_n, \beta_n\}$

$$\alpha_n = \frac{H_{n+1}(c_0)}{H_n(c_0)} \frac{H_{n-1}(c_1)}{H_n(c_1)} + \frac{H_{n+1}(c_1)}{H_n(c_1)} \frac{H_n(c_0)}{H_{n+1}(c_0)} \tag{46.13}$$

$$\beta_n^2 = \frac{H_{n-1}(c_0) H_{n+1}(c_0)}{[H_n(c_0)]^2}. \tag{46.14}$$

Here, β_n^2 is given exclusively in terms of the determinant of the Schrödinger overlap matrix $H_m(c_0) = \det \mathbf{S}_m$. Recursive numerical computations of the Hankel determinants $H_m(c_0)$ and $H_m(c_1)$ from (46.13) can be carried out by Gordon's [170] PD algorithm which is accurate, efficient and robust with only $\sim n^2$ multiplications relative to some formidable $n!$ multiplications in the Cramer rule via direct evaluations. Moreover, the same PD recursion can also be used for the CF coefficients $\{a_n\}$ [142].

Chapter 47

The Lanczos continued fractions and contracted continued fractions

Employing the relationships from (46.12), we can rewrite the Lanczos continued fraction (41.7) as

$$\mathcal{R}^{LCF}(u) = \frac{a_1}{u-a_2} - \frac{a_2 a_3}{u-(a_3+a_4)} - \frac{a_4 a_5}{u-(a_5+a_6)} - \frac{a_6 a_7}{u-(a_7+a_8)} - \cdots \quad (47.1)$$

A truncation of this expression at the order n will give the nth order LCF approximant

$$\mathcal{R}_n^{LCF}(u) = \{\mathcal{R}^{LCF}(u)\}_{a_m=0, m \geq 2n+1} \quad (n = 1, 2, 3, \ldots). \quad (47.2)$$

For example

$$\mathcal{R}_1^{LCF}(u) = \frac{a_1}{u-a_2} \qquad \mathcal{R}_2^{LCF}(u) = \frac{a_1}{u-a_2} - \frac{a_2 a_3}{u-(a_3+a_4)}$$

$$\mathcal{R}_3^{LCF}(u) = \frac{a_1}{u-a_2} - \frac{a_2 a_3}{u-(a_3+a_4)} - \frac{a_4 a_5}{u-(a_5+a_6)}. \quad (47.3)$$

The equivalent Padé forms of these equations are given by

$$\mathcal{R}_1^{LCF}(u) = \frac{a_1}{u-a_2}$$

$$\mathcal{R}_2^{LCF}(u) = a_1 \frac{u-(a_3+a_4)}{u^2 - (a_2+a_3+a_4)u + a_2 a_4}$$

$$\mathcal{R}_3^{LCF}(u) = a_1 \frac{u^2 - (a_3+a_4+a_5+a_6)u + [a_3(a_5+a_6) + a_4 a_6]}{u^3 - (a_2+a_3+a_4+a_5+a_6)u^2 + A'u - a_2 a_4 a_6} \quad (47.4)$$

where A' is given in (43.6). Comparing (43.5) and (47.4) reveals that

$$\mathcal{R}_1^{LCF}(u) = \mathcal{R}_2^{CF}(u) \qquad \mathcal{R}_2^{LCF}(u) = \mathcal{R}_4^{CF}(u) \qquad \mathcal{R}_3^{LCF}(u) = \mathcal{R}_6^{CF}(u). \quad (47.5)$$

Therefore, for an arbitrary positive integer n the following equivalence is established

$$\mathcal{R}_n^{\text{LCF}}(u) = \mathcal{R}_{2n}^{\text{CF}}(u) \qquad (n = 1, 2, 3, \ldots). \tag{47.6}$$

This implies, according to $\mathcal{R}_n^{\text{LCF}}(u) = \mathcal{R}_n^{\text{PLA}}(u)$ from (41.14), that we have, in general

$$\mathcal{R}_n^{\text{PLA}}(u) = \mathcal{R}_{2n}^{\text{CF}}(u) \qquad (n = 1, 2, 3, \ldots). \tag{47.7}$$

Let us recall the definition of the contracted continued fraction (CCF). A general CF, say $\tilde{C}(z)$, whose approximants coincide with a *subset* of the approximants of another continued fraction $C(z)$ is called a *contraction* of $C(z)$ [186]. The result (47.6) shows that the approximant $\mathcal{R}_n^{\text{LCF}}(u)$ of order n of the Lanczos continued fraction $\mathcal{R}^{\text{LCF}}(u)$ is equal exactly to the approximant $\mathcal{R}_{2n}^{\text{CF}}(u)$ of order $2n$ of the continued fraction $\mathcal{R}^{\text{CF}}(u)$ from (43.1) of the series (38.2) of autocorrelation functions. Thus, $\mathcal{R}^{\text{CCF}}(u)$ is a contraction of $\mathcal{R}^{\text{CF}}(u)$. Therefore the PLA is equal to the LCF and to the CCF.

Chapter 48

Exact retrieval of time signals from the Padé polynomial coefficients

In this chapter we shall devise a procedure with the goal of checking the internal consistency of spectral analysis within the PLA. For this purpose, we shall adapt the Longman and Sharir [270] method which has originally been designed for the Laplace inversion of rational response functions. If we have already obtained the roots and residues $\{u_k, d_k\}_{k=1}^{K}$ of the denominator polynomial $Q_K(u)$, we could simply recompute the elements of the auto-correlation functions, or equivalently, time signals $\{C_n\} = \{c_n\}$ by using the decomposition $c_n = \sum_{k=1}^{K} d_k u_k^n$ from (5.11). This would constitute one of the possible tests of the internal consistency of the PLA. Such a procedure would be feasible if we were doing parametric estimation which necessitates the set $\{u_k, d_k\}_{k=1}^{K}$. The PLA is both a parametric and a non-parametric estimator of spectra. When used as a non-parametric estimator, the PLA does not provide the pair $\{u_k, d_k\}_{k=1}^{K}$, as here the interest is limited only to the shape of a spectrum. Then the question arises as to whether is it still possible to check the internal consistency of the PLA by using (5.11)? The reason why it is important, in principle, to retrieve exactly the set $\{c_n\}$ in both parametric and non-parametric estimations, is that the only input data to spectral analysis are the signal points $\{c_n\}$. Exact reconstruction of the set of the initial data $\{c_n\}$ is essential for a retrospective validation of the performed spectral analysis. Such a reconstruction should yield accurate values for the quantities $\{c_n\}$ in an efficient and numerically stable way. A computational algorithm of this type will be analysed in the present chapter.

We assume that both the numerator and denominator polynomials $P_K(u)$ and $Q_K(u)$ of the PLA have already been generated. In particular, it will be supposed that the expansion coefficients $\{p_{K,K-r}, q_{K,K-r'}\}_{r,r'=0}^{K-1,K}$ from the power series representation (26.1) of these polynomials are available. Equivalently, the algorithm also works by supposing that instead of the coefficients $\{p_{K,K-r}, q_{K,K-r}\}$, we have the precomputed couplings $\{\alpha_n, \beta_n\}$. By contrast, the spectral parameters $\{u_k, d_k\}$ will not be required. Yet our analysis

Exact retrieval of time signals from the Padé polynomial coefficients 267

will rely upon the decomposition (5.11) for the discrete signal points $\{c_n\}$. The idea is first to use (5.11) only as the starting point for the definition of the Lorentzian model for c_n, and then try to eliminate altogether both sets $\{u_k\}$ and $\{d_k\}$ from the algorithm without any approximation. This is possible as will be analysed in this chapter. We begin by employing the Cauchy residue theorem within the PLA to derive (5.11)

$$\frac{1}{2\pi i}\oint_C du\, u^n \mathcal{R}_K^{PLA}(u) = \frac{c_0}{\beta_1}\frac{1}{2\pi i}\oint_C du\, u^n \frac{P_K(u)}{Q_K(u)}$$
$$= \frac{c_0}{\beta_1}\sum_{k=1}^{K} u_k^n \frac{P_K(u_k)}{Q'_K(u_k)} = \sum_{k=1}^{K} d_k u_k^n = c_n \quad (48.1)$$

with $0 \leq n \leq N-1$ and where $d_k = (c_0/\beta_1)P_K(u_k)/Q'_K(u_k)$ as in (36.10). This line integral in the complex u-plane is carried out along the circle $C : |u| = R$ of a sufficiently large radius R, which tends to infinity to include all the zeros $\{u_k\}_{k=1}^{K}$ of polynomial $Q_K(u)$. An important role in the elimination of the roots $\{u_k\}$ from the sought representation of signal points is played by the characteristic equation (23.12) itself which defines these roots. Multiplying (23.12) by the term u_k^{n-K}, we have

$$q_{K,K}u_k^{n-K} + q_{K,K-1}u_k^{1+n-K} + \cdots + q_{K,1}u_k^{n-1} + q_{K,0}u_k^n = 0.$$

Extracting the term u_k^n from this equation, it follows

$$u_k^n = -\frac{1}{q_{K,0}}\sum_{r=1}^{K} q_{K,r}u_k^{n-r}. \quad (48.2)$$

Inserting the power function u_k^n from (48.2) into (48.1) yields

$$c_n = -\frac{1}{q_{K,0}}\sum_{r=1}^{K} q_{K,r}c_{n-r} \quad (n \geq K). \quad (48.3)$$

This is a recursion relation for the time signal points $\{c_n\}_{n=0}^{N-1}$ ($n = K, K+1, K+2, \ldots$). However, for a given K the relationship (48.3) is valid only for $n \geq K$, since the subscript n in c_n counts time and, as such, must be non-negative. Therefore, the recursion (48.3) must be supplied by the initial condition through which the missing set $\{c_n\}_{n=0}^{K-1}$ will become available. At this point in the analysis, we need to represent the signal c_n in the form (48.1), where the residue d_k is defined in (36.10). The parameter d_k can be eliminated from (48.1) by using (26.1) and (36.10). This results in

$$c_n = \sum_{k=1}^{K} d_k u_k^n$$

$$= \sum_{k=1}^{K} \left\{ \frac{c_0}{\beta_1} \frac{P_K(u_k)}{Q'_K(u_k)} \right\} u_k^n = \sum_{k=1}^{K} \frac{c_0}{\beta_1} \frac{1}{Q'_K(u_k)} u_k^n \sum_{r=0}^{K-1} p_{K,K-r} u_k^r$$

$$= \sum_{r=0}^{K-1} p_{K,K-r} \left\{ \sum_{k=1}^{K} \frac{c_0}{\beta_1} \frac{1}{Q'_K(u_k)} u_k^{n+r} \right\}$$

$$= \sum_{r=1}^{K} p_{K,r} \left\{ \sum_{k=1}^{K} \frac{c_0}{\beta_1} \frac{1}{Q'_K(u_k)} u_k^{n+K-r} \right\} \equiv \sum_{r=1}^{K} p_{K,r} \tilde{c}_{n+K-r}$$

$$c_n = \sum_{r=1}^{K} p_{K,r} \tilde{c}_{n+K-r} \qquad (0 \leq n \leq N-1) \qquad (48.4)$$

$$\tilde{c}_n \equiv \sum_{k=1}^{K} \tilde{d}_k u_k^n \qquad (0 \leq n \leq N-1) \qquad (48.5)$$

$$\tilde{d}_k = \frac{c_0}{\beta_1} \frac{1}{Q'_K(u_k)}. \qquad (48.6)$$

The decomposition (48.4) is free from the unknown sequence $\{d_k\}$ of the residues and this is valid for any non-negative integer n ($n = 0, 1, 2, \ldots$). The finding (48.5) still contains the other set of the unknown parameters $\{u_k\}$ that, as such, should be eliminated. Notice that the quantity $\tilde{d}_k$ from (48.6) is the residue of the spectrum in the AR model, or equivalently, the 'all pole model' of Kth order

$$\tilde{\mathcal{R}}_K(u) \equiv \mathcal{R}_K^{AR}(u) = \frac{c_0}{\beta_1} \frac{1}{Q_K(u)}. \qquad (48.7)$$

The usual spectrum in the AR model is proportional to the inverse of the characteristic polynomial $\sim 1/Q_K(u)$. We multiply the latter quantity by the constant c_0/β_1 in order to make an easy link with the PLA, which is of the type of the 'all-zero-all-pole models', i.e. ARMA. Just like (48.1), the representation (48.5) can be obtained directly from the Cauchy residue theorem as

$$\frac{1}{2\pi i} \oint_C du\, u^n \mathcal{R}_K^{AR}(u) = \frac{c_0}{\beta_1} \frac{1}{2\pi i} \oint_C du\, u^n \frac{1}{Q_K(u)} = \sum_{k=1}^{K} \tilde{d}_k u_k^n = \tilde{c}_n \qquad (48.8)$$

Exact retrieval of time signals from the Padé polynomial coefficients   269

with $0 \leq n \leq N - 1$. Substituting the term u_k^n from (48.2) into (48.5) yields at once

$$\tilde{c}_n = -\frac{1}{q_{K,0}} \sum_{r=1}^{K} q_{K,r} \tilde{c}_{n-r} \qquad (n \geq K). \qquad (48.9)$$

It then follows that this recursion for the set $\{\tilde{c}_n\}$ is identical in its form with (48.3) for $\{\tilde{c}_n\}$. Hence, the same limitation $n \geq K$ also applies to (48.9) since only non-negative subscripts are allowed in any of the elements $\tilde{c}_n$. Thus the problem of finding the initializing set $\{c_n\}_{n=0}^{K-1}$ is considered as solved, if one could determine the initialization for $\{\tilde{c}_n\}$, i.e. the subset $\{\tilde{c}_n\}_{n=0}^{K-1}$. This is because the entries $\{c_n\}$ are the sums of the elements $\{\tilde{c}_n\}$, according to (48.4). Although (48.4) is not limited to $n \geq K$, its usefulness depends upon the availability of the elements of the whole set $\{\tilde{c}_n\}_{n=0}^{N-1}$. To find the initial conditions to the recursion (48.9), i.e. to obtain the elements of the sequence $\{\tilde{c}_n\}$ $(0 \leq n \leq K - 1)$, we shall use (48.8). We write the complex variable u in its polar form

$$u = Re^{i\theta} \qquad (R, \theta \text{ real}) \qquad (48.10)$$

such that the differential du is given by $du = iu\, d\theta$. Then the power series representation (26.1) gives the relations

$$-i \lim_{R\to\infty} u^n \tilde{\mathcal{R}}_K(u) \frac{du}{d\theta} = -i \frac{c_0}{\beta_1} \lim_{R\to\infty} \frac{u^n}{Q_K(u)} \frac{du}{d\theta}$$

$$= \frac{c_0}{\beta_1} \lim_{R\to\infty} \frac{u^{n+1}}{q_{K,0} u^K + \sum_{r=0}^{K-1} q_{K,K-r} u^r}$$

and, in the limit $R \longrightarrow \infty$, the result of this quotient is equal to zero or $c_0/(\beta_1 q_{K,0})$, depending upon whether $0 \leq n \leq K - 2$ or $n = K - 1$, respectively [270]

$$\lim_{R\to\infty} \frac{1}{2\pi i} \oint_C du \frac{c_0}{\beta_1} \frac{u^n}{Q_K(u)} = 0 \qquad (0 \leq n \leq K - 2) \qquad (48.11)$$

with $c_0/\beta_1 \neq 0$ and

$$\lim_{R\to\infty} \frac{1}{2\pi i} \oint_C du \frac{c_0}{\beta_1} \frac{u^{K-1}}{Q_K(u)} = \frac{c_0}{p_{K,1}} \qquad (n = K - 1) \qquad (48.12)$$

where the relation $p_{K,1}/q_{K,0} = \beta_1$ from (26.9) is utilized. Therefore, the values of the elements from the initializing set $\{\tilde{c}_n\}$ for $n \leq K - 1$ are

$$\tilde{c}_n = \begin{cases} 0 & 0 \leq n \leq K - 2 \\ \dfrac{c_0}{p_{K,1}} & n = K - 1 \end{cases} \qquad (n \leq K - 1). \qquad (48.13)$$

This concludes the derivation of the presented algorithm for the exact reconstruction of the input data that are auto-correlation functions or time signal points $\{C_n\} = \{c_n\}$, without having to compute the set $\{u_k, d_k\}$ of the eigenvalues and residues. To recapitulate, the described algorithm consists of using (48.3) for $n \geq K$ and (48.4), (48.9) and (48.13) for $n \leq K - 1$. The outlined procedure assumes that the expansion coefficients $\{p_{K,K-r}, q_{K,K-r}\}$ of the numerator and denominator polynomials $\{P_K(u), Q_K(u)\}$ of the PLA are available. Here it is tempting to bypass (48.3) altogether, since all the signal prints $\{c_n\}$ are retrievable from (48.4), (48.9) and (48.13). Recall that (48.4) is valid for any non-negative time ($n = 0, 1, 2, \ldots, K, K+1, \ldots$). Nevertheless, the use of (48.3) for $n \geq K$ is more economical, since $c_{n \geq K}$ are computed directly from the set $\{c_n\}_{n=0}^{K-1}$ without the need for the additional recursion on $\tilde{c}_n$.

For checking purposes, we used the previously obtained particular values of the expansion coefficients $\{p_{K,K-r}, q_{K,K-r}\}$ for, e.g. $K \leq 3$ from (26.6), and found the identity $c_n \equiv c_n$ by exploring the rhs of (48.3) or (48.4). For example, for $K = 2$ we employ (48.9) and (48.13) to compute $\tilde{c}_{n \leq 3}$ with the results

$$\tilde{c}_0 = 0 \qquad \tilde{c}_1 = c_0 \beta_2 \qquad \tilde{c}_2 = c_0(\alpha_0 + \alpha_1)\beta_2$$
$$\tilde{c}_3 = c_0[(\alpha_0 + \alpha_1)^2 - (\alpha_0 \alpha_1 - \beta_1^2)]\beta_2. \tag{48.14}$$

Substituting these expressions into (48.4) for $n = 0$ and $n = 1$ ($n \leq K - 1 = 1$) yields the formulae

$$c_0 = \sum_{r=1}^{2} p_{2,r} \tilde{c}_{2-r} = p_{2,1} \tilde{c}_1 + p_{2,2} \tilde{c}_0 = p_{2,1} \tilde{c}_1$$
$$= \frac{1}{\beta_2}(c_0 \beta_2) = c_0 \qquad \text{(QED)} \tag{48.15}$$

$$c_1 = \sum_{r=1}^{2} p_{2,r} \tilde{c}_{3-r} = p_{2,1} \tilde{c}_2 + p_{2,2} \tilde{c}_1$$
$$= \frac{1}{\beta_2}(\alpha_0 + \alpha_1)(c_0 \beta_2) - \frac{\alpha_1}{\beta_2}(c_0 \beta_2)$$
$$= \alpha_0 c_0 = \frac{c_1}{c_0} c_0 = c_1 \qquad \text{(QED)}. \tag{48.16}$$

As emphasized above, at the values $n \geq K$ we should always switch to (48.3). Thus for, e.g. $n = 2$ the following is true

$$c_2 = -\frac{1}{q_{2,0}} \sum_{r=1}^{2} q_{2,r} c_{2-r} = -\frac{q_{2,1} c_1 + q_{2,2} c_0}{q_{2,0}}$$
$$= (\alpha_0 + \alpha_1) c_1 - (\alpha_0 \alpha_1 - \beta_1^2) c_0 = \frac{c_0 c_3 - c_1 c_2}{c_0 c_2 - c_1^2} c_1 - \frac{c_1 c_3 - c_2^2}{c_0 c_2 - c_1^2} c_0$$

Exact retrieval of time signals from the Padé polynomial coefficients   271

$$= \frac{(c_0c_3 - c_1c_2)c_1 - (c_1c_3 - c_2^2)c_0}{c_0c_2 - c_1^2}$$

$$= c_2 \frac{c_0c_2 - c_1^2}{c_0c_2 - c_1^2} = c_2 \quad \text{(QED)} \tag{48.17}$$

where (22.6) and (26.6) are used. We pointed out that (48.4) is less efficient for $n \geq K$ than (48.3) in exhaustive numerical computations, but for a low value of rank, such as $K = 2$, the difference in the effort is, of course, negligible. Thus with the help of (26.6), (26.12) and (48.4) for $n = K = 2$ we find

$$c_2 = \sum_{r=1}^{2} p_{2,r}\tilde{c}_{4-r} = p_{2,1}\tilde{c}_3 + p_{2,2}\tilde{c}_2$$

$$= \frac{(\alpha_0 + \alpha_1)^2 - (\alpha_0\alpha_1 - \beta_1^2)}{\beta_2}(c_0\beta_2) - \frac{\alpha_1(\alpha_0 + \alpha_1)}{\beta_2}(c_0\beta_2)$$

$$= \frac{(\alpha_0 + \alpha_1)^2 - (\alpha_0\alpha_1 - \beta_1^2) - \alpha_1(\alpha_0 + \alpha_1)}{\beta_2}(c_0\beta_2)$$

$$= c_0(\alpha_0^2 + \beta_1^2) = c_0\left(\frac{c_1^2}{c_0^2} + \frac{c_0c_2 - c_1^2}{c_0^2}\right)$$

$$= c_0\frac{c_0c_2}{c_0^2} = c_2 \quad \text{(QED)} \tag{48.18}$$

where (22.6) is used. The same type of calculation for $K = 3$ based upon (48.3) and (48.4) leads again to the identity $c_n = c_n$. Here, we give the details of the calculation using, e.g. only (48.3) for $n = K = 3$

$$c_3 = -\frac{1}{q_{3,0}}\sum_{r=1}^{3} q_{3,r}c_{3-r} = -\frac{q_{3,1}c_2 + q_{3,2}c_1 + q_{3,3}c_0}{q_{3,0}}$$

$$= (\alpha_0 + \alpha_1 + \alpha_2)c_2 - [\alpha_0\alpha_1 + \alpha_2(\alpha_0 + \alpha_1) - \beta_1^2 - \beta_2^2]c_1$$
$$\quad - \left[\alpha_0\beta_2^2 - \alpha_2(\alpha_0\alpha_1 - \beta_1^2)\right]c_0$$

$$= -\frac{q_1c_2 + q_2c_1 + q_3c_0}{q_0}$$

$$= q_0^{-1}[(c_0c_2c_5 - c_0c_3c_4 + c_1c_2c_4 + c_1c_3^2 - c_1^2c_5 - c_2^2c_3)c_2$$
$$\quad - (c_0c_3c_5 + c_1c_3c_4 - c_1c_2c_5 - c_0c_4^2 - c_2c_3^2 + c_2^2c_4)c_1$$
$$\quad + (c_1c_3c_5 - c_3^3 + 2c_2c_3c_4 - c_1c_4^2 - c_2^2c_5)c_0]$$

$$= \frac{2c_1c_2c_3^2 + c_0c_2c_3c_4 - c_0c_3^3 - c_1^2c_3c_4 - c_2^2c_3}{q_0}$$

$$= c_3\frac{2c_1c_2c_3 + c_0c_2c_4 - c_0c_3^2 - c_1^2c_4 - c_2^3}{q_0}$$

$$= c_3 \frac{q_0}{q_0} = c_3 \quad \text{(QED)} \tag{48.19}$$

where (26.6) and (26.12) are employed. Of course, a similar calculation with (48.4) also gives the identities, $c_1 = c_1, c_2 = c_2, c_3 = c_3$, etc. Notice that in this illustration, all the 'normalized' auto-correlation functions $\{c_n/c_0\}$ [271] are seen to depend exclusively upon the coupling parameters $\{\alpha_n, \beta_n\}$ viz

$$\frac{c_1}{c_0} = \alpha_0 \qquad \frac{c_2}{c_0} = \alpha_0^2 + \beta_1^2 \qquad \frac{c_3}{c_0} = \alpha_0(\alpha_0^2 + \beta_1^2) + (\alpha_0 + \alpha_1)\beta_1^2 \tag{48.20}$$

and similarly for higher values of the integer n. In other words, the outlined procedure of self-consistency of the PLA can be used equally well to generate the auto-correlation functions $\{C_n\} = \{c_n\}$ from the precomputed set of the couplings $\{\alpha_n, \beta_n\}$ or expansion coefficients $\{p_{n,n-r}, q_{n,n-r}\}$ of the Lanczos polynomials $\{P_n(u), Q_n(u)\}$, respectively. In the case that this is chosen as a way of computing C_n, the parameters $\{\alpha_n, \beta_n\}$, or equivalently, the expansion coefficients $\{p_{n,n-r}, q_{n,n-r}\}$ should be generated by using the operator $\hat{\Omega}$ in (22.1) instead of $\hat{U}(\tau)$.

Chapter 49

Auto-correlation functions at large times

The flexibility of the chain model is quite transparent already at its starting point. The studied problem can be specified in two major ways. One, in which both the initial state $|\Phi_0\rangle$ and the dynamical operator $\hat{\Omega}$ are explicitly given, as is the case in, e.g. quantum mechanics. Second, in which neither $|\Phi_0\rangle$ or $\hat{\Omega}$ is known, but instead the matrix elements of $\hat{\Omega}$ or $\hat{U}$ are available. In the former case, one would proceed by first applying the so-called discrete variable representation (DVR) [204] to discretize all the dynamic variables of the known input pair $\{|\Phi_0\rangle, \hat{\Omega}\}$ and generate afterwards the Lanczos states $\{|\psi_n\rangle\}$ from the recursion similar to (22.1), except that $\hat{U}(\tau)$ is replaced by $\hat{\Omega}$

$$\beta_{n+1}|\psi_{n+1}\rangle = \{\hat{\Omega} - \alpha_n\hat{1}\}|\psi_n\rangle - \beta_n|\psi_{n-1}\rangle \tag{49.1}$$

$$\alpha_n = \frac{\langle\psi_n|\hat{\Omega}|\psi_n\rangle}{\langle\psi_n|\psi_n\rangle} \qquad \beta_n = \frac{\langle\psi_{n-1}|\hat{\Omega}|\psi_n\rangle}{\langle\psi_{n-1}|\psi_{n-1}\rangle} \qquad \beta_0 = 0 \tag{49.2}$$

where $\langle\psi_m|\psi_n\rangle = c_0\delta_{n,m}$ holds true, as in (22.9). No confusion should arise while still keeping the same labels $\{|\psi_n\rangle\}$ for the Lanczos states and coupling constants $\{\alpha_n, \beta_n\}$, despite the replacement of $\hat{U}$ by $\hat{\Omega}$. This would lead to the Jacobi matrix **J**, which is the same as in (22.10), bearing in mind that now the constants $\{\alpha_n, \beta_n\}$ are redefined by (49.2). Then diagonalization of **J** would lead to the eigenvalues and residues $\{\omega_k, d_k\}$. With this at hand, the auto-correlation functions $\{C_n\}$ are obtained from (5.9). This is the current practice in the literature, but the procedure experiences serious difficulties when the studied system requires wave packet propagations at large times. An alternative method of obtaining the information at large times with data acquired at short times is analysed in this chapter *without the need to solve any eigenvalue problems.* Instead, we rely only upon the availability of the coupling or recurrence parameters $\{\alpha_n, \beta_n\}$. In the special case $\hat{\Omega} = \hat{H}$, where $\hat{H}$ is a Hermitian Hamiltonian of the studied system, the coupling parameters $\{\alpha_n, \beta_n\}$ have a direct physical interpretation as discussed in chapter 22.

The auto-correlation function $C(t)$ from (4.9) and the corresponding spectrum $\mathcal{F}(\omega)$ as given by the expression (4.10) are interrelated via the Fourier integral. The function $C(t)$ agrees with the time signal $c(t)$ from (5.9) and this is seen in the equivalence from (5.12) which applies to Lorentzian spectra. The associated representation for the time signal $c(t) = \sum_{k=1}^{K} d_k \exp(-i\omega_k t)$ from (5.10) requires that the whole sets of the eigenvalues $\{\omega_k\}$ and residues $\{d_k\}$ are precomputed. By contrast, no diagonalization is needed to compute the Lorentzian spectrum at any value of ω

$$\mathcal{F}(\omega) \equiv \mathcal{R}_K^{\text{PLA}}(\omega) = \frac{c_0}{\beta_1} \frac{P_K(\omega)}{Q_K(\omega)}. \tag{49.3}$$

This is the PLA to the Green function $\mathcal{G}(\omega)$ whose exact expression is given by (3.2). The result (49.3) follows immediately from the analogy to the corresponding spectrum (36.5) in the PLA for the variable u and the associated Green function (5.22). This asymmetry in the demanded computational effort for $\mathcal{F}(\omega)$ and $C(t)$ is not expected, in view of the fact that the two quantities are interrelated by the simple Fourier integral (4.8). This is an apparent paradox which can be resolved with a bonus of ending up with a method which can accelerate generation of auto-correlation functions at large times. As is clear from our analysis in chapter 36, all that is needed for the Green function $-i\mathcal{F}(\omega) = \mathcal{G}(\omega)$ is the set of coupling constants $\{\alpha_n, \beta_n\}$ to create a shape spectrum. The same information must also be sufficient to obtain C_n. This would obviate the need to root the characteristic polynomial $Q_K(u)$, or equivalently, to solve the corresponding eigenvalue problem. When using $\hat{\Omega}$ as the generator of the Lanczos states as in (49.2), the analysis and the results of chapter 22 remain the same, provided that appropriate care is taken to replace the triple $\{\hat{U}, u, u_k\}$ by $\{\hat{\Omega}, \omega, \omega_k\}$ throughout, on top of having the new definitions (49.2) of the coupling parameters $\{\alpha_n, \beta_n\}$. In particular, using the Green function $\mathcal{R}_K^{\text{PLA}}(\omega)$ from (36.5), we can write down at once the final expression for the PLA

$$\mathcal{R}_K^{\text{PLA}}(\omega) = \frac{c_0}{\beta_1} \frac{P_K(\omega)}{Q_K(\omega)} = \sum_{k=1}^{K} \frac{d_k}{\omega - \omega_k} \qquad d_k = \frac{c_0}{\beta_1} \frac{P_K(\omega_k)}{Q_K'(\omega_k)}. \tag{49.4}$$

The Lanczos orthogonal polynomials $P_K(\omega)$ and $Q_K(\omega)$ are defined recursively in (23.16) and (23.20), respectively, where the variable u is now the frequency ω. Obviously, the auto-correlation function $C(t)$ should possess an expression which is alternative to (4.9), such that only the coupling constants $\{\alpha_n, \beta_n\}$ are needed and *not* the pair $\{\omega_k, d_k\}$. To this end, we shall use the explicit representations of the numerator and denominator polynomials in the PLA as the expansions in terms of powers of ω in the analogy with (26.1)

$$P_K(\omega) = \sum_{r=0}^{K-1} p_{K,K-r} \omega^r \qquad Q_K(\omega) = \sum_{r=0}^{K} q_{K,K-r} \omega^r \tag{49.5}$$

where the polynomial coefficients are defined in (26.2) and (26.3). As noted before, the so-called 'all-pole model', or equivalently, the AR model, is a simplification of the PLA. This simplification consists of setting the numerator polynomial of the PLA in (49.4) to a constant, say $P_K(\omega) = 1$

$$X(\omega) = \frac{c_0}{\beta_1} \frac{1}{Q_K(\omega)} = \sum_{k=1}^{K} \frac{d_{k,x}}{\omega - \omega_k} \qquad d_{k,x} = \frac{c_0}{\beta_1} \frac{1}{Q'_K(\omega_k)} \qquad (49.6)$$

where $Q'_K(\omega) = dQ_K(\omega)/d\omega$. Here, we used non-degeneracy of the denominator polynomial $Q_K(\omega)$ which has no multiple zeros $\{\omega_k\}$, as discussed in chapter 22 and, therefore, only the first-order or simple poles appear in the spectrum (49.6) of the AR modelling via the Lanczos polynomial $Q_K(\omega)$. Nevertheless, this is not a limitation of the analysis, since the upcoming procedure can be readily reinterpreted to encompass polynomials that may possess multiple zeros, as will be indicated later on. The inverse Fourier transform of function $X(\omega)$ denoted by $C_x(t)$ is given by

$$iC_x(t) = \sum_{k=1}^{K} d_{k,x} e^{-i\omega_k t} \qquad \text{Im}(\omega_k) < 0. \qquad (49.7)$$

An equivalent representation of $C_x(t)$ can be written through the replacement of the exponential $\exp(-i\omega_k t)$ with its Maclaurin series $\exp(-i\omega_k t) = \sum_{\ell=0}^{\infty}(-i\omega_k t)^\ell/\ell!$ which converges uniformly for any value of $\omega_k t$

$$iC_x(t) = \sum_{\ell=0}^{\infty} x_\ell \frac{(-it)^\ell}{\ell!} \qquad x_\ell = \sum_{k=1}^{K} d_{k,x} \omega_k^\ell \qquad (49.8)$$

where $x_{-n} = 0$ for $n > 0$. Likewise for $C(t) = \sum_{k=1}^{K} d_k e^{-i\omega_k t}$ from (4.9) we have

$$C(t) = \sum_{\ell=0}^{\infty} y_\ell \frac{(-it)^\ell}{\ell!} \qquad y_\ell = \sum_{k=1}^{K} d_k \omega_k^\ell. \qquad (49.9)$$

The same moments $\{x_\ell, y_\ell\}$ as in (49.8) and (49.9) can also be obtained from the Cauchy residue theorem

$$\frac{c_0}{\beta_1} \frac{1}{2\pi i} \oint_{|z|=R} dz \frac{z^\ell}{Q_K(z)} = \sum_{k=1}^{K} d_{k,x} \omega_k^\ell = x_\ell$$

$$\frac{c_0}{\beta_1} \frac{1}{2\pi i} \oint_{|z|=R} dz \frac{z^\ell P_K(z)}{Q_K(z)} = \sum_{k=1}^{K} d_k \omega_k^\ell = y_\ell. \qquad (49.10)$$

Carrying out the contour integral from (49.10) in the same way as done in (48.11) and (48.12) will produce the following result for $\ell < K$

$$x_\ell = \begin{cases} 0 & 0 \leq \ell \leq K-2 \\ \dfrac{c_0}{p_{K,1}} & \ell = K-1 \end{cases} \qquad \ell \leq K-1. \qquad (49.11)$$

On the other hand, for $\ell \geq K$, we use the equivalent of (25.10) and (26.1), namely, $Q_K(\omega_k) = 0 = \sum_{r=0}^{K} q_{K,K-r} \omega_k^r$. Multiplying this characteristic equation by $\omega_k^{\ell-K}$, we can extract the term ω_k^ℓ in the form, $\omega_k^\ell = -(1/q_{K,0}) \sum_{r=1}^{K} q_{K,r} \omega_k^{\ell-r}$. When this term is inserted into (49.8), the following recursion is obtained

$$x_\ell = -\frac{1}{q_{K,0}} \sum_{r=1}^{K} q_{K,r} x_{\ell-r} \qquad \ell \geq K \qquad (49.12)$$

with the starting values provided by (49.11). Having obtained coefficients $\{x_\ell\}$ from recursion (49.12), the auto-correlation function $C_x(t)$ in the all-pole model can be computed from (49.7) via

$$C_x(t) = \sum_{\ell=0}^{\infty} z_{\ell,x} t^\ell \qquad z_{\ell,x} = \frac{(-i)^{\ell+1}}{\ell!} x_\ell. \qquad (49.13)$$

Substituting the power representation of the numerator polynomial $P_K(\omega)$ from (49.5) into (49.9) with d_k taken from (49.4), we have

$$y_\ell = \sum_{r=1}^{K} p_{K,r} x_{K+\ell-r} \qquad \ell \geq K. \qquad (49.14)$$

Another equivalent expression can be derived for y_ℓ, if we insert (49.12) into the rhs of (49.14) so that

$$y_\ell = -\frac{1}{q_{K,0}} \sum_{r=1}^{K} q_{K,r} y_{\ell-r} \qquad \ell \geq K \qquad (49.15)$$

which is a recurrence relation of the type similar to (49.12). The initial values $\{y_\ell\}$ for $\ell \leq K-1$ to the recursion (49.15) are readily generated by using (49.11) in the rhs of (49.14) with the results

$$y_\ell = \begin{cases} \sum_{m=1}^{\ell+1} p_{K,m} x_{\ell+K-m} & 0 \leq \ell \leq K-2 \\ p_{K,1} x_0 & \ell = K-1 \end{cases} \qquad \ell \leq K-1. \qquad (49.16)$$

Once the set $\{y_\ell\}$ becomes available from recurrence (49.15), the corresponding auto-correlation function $C(t)$ in the PLA, can be deduced from (49.9) with the final result as the following power series expansions

$$C(t) = \sum_{\ell=0}^{\infty} z_\ell t^\ell \qquad z_\ell = \frac{1}{\beta_1} \frac{(-i)^{\ell+1}}{\ell!} y_\ell. \qquad (49.17)$$

The result (49.17) is the sought expression for the auto-correlation function $C(t)$, which is independent of the eigenvalues and residues $\{\omega_k, d_k\}$ of the operator

$\hat{\Omega}$. This method has been introduced by Longman and Sharir [270] for inversion of the PA by the Laplace transform. Their method is used in chapter 48 for the exact retrieval of time signals from the Padé polynomial coefficients. The series in (49.17) is convergent, since the Maclaurin expansion of an exponential e^z always converges for any value of z. Convergence of series (49.17) might be slow at large values of time t. Nevertheless, this is not a problem, since slowly converging series can efficiently be accelerated by using, e.g. the one-dimensional Wynn ε-recursive algorithm (18.14). Thus (49.17) represents a fast method of computing the auto-correlation functions and this should be of particular importance for systems displaying detailed structures over a long period of time, as is actually the case with, e.g. the ozone photodissociation process and many other collisional systems [181]. In this way, explicit propagations of a wave packet to very large times are avoided, as the Wynn algorithm can effectively reach the asymptotic t-region by using relatively small sets of the Lanczos coupling constants $\{\alpha_n, \beta_n\}$. This achieves a huge reduction in matrix-vector multiplications that represent the bottleneck of any Lanczos scheme. The above analysis concerns the non-degenerate case, where all the poles $\{\omega_k\}$ are distinct. However, the end results (49.13) and (49.17) also retain the same forms for a degenerate spectrum, where some ω_k coincide with each other, provided that the coefficients $z_{\ell,x}$ and z_ℓ are appropriately redefined. An inspection shows that these changes should be such that (49.11) and (49.12) should now be conceived as the defining expression for the new coefficients $\{z_{\ell,x}\}$. Similarly the new expansion coefficients $\{z_\ell\}$ for a degenerate spectrum are taken as being defined by (49.15) and (49.16) [270].

Chapter 50

The power moment problem in celestial mechanics

Consider a physical system or a body or an object of known total finite mass $m < \infty$. We want to determine how the mass m is spatially distributed across the body. Specifically, assuming that the investigated system possesses K constituents, each of which has the elementary mass $m_k (1 \leq k \leq K)$, we want to find out how these mass points are distributed by preserving total mass m [159, 160]

$$m = \sum_{k=1}^{K} m_k. \tag{50.1}$$

To this end, it is sufficient to consider the gravitational potential V due to the mass m located at certain point $\vec{r}$. This potential is felt at another 'test' point $\vec{R} \neq \vec{r}$ fixed in space with the intensity or strength given by

$$V = \frac{m}{|\vec{r} - \vec{R}|} \tag{50.2}$$

where, for convenience, the gravitational constant is set to unity. For our further analysis, it proves convenient to consider the Laplace expansion for the potential (50.2) in terms of the Legendre polynomials $P_n(\chi)$

$$V = m \sum_{n=0}^{\infty} \frac{r_<^n}{r_>^{n+1}} P_n(\chi) \qquad \chi = \hat{\vec{r}} \cdot \hat{\vec{R}} \tag{50.3}$$

$$r_< = \begin{cases} r & r < R \\ R & R < r \end{cases} \qquad r_> = \begin{cases} R & R > r \\ r & r > R. \end{cases} \tag{50.4}$$

A relatively simple, algebraic expression also exists which can be written in a more general form than (50.3) by avoiding consideration of two separate cases

$r > R$ and $r < R$. However, our concern here is to use the simplest arguments to arrive straight to the main point of emergence of the moment problem for determining the distribution of m_k in the studied system. For this purpose, we shall simplify the analysis by further assuming that the examined object has rotational symmetry. In such a case, working in spherical coordinates (r, ϑ, φ), it becomes immediately apparent that V is independent of the azimuthal angle φ so that $V = V(r, \vartheta)$. For simplicity, from now on we shall consider only the case $r > R$

$$V(r, \vartheta) = \sum_{n=0}^{\infty} c_n r^{-n-1} P_n(\cos \vartheta) \qquad r > R \qquad (50.5)$$

where the expansion coefficients c_n are fixed, i.e. known and given by

$$c_n = m R^n. \qquad (50.6)$$

Let us suppose that mass m_k is located at the spatial point $M_k(x_k, y_k, z_k)$ and let the Descartes coordinates of the centre of gravity point r be $M(x, y, z)$, such that ℓ_k represents the distance between M and M_k

$$\ell_k = \sqrt{(x - x_k)^2 + (y - y_k)^2 + (z - z_k)^2}. \qquad (50.7)$$

Rotational symmetry implies that all the masses $\{m_k\}$ are distributed along the rotation axis. If this latter axis is taken as the Z-axis in the XYZ-coordinate system, we have

$$\ell_k = \sqrt{r^2 + z_k^2 - 2r z_k \cos \vartheta} \qquad r^2 = x^2 + y^2 + z^2$$
$$x_k = 0 \qquad y_k = 0 \qquad z = r \cos \vartheta. \qquad (50.8)$$

We will now represent the actual potential V by a plausible model of a sum of gravitational potentials $V_k(r, \vartheta) = m_k/\ell_k$ stemming from the elementary masses m_k, as encountered in Newtonian mechanics and used extensively in celestial physics

$$V(r, \vartheta) = \sum_{k=1}^{K} V_k(r, \vartheta) \equiv \sum_{k=1}^{K} \frac{m_k}{\ell_k}. \qquad (50.9)$$

Setting $\xi = z_k/r$ and $\zeta = \cos \vartheta$, the ansatz $\ell_k = r\sqrt{1 + \xi^2 - 2\xi\zeta}$ needed in (50.9) can be recognized as the generating function of the Legendre polynomial $P_n(\zeta)$

$$\frac{1}{\sqrt{1 + \xi^2 - 2\xi\zeta}} = \sum_{n=0}^{\infty} \xi^n P_n(\zeta) \qquad 1 \leq \zeta \leq 1 \qquad |\xi| < 1. \qquad (50.10)$$

This implies

$$\frac{1}{\ell_k} = \sum_{n=0}^{\infty} z_k^n r^{-n-1} P_n(\cos \vartheta) \qquad (50.11)$$

so that upon inserting (50.11) into (50.9), it follows

$$V(r, \vartheta) = \sum_{n=0}^{\infty} \mu_n r^{-n-1} P_n(\cos \vartheta) \qquad (50.12)$$

where μ_n is the power moment of the order n defined by

$$\mu_n = \sum_{k=1}^{K} m_k z_k^n. \qquad (50.13)$$

Using the theorem of uniqueness of the power expansion of a given function, we can compare (50.5) and (50.13) and deduce

$$\mu_n = c_n. \qquad (50.14)$$

In other words, the lhs of (50.13), i.e. μ_n is known, since c_n is given by (50.6). With this, all the ingredients of the power moment problem are in place so that we can formulate it as follows: given the set $\{\mu_n\}$ ($0 \leq n \leq 2K - 1$), find the pair of the unknown $\{z_k, m_k\}$ ($1 \leq k \leq K$).

To solve the moment problem, we first provisionally assume that we already know all the locations $\{z_k\}$ ($1 \leq z_k \leq K$). Of course, we are free to consider these given points $\{z_k\}$ as the K zeros of a polynomial $B_K(z)$ of degree K, which can be constructed in the form

$$B_K(z) = b_K(z - z_1)(z - z_2) \cdots (z - z_K) \qquad (50.15)$$

where b_K is the expansion coefficient of the highest power in the series representation of $B_K(z)$

$$B_K(z) = \sum_{r=0}^{K} b_r z^r \qquad b_0 = 1. \qquad (50.16)$$

If the polynomial $B_K(z)$ is to be of degree K, its leading coefficient b_K must be non-zero

$$b_K \neq 0. \qquad (50.17)$$

The characteristic equation associated with $B_K(z)$ reads as

$$B_K(z_k) = 0 \qquad 1 \leq k \leq K. \qquad (50.18)$$

The polynomial $B_K(z)$ is real, i.e. z and $\{b_r\}$ ($1 \leq r \leq K$) are all real, since potential V is the real valued gravitational interaction. Based upon (50.18) the following relation is obviously valid for any continuous function $f(z)$

$$\sum_{k=1}^{K} m_k B_K(z_k) f(z_k) = 0. \qquad (50.19)$$

The power moment problem in celestial mechanics   281

Unlike (50.15), the form (50.16) does not require knowledge of the roots $\{z_k\}$ of $B_K(z)$, but instead the coefficients $\{b_r\}$ must be given. These expansion coefficients $\{b_r\}$ can be determined without knowing z_k, provided that the power moments $\{\mu_n\}$ are given and that their representation is prescribed by (50.13). To show this, we set $f(z_k) = z_k^n$ in (50.19) and use (50.13) together with (50.16)

$$0 = \sum_{k=1}^{K} B_K(z_k) m_k z_k^n = \sum_{k=1}^{K} \left(\sum_{r=0}^{K} b_r z_k^r \right) m_k z_k^n$$

$$= \sum_{r=0}^{K} b_r \left(\sum_{k=1}^{K} m_k z_k^{n+r} \right) = \sum_{r=0}^{K} b_r \mu_{n+r}$$

$$\sum_{r=0}^{K} b_r \mu_{n+r} = 0. \qquad (50.20)$$

If in (50.20) we vary index n in the range $0 \leq n \leq K$, we shall obtain a system of linear equations from which all the K unknown coefficients $\{b_r\}$ can be uniquely found. Once the coefficients $\{b_r\}$ are determined, the polynomial $B_K(z)$ from (50.16) can be constructed. Therefore, our provisional request for knowing z_k before we are able to set up the polynomial $B_K(z)$ from (50.18) can now be dropped. Of course, when the set $\{b_r\}$ has been found, we can solve the eigenequation $B_K(z) = 0$ as in (50.18) to obtain the K roots $\{z_k\}$ $(1 \leq k \leq K)$. These roots are given by the known algebraic expressions for $K \leq 4$, whereas for $K \geq 5$ numerical algorithms should be employed [252].

In order to determine the elementary masses $\{m_k\}$ $(1 \leq k \leq K)$, the characteristic polynomial $B_K(z)$ becomes again pivotal. We know that m_k should be located at the position $(0, 0, z_k)$. Moreover, an inspection of (50.15) would reveal that the ansatz $B_K(z)/(z - z_{k'})$ evaluated at $z = z_{k'}$ is non-zero. Therefore, it should be possible to extract the kth mass m_k from the product $m_{k'}[B_K(z)/(z - z_{k'})]_{z=z_k}$ summed over k' from 1 to K. With this goal, we insert (50.15) into $[B_K(z)/(z - z_{k'})]_{z=z_k}$ to arrive at

$$\left\{ \frac{B_K(z)}{z - z_{k'}} \right\}_{z=z_k} = b_K \left\{ \frac{(z-z_1)(z-z_2)\cdots(z-z_k)\cdots(z-z_K)}{z - z_{k'}} \right\}_{z=z_k}$$

$$= b_K \frac{(z_k - z_1)(z_k - z_2)\cdots(z_k - z_k)\cdots(z_k - z_K)}{z_k - z_{k'}}$$

$$= \begin{cases} 0 & k' \neq k \\ \neq 0 & k' = k \end{cases}$$

$$\therefore \quad \left\{ \frac{B_K(z)}{z - z_{k'}} \right\}_{z=z_k} = \begin{cases} 0 & k' \neq k \\ L_{K,k} & k' = k \end{cases} \qquad (50.21)$$

$$L_{K,k} = b_K(z_k - z_1)(z_k - z_2) \cdots (z_k - z_{k-1})(z_k - z_{k+1})$$
$$\cdots (z_k - z_K) \qquad (50.22)$$

$$\left\{\frac{B_K(z)}{z-z_{k'}}\right\}_{z=z_k} = \left\{\frac{B_K(z)}{z-z_k}\right\}_{z=z_k} \delta_{k,k'} \qquad (50.23)$$

where $\delta_{k,k'}$ is the Kronecker δ-symbol (3.7). Next we multiply both sides of (50.23) by $m_{k'}$ and sum over $1 \leq k' \leq K$ in which case the term $\delta_{k,k'}$ will reduce the whole sum to a single member

$$\sum_{k'=1}^{K} m_{k'} \left\{\frac{B_K(z)}{z-z_{k'}}\right\}_{z=z_k} = m_k \left\{\frac{B_K(z)}{z-z_k}\right\}_{z=z_k}. \qquad (50.24)$$

The rhs of (50.24) becomes an undetermined expression of the type 0/0 in the limit $z \longrightarrow z_k$, in which case the l'Hôpital rule applies, thus yielding $m_k[(d/dz)B_K(z)]_{z=z_k}$

$$\sum_{k'=1}^{K} m_{k'} \left\{\frac{B_K(z)}{z-z_{k'}}\right\}_{z=z_k} = m_k B'_K(z_k) \qquad (50.25)$$

where $B'_K(z) = (d/dz)B_K(z)$. Using (50.21), the first derivative $B'_K(z_k)$ from (50.25) can be written as

$$B'_K(z_k) = b_K \prod_{k'=1\,(k'\neq k)}^{K} (z_k - z_{k'}). \qquad (50.26)$$

From here it immediately follows for non-degenerate roots

$$B'_K(z_k) \neq 0 \qquad z_{k'} \neq z_k \qquad (k' \neq k). \qquad (50.27)$$

Clearly the rhs of (50.25) could yield m_k if we could somehow eliminate all the unknown terms $m_{k'}$ ($1 \leq k' \leq K$) from the lhs of the same equation. We could get rid of the terms $m_{k'}$ ($1 \leq k' \leq K$) only if we would be able to collect them as the ansatz $\sum_{k'=1}^{K} m_{k'} z_{k'}^n$ which is then substituted by the moment μ_n according to (50.13). Such a task can be accomplished by finding a series representation of $[B_K(z)/(z-z_{k'})]_{z=z_k}$ in powers of $z_{k'}$ with the expansion coefficients which are independent of k'. When such a power series representation is inserted in the lhs of (50.25) for $[B_K(z)/(z-z_{k'})]_{z=z_k}$, the term $\sum_{k'=1}^{K} m_{k'} z_{k'}^n$ would pop out to produce μ_n and with this step achieved, the unknowns $m_{k'}(1 \leq k \leq K)$ will be gone. This sketched proof suggests the departure from the power series (50.16) for polynomial $B_K(z)$ from which we calculate

$$B_K(z) = \sum_{r=0}^{K} b_r z^r$$

$$= \left[\sum_{r=0}^{K} b_{r+s} z_{k'}^s z^r \right]_{s=0}$$

$$= \left\{ \left[\sum_{r=0}^{K} b_{r+s} z_{k'}^s z^r \right]_{s=0} + \sum_{r=0}^{K} \left[\sum_{s=1}^{K-r} b_{r+s} z_{k'}^s \right] z^r \right\}$$

$$- \sum_{r=0}^{K} \left[\sum_{s=1}^{K-r} b_{r+s} z_{k'}^s \right] z^r$$

$$= \sum_{r=0}^{K} \left[\sum_{s=0}^{K-r} b_{r+s} z_{k'}^s \right] z^r - \sum_{r=0}^{K} \left[\sum_{s=1}^{K-r} b_{r+s} z_{k'}^s \right] z^r$$

$$= \left\{ \left[\sum_{s=0}^{K-r} b_{r+s} z_{k'}^s \right]_{r=0} + \sum_{r=1}^{K} \left[\sum_{s=0}^{K-r} b_{r+s} z_{k'}^s \right] z^r \right\}$$

$$- \sum_{r=0}^{K} \left[\sum_{s=1}^{K-r} b_{r+s} z_{k'}^s \right] z^r$$

$$= B_K(z_{k'}) + \sum_{r=1}^{K} \left[\sum_{s=0}^{K-r} b_{r+s} z_{k'}^s \right] z^r - \sum_{r=0}^{K} \left[\sum_{s=1}^{K-r} b_{r+s} z_{k'}^s \right] z^r$$

$$= \sum_{r=1}^{K} \left[\sum_{s=0}^{K-r} b_{r+s} z_{k'}^s \right] z^r - \sum_{r=0}^{K} \left[\sum_{s=1}^{K-r} b_{r+s} z_{k'}^s \right] z^r$$

$$= \sum_{r=1}^{K} \left[\sum_{s=0}^{K-r} b_{r+s} z_{k'}^s \right] z^r - \sum_{r=0}^{K-1} \left[\sum_{s=1}^{K-r} b_{r+s} z_{k'}^s \right] z^r$$

$$= \sum_{r=0}^{K-1} \left[\sum_{s=0}^{K-r-1} b_{r+s+1} z_{k'}^s \right] z^{r+1} - \sum_{r=0}^{K-1} \left[\sum_{s=0}^{K-r-1} b_{r+s+1} z_{k'}^{s+1} \right] z^r$$

$$= (z - z_{k'}) \sum_{r=0}^{K-1} \left[\sum_{s=0}^{K-r-1} b_{r+s+1} z_{k'}^s \right] z^r$$

where the characteristic equation $B_K(z_{k'}) = 0$ is used. Therefore, the final result of this calculation can be written in the following compact form

$$\frac{B_K(z)}{z - z_{k'}} = \tilde{A}_{K-1}(z, z_{k'}) \tag{50.28}$$

where $\tilde{A}_{K-1}(z, z_{k'})$ is an auxiliary polynomial of degree $K - 1$ in variable z for a fixed value $z_{k'}$

$$\tilde{A}_{K-1}(z, z_{k'}) = \sum_{r=0}^{K-1} \left\{ \sum_{s=0}^{K-r-1} b_{r+s+1} z_{k'}^s \right\} z^r. \tag{50.29}$$

The general expansion coefficient $\sum_{s=0}^{K-r-1} b_{s+r+1} z_{k'}^s$ of polynomial (50.29) is itself a polynomial in variable $z_{k'}$. The result (50.28) is valid for any z including the eigenroot $z = z_k$, as required in (50.25). The whole dependence upon index k' on the rhs of (50.28) is contained in the power function $z_{k'}^s$, as seen in (50.29). This latter feature leads to a very important sum rule for the set of polynomial values $\{\tilde{A}_{K-1}(z, z_{k'})\}$ ($1 \le k' \le K$) multiplied with $m_{k'}$ and summed over k'

$$\sum_{k'=1}^{K} m_{k'} \tilde{A}_{K-1}(z, z_{k'}) = \sum_{r=0}^{K-1} \sum_{s=0}^{K-r-1} b_{r+s+1} z^r \left\{ \sum_{k'=1}^{K} m_{k'} z_{k'}^s \right\}$$

$$= \sum_{r=0}^{K-1} \sum_{s=0}^{K-r-1} b_{r+s+1} z^r \mu_s$$

$$= \sum_{r=0}^{K-1} \left\{ \sum_{s=0}^{K-r-1} b_{r+s+1} \mu_s \right\} z^r \equiv A_{K-1}(z). \tag{50.30}$$

In (50.30), the sum over k' is explicitly carried out to give μ_s according (50.13). Consequently, the K products $m_{k'} \tilde{A}_{K-1}(z, z_{k'})$ summed over k' retains the structure of the polynomial $\tilde{A}_{K-1}(z, z_{k'})$ in which μ_s plays the role of $z_{k'}^s$. For a simpler notation, the result of the sum rule (50.30) is relabelled as $A_{K-1}(z)$ to denote a polynomial of degree $K - 1$ in variable z

$$A_{K-1}(z) = \sum_{r=0}^{K-1} a_r z^r \tag{50.31}$$

$$a_r = \sum_{s=0}^{K-r-1} b_{s+r+1} \mu_s. \tag{50.32}$$

By evaluating (50.28) at $z = z_k$ with the subsequent usage of the sum rule (50.30) and relation (50.25), we arrive at the following result

$$\sum_{k'=1}^{K} m_{k'} \left\{ \frac{B_K(z)}{z - z_{k'}} \right\}_{z=z_k} = A_{K-1}(z_k) = m_k B'_K(z_k) \tag{50.33}$$

$$m_k = \frac{A_{K-1}(z_k)}{B'_K(z_k)} \tag{50.34}$$

where $B'_K(z_k)$ is different from zero according to (50.27). The result (50.34) is recognized as the corresponding prediction of the Padé approximant (PA) [19].

Yet nowhere have we used the PA in the above derivation. Still the PA popped out nearly everywhere, which is expected, as this method is naturally ingrained in the moment problem. For example, in (50.20) and (50.32) we can immediately recognize the two systems of linear equations for the coefficients $\{b_r\}$ and $\{a_r\}$ of the denominator $B_K(z)$ and numerator $A_{K-1}(z)$ polynomial, respectively, from the paradiagonal PA of order $(K-1, K)$. As a byproduct of the above analysis, the obtained intermediate result (50.28) is particularly interesting, as it represents a new expansion for the the binomial $1/(z - z_k)$

$$\frac{1}{z - z_k} = \frac{\tilde{A}_{K-1}(z, z_k)}{B_K(z)} \qquad (50.35)$$

with z_k being the root of $B_K(z)$ and where $\tilde{A}_{K-1}(z, z_k)$ is given in (50.29). The polynomial $\tilde{A}_{K-1}(z, z_k)$ depends upon the input power moments $\{\mu_n\}$ through b_r, as is clear from (50.20). It also depends upon the given fixed root z_k. Thus, the usefulness of the new binomial expansion (50.35) is considerable since, through a separate computation, it could reliably check the accuracy of each found *individual* root z_k. Likewise, once all the masses m_k ($1 \le k \le K$) have been obtained, their sum could be checked against the total mass of the body (50.1), i.e. $m = \sum_{k=1}^{K} A_{K-1}(z_k)/B'_K(z_k)$. Of course, the found pair $\{z_k^n, m_k\}$ could be multiplied with each other as in (50.13) to check whether all the known power moments $\{\mu_n\}$ are retrieved with fidelity

$$m = \sum_{k=1}^{K} \frac{A_{K-1}(z_k)}{B'_K(z_k)} \qquad \mu_n = \sum_{k=1}^{K} \frac{A_{K-1}(z_k)}{B'_K(z_k)} z_k^n. \qquad (50.36)$$

Chapter 51

Mass positivity and the power moment problem

With the real potential V from chapter 50, all the calculated, i.e. predicted masses $\{m_k\}$ must be positive and, moreover, all the roots $\{z_k\}$ ought to be real and located within the segment (a, b). This will be the case with certainty if the following theorem could be proved [159].

Theorem: In order that for any integer $K > 0$, all the K parameters $\{z_k\}$ appearing as the unique solution of the power moment problem are real, distinct and located within the given real segment (a, b) such that all the K corresponding masses $\{m_k\}$ are unique as well as positive, it is necessary and sufficient that the input numbers $\{\mu_n\}$ ($n \leq 2K - 1$) represent the power moments defined as the Stieltjes integral

$$\mu_n = \int_a^b z^n \mathrm{d}w(z) \tag{51.1}$$

where $w(z)$ is a positive, non-decreasing, bounded function in the interval $[a, b]$.

Proof (necessity): Necessity means that if $m_k > 0$ and $a < z_k < b$ ($1 \leq k \leq K$), then the power moment μ_n must be given by (51.1). To prove this, we introduce a weight function $\sigma_K(z)$ which is real and positive for $z \in (a, b)$

$$\sigma_K(z) = \sum_{k=1}^{K} m_k \theta(z - z_k) \tag{51.2}$$

where $\theta(z - z_k)$ is the Heaviside step or jump function (9.20). The associated distribution $\mathrm{d}\sigma_K(z)$ is obtained from (51.2) using the property $(\mathrm{d}/\mathrm{d}z)\theta(z - z_k) = \delta(z - z_k)$, where $\delta(z - z_k)$ is the Dirac δ-function

$$\mathrm{d}\sigma_K(z) = \sum_{k=1}^{K} m_k \delta(z - z_k) \mathrm{d}z. \tag{51.3}$$

Thus (51.1) becomes

$$\mu_n = \int_a^b z^n dw_K(z) \qquad n \leq 2K - 1 \qquad (51.4)$$

$$w_K(z) = \int_a^z \sigma_K(v) dv. \qquad (51.5)$$

In other words, we replace the true distribution $w(z)$ from (51.1) by the approximate sequence of functions $\{w_K(z)\}$ ($K = 1, 2, 3, \ldots$) hoping that the relation $\lim_{K \to \infty} w_K(z) = w(z)$ will exist. In so doing, we require that each approximation $\{w_K(z)\}$ to $w(z)$ is built from the weight $\sigma_K(z)$ in the form of a linear combination of the Heaviside step functions with K points of increase such that its first $2K$ moments μ_n agree with those of the exact distribution $w(z)$. The Kth approximation $w_K(z)$ is fully determined by the $2K$ numbers that are: (1) the K values of the independent variable $\{z_k\}_{k=1}^K$ at which $w_K(z)$ is discontinuous, and (2) the K elementary masses $\{m_k\}_{k=1}^K$ which are the magnitudes of the discontinuities of $w_K(z)$. Therefore, these two sets of numbers $\{z_k, m_k\}_{k=1}^K$ must satisfy the $2K$ relations

$$\mu_n = \int_a^b z^n dw(z) = \int_a^b z^n dw_K(z) = \sum_{k=1}^K m_k \int_a^b z^n \delta(z - z_k) dz = \sum_{k=1}^K m_k z_k^n$$

$$\mu_n = \sum_{k=1}^K m_k z_k^n \qquad (51.6)$$

where $0 \leq n \leq 2K - 1$ as in (50.13). If in (51.5) we vary $w_K(z)$ along the segment (a, b), we would obtain the total mass $m < \infty$, and this signifies that $w_K(z)$ is a bounded function for $z \in (a, b)$

$$w_K(z) = \int_a^z \sigma_K(v) dv = \sum_{k=1}^K m_k \int_a^z \delta(v - z_k) dv = \sum_{k=1}^K m_k = m < \infty. \qquad (51.7)$$

Thus, the weight $w_K(z)$ is a non-decreasing function of bounded variation throughout the strip $z \in (a, b)$. The clause 'non-decreasing' is due to the fact that $w_K(z)$ changes only by amounts $m_k > 0$ at the points $z = z_k$. Boundedness of $w_K(z)$ in (a, b) is implied by finiteness of the total mass m, as per (51.7). It can be shown that the set of uniformly bounded non-decreasing functions $\{w_K(z)\}_{K=1}^\infty$ converges to the non-decreasing bounded function $w(z)$ as $K \to \infty$. Therefore, the limit $K \to \infty$ will transform (51.5) into (51.1) and this proves the sought necessity in the Theorem

$$\mu_n = \lim_{K \to \infty} \int_a^b z^n dw_K(z) = \int_a^b z^n dw(z) \qquad \text{(QED)}. \qquad (51.8)$$

Proof (sufficiency): Sufficiency signifies that if the power moments $\{\mu_n\}$ are given by (51.1), then for $1 \leq k \leq K$ all the roots $\{z_k\}$ must be distinct, as well as $a < z_k < b$ and, moreover $m_k > 0$. To prove this, we use the set $\{z_k\}$ to first construct the characteristic polynomial $B_K(z)$ as in (50.15). This polynomial of degree K must have K distinct roots $\{z_k\}$ all contained within the open interval (a, b) if the Theorem is to hold. Suppose that the opposite is true, i.e. that $B_K(z)$ has $M < K$ real roots $\{z_k\}$ with, e.g. odd multiplicity and $z_k \in (a, b)$ for $1 \leq k \leq M$. The remaining $K - M$ roots can be either outside the interval (a, b) or complex and, hence, appearing always as a pair of a complex root and its complex conjugate due to realness of $B_K(z)$. Then the product of the two polynomials $B_K(z)$ and $C_K(z) = \prod_{k=1}^{M}(z - z_k)^{r_k}$ is a new polynomial of degree $K + M < 2K$ with a constant sign for $z \in [a, b]$. Let $g(z)$ be the product of all other divisors of $B_K(z)$. The function $g(z)$ has the same sign throughout (a, b). Then $B_K(z)$ can be cast into the form

$$B_K(z) = C_K(z)g(z) \qquad C_K(z) = \prod_{k=1}^{M}(z - z_k)^{r_k} \qquad (51.9)$$

where the number r_k denotes the odd multiplicity of the roots $\{z_k\}$ ($1 \leq k \leq M$). We can rewrite (51.9) equivalently as

$$B_K(z) = \tilde{B}_K(z)h(z) \qquad \tilde{B}_K(z) = \prod_{k=1}^{M}(z - z_k)$$

$$h(z) = \prod_{k=1}^{M}(z - z_k)^{r_k - 1}g(z). \qquad (51.10)$$

Since $r_k - 1$ is even for r_k odd, the polynomial $\prod_{k=1}^{M}(z - z_k)^{r_k - 1}$ from $h(z)$ is always positive since, by the mentioned assumption, all the roots $\{z_k\}$ with $1 \leq k \leq M$ belong to (a, b). Therefore, just like $g(z)$ itself, the function $h(z)$ does not change its sign in (a, b). The relation $M = K$ must be proven. If we assume the opposite, i.e. $M < K$ then the moments μ_n relative to the distributions $w_K(z)$ and $w(z)$ are the same up to the order $2K - 1$ and, therefore

$$\int_a^b B_K(z)\tilde{B}_K(z)\mathrm{d}w(z) = \int_a^b B_K(z)\tilde{B}_K(z)\mathrm{d}w_K(z)$$

$$= \int_a^b \tilde{B}_K^2(z)h(z)\mathrm{d}w_K(z) \neq 0. \qquad (51.11)$$

The result of the last integral in (51.11) is non-zero because $\tilde{B}_K^2(z) > 0$ and the function $h(z)$ has one sign in the whole range (a, b). However, this non-zero result is a contradiction, since the integral $\int_a^b B_K(z)\tilde{B}_K(z)\mathrm{d}w_K(z)$ must yield zero according to (50.19)

$$\int_a^b B_K(z)\tilde{B}(z)dw_K(z) = \sum_{k=1}^{K} m_k \int_a^b B_K(z)\tilde{B}(z)\delta(z-z_k)dz$$

$$\sum_{k=1}^{K} m_k B_K(z_k) f(z_k) = 0 \qquad f(z) = \tilde{B}_K(z). \qquad (51.12)$$

Therefore, the opposite of the above supposition is true, i.e. on the open interval (a, b) the characteristic polynomial $B_K(z)$ has exactly K distinct zeros (QED). Finally, it remains to show that all the masses m_k are real and positive. This can be proven by observing that the square of the derivative polynomial $B'_K(z)$ is a polynomial of order $2K-2$ and, therefore, we can use the equality of the moments for two distributions $w(z)$ and $w_K(z)$ in a fashion similar to (51.11)

$$\int_a^b \left[\frac{B_K(z)}{z-z_k}\right]^2 dw(z) = \int_a^b \left[\frac{B_K(z)}{z-z_k}\right]^2 dw_K(z) = m_k \prod_{k'=1\ (k'\neq k)}^{K} (z_k - z_{k'})^2. \qquad (51.13)$$

This, together with (50.17) and (50.27), as well as the fact that $w_K(z)$ is a positive function throughout (a, b), finally gives the sought result

$$m_k = b_K^2 \int_a^b \left[\frac{B_K(z)}{(z-z_k)B'_K(z_k)}\right]^2 dw_K(z) > 0 \qquad \text{(QED)} \qquad (51.14)$$

and, with this accomplished, the proof of the entire Theorem is completed (QED). Note that positivity of m_k and the relationship (50.34) imply that the polynomials $A_{K-1}(z_k)$ and $Q_K(z_k)$ are of the same sign. The above Theorem can also be equivalently stated in terms of the following plausible physical formulation [159].

Potential V exerted by a rotating body on a fixed point can be approximated, for any positive integer K, by the gravitational potential originating from K positive point masses distributed on the finite segment (a, b) along the symmetry axis, if and only if V coincides with the mentioned segment's potential when weighted with the non-negative linear density function.

Chapter 52

The power moment problem and Gaussian numerical quadratures

Comparing (50.13) and (51.1) establishes directly the following important relationship

$$\int_a^b z^n \mathrm{d}w(z) = \sum_{q=1}^K m_q z_q^n \qquad 0 \le n \le 2K - 1. \tag{52.1}$$

If we multiply (52.1) by the constant coefficients $\{p_n\}$ ($1 \le n \le L$), where $L \le 2K - 1$, and sum the results over n, we shall obtain a more general result

$$\int_a^b \mathcal{P}_L(z) \mathrm{d}w(z) = \sum_{q=1}^K m_q \mathcal{P}_L(z_q) \tag{52.2}$$

$$\mathcal{P}_L(z) = \sum_{n=0}^L p_n z^n \qquad L \le 2K - 1. \tag{52.3}$$

Here (52.2) is recognized as a Gaussian quadrature formula which represents the exact result for the numerical integration of the integral of any polynomial $\mathcal{P}_L(z)$ of degree $L \le 2K - 1$, defined for $z \in (a, b)$. For example, taking $\mathcal{P}_L(z)$ to be a polynomial of order $2K - 2$ in the form

$$\mathcal{P}_L(z) = b_K^2 \left[\frac{B_K(z)}{(z - z_k) B_K'(z_k)} \right]^2 \tag{52.4}$$

we can use (3.7), (50.23) and (50.33) to find

$$\mathcal{P}_L(z_q) = \delta_{k,q}. \tag{52.5}$$

Inserting this result into the rhs of (52.2), the whole sum reduces to m_k by means of the $\delta_{k,q}$-symbol from (3.7), so that

$$\int_a^b \mathcal{P}_L(z) \mathrm{d}w(z) = b_K^2 \int_a^b \left[\frac{B_K(z)}{(z - z_k) B_K'(z_k)} \right]^2 \mathrm{d}w(z) = m_k \tag{52.6}$$

in agreement with the result (51.14) from the previous calculation (QED). More generally, the Gaussian quadrature rule (52.2) will be valid *approximately* for any other analytical function $F(z)$ in the sector (a, b)

$$\int_a^b F(z)\mathrm{d}w(z) \approx \sum_{q=1}^K m_q F(z_q). \tag{52.7}$$

Chapter 53

The power moment problem of the generalized Stieltjes type

Let us introduce the following Stieltjes integral as the complex Hilbert transform

$$S(u) = \frac{1}{2\pi i} \oint_C \frac{d\sigma_0(z)}{u - z} \qquad (53.1)$$

where $d\sigma_0(z)$ is the complex valued Lebesgue measure given in (30.5) for the Lanczos vector space $\mathcal{L}$ from (30.1). Here the contour C is the same as in (30.3) and it is set up to encircle all the eigenroots $\{u_k\}$ ($1 \le k \le K$) of $Q_K(u)$. Inserting the measure (30.5) into the integral (53.1) yields the result which is the spectral representation (5.23) for the Green function

$$S(u) = \sum_{k=1}^{K} \frac{d_k}{u - u_k}. \qquad (53.2)$$

If the function $S(u)$ from (53.1) is expanded as a formal power series, it will follow

$$S(u) = \sum_{n=0}^{\infty} \mu_n u^{-n-1} \qquad (53.3)$$

where μ_n is the complex valued power moment defined as[1]

$$\mu_n = \frac{1}{2\pi i} \oint_C z^n d\sigma_0(z) \qquad (53.4)$$

provided that the integral exists for all the non-negative finite values of the integer n ($n = 0, 1, 2, \ldots$). Inserting the measure (30.5) into (53.4), we shall have

$$\mu_n = \sum_{k=1}^{K} d_k u_k^n. \qquad (53.5)$$

---
[1] Recall that the standard Stieltjes moment problem operates with real valued quantities (functions, distributions, etc) [166, 167], as analysed in chapters 50 and 51.

According to (5.9) the rhs of (53.5) is equal to the auto-correlation function C_n, so that

$$\mu_n = C_n. \tag{53.6}$$

On the other hand, the equivalence (5.12) already exists between the auto-correlation functions $\{C_n\}$ from (5.9) and time signals $\{c_n\}$ from (5.11). Thus (53.6) extends (5.12) to encompass the triple equivalence among the three quantities, the power moment μ_n from (53.4), the auto-correlation function C_n from (5.9) and the signal point c_n from (5.11)

$$\mu_n = C_n = c_n. \tag{53.7}$$

The 'power moment problem', which is most frequently used in statistics and the theory of probability [167], is defined as follows: given the set of power moments $\{\mu_n\}$ the task is to find the building constituents $\{K, u_k, d_k\}$. Hence, the previously analysed 'harmonic inversion' coincides precisely with the definition of the power moment problem from (53.4). Using (53.4) and the equivalence (53.7), we can write the following integral representation for the time signal

$$c_n = \frac{1}{2\pi i} \oint_C z^n d\sigma_0(z). \tag{53.8}$$

For any analytic function $f(\hat{\Omega})$ of the operator $\hat{\Omega}$, the following contour integral is valid

$$f(\hat{\Omega}) = \frac{1}{2\pi i} \oint_C d\sigma_0(z)|\Upsilon(z))f(z)(\Upsilon(z)| = \sum_{k=1}^{K} |\Upsilon_k)f(u_k)(\Upsilon_k| \tag{53.9}$$

where C is again the same contour as in (30.3). Finally, the key question to ask here, while working in the Lanczos space $\mathcal{L}$ defined in (30.1) is: given the power moments $\{\mu_n\}$, or equivalently, the exponentially attenuated signal points $\{c_n\}$, can we compute accurately the coupling parameters $\{\alpha_n, \beta_n\}$ that are needed to construct the basis of the characteristic polynomials $\{Q_n(u)\}$? The answer would be in the affirmative, if we use the definitions of the coupling constants $\alpha_n = (Q_n(u)|u|Q_n(u))/(Q_n(u)|Q_n(u))$ and $\beta_n = (Q_{n-1}(u)|u|Q_n(u))/(Q_{n-1}(u)|Q_{n-1}(u))$ from (32.4), provided that we know how to evaluate the indicated inner products. The scalar product in $\mathcal{L}$ is defined in (30.8) which, however, invokes the unknown spectral parameters $\{u_k, d_k\}$. Nevertheless, this obstacle can be circumvented by using the duality of the Lanczos representation $\{Q_n(u), |\psi_n)\}$ from chapter 32. According to this dual representation, the couplings $\{\alpha_n, \beta_n\}$ can be computed by switching from the scalar products over the polynomial basis $\{Q_n(u)\}$ to the scalar product over the state vectors $\{|\psi_n)\}$, as prescribed by (32.5). In other words, we return to the original Lanczos algorithm (22.1) for state vectors and, here, as we know the parameters $\{\alpha_n, \beta_n\}$ are generated in the course of the recursive construction of the basis set $\{|\psi_n)\}$. Having obtained the couplings $\{\alpha_n, \beta_n\}$ in this way from (22.2), the Lanczos polynomial recursion (23.16) is used to generate the set $\{Q_n(u)\}$.

Chapter 54

Matrix elements $(u^m|Q_n(u))$ and the Stieltjes integral

In general, the Stieltjes integral (53.1) and Stieltjes series (53.3) are formally equal to each other

$$\frac{1}{2\pi i}\oint_C \frac{d\sigma_0(z)}{u-z} = \sum_{n=0}^{\infty} \mu_n u^{-n-1} \tag{54.1}$$

if $|u| < 1$ and $|u| \neq 0$, where we assume that all the moments $\{\mu_n\}$ exist and are finite. The series on the rhs of (54.1) is divergent for $|u| > 1$, in which case the concept of analytical continuation is used for resummation via anti-limits [72], as in chapter 18. When the first $2n$ power moments are available, the PLA of the order $(n-1, n)$ for the Stieltjes series on the rhs of (54.1) will be completely determined, if the function $(2\pi i)^{-1} \oint_C d\sigma_0(z)/(u-z) - c_0 P_n(u)/[\beta_1 Q_n(u)]$ expanded in powers of u^{-1} does not contain the terms multiplying u^{-m} with $m \leq 2n$

$$\frac{1}{2\pi i}\oint_C \frac{d\sigma_0(z)}{u-z} - \frac{c_0}{\beta_1}\frac{P_n(u)}{Q_n(u)} = \mathcal{O}(u^{-2n-1}). \tag{54.2}$$

Stated equivalently, the series expansion of the difference of the following two terms $\beta_1 Q_n(u)(2\pi i)^{-1}\oint_C d\sigma_0(z)/(u-z)$ and $c_0 P_n(u)$ in powers of u^{-1} should begin with the term u^{-n-1}

$$\beta_1 Q_n(u) \frac{1}{2\pi i}\oint_C \frac{d\sigma_0(z)}{u-z} - c_0 P_n(u) = \mathcal{O}(u^{-n-1}). \tag{54.3}$$

This fact, together with the identity $Q_n(u)d\sigma_0(z) = Q_n(z)d\sigma(z) + [Q_n(u) - Q_n(z)]d\sigma_0(z)$, leads to

$$\frac{\beta_1}{c_0}\frac{1}{2\pi i}\oint_C d\sigma_0(z)\frac{Q_n(z)}{u-z} + \left\{\frac{\beta_1}{c_0}\oint_C d\sigma_0(z)\frac{Q_n(u)-Q_n(z)}{u-z} - P_n(z)\right\}$$
$$= \mathcal{O}(u^{-n-1}). \tag{54.4}$$

Here, the term in the curly brackets is exactly zero according to the definition of the Lanczos polynomial $P_n(u)$ of the second kind and degree $n-1$ in the usual form (36.2), i.e. $P_n(u) = [\beta_1/(2\pi i c_0)] \oint_C d\sigma_0(z)[Q_n(u) - Q_n(z)]/(u-z)$, so that

$$\frac{\beta_1}{c_0} \frac{1}{2\pi i} \oint_C d\sigma_0(z) \frac{Q_n(z)}{u-z} = \mathcal{O}(u^{-n-1}). \tag{54.5}$$

Therefore, the function $(2\pi i)^{-1} \oint_C d\sigma_0(z) Q_n(z)/(u-z)$ developed in powers of u^{-1} does not involve any expansion terms multiplying $\{u^{-m-1}\}$ with $0 \leq m \leq n$. This condition, which will hereafter be called the Padé condition, means that the following n equations ought to be satisfied [167]

$$\frac{1}{2\pi i} \oint_C d\sigma_0(z) z^m Q_n(z) = 0 \quad (m = 0, 1, 2, \ldots, n-1). \tag{54.6}$$

Multiplying this equation by the constant coefficient $v_{m,m-r}$ and summing up the result over s from 0 to m, a generalization of (54.6) is obtained viz

$$\frac{1}{2\pi i} \oint_C d\sigma_0(z) V_m(z) Q_n(z) = 0 \quad (m = 0, 1, 2, \ldots, n-1) \tag{54.7}$$

where $V_m(z)$ is any polynomial of degree $m \leq n-1$

$$V_m(z) = \sum_{r=0}^{m} v_{m,m-r} z^r \quad (m = 0, 1, 2, \ldots, n-1). \tag{54.8}$$

The contour integral on the lhs of (54.6) can be discretized with the help of the complex valued Dirac measure (30.4), thus yielding for any m

$$\frac{1}{2\pi i} \oint_C d\sigma_0(z) z^m Q_n(z) = \sum_{k=1}^{K} d_k u_k^m Q_n(u_k) \equiv (u^m | Q_n(u)). \tag{54.9}$$

Therefore, when $1 \leq m \leq n-1$, the discrete counterpart of the condition (54.6) becomes

$$(u^m | Q_n(u)) = \sum_{k=1}^{K} d_k u_k^m Q_n(u_k) = 0 \quad (m = 0, 1, 2, \ldots, n-1). \tag{54.10}$$

In the above expressions, we used the inner product (30.8) in the Lanczos vector space $\mathcal{L}$. The Padé condition (54.6), in fact, *defines* the polynomial $Q_n(u)$ up to an overall multiplicative constant. Conversely, if there is a polynomial $Q_n(u)$ which satisfies the Padé condition (54.6), such that (36.2) defines the polynomial $P_n(u)$ of degree $n-1$, then the PLA is uniquely given by the quotient $c_0 P_n(u)/[\beta_1 Q_n(u)]$. To determine explicitly the polynomial $Q_n(u)$, we insert the power series representation (26.1) into (54.6) and use the definition (53.4) to identify the power moments

$$\frac{1}{2\pi i} \oint_C d\sigma_0(z) z^m Q_n(z) = \sum_{r=0}^{n} q_{n,n-r} \left\{ \frac{1}{2\pi i} \oint_C d\sigma_0(z) z^{m+r} \right\}$$

$$= \sum_{r=0}^{n} q_{n,n-r} \mu_{m+r} = 0.$$

Hence, the following system of equations allows determination of the expansion coefficients $\{q_{n,n-r}\}$

$$\sum_{r=0}^{n} q_{n,n-r} \mu_{m+r} = 0. \tag{54.11}$$

An alternative form of this system of linear equations is obtained by using the equivalence (53.7)

$$\sum_{r=0}^{n} q_{n,n-r} c_{m+r} = 0 \quad (m = 0, 1, 2, \ldots, n-1) \tag{54.12}$$

or explicitly [167]

$$c_0 q_{n,n} + c_1 q_{n,n-1} + c_2 q_{n,n-2} + \cdots + c_{n-1} q_{n,1} = -c_n q_{n,0}$$
$$c_1 q_{n,n} + c_2 q_{n,n-1} + c_3 q_{n,n-2} + \cdots + c_n q_{n,1} = -c_{n+1} q_{n,0}$$
$$c_2 q_{n,n} + c_3 q_{n,n-1} + c_4 q_{n,n-2} + \cdots + c_{n+1} q_{n,1} = -c_{n+2} q_{n,0} \tag{54.13}$$
$$\vdots$$
$$c_{n-1} q_{n,n} + c_n q_{n,n-1} + c_{n+1} q_{n,n-2} + \cdots + c_{2n-2} q_{n,1} = -c_{2n-1} q_{n,0}.$$

The determinant of the system (54.12) of n inhomogeneous linear equations is the Hankel determinant $H_n(c_0)$ from (5.15), which is also the determinant $\det \mathbf{S}_n$ of the overlap matrix $\mathbf{S}_n$ in the Schrödinger basis $\{|\Phi_n\rangle\}$. The system (54.12) will have the unique solution $\{q_{n,n-r}\}$ if

$$\det \mathbf{S}_n \equiv H_n(c_0) \neq 0. \tag{54.14}$$

In [142] the PD algorithm [170] has been used for computations of the general Hankel determinant $H_m(c_n)$ and this can be applied to (54.14) (see also chapter 68). This is an error-free algorithm whenever the signal $\{c_n\}$ is comprised of integers as, e.g. in MR physics. The nth leading coefficient $q_{n,0}$ from the power series representation (26.1) can be determined from the normalization of $Q_n(u)$. The general element $q_{n,n-r}$ of the set of the solutions $\{q_{n,n-r}\}$ of the system of equations (54.12) rewritten as $\sum_{r=0}^{n-1} q_{n,n-r} c_{m+r} = -c_n q_{n,0}$

($m = 0, 1, 2, \ldots, n-1$) is given by the Cramer rule via the following quotient of the determinants

$$q_{n,n-r} = \frac{H_{n,r}(c_0)}{H_n(c_0)} \qquad (54.15)$$

where $H_n(c_0)$ is the Hankel determinant from (5.15). The determinant $H_{n,r}(c_0)$ from (54.15) is obtained by replacing the rth column $\{c_r, c_{r+1}, c_{r+2}, \ldots, c_{n-1+r}\}$ in $H_n(c_0)$ by the column $\{-\tilde{c}_n, -\tilde{c}_{n+1}, -\tilde{c}_{n+2}, \ldots, -\tilde{c}_{2n-1}\}$ where $\tilde{c}_n = c_n q_{n,0}$. If the input set $\{c_n\}$ represents a geometric sequence (19.2) or auto-correlation functions (5.9) or time signals (5.11), then it follows from the explicit analytical expression (7.26) for the general Hankel determinant that $H_n(c_m) \neq 0$ [142] (see also Appendices B and C). In such a case, the condition (54.14) is satisfied and this guarantees that the system of equations (54.12) has the unique solution $\{q_{n,n-r}\}$ given by (54.15). The exact determinantal quotient (54.15) is not practical in applications. However, in [142] the efficient recursive PD algorithm [170] is employed for computation of any coefficient from the set $\{q_{n,n-r}\}$. This algorithm is error-free, i.e. it can be performed in the exact arithmetic using symbolic computations [102], e.g. MAPLE [91, 216] whenever the corresponding c_n are either integers or rational numbers as in MR physics.

Chapter 55

The modified moment problem

The analysis from chapter 22, which deals with the nearest neighbour approximation, shows that the three-term recursive construction (22.1) of the Lanczos orthonormal basis set $\{|\psi_n\rangle\}$ necessitates knowledge of the coupling constants $\{\alpha_n, \beta_n\}$. All the classical polynomials [180] satisfy the same type of contiguous recursions and have their constants $\{\alpha_n, \beta_n\}$ determined in advance. However, this is not the case with non-classical Lanczos polynomials $\{P_n(u), Q_n(u)\}$ for which the parameters $\{\alpha_n, \beta_n\}$ are unknown prior to computation and, as such, must be generated in a self-consistent manner in the process of the Lanczos algorithm for the wavefunctions $\{|\psi_n\rangle\}$. The polynomial pair $\{Q_n(u), P_n(u)\}$, as well as the accompanying coupling parameters $\{\alpha_n, \beta_n\}$, are also created recursively via the Lanczos algorithms (23.16), (23.20), (26.2) and (26.3). Particular importance is attached to the characteristic polynomial $Q_K(u)$ whose zeros $Q_K(u_k) = 0$ yield precisely the eigenvalues $\{u_k\}$ of the evolution operator $\hat{U}(\tau)$ from (8.1). This whole process is done through a twofold recursion (22.1) and (23.16), where the former algorithm produces the couplings $\{\alpha_n, \beta_n\}$, whereas the latter yields the sequence of polynomials $\{Q_n(u)\}$. Computations of the expansion coefficients $\{q_{n,n-r}\}$ of the power series representation (26.1) for the polynomials $\{Q_n(u)\}$ are also done recursively via (26.3). This entire computational strategy depends upon the availability of the coupling parameters $\{\alpha_n, \beta_n\}$. The final results of this procedure are the Lanczos states $\{|\psi_n\rangle\}$ that are also the building blocks of the total wavefunction $|\Upsilon_k\rangle$, according to (22.12). However, in the cases that do not require knowledge of the complete Schrödinger wavefunction $|\Upsilon_k\rangle$ computations of the Lanczos state vectors $\{|\psi_n\rangle\}$ seem to represent an unnecessary burden which, as such, should be alleviated altogether. Additionally, practice has shown that the finite precision numerical arithmetic in the recursion (22.1) leads to loss of orthogonalization and linear independence of the created basis set $\{|\psi_n\rangle\}$ for larger values of n. This is usually compensated by the explicit Gram–Schmidt re-orthogonalization of the basis $\{|\psi_n\rangle\}$. One of the obvious drawbacks of such a procedure is the need to keep all the generated Lanczos states $\{|\psi_n\rangle\}$. With this, the chief storage advantage of the original Lanczos algorithm is lost. Here, it is pertinent to recall that the recursion (22.1) needs to retain *only two* Lanczos states at a time. In order to bypass an explicit generation of the basis set $\{|\psi_n\rangle\}$, we shall introduce the matrix elements $\{\mu_{n,m}\}$ of the power function or a monomial u^m and the polynomial $Q_n(u)$ via

$$\mu_{n,m} = \frac{(u^m|Q_n(u))}{c_0}. \tag{55.1}$$

The quantity $\mu_{n,m}$ is known as the *modified moment* [171]. According to (54.6) or (54.10), the following key property of $\mu_{n,m}$ holds true

$$\mu_{n,m} = 0 \quad (m = 0, 1, 2, \ldots, n-1). \tag{55.2}$$

In other words, the modified moments fill in an upper triangular matrix $\boldsymbol{\mu} = \{\mu_{i,j}\}$ in which all the elements below the main diagonal are equal to zero [176]

$$\boldsymbol{\mu} = \begin{pmatrix} \mu_{0,0} & \mu_{0,1} & \mu_{0,2} & \mu_{0,3} & \mu_{0,4} & \cdots & \mu_{0,n} \\ 0 & \mu_{1,1} & \mu_{1,2} & \mu_{1,3} & \mu_{1,4} & \cdots & \mu_{1,n} \\ 0 & 0 & \mu_{2,2} & \mu_{2,3} & \mu_{2,4} & \cdots & \mu_{2,n} \\ 0 & 0 & 0 & \mu_{3,3} & \mu_{3,4} & \cdots & \mu_{3,n} \\ \vdots & \vdots & \vdots & \vdots & \vdots & \ddots & \vdots \\ 0 & 0 & 0 & 0 & 0 & \cdots & \mu_{n,n} \end{pmatrix}. \tag{55.3}$$

By definition (54.9) of the inner product, the matrix element (55.1) is

$$(u^m|Q_n(u)) = \sum_{k=1}^{K} d_k u_k^m Q_{n,k} \tag{55.4}$$

where $Q_{n,k} \equiv Q_n(u_k)$, $d_k = (\Phi_0|\Upsilon_k)^2$ and u_k is the eigenvalue of the evolution operator $\hat{U}(\tau)$, i.e. $\hat{U}|\Upsilon_k) = u_k|\Upsilon_k)$, as in (8.1). In the special case $n = 0$, we have $\mu_{0,m} = (1/c_0)\sum_{k=1}^{K} u_k^m d_k = \mu_m$, according to (53.5), so that

$$\mu_{0,m} = \mu_m \tag{55.5}$$

where μ_m is the mth power moment from (53.4). Furthermore, for $n = 0 = m$, it follows $\mu_{0,0} = (1/c_0)\sum_{k=1}^{K} d_k = 1$, where the sum rule (5.20) is used with $c_0 \neq 0$, so that

$$\mu_{0,0} = 1. \tag{55.6}$$

Comparing (53.5) and (55.1), it follows that the modified moment $\mu_{n,m}$ is of the form of the power moment μ_n, but with a polynomial weight function $w_{n,k} \equiv Q_n(u_k)$

$$\mu_{n,m} = \frac{1}{c_0} \sum_{k=1}^{K} Q_{n,k} \{d_k u_k^m\}. \tag{55.7}$$

The content in the curly brackets on the rhs of (55.7) is also present in the corresponding innermost part of the power moment (53.5). Inserting the scalar product (55.4) into (55.1) and subsequently using the power series representation (26.1) for $Q_n(u_k)$, we obtain

$$c_0 \mu_{n,m} = (u^m | Q_n(u)) = \sum_{k=1}^{K} d_k u_k^m Q_{n,k} = \sum_{r=0}^{n} q_{n,n-r} \left\{ \sum_{k=1}^{K} d_k u_k^{m+r} \right\}$$

$$= \sum_{r=0}^{n} q_{n,n-r} c_{m+r}.$$

Here, the term in the curly brackets is identified as the $(m+r)$th signal point or the auto-correlation function $c_{m+r} = C_{m+r}$, according to (5.9), (5.11) and (5.12), and thus

$$\mu_{n,m} = \frac{1}{c_0} \sum_{r=0}^{n} c_{m+r} q_{n,n-r}. \tag{55.8}$$

If for $m = n$, the product $u^n Q_n(u)$ from (55.1) is rewritten as $u^{n-1}[u Q_n(u)]$ with the subsequent elimination of the term $u Q_n(u)$ from the Lanczos recursion (23.16) for the polynomials $\{Q_n(u)\}$, then the following result emerges

$$c_0 \mu_{n,n} = (u^{n-1} | u Q_n(u))$$
$$= \beta_{n+1}(u^{n-1}|Q_{n+1}(u)) + \alpha_n(u^{n-1}|Q_n(u)) + \beta_n(u^{n-1}|Q_{n-1}(u))$$
$$= c_0 \beta_{n+1} \mu_{n+1,n-1} + c_0 \alpha_n \mu_{n,n-1} + c_0 \beta_n \mu_{n-1,n-1}$$
$$= c_0 \beta_n \mu_{n-1,n-1} \tag{55.9}$$

where the property (55.1) is used via $\mu_{n+1,n-1} = 0 = \mu_{n,n-1}$, so that

$$\beta_n = \frac{\mu_{n,n}}{\mu_{n-1,n-1}} \qquad n \geq 1 \tag{55.10}$$

with the initializations $\beta_0 = 0$, as in (22.6). Once the whole set $\{\beta_n\}$ becomes available in this way, we can find an explicit relationship between $\mu_{n,n}$ and all the coupling parameters β_n. We rewrite (55.10) as $\mu_{n,n} = \beta_n \mu_{n-1,n-1}$ which is a simple recursion for $\mu_{n,n}$. This latter recursion can be easily solved, if the term $\mu_{n-1,n-1}$ is replaced by $\beta_{n-1}\mu_{n-2,n-2}$ and likewise $\mu_{n-2,n-2}$ by $\beta_{n-2}\mu_{n-3,n-3}$. Continuing such a chaining until $n = 0$, we have $\mu_{n,n} = \beta_n \mu_{n-1,n-1} = \beta_n(\beta_{n-1}\mu_{n-2,n-2}) = \beta_n \beta_{n-1} \beta_{n-2} \cdots \beta_1 \mu_{0,0}$

$$\mu_{n,n} = \beta_n \beta_{n-1} \beta_{n-2} \cdots \beta_1 = \bar{\beta}_n \tag{55.11}$$

where $\mu_{0,0} = 1$ is used from (55.6) and $\bar{\beta}_n$ is the same as in from (22.4). The next goal is to see whether an analytical expression could also be obtained for the remaining set of the coupling constants $\{\alpha_n\}$. To solve this task, we take $\mu_{n,n+1}$

and $\mu_{n,n}$ from (55.1) to arrive at the quotient $\mu_{n,n+1}/\mu_{n,n}$ and finally abbreviate the result by λ_{n+1}

$$\lambda_{n+1} \equiv \frac{\mu_{n,n+1}}{\mu_{n,n}} \qquad \lambda_0 = 0. \qquad (55.12)$$

Here, we follow the derivation in (55.10) by first expressing the product of the type $u^{n+1} Q_n(u)$ as $u^n [u Q_n(u)]$ and then eliminating the term $u Q_n(u)$ by means of the recurrence (23.16), so that

$$\begin{aligned}
c_0 \mu_{n,n} \lambda_{n+1} &= c_0 \mu_{n,n+1} = (u^{n+1} | Q_n(u)) = (u^n | u Q_n(u)) \\
&= \beta_{n+1}(u^n | Q_{n+1}(u)) + \alpha_n (u^n | Q_n(u)) + \beta_n(u^n | Q_{n-1}(u)) \\
&= c_0 \beta_{n+1} \mu_{n+1,n} + c_0 \alpha_n \mu_{n,n} + c_0 \beta_n \mu_{n-1,n} \\
&= c_0 \alpha_n \mu_{n,n} + c_0 \beta_n \lambda_n \mu_{n-1,n-1} \\
&= c_0 \alpha_n \mu_{n,n} + c_0 \frac{\mu_{n,n}}{\mu_{n-1,n-1}} \lambda_n \mu_{n-1,n-1} \\
&= c_0 \mu_{n,n} (\alpha_n + \lambda_n)
\end{aligned}$$

where the relations $\mu_{n+1,n} = 0$ and $\mu_{n-1,n} = \lambda_n \mu_{n-1,n-1}$ from (55.1) and (55.12) are employed, respectively. Thus, the final result for the couplings $\{\alpha_n\}$ is

$$\alpha_n = \lambda_{n+1} - \lambda_n \qquad \alpha_0 = \frac{c_1}{c_0}. \qquad (55.13)$$

This result can be written in the following equivalent form using (55.12)

$$\alpha_n = \frac{\mu_{n,n+1}}{\mu_{n,n}} - \frac{\mu_{n-1,n}}{\mu_{n-1,n-1}}. \qquad (55.14)$$

Given the set $\{\alpha_n, \beta_n\}$, the expansion coefficients $\{q_{n,n-r}\}$ for $Q_n(u)$ are obtained from (26.3). The first few values from the set $\{\mu_{n,m}, \lambda_n\}$ are

$$\mu_{0,1} = \alpha_0 \qquad \mu_{1,1} = \frac{c_0 c_2 - c_1^2}{c_0^2 \beta_1} \qquad \mu_{1,2} = \frac{c_0 c_3 - c_1 c_2}{c_0^2 \beta_1}$$

$$\mu_{2,2} = \frac{q_0}{c_0^3 \beta_1^3 \beta_2} \qquad \mu_{2,3} = -\frac{q_1}{c_0^3 \beta_1^3 \beta_2} \qquad (55.15)$$

$$\lambda_1 = \frac{c_1}{c_0} \qquad \lambda_2 = \frac{c_0 c_3 - c_1 c_2}{c_0 c_2 - c_1^2} \qquad \lambda_3 = -\frac{q_1}{q_0} \qquad (55.16)$$

where the constants q_0 and q_1 are given in (26.14). Finally, using (55.10) and (55.13)–(55.16), the same results as in (22.6) for $\{\alpha_1, \alpha_2\}$ and $\{\beta_1, \beta_2\}$ are retrieved as a check of the current derivation. However, the advantage is that this new calculation is much simpler than the derivation of the results from (22.6) via the original Lanczos prescription (22.2). The above method within the modified moment problem using the steps (55.1)–(55.13) for generating the coupling constants $\{\alpha_n, \beta_n\}$, alongside the expansion coefficients $\{q_{n,n-r}\}$ of the polynomial $Q_n(u)$, via the recursion (26.3), is usually called the Gragg algorithm [171]. However, it should be noted that prior to Gragg [171], the modified moment problem has been used by Gautschi [172], as well as by Sack and Donovan [173].

Gragg [171] gave a symbolic formulation of this algorithm by introducing a linear functional $c_\downarrow = \mu_\downarrow$ acting on the power function u^n to produce the normalized signal point c_n/c_0 [271] as per (17.7). Similarly, for matrix elements we use the following symbolic rule which was, as mentioned in chapter 17, first put forward by Prony [57]

$$\mu_\downarrow(u^m|u^n) = \frac{\mu_{n+m}}{\mu_0} \tag{55.17}$$

or, in view of the equivalence (53.7)

$$c_\downarrow(u^m|u^n) = \frac{c_{n+m}}{c_0}. \tag{55.18}$$

Note that the matrix element $(u^m|u^n)$ could also be calculated using (5.11) and (54.9) with the result $(u^m|u^n) = \sum_{k=1}^{K} d_k u_k^{n+m} = c_{n+m}$. The main property of the functional $c_\downarrow$ is its linearity

$$c_\downarrow(au^n + bu^m) = \frac{ac_n + bc_m}{c_0} \tag{55.19}$$

where a and b are constants that do not depend upon n. Applying $c_\downarrow$ to the product $u^m Q_n(u)$ and using (26.1) and (55.18) together with the linearity property (55.19), we obtain

$$c_\downarrow(u^m|Q_n(u)) = \sum_{r=0}^{n} q_{n,n-r} c_\downarrow(u^m|u^r) = \sum_{r=0}^{n} q_{n,n-r} \frac{c_{m+r}}{c_0}$$

$$c_\downarrow(u^m|Q_n(u)) = \frac{1}{c_0} \sum_{r=0}^{n} q_{n,n-r} c_{m+r} = \mu_{n,m} \tag{55.20}$$

as in (55.8). Thus (55.10) and (55.13) can equivalently be written as

$$\alpha_n = \frac{c_\downarrow(u^{n+1}|Q_n(u))}{c_\downarrow(u^n|Q_n(u))} - \frac{c_\downarrow(u^n|Q_{n-1}(u))}{c_\downarrow(u^{n-1}|Q_{n-1}(u))}$$

$$\beta_n = \frac{c_\downarrow(u^n|Q_n(u))}{c_\downarrow(u^{n-1}|Q_{n-1}(u))}. \tag{55.21}$$

The explicit computation based upon (55.20) and (55.21) for the first few elements of the sequences $\{\alpha_n, \beta_n\}$ leads again to the previously derived expressions from (22.6). From the computational complexity viewpoint, the Gragg algorithm requires $\propto N^2$ multiplications and divisions and, therefore, it is an $\mathcal{O}(N^2)$ process for the given time signal $\{c_n\}$ of length N.

The modified moment problem

If the expansion coefficients $\{q_{n,n-r}\}$ from (26.3), as the main constituent of the power series representation (26.1), are not required in the analysis, then the Gragg algorithm can be simplified considerably. To this end, the ansatz $u^m Q_n(u)$ from (55.1) is first expressed through the identity $u^m Q_n(u) \equiv u^{m-1}[u Q_n(u)]$ and then the product $u Q_n(u)$ is eliminated by using the recursion (23.16). This is an extension of (55.9)

$$\begin{aligned} c_0 \mu_{n,m} &= (u^{m-1} | u Q_n(u)) \\ &= \beta_{n+1}(u^{m-1}|Q_{n+1}(u)) + \alpha_n(u^{m-1}|Q_n(u)) + \beta_n(u^{m-1}|Q_{n-1}(u)) \\ &= c_0 \beta_{n+1} \mu_{n+1,m-1} + c_0 \alpha_n \mu_{n,m-1} + c_0 \beta_n \mu_{n-1,m-1} \end{aligned}$$

so that when setting here $m' = m - 1$ and dividing both sides by $c_0 \neq 0$, we shall have

$$\beta_{n+1} \mu_{n+1,m} = \mu_{n,m+1} - \alpha_n \mu_{n,m} - \beta_n \mu_{n-1,m}, \qquad \mu_{0,0} = 1. \quad (55.22)$$

This recursion combined with (55.10) and (55.13) yields both the set $\{\mu_{n,m}\}$ and the $2N$ coupling parameters $\{\alpha_n, \beta_n\}$ for the given data of $2N$ power moments or signal points $\{\mu_n\} = \{c_n\}$. The outlined modification via avoiding the computation of the coefficients $\{q_{n,n-r}\}$, whenever they are not needed, is hereafter called the Chebyshev–Gragg algorithm for the modified moment problem (55.1).

Chapter 56

The mixed moment problem

The key in the above derivation of the Gragg and the Chebyshev–Gragg algorithms for the modified moment problem is the observation that the matrix element $(u^m|Q_n(u))$ is equal to zero for $n > m$, as stated in (55.1). Moreover, according to (54.7), the equality $(V_m(u)|Q_n(u)) = 0$ is valid for any polynomial $V_m(u)$ of degree m with $m < n$. Therefore we can now introduce a generalization of the two algorithms from chapter 55 to encompass a more general case of the moment problem, where the power function u^m is replaced by a known, i.e. given polynomial $V_m(u)$. This amounts to considering a mixed matrix element $(V_m(u)|Q_n(u))$ instead of $(u^m|Q_n(u))$ [174]

$$M_{n,m} = \frac{(V_m(u)|Q_n(u))}{c_0} \tag{56.1}$$

$$M_{n,m} = 0 \qquad n > m \tag{56.2}$$

where $c_0 = (\Phi_0|\Phi_0) \neq 0$, as in (4.5). Similarly to (55.3), the mixed moments also appear to form an upper triangular matrix $\mathbf{M} = \{M_{n,m}\}$ filled with zero elements below the main diagonal

$$\mathbf{M} = \begin{pmatrix} M_{0,0} & M_{0,1} & M_{0,2} & M_{0,3} & M_{0,4} & \cdots & M_{0,m} \\ 0 & M_{1,1} & M_{1,2} & M_{1,3} & M_{1,4} & \cdots & M_{1,m} \\ 0 & 0 & M_{2,2} & M_{2,3} & M_{2,4} & \cdots & M_{2,m} \\ 0 & 0 & 0 & M_{3,3} & M_{3,4} & \cdots & M_{3,m} \\ \vdots & \vdots & \vdots & \vdots & \vdots & \ddots & \vdots \\ 0 & 0 & 0 & 0 & 0 & \cdots & M_{m,m} \end{pmatrix}. \tag{56.3}$$

It will also prove convenient to introduce the following auxiliary matrix element

$$M_{n,m}^{(s)} = \frac{(V_m(u)|u^s|Q_n(u))}{c_0} \tag{56.4}$$

such that, according to (54.10), one has

$$M_{n,m}^{(s)} = 0 \qquad n > m + s \tag{56.5}$$

$$M_{n,m}^{(0)} = M_{n,m} \qquad M_{n,0}^{(s)} = \mu_{n,s} \qquad M_{0,0}^{(s)} = \mu_s. \tag{56.6}$$

In (56.1) and (56.4), the set $\{Q_n(u)\}$ represents a sequence of orthogonal polynomials $Q_n(u)$ each of which satisfies the three-term contiguous recursive relation (23.16). In (56.4) we shall consider a general set $\{V_m(u)\}$ of polynomials $V_m(u)$, and will analyse separately two cases without and with the known recursion, yielding two algorithms for the mixed moment (56.1). By construction the mixed moment (56.1) is a linear combination of the modified moments (55.1).

First, we shall assume that we are given the polynomial set $\{V_m(u)\}$ through the power series representation $V_m(u) = \sum_{r=0}^{m} v_{m,m-r} u^r$ from (54.8), where the expansion coefficients $\{v_{m,m-r}\}$ are known. Using the general definition (30.8) of the scalar product, we substitute the power series representations (26.1) and (54.8) into the matrix element (56.4) to calculate

$$c_0 M_{n,m}^{(s)} = (V_m(u)|u^s|Q_n(u)) = \sum_{k=1}^{K} d_k V_m(u_k) u_k^s Q_n(u_k)$$

$$= \sum_{r=0}^{n} \sum_{r'=0}^{m} q_{n,n-r} v_{m,m-r'} \left\{ \sum_{k=1}^{K} d_k u_k^{r+r'+s} \right\}$$

$$= \sum_{r=0}^{n} \sum_{r'=0}^{m} q_{n,n-r} v_{m,m-r'} c_{r+r'+s}$$

where (5.11) is utilized to identify the signal point $c_{r+r'+s}$

$$M_{n,m}^{(s)} = \frac{1}{c_0} \sum_{r=0}^{n} \sum_{r'=0}^{m} q_{n,n-r} v_{m,m-r'} c_{r+r'+s}. \tag{56.7}$$

The double sum in (56.7) can be simplified in the same fashion as previously done in (35.10) for the Lanczos coupling constants $\{\alpha_n, \beta_n\}$, so that

$$M_{n,m}^{(s)} = \frac{1}{c_0} \sum_{n'=0}^{n+m} \eta_{n'}^{(n,m)} c_{n'+s} \tag{56.8}$$

$$\eta_{n'}^{(n,m)} = \sum_{r=r_1}^{r_2} q_{n,n-r} v_{m,m-n'+r}$$

$$r_1 = \max\{0, n'-m\} \qquad r_2 = \min\{n', m\}. \tag{56.9}$$

In addition to the results from (32.4) and (32.5), we can derive two other representations for the parameters α_n and β_n. Namely, using the identity $(V_m(u)|u^s|Q_n(u)) = (V_m(u)|u^{s-1}|uQ_n(u))$ with the product $uQ_n(u)$ extracted from (23.16), we will have

$$(V_m(u)|u^s|Q_n(u)) = \beta_{n+1}(V_m(u)|u^{s-1}|Q_{n+1}(u)) + \alpha_n(V_m(u)|u^{s-1}|Q_n(u))$$
$$+ \beta_n(V_m(u)|u^{s-1}|Q_{n-1}(u))$$
$$= c_0[\beta_{n+1} M_{n+1,m}^{(s-1)} + \alpha_n M_{n,m}^{(s-1)} + \beta_n M_{n-1,m}^{(s-1)}]$$

$$\beta_{n+1} M_{n+1,m}^{(s-1)} = M_{n,m}^{(s)} - \alpha_n M_{n,m}^{(s-1)} - \beta_n M_{n-1,m}^{(s-1)}. \tag{56.10}$$

Setting $m = n$ and $s = 0$ in (56.10) will yield the relation

$$0 = \beta_{n+1} M_{n+1,n}^{(-1)} = M_{n,n}^{(0)} - \alpha_n M_{n,n}^{(-1)} - \beta_n M_{n-1,n}^{(-1)} = M_{n,n}^{(0)} - \beta_n M_{n-1,n}^{(-1)}$$

where $M_{n+1,n}^{(-1)} = 0 = M_{n,n}^{(-1)}$, as in (56.5). Thus, the coupling parameter β_n becomes

$$\beta_n = \frac{M_{n,n}}{M_{n-1,n}^{(-1)}} \tag{56.11}$$

where $M_{n,n}^{(0)} \equiv M_{n,n}$. Letting $m = n$ and $s = 1$ in (56.10) will produce the formula

$$M_{n,n}^{(1)} = \beta_{n+1} M_{n+1,n}^{(0)} + \alpha_n M_{n,n}^{(0)} + \beta_n M_{n-1,n}^{(0)}$$
$$= \beta_{n+1} M_{n+1,n} + \alpha_n M_{n,n} + \beta_n M_{n-1,n} = \alpha_n M_{n,n} + \beta_n M_{n-1,n}$$

with $M_{n+1,n} = 0$, according to (56.5), so that

$$\alpha_n = \frac{M_{n,n}^{(1)}}{M_{n,n}} - \beta_n \frac{M_{n-1,n}}{M_{n,n}}. \tag{56.12}$$

Substituting (56.11) for β_n into (56.12), we obtain the following result for the parameters $\{\alpha_n\}$

$$\alpha_n = \frac{M_{n,n}^{(1)}}{M_{n,n}} - \frac{M_{n-1,n}}{M_{n-1,n}^{(-1)}}. \tag{56.13}$$

Moreover, it is possible to derive the expressions for $\{\alpha_n, \beta_n\}$ containing only the matrix elements $M_{n,m}$ and $M_{n,m}^{(1)}$, but not $M_{n,m}^{(-1)}$. Thus, setting $m = n - 1$ and $s = 1$ into (56.10) yields

$$M_{n,n-1}^{(1)} = \beta_{n+1} M_{n+1,n-1}^{(0)} + \alpha_n M_{n,n-1}^{(0)} + \beta_n M_{n-1,n-1}^{(0)}$$
$$= \beta_{n+1} M_{n+1,n-1} + \alpha_n M_{n,n-1} + \beta_n M_{n-1,n-1} = \beta_n M_{n-1,n-1}$$

where $M_{n+1,n-1} = 0 = M_{n,n-1}$, as in (56.2). The parameter β_n can now be written in the form

$$\beta_n = \frac{M_{n,n-1}^{(1)}}{M_{n-1,n-1}} \tag{56.14}$$

or, equivalently, according to the definition (56.1)

$$\beta_n = \frac{(V_{n-1}(u)|u|Q_n(u))}{(V_{n-1}(u)|Q_{n-1}(u))} \tag{56.15}$$

and this generalizes (32.4). Inserting (56.14) into (56.12) yields the parameter α_n as follows

$$\alpha_n = \frac{M^{(1)}_{n,n}}{M_{n,n}} - \frac{M_{n-1,n}}{M_{n-1,n-1}} \frac{M^{(1)}_{n,n-1}}{M_{n,n}}. \tag{56.16}$$

Hence, unlike (56.11) and (56.13), the alternative formulae (56.14) and (56.16) for $\{\alpha_n, \beta_n\}$ do not require the computations of the matrix element $M^{(-1)}_{n,m}$. In a special case $V_m(u) \equiv u^m$, we have $M^{(s)}_{n,m} = \mu_{n,m+s}$, so that (56.16) becomes $\alpha_n = (\mu_{n,n+1}/\mu_{n,n}) - (\mu_{n-1,n}/\mu_{n-1,n-1})$ which agrees with the result (55.14) for the modified moment, as it should. The outlined method of generating the Lanczos parameters $\{\alpha_n, \beta_n\}$ by using the mixed moment problem is hereafter called the Chebyshev–Wheeler algorithm [174].

The above procedure is a direct way of tackling the mixed moment problem when only the power series expansion (54.8) of the potential $V_m(u)$ is given. An indirect way is also possible by expressing the matrix element $M_{n,m}$ as a linear combination of the modified moments $\{\mu_{n,r}\}_{r=0}^{m}$. Then, substituting (54.8) into the defining expression (56.1), we can write

$$M_{n,m} = \sum_{r=0}^{m} v_{m,m-r} \mu_{n,r} \tag{56.17}$$

where (55.1) is used. As we saw in chapter 55, the modified moment problem itself does not proceed by generating $\mu_{n,m}$ at once for only one fixed value of m. Rather, the needed set $\{\mu_{n,m}\}$ is constructed through recursions by forming a string of the modified moments $\{\mu_{n,r}\}_{r=0}^{m}$. This latter sequence is all that is needed in (56.17). Once the matrix elements $\{M_{n,m}\}$ are available, the parameters $\{\alpha_n, \beta_n\}$ can be computed from (56.14) and (56.16). Such a variant of the Wheeler algorithm can provide the parameters $\{\alpha_n, \beta_n\}$ without computing the expansion coefficients $\{q_{n,n-r}\}$ for the polynomials $\{Q_n(u)\}$. This is done by first using the Chebyshev–Gragg algorithm to obtain the modified moments $\{\mu_{n,m}\}$ that are afterwards imported into $M_{n,m}$ as intact via (56.17).

Finally, we shall consider the case when the polynomial set $\{V_m(u)\}$ forms an orthogonal sequence. In such a case, the members of the set $\{V_m(u)\}$ satisfy the three-term recursion relations of the type (23.16) with the known coefficients $\{a_n, b_n\}$ that are, of course, different from $\{\alpha_n, \beta_n\}$

$$b_{m+1} V_{m+1}(u) = (u - a_m) V_m(u) - b_m V_{m-1}(u)$$
$$V_{-1}(u) = 0 \qquad V_0(u) = 1. \tag{56.18}$$

This case is very important since all the classical polynomials [180] are orthogonal and, as such, are the solutions of the difference equation (56.18), where all the coefficients $\{a_m, b_m\}$ are fixed in advance through, e.g. the known analytical expressions. This is opposed to the couplings $\{\alpha_n, \beta_n\}$ in the set $\{Q_n(u)\}$ that need to be generated simultaneously with the recursive construction (23.16) of the polynomials $Q_n(u)$ themselves. In such a setting, the mixed moment problem consists of determining the coupling parameters $\{\alpha_n, \beta_n\}$ for the given sequence of parameters $\{a_m, b_m\}$ of the polynomials $\{V_m(u)\}$. We begin by projecting (56.18) and (23.16) onto $Q_n(u)$ and $V_m(u)$, respectively, through forming their inner products of the type $(V_j(u)|Q_{j'}(u))$ yielding

$$b_{m+1}M_{n,m+1} = M_{n,m}^{(1)} - a_m M_{n,m} - b_m M_{n,m-1} \tag{56.19}$$

$$\beta_{n+1}M_{n+1,m} = M_{n,m}^{(1)} - \alpha_n M_{n,m} - \beta_n M_{n-1,m} \tag{56.20}$$

where (56.1) and (56.4) are used to identify $M_{j,j'}$ and $M_{j,j'}^{(s)}$, respectively. Setting $m = n - 1$ in (56.19) and utilizing the parameter β_n with the help of the relation (56.14), it follows

$$b_n M_{n,n} = \beta_n M_{n-1,n-1} - a_{n-1} M_{n,n-1} - b_{n-1} M_{n,n-2}$$

so that after using $M_{n,n-1} = 0 = M_{n,n-2}$ from (56.2), we have

$$\beta_n = \frac{M_{n,n}}{M_{n-1,n-1}} b_n. \tag{56.21}$$

This determines completely the parameters $\{\beta_n\}$ in terms of the known quantities $\{b_n, M_{n,m}\}$. The common matrix elements $M_{n,m}^{(1)}$ from (56.19) and (56.20) can be eliminated by subtracting these two equations from each other

$$\beta_{n+1}M_{n+1,m} = (a_m - \alpha_n)M_{n,m} + b_m M_{n,m-1} - \beta_n M_{n-1,m} + b_{m+1}M_{n,m+1}. \tag{56.22}$$

For $m = n$ in (56.22) we have $M_{n+1,n} = 0 = M_{n,n-1}$, according to (56.2) and this leads to

$$b_{n+1}M_{n,n+1} = (\alpha_n - a_n)M_{n,n} + \beta_n M_{n-1,n}. \tag{56.23}$$

Inserting the formula (56.21) for β_n into (56.23) and dividing the result by $M_{n,n} \neq 0$, we find

$$\alpha_n = a_n + b_{n+1}\frac{M_{n,n+1}}{M_{n,n}} - b_n \frac{M_{n-1,n}}{M_{n-1,n-1}}. \tag{56.24}$$

This result determines completely the couplings $\{\alpha_n\}$ as a function of the known quantities $\{a_m, b_m, M_{n,m}\}_{n=0}^{N-1}$ ($0 \leq m \leq n - 1$). The first pair $\{a_m, b_m\}$ are the recurring constants of the known set $\{V_m(u)\}$ and, as such, they are given in advance in, e.g. their analytical forms. The initial values for computing the parameters α_n and β_n are α_0 and β_0. The value β_0 is fixed by the convention as $\beta_0 = 0$. On the other hand, using the convention $M_{-1,0}/M_{-1,-1} \equiv 0$, we can fix the parameter α_0 from the recursion (56.24) itself as $\alpha_0 = a_0 + b_1 M_{0,1}/M_{0,0}$

$$\alpha_0 = a_0 + b_1 \frac{\nu_1}{\nu_0} \tag{56.25}$$

$$\nu_m \equiv M_{0,m}. \tag{56.26}$$

The mixed moments $\{M_{n,m}\}$ themselves are obtainable from (56.22), where we set $n' = n + 1 \equiv n$ and write

$$\beta_n M_{n,m} = b_{m+1} M_{n-1,m+1} - (\alpha_{n-1} - a_m) M_{n-1,m}$$
$$- \beta_{n-1} M_{n-2,m} + b_m M_{n-1,m-1}. \quad (56.27)$$

This recursion for the sequence $\{M_{n,m}\}$ needs the initial conditions for $n = 0$ that are $\{M_{0,m}\} \equiv \{v_m\}$. Inserting the given power series expansion (54.8) for $V_m(u)$ into (56.26), we obtain

$$v_m \equiv M_{0,m} = (V_m(u)|1) = \sum_{k=1}^{K} d_k V_m(u_k)$$

$$= \sum_{r=0}^{m} v_{m,m-r} \left\{ \sum_{k=1}^{K} d_k u_k^r \right\} = \sum_{r=0}^{m} v_{m,m-r} c_r$$

$$v_m = \sum_{r=0}^{m} v_{m,m-r} c_r \qquad (m = 0, 1, 2, \ldots, n-1). \quad (56.28)$$

Past experience has shown [175, 176] that the Wheeler algorithm [174] is an extremely accurate, stable and robust algorithm for computation of the tight binding constants/couplings $\{\alpha_n, \beta_n\}$.

Chapter 57

Proof for zero-valued amplitudes of spurious resonances

Suppose that we are given a physical time signal c_n of the form $c_n = \sum_{k=1}^{K} d_k u_k^n$ from (5.11) with exactly K genuine resonances described by the spectral parameters $\{u_k, d_k\}$, where $u_k \neq 0, u_k \neq 1 (1 \leq k \leq K)$. Here, the adjective 'physical' means that all the harmonics $\{u_k^n\} = \{\exp(-in\omega_k \tau)\}$ ($k = 1, \ldots, K$) will decay to zero when $n \longrightarrow \infty$. The points $u = 0$ and $u = 1$ are also excluded from the corresponding exact frequency spectrum $\mathcal{R}(u) = \sum_{n=0}^{\infty} c_n u^{-n-1}$ which is the Green function (38.1) in the harmonic variable $u = \exp(-i\omega\tau)$. We saw in chapters 7 and 21 that the action of the Shanks transform $e_K(c_n)$ on the signal $c_n = \sum_{k=1}^{K} d_k u_k^n$ gave exactly zero through the process of the explicit filtering out of all the K transients $\{d_k u_k\}$ from the signal c_n, so that $e_K(c_n) = 0$, as in (7.53). In the spectral problem, e.g. the HIP, one of the key unknowns is the number K. However, if the signal c_n indeed has K harmonics, then $c_n = \sum_{k=1}^{K} d_k u_k^n$ can be converted to the Dirichlet series

$$c_n = \sum_{k=1}^{\infty} d_k u_k^n \qquad (57.1)$$

provided that

$$d_k = 0 \qquad k > K. \qquad (57.2)$$

In other words, all the amplitudes $\{d_k\}$ of higher harmonics $k > K$ in (57.1) must be equal to zero, when the signal is sampled by invoking only K harmonics as in chapter 21. For convenience, we also set in (57.1)

$$u_k = 1 \qquad k > K. \qquad (57.3)$$

The problem posed in this chapter is to find an explicit expression for the amplitudes $\{d_k\}$ for these higher harmonics $k > K$ from which (57.2) will be deduced rather than imposed. We shall solve this problem in two different ways. First, we will use (17.5) for d_k associated with c_n from (5.11). The definition (17.5) for d_k is convenient for the present purpose, since it permits that the subscript k takes the values greater than K. Thus (17.5) gives for, e.g. $K = 1$

$$d_1 = c_0 \qquad d_2 = \frac{c_1 - u_1 c_0}{1 - u_1} \qquad K = 1. \tag{57.4}$$

The value of u_1 found from the characteristic equation $Q_1(u) = q_{1,1} + q_{1,0} u = 0$ is given by $u_1 = -q_{1,1}/q_{1,0}$, where $\{q_{r,K-r}\}$ are the coefficients of the characteristic polynomial $Q_K(u)$ from the power expansion (26.1). Thus, inserting $u_1 = -q_{1,1}/q_{1,0}$ into the rhs of (57.4) yields

$$d_2 = \beta_1 \frac{\sum_{r=0}^{1} c_r q_{1,r-1}}{1 - u_1} \qquad K = 1. \tag{57.5}$$

This can equivalently be written as follows

$$d_2 = (-1)^1 c_0 \frac{\mu_{1,0}}{Q_1(1)} \qquad K = 1 \tag{57.6}$$

where $\mu_{n,m}$ is the modified moment from (55.1) which is, by definition (55.2), equal to zero when $m > n$. Therefore, $\mu_{1,0} = 0$ which implies

$$d_2 = 0 \qquad K = 1 \qquad \text{(QED)}. \tag{57.7}$$

In (57.6) we have

$$Q_k(1) = \frac{W_k}{\bar{\beta}_k} = \frac{\tilde{Q}_k(1)}{\bar{\beta}_k} \neq 0 \tag{57.8}$$

where $\bar{\beta}_k = \beta_1 \beta_2 \cdots \beta_k \neq 0$ and $W_k = \prod_{j=1}^{k}(u_j - 1) = \tilde{Q}_k(1) \neq 0$, as in (22.4) and (7.50), respectively. Similarly, for $K = 2$ it follows from (17.6)

$$d_1 = \frac{c_{\downarrow} \prod_{j=1, j \neq 1}^{2}(u - u_j)}{\prod_{j=1, j \neq 1}^{2}(u_1 - u_j)} = \frac{c_{\downarrow}(u - u_2)}{u_1 - u_2} = c_0 \frac{u_2 - c_1/c_0}{u_2 - u_1}$$

$$\therefore \quad d_1 = c_0 \frac{u_2 - \alpha_0}{u_2 - u_1} \qquad K = 2. \tag{57.9}$$

A verification of this expression can be carried out by using the defining relation (36.10) for d_k in the PLA

$$d_1 = \frac{c_0}{\beta_1} \frac{P_2(u_1)}{Q'_2(u_1)} = c_0 \frac{u_1 - \alpha_1}{-(\alpha_0 + \alpha_1) + 2u_1} = c_0 \frac{u_1 - \alpha_1}{-(u_1 + u_2) + 2u_1}$$

$$= c_0 \frac{u_1 - \alpha_1}{u_1 - u_2} = c_0 \frac{u_2 - \alpha_0}{u_2 - u_1}$$

where $u_1 + u_2 = \alpha_0 + \alpha_1$ so that

$$\therefore \quad d_1 = c_0 \frac{u_2 - \alpha_0}{u_2 - u_1} \quad K = 2 \quad \text{(QED)} \quad (57.10)$$

in agreement with (57.9). For $K = 2$ and $k = 2$ we have from (17.6)

$$d_2 = \frac{c_\downarrow \prod_{j=1, j \neq 2}^2 (u - u_j)}{\prod_{j=1, j \neq 2}^2 (u_2 - u_j)} = \frac{c_\downarrow (u - u_1)}{u_2 - u_1} = \frac{c_1 - u_1 c_0}{u_2 - u_1}$$

$$\therefore \quad d_2 = c_0 \frac{\alpha_0 - u_1}{u_2 - u_1} \quad K = 2. \quad (57.11)$$

To check this result we use (36.10) to calculate

$$d_2 = \frac{c_0}{\beta_1} \frac{P_2(u_2)}{Q_2'(u_2)} = c_0 \frac{u_2 - \alpha_1}{-(\alpha_0 + \alpha_1) + 2u_2}$$

$$= c_0 \frac{u_2 - \alpha_1}{-(u_1 + u_2) + 2u_2} = c_0 \frac{u_2 - \alpha_1}{u_2 - u_1} = c_0 \frac{\alpha_0 - u_1}{u_2 - u_1}$$

$$\therefore \quad d_2 = c_0 \frac{\alpha_0 - u_1}{u_2 - u_1} \quad K = 2 \quad \text{(QED)} \quad (57.12)$$

in agreement with (57.11). Setting $K = 2$ and $k = 3$ into (17.6) yields

$$d_3 = \frac{c_\downarrow \prod_{j=1}^2 (u_j - u)}{\prod_{j=1}^2 (u_j - u_3)} = \frac{c_\downarrow (u_1 - u)(u_2 - u)}{(u_1 - 1)(u_2 - 1)}$$

$$= \frac{u_1 u_2 c_0 - (u_1 + u_2) c_1 + c_2}{(u_1 - 1)(u_2 - 1)}$$

$$\therefore \quad d_3 = \frac{u_1 u_2 c_0 - (u_1 + u_2) c_1 + c_2}{(u_1 - 1)(u_2 - 1)} \quad K = 2 \quad (57.13)$$

where the simplified notation $\prod_{j=1}^2 (u_j - u)$ is used for $\prod_{j=1, j \neq 3}^2 (u_j - u)$, since the constraint $j \neq 3$ is obviously superfluous. The roots u_1 and u_2 are the solutions of the characteristic equation $\beta_1 \beta_2 Q_2(u) = u^2 - (\alpha_0 + \alpha_1) u + (\alpha_0 \alpha_1 - \beta_1^2) = 0$ so that $u_{1,2} = [(\alpha_0 + \alpha_1) \pm \sqrt{(\alpha_0 + \alpha_1)^2 - 4(\alpha_0 \alpha_1 - \beta_1^2)}]/2$. Thus $u_1 + u_2 = \alpha_0 + \alpha_1$ and $u_1 u_2 = \alpha_0 \alpha_1 - \beta_1^2$ so that

$$d_3 = \frac{(\alpha_0 \alpha_1 - \beta_1^2) c_0 - (\alpha_0 + \alpha_1) c_1 + c_2}{(u_1 - 1)(u_2 - 1)}$$

$$= \beta_1 \beta_2 \frac{q_{2,2} c_0 + q_{2,1} c_1 + q_{2,0} c_2}{(u_1 - 1)(u_2 - 1)} \quad (57.14)$$

where $q_{2,0}, q_{2,1}$ and $q_{2,2}$ are the coefficients of the polynomial $Q_2(u)$ and they are given by (26.6). Using (17.6) we have

$$d_3 = c_0\beta_1\beta_2 \frac{1}{(u_1-1)(u_2-1)} \frac{1}{c_0} \sum_{r=0}^{2} c_r q_{2,2-r} = c_0\beta_1\beta_2 \frac{\mu_{2,0}}{(u_1-1)(u_2-1)}$$

$$\therefore \quad d_3 = c_0(-1)^2 \frac{\mu_{2,0}}{Q_2(1)} \qquad K=2. \qquad (57.15)$$

Since $\mu_{2,0} = 0$, as per (55.2), it follows

$$d_3 = 0 \qquad K=2 \qquad (QED). \qquad (57.16)$$

Our next example is $K=3$ for which we calculate from (17.6)

$$d_4 = \frac{c_\downarrow \prod_{j=1}^{3}(u_j-u)}{\prod_{j=1}^{3}(u_j-u_4)} = \frac{c_\downarrow[(u_1-u)(u_2-u)(u_3-u)]}{(u_1-u_4)(u_2-u_4)(u_3-u_4)}$$

$$= \frac{u_1u_2u_3c_0 - (u_1u_2+u_1u_3+u_2u_3)c_1 + (u_1+u_2+u_3)c_2 - c_3}{(u_1-1)(u_2-1)(u_3-1)}$$

$$\therefore \quad d_4 = \frac{u_1u_2u_3c_0 - (u_1u_2+u_1u_3+u_2u_3)c_1 + (u_1+u_2+u_3)c_2 - c_3}{(u_1-1)(u_2-1)(u_3-1)}$$

$$(57.17)$$

where we wrote $\prod_{j=1}^{3}(u_j-u)$ instead of $\prod_{j=1, j\neq 4}^{3}(u_j-u)$, since $1 \leq j \leq 3$. The polynomial $Q_3(u)$ can be written via its zeros as

$$Q_3(u) = q_{3,0}(u-u_1)(u-u_2)(u-u_3) = q_{3,0}[-u_1u_2u_3$$
$$+ (u_1u_2+u_1u_3+u_2u_3)u - (u_1+u_2+u_3)u^2 + u^3]$$
$$\equiv q_{3,3} + q_{3,2}u + q_{3,1}u^2 + q_{3,0}u^3. \qquad (57.18)$$

By comparing (26.8) with (57.18), we arrive at

$$u_1+u_2+u_3 = -\frac{q_{3,1}}{q_{3,0}} \qquad u_1u_2+u_1u_3+u_2u_3 = \frac{q_{3,2}}{q_{3,0}}$$

$$u_1u_2u_3 = -\frac{q_{3,3}}{q_{3,0}} \qquad (57.19)$$

where, in particular, $q_{3,0} = 1/(\beta_1\beta_2\beta_3)$, as in (26.6). Then, the usage of these expressions in (57.17) gives

$$d_4 = (-1)^3 \frac{\beta_1\beta_2\beta_3}{(u_1-1)(u_2-1)(u_3-1)} c_0 \left[\frac{1}{c_0} \sum_{r=0}^{3} c_r q_{3,3-r} \right]$$

$$= (-1)^3 \frac{c_0\beta_1\beta_2\beta_3}{(u_1-1)(u_2-1)(u_3-1)} \mu_{3,0}$$

$$\therefore \quad d_4 = (-1)^3 c_0 \frac{\mu_{3,0}}{Q_3(1)} \qquad K=3. \qquad (57.20)$$

Since $\mu_{3,0} = 0$ according to (55.2), we can write

$$d_4 = 0 \qquad K = 3 \qquad \text{(QED)}. \tag{57.21}$$

Obviously, from the particular results (57.6), (57.15) and (57.20) we shall have in the general harmonic inversion problem with K resonances

$$d_{K+1} = (-1)^K c_0 \frac{\mu_{K,0}}{Q_K(1)} \tag{57.22}$$

$$\therefore \qquad d_{K+1} = 0 \qquad \text{(QED)} \tag{57.23}$$

since $\mu_{K,0} = 0$. Therefore, to check whether the rank K for the given signal c_n is correct, we must have

$$\sum_{r=0}^{K} c_r q_{K,K-r} = 0. \tag{57.24}$$

Let us now define an auxiliary time signal $\{\tilde{c}_n\}$ $(0 \le n \le N-1)$

$$\tilde{c}_n = \sum_{k=0}^{K+1} d_k u_k^n \qquad \tilde{c}_n = d_{K+1} + c_n \qquad u_{K+1} \equiv 1. \tag{57.25}$$

By means of the equation $e_K(c_n) = 0$ from (7.53), it follows $e_K(\tilde{c}_n) = e_K(d_{K+1} + c_n) = d_{K+1} + e_K(c_n) = d_{K+1}$ so that

$$\therefore \qquad e_K(\tilde{c}_n) = d_{K+1}. \tag{57.26}$$

On the other hand, we have $e_K(c_n) = H_{K+1}(c_n)/H_K(\Delta^2 c_n)$ from (7.52) where $\Delta^2 c_n = \Delta c_{n+1} - \Delta c_n$ and $\Delta c_n = c_{n+1} - c_n$. Hence, if the signal c_n has exactly K harmonics, it ought to be

$$e_K(\tilde{c}_n) = e_K(c_n) \tag{57.27}$$

which implies $d_{K+1} = 0$ as in (57.23). The transform $e_K(c_n)$ is the exact filter which is capable of filtering out all the K transients $\sum_{k=1}^{K} d_k u_k^n$ and thus arriving straight at the equilibrium value d_{K+1} which is zero, as in (57.23). We know from chapter 7 that (57.23) is satisfied if the two conditions (7.11) and (7.12) are fulfilled simultaneously. In other words, $e_K(c_n) = 0$ if and only if $H_K(c_n) \ne 0$ and $H_{K+1}(c_n) = 0$ for $n \in [0, N-1]$. Also from the definition (7.52) of the Shanks transform we have

$$e_K(c_n) = 0 \iff \begin{cases} H_{K+1}(c_n) = 0 \\ H_K(\Delta^2 c_n) \ne 0. \end{cases} \tag{57.28}$$

We have already shown in chapter 7 that $H_K(c_n) = V_K^2 \bar{d} \bar{u}^n \ne 0$ as in (7.26). In the same fashion, we can easily derive the relation

$$H_K(\Delta^m c_n) = \tilde{Q}_K^m(1) H_K(c_n) \tag{57.29}$$

$$\Delta^m c_n = \sum_{k=1}^{K} d_k u_k^n (u_k - 1)^m. \tag{57.30}$$

Since $Q_K^m(1) \neq 0$ and $H_K(c_n) \neq 0$, it follows from (57.28)

$$\therefore \quad H_K(\Delta^m c_n) \neq 0 \quad (m = 1, 2, 3, \ldots) \tag{57.31}$$

as required in (57.28). Moreover, we have

$$H_{K+1}(c_n) = d_{K+1}\bar{d}\bar{u}^n V_{K+1}^2 \quad V_{K+1} = Q_K(1)V_K$$
$$\therefore \quad H_{K+1}(c_n) = d_{K+1}\tilde{Q}_K^2(1)H_K(c_n). \tag{57.32}$$

Using (57.28), we obtain

$$e_K(c_n) = \frac{H_{K+1}(c_n)}{H_K(\Delta^2 c_n)} = \frac{d_{K+1}\tilde{Q}_K^2(1)H_K(c_n)}{\tilde{Q}_K^2(1)H_K(c_n)} = d_{K+1}$$
$$\therefore \quad e_K(c_n) = d_{K+1}. \tag{57.33}$$

Thus, it follows from (57.26) and (57.33) that $e_K(\tilde{c}_n) = e_K(c_n)$ as stated in (57.27). The results (7.53) and (57.32) lead to $d_{K+1} = 0$, as in (57.23). Since $Q_K(1) \neq 0$ and $H_K(c_n) \neq 0$, it follows from (57.32)

$$d_{K+1} = 0 \iff H_{K+1}(c_n) = 0 \tag{57.34}$$

which is required in (57.28) where $H_K(\Delta^2 c_n) \neq 0$, as per (57.31). Moreover, it is easy to see in the course of the performed derivation via (17.6) that not only d_{K+1} is equal to zero for the signal $c_n = \sum_{k=1}^{K} d_k u_k^n$, but also

$$\left.\begin{array}{l} d_{K+m} = 0 \\ e_K(c_n) = d_{K+m} \end{array}\right\} \quad (m = 1, 2, \ldots) \tag{57.35}$$

as in (7.43). In other words, if we have a sampled signal $\{c_n\}$ with K harmonics (K is unknown prior to analysis), then all the transients $\{d_k u_k\}$ for $k > K$ will have zero amplitudes exactly as required by (57.2). Therefore, they can be considered as spurious.

Let us re-emphasize that a time signal $\{c_n\}_{n=0}^{N-1}$, will be uniquely given by a linear combination of K damped complex exponentials $c_n = \sum_{k=1}^{K} d_k \exp(-i\omega_k n\tau)$, or equivalently, the corresponding spectrum will have exactly K resonances, if and only if the following two conditions are satisfied simultaneously

$$\left.\begin{array}{l} H_K(c_n) \neq 0 \\ H_{K+1}(c_n) = 0 \end{array}\right\} \quad \forall n \in [0, N-1] \tag{57.36}$$

as in (7.11) and (7.12). These conditions have been derived from the existence condition of the Padé approximant. However, (57.36) ought to be in the nature of the harmonic inversion problem and, as such, independent of the processing method used. Indeed, a difference equation of order K with constant coefficients and with K known initial conditions (a direct problem) will have the unique solution given by a linear combination of K damped exponentials, provided that (57.36) is satisfied as stated in (7.11) and (7.11). Such a difference equation is the characteristic equation

$$Q_K(u) = 0 \Longrightarrow \{u_k\} \qquad (1 \le k \le K). \tag{57.37}$$

If we take the scalar product of $Q_K(u)$ with $u^0 = 1$, we shall have $0 = (1|Q_K(u)) = (1|\sum_{r=0}^{K} q_{K,K-r} u^r) = \sum_{r=0}^{K} q_{K,K-r}(1|u^r) = \sum_{r=0}^{K} q_{K,K-r} c_r = \mu_{K,0}$ where $(u^n|u^m) = c_{n+m}$, so that

$$\sum_{r=0}^{K} c_r q_{K,K-r} = \mu_{K,0} = (1|Q_K(u)) = 0 \tag{57.38}$$

as in (57.24). Hence, the relation (57.38) is nothing but an alternative way of writing the characteristic equation (57.37). Thus, the result (57.22) can be rewritten as

$$d_{K+1} = (-1)^K c_0 \frac{(1|Q_K(u))}{Q_K(1)}. \tag{57.39}$$

The inner product is defined by $(g(u)|f(u)) = \sum_{k=0}^{K} d_k f(u_k) g(u_k)$ as in (30.8), so that

$$\therefore \qquad \mu_{K,0} = (1|Q_K(u)) = \sum_{k=0}^{K} d_k Q_K(u_k) = 0 \tag{57.40}$$

where (57.37) is used via $Q_K(u_k) = 0$. The relation $d_{K+1} = 0$ is equivalent to (57.36). This implies that the exact number K of resonances can be extracted by using the whole signal to verify whether (57.36) is valid. In practice, since we do not know K in advance, we compute $H_M(c_n)$ for each $n \in [0, N-1]$ by setting $M = K_0 + \kappa$ ($\kappa = 1, 2, \ldots, p$), where K_0 is an initial guess (for K) taken to be equal to the number of major peaks seen in, e.g. the magnitude Fourier spectrum. Then, the sought value of the integer p will be the smallest κ for which we have

$$H_{K_0+p}(c_n) \ne 0 \qquad H_{K_0+p+1}(c_n) = 0 \qquad \forall n \in [0, N-1] \tag{57.41}$$
$$\therefore \qquad K = K_0 + p. \tag{57.42}$$

Thus, we have shown explicitly that the Shanks transform $e_K(c_n)$ for the signal (5.11) is given exactly by $e_K(c_n) = d_{K+m}$, as per (57.35). Since $e_K(c_n) = 0$ for the signal of the form $c_n = \sum_{k=1}^{K} d_k u_k^n$, it follows that all the spurious resonances will have zero amplitudes $d_k = 0$ ($k > K$) as in (57.2) which was set up to prove (QED) (see also chapter 21).

Chapter 58

Mapping from monomials u^n to polynomials $Q_n(u)$

The orthogonal characteristic polynomials or eigenpolynomials $\{Q_n(u)\}$ play one of the central roles in spectral analysis, since they form a basis due to the completeness relation (31.20). They can be computed either via the Lanczos recursion (23.16) or from the power series representation (26.1). The latter method generates the expansion coefficients $\{q_{n,n-r}\}$ through the recursion (26.3). Alternatively, these coefficients can be deduced from the Lanczos recursion (24.4) for the rth derivative $Q_{n,r}(0)$, since we have $q_{n,n-r} = (1/r!)Q_{n,r}(0)$, as in (26.10). The polynomial set $\{Q_n(u)\}$ is the basis comprised of scalar functions in the Lanczos vector space $\mathcal{L}$ from (30.1). In (30.1), the definition (30.8) of the inner product implies that the polynomials $Q_n(u)$ and $Q_m(u)$ are orthogonal to each other (for $n \neq m$) with respect to the complex weight function d_k as per (31.23). The completeness (31.20) of the set $\{Q_n(u)\}$ enables expansion of every function $f(u) \in \mathcal{L}$ in a series in terms of the polynomials $\{Q_n(u)\}$, as shown in (31.24). Doing this with the monomial $f(u) = u^m$ yields

$$u^m = \sum_{r=0}^{m} \mu_{r,m} Q_r(u). \qquad (58.1)$$

The unknown set $\{\mu_{r,m}\}$ can be found by using (31.23) in the scalar product of u^m with $Q_n(u)$

$$(u^m|Q_r(u)) = \sum_{n=0}^{m} \mu_{n,m}(Q_n(u)|Q_r(u)) = \sum_{n=0}^{m} \mu_{n,m}\{c_0 \delta_{r,n}\} = c_0 \mu_{r,m}$$

$$\mu_{r,m} = \frac{(u^m|Q_r(u))}{c_0}. \qquad (58.2)$$

Thus, the expansion coefficients $\{\mu_{r,m}\}$ are the modified moments (55.1) with the feature $\mu_{r,m} = 0$ for $r > m$, as in (55.2). Taking the inner product (30.8) of u^n

with u^m and using (5.11) gives

$$(u^m|u^n) = \sum_{k=1}^{K} d_k u_k^m u_k^n = \sum_{k=1}^{K} d_k u_k^{n+m} = c_{n+m}$$

$$(u^m|u^n) = c_{n+m}. \tag{58.3}$$

Likewise, computing the same scalar product $(u^m|u^n)$ by using the expansion (58.1) for both u^m and u^n together with the orthogonality (31.23) implies

$$(u^m|u^n) = \sum_{r=0}^{n} \sum_{r'=0}^{m} \mu_{r,n} \mu_{r',m} (Q_{r'}(u)|Q_r(u))$$

$$= \sum_{r=0}^{n} \sum_{r'=0}^{m} \mu_{r,n} \mu_{r',m} \{c_0 \delta_{r,r'}\} = c_0 \sum_{r=0}^{n} \mu_{r,n} \mu_{r,m}$$

$$(u^m|u^n) = c_0 \sum_{r=0}^{n} \mu_{r,n} \mu_{r,m}. \tag{58.4}$$

On comparing (58.3) and (58.4), we deduce

$$\therefore \quad \sum_{r=0}^{n} \mu_{r,n} \mu_{r,m} = \frac{c_{n+m}}{c_0}. \tag{58.5}$$

If (26.1) is inserted into (55.1), the following result is obtained using (58.3)

$$\mu_{n,m} = \frac{(u^m|Q_n(u))}{c_0} = \sum_{r=0}^{n} q_{n,n-r}(u^m|u^r) = \frac{1}{c_0} \sum_{r=0}^{n} q_{n,n-r} c_{n+r}$$

$$\mu_{n,m} = \frac{1}{c_0} \sum_{r=0}^{n} c_{n+r} q_{n,n-r} \tag{58.6}$$

as in (55.20). When (55.2) is used in (58.6), the following system of linear equations for the coefficients $\{q_{n,n-r}\}$ of the polynomial $Q_n(u)$ is obtained, as typical for the PA

$$\sum_{r=0}^{n} c_{m+r} q_{n,n-r} = 0 \quad (m = 0, 1, 2, \ldots, n-1). \tag{58.7}$$

Chapter 59

Mapping of state vectors: Schrödinger ⟷ Lanczos

In (35.2) we give the mapping $|\psi_n\rangle = \sum_{r=0}^{n} q_{n,n-r}|\Phi_r\rangle$ between the two different representations $\{|\psi_n\rangle\}$ and $\{|\Phi_n\rangle\}$. Here, the expansion coefficients $\{q_{n,n-r}\}$ play the role of the elements of the direct transformation $|\psi_n\rangle \longrightarrow |\Phi_n\rangle$ between the Lanczos and the Schrödinger states. The result (35.2) shows that each Lanczos state $|\psi_n\rangle$ is a linear combination of n Schrödinger states $\{|\Phi_m\rangle\}(0 \leq m \leq n)$. To proceed further in this direction, we need the inverse transformation $|\Phi_n\rangle \longrightarrow |\psi_n\rangle$. This could be obtained using (5.6) as the definition of the Schrödinger state $|\Phi_n\rangle = \hat{U}^n(\tau)|\Phi_0\rangle$, if we have $\hat{U}^n$ as a linear combination of the polynomials $\{Q_r(\hat{U})\}$, such that the ansatz $Q_r(\hat{U})|\Phi_0\rangle$ could be used to identify the Lanczos state $|\psi_r\rangle$ via (23.19). The mapping $\hat{U}^n \longrightarrow Q_n(\hat{U})$ can be obtained as the operator counterpart of (58.1) by resorting to the Cayley–Hamilton theorem [228]

$$\hat{U}^n = \sum_{r=0}^{n} \mu_{r,n} Q_r(\hat{U}). \tag{59.1}$$

Substituting this expression into $|\Phi_n\rangle = \hat{U}^n(\tau)|\Phi_0\rangle$ from (5.6), we have

$$|\Phi_n\rangle = \sum_{r=0}^{n} \mu_{r,n} Q_r(\hat{U})|\Phi_0\rangle = \sum_{r=0}^{n} \mu_{r,n}|\psi_r\rangle$$

where the definition $Q_r(\hat{U})|\Phi_0\rangle = |\psi_r\rangle$ from (23.19) is employed, so that

$$|\Phi_n\rangle = \sum_{r=0}^{n} \mu_{r,n}|\psi_r\rangle. \tag{59.2}$$

Hence, the inverse mapping $|\Phi_n\rangle \longrightarrow |\psi_n\rangle$ between the Schrödinger and Lanczos states is carried out by means of the matrix $\boldsymbol{\mu}$ of modified moments $\{\mu_{i,j}\}$

from (55.3). The scalar product between two Schrödinger states $(\Phi_m|$ and $|\Phi_n\rangle$ from (59.2) followed by the use of (22.9) will give

$$(\Phi_m|\Phi_n) = \sum_{r=0}^{n}\sum_{r'=0}^{m} \mu_{r,n}\mu_{r',m}(\psi_{r'}|\psi_r)$$

$$= \sum_{r=0}^{n}\sum_{r'=0}^{m} \mu_{r,n}\mu_{r',m}\{c_0\delta_{r',r}\} = c_0\sum_{r=0}^{n}\mu_{r,n}\mu_{r,m}$$

$$\therefore \quad \sum_{r=0}^{n}\mu_{r,n}\mu_{r,m} = \frac{(\Phi_m|\Phi_n)}{c_0}. \tag{59.3}$$

According to (5.13), we have $(\Phi_m|\Phi_n) = c_{n+m}$ which is the general element $S_{n,m}$ of the Schrödinger overlap matrix $\mathbf{S} = \{S_{n,m}\} = \{c_{n+m}\}$, so that

$$\sum_{r=0}^{n}\mu_{r,n}\mu_{r,m} = \frac{c_{n+m}}{c_0} = \frac{S_{n,m}}{c_0} \tag{59.4}$$

as in (58.5). Our formula (59.4) can also be verified in the opposite direction by using (55.1) for both $\mu_{r,m}$ and $\mu_{r,n}$ together with (30.8) and (31.20)

$$c_0\sum_{r=0}^{n}\mu_{r,m}\mu_{r,n} = c_0\sum_{r=0}^{n}\left\{\frac{1}{c_0}(u^m|Q_r(u))\right\}\left\{\frac{1}{c_0}(u^n|Q_r(u))\right\}$$

$$= \frac{1}{c_0}\sum_{r=0}^{n}\left\{\sum_{k=1}^{K}d_k u_k^n Q_r(u_k)\right\}\left\{\sum_{k'=1}^{K}d_{k'} u_{k'}^m Q_r(u_{k'})\right\}$$

$$= \frac{1}{c_0}\sum_{k=1}^{K}\sum_{k'=1}^{K}d_k d_{k'} u_k^n u_{k'}^m \left\{\sum_{r=0}^{n}Q_r(u_k)Q_r(u_{k'})\right\}$$

$$= \frac{1}{c_0}\sum_{k=1}^{K}\sum_{k'=1}^{K}d_k d_{k'} u_k^n u_{k'}^m \left\{c_0\frac{\delta_{k,k'}}{d_k}\right\} = \sum_{k=1}^{K}d_k u_k^{n+m} = c_{n+m}$$

$$c_0\sum_{r=0}^{n}\mu_{r,m}\mu_{r,n} = \sum_{k=1}^{K}d_k u_k^{n+m} = c_{n+m} = (\Phi_m|\Phi_n) = S_{n,m} \quad \text{(QED)}$$

$$\tag{59.5}$$

in agreement with (5.13) and (59.4). In the above checking of (59.4), we used the completeness relation (31.20) in the form $\sum_{r=0}^{n} Q_r(u_k)Q_r(u_{k'}) = c_0\delta_{k,k'}/d_k$ with the underlying characteristic equation which is here defined by $Q_n(u_k) = 0$. The final result (59.4) is, of course, independent of the eigenvalues $\{u_k\}$. This, in turn, proves the correctness of the inverse transformation (59.2) between the Schrödinger and Lanczos states. Likewise, we can verify the direct transformation (35.2) between the Lanczos and Schrödinger states by retrieving

Mapping of state vectors: Schrödinger ⟷ Lanczos

the orthonormalization (22.9). With this goal, we project $|\psi_n)$ onto $(\psi_m|$ and employ (35.2) for both Lanczos states, together with the subsequent usage of (5.11), (5.13), (26.1) and (31.23) to arrive at

$$(\psi_m|\psi_n) = \sum_{r=0}^{n}\sum_{r'=0}^{m} q_{n,n-r} q_{m,m-r'} (\Phi_{r'}|\Phi_r)$$

$$= \sum_{r=0}^{n}\sum_{r'=0}^{m} q_{n,n-r} q_{m,m-r'} c_{r+r'}$$

$$= \sum_{r=0}^{n}\sum_{r'=0}^{m} q_{n,n-r} q_{m,m-r'} \sum_{k=1}^{K} d_k u_k^{r+r'}$$

$$= \sum_{k=1}^{K} d_k \left\{\sum_{r=0}^{n} q_{n,n-r} u_k^r\right\}\left\{\sum_{r'=0}^{m} q_{m,m-r'} u_k^{r'}\right\}$$

$$= \sum_{k=1}^{K} d_k Q_n(u_k) Q_m(u_k) = c_0 \delta_{n,m}$$

$$\therefore \quad (\psi_m|\psi_n) = c_0 \delta_{n,m} \quad \text{(QED)} \tag{59.6}$$

as in (22.9) for the Lanczos states. This establishes the correctness of (35.2). Moreover, in a mixed scalar product between a Lanczos and a Schrödinger state, we project $|\psi_n)$ onto $(\Phi_m|$ and use (22.9) and (59.2) to deduce

$$(\Phi_m|\psi_n) = \sum_{r=0}^{m} \mu_{r,m}(\psi_r|\psi_n) = \sum_{r=0}^{m} \mu_{r,m}\{c_0 \delta_{r,n}\} = c_0 \mu_{n,m}$$

$$\therefore \quad \frac{(\Phi_m|\psi_n)}{c_0} = \mu_{n,m}. \tag{59.7}$$

Thus, the scalar product $(\Phi_m|\psi_n)/c_0$ between the Schrödinger $(\Phi_m|$ and Lanczos $|\psi_n)$ states is equal to the modified moment $\mu_{n,m}$, which is equivalently defined by (55.1). Using the property (55.2) of $\mu_{n,m}$ for $(m < n)$, it follows

$$(\Phi_m|\psi_n) = 0 \quad (m = 0, 1, 2, \ldots, n-1). \tag{59.8}$$

Hence, every Lanczos state $|\psi_n)$ is orthogonal to each Schrödinger state $(\Phi_m|$ for $0 \leq m \leq n-1$ and this generalizes $(\Phi_0|\psi_n) = 0$ from (34.7). Similarly, inserting (35.2) into $(\Phi_m|\psi_n)$, it follows

$$(\Phi_m|\psi_n) = \sum_{r=0}^{n} q_{n,n-r}(\Phi_m|\Phi_r) = \sum_{r=0}^{n} q_{n,n-r} c_{m+r}$$

$$(\Phi_m|\psi_n) = \sum_{r=0}^{n} c_{m+r} q_{n,n-r}. \tag{59.9}$$

Comparison between (59.7) and (59.9) yields the expression

$$\mu_{n,m} = \frac{1}{c_0} \sum_{r=0}^{n} c_{m+r} q_{n,n-r} \qquad \text{(QED)} \tag{59.10}$$

which confirms the previously obtained formula (58.6).

Chapter 60

The Gram–Schmidt *versus* the Lanczos orthogonalization

As mentioned, the Krylov set $\{|\Phi_n)\}$ is not orthogonal. However, using the standard Gram–Schmidt orthonormalization (GSO), the set $\{|\Phi_n)\}$ can be orthonormalized. Denoting by $\{|\Psi_n)\}$ the new orthonormal set, the GSO applied to $\{|\Phi_n)\}$ gives the results

$$|\Psi_0) = \frac{|\Phi_0)}{(\Phi_0|\Phi_0)^{1/2}}$$

$$|\Psi_1) = \frac{|\Phi_1) - c_0^{-1}(\Psi_0|\Phi_1)|\Psi_0)}{\{c_0^{-1}(\Phi_1|\Phi_1) - c_0^{-2}(\Psi_0|\Phi_1)^2\}^{1/2}}$$

$$|\Psi_2) = \frac{|\Phi_2) - c_0^{-1}[(\Psi_1|\Phi_2)|\Psi_1) + (\Psi_0|\Phi_2)|\Psi_0)]}{\{c_0^{-1}(\Phi_2|\Phi_2) - c_0^{-2}[(\Psi_1|\Phi_2)^2 + (\Psi_0|\Phi_2)^2]\}^{1/2}}. \quad (60.1)$$

Then the nth Gram–Schmidt state, which is the general term $|\Psi_n)$ of the sequence $\{|\Psi_n)\}$ ($n = 0, 1, 2, \ldots$), reads as

$$|\Psi_n) = \tilde{N}_n|\tilde{\Psi}_n) \quad (60.2)$$

$$|\tilde{\Psi}_n) = |\Phi_n) - \sum_{k=0}^{n-1} \tilde{\mu}_{n,k}|\tilde{\Psi}_k) \quad (60.3)$$

$$\tilde{N}_n^{-2} \equiv \|\Psi_n\| = (\Psi_n|\Psi_n) = \frac{c_{2n}}{c_0} - \sum_{k=0}^{n-1} \tilde{\mu}_{n,k}^2 \quad (60.4)$$

$$\tilde{\mu}_{n,m} = \frac{(\Phi_m|\Psi_n)}{c_0} \quad (60.5)$$

where we used (3.3) and (5.13) with $(\Psi_n|\Phi_n) = (\Phi_n|\Psi_n)$ and $(\Phi_n|\Phi_n) = c_{2n}$. In (60.3) the sum $\sum_{k=0}^{n-1} \tilde{\mu}_{n,k}|\tilde{\Psi}_k)$ measures the departure of the non-orthogonal Schrödinger orbital $|\Phi_n)$ from its orthogonalized counterpart $|\tilde{\Psi}_n)$. Specifically, the difference $|\Phi_n\rangle - |\tilde{\Psi}_n\rangle$ is given by a linear combination of n orbitals $\{|\tilde{\Psi}_k\rangle\} (0 \leq k \leq n-1)$ with weights $\tilde{\mu}_{n,k}$ that are proportional to the overlap of $|\Phi_n\rangle$ with $|\tilde{\Psi}_k\rangle$, as per (60.5). For an illustration, it suffices to calculate $\tilde{\mu}_{n,k}$ for a few subscripts only, with the simple results

$$\tilde{\mu}_{0,1} = \frac{c_1}{c_0} \qquad \tilde{\mu}_{0,2} = \frac{c_2}{c_0} \qquad \tilde{\mu}_{1,2} = \frac{c_0 c_3 - c_1 c_2}{c_0 (c_0 c_2 - c_1^2)^{1/2}}. \tag{60.6}$$

Having (60.6), the numerator and the denominator in (60.1) become

$$|\tilde{\Psi}_0\rangle = |\Phi_0\rangle \qquad |\tilde{\Psi}_1\rangle = |\Phi_1\rangle - \frac{c_1}{c_0}|\Phi_0\rangle$$

$$|\tilde{\Psi}_2\rangle = |\Phi_2\rangle - \frac{c_0 c_3 - c_1 c_2}{c_0 c_2 - c_1^2}|\Phi_1\rangle + \frac{c_1 c_3 - c_2^2}{c_0 c_2 - c_1^2}|\Phi_0\rangle. \tag{60.7}$$

It is convenient to set up the auxiliary parameters $\tilde{\alpha}_0, \tilde{\alpha}_1, \tilde{\beta}_1^2$ and $\tilde{\beta}_2^2$ via

$$\tilde{\alpha}_0 = \frac{c_1}{c_0} \qquad \tilde{\alpha}_1 = \frac{c_0^2 c_3 - 2 c_0 c_1 c_2 + c_1^3}{c_0(c_0 c_2 - c_1^2)}$$

$$\tilde{\beta}_1^2 = \frac{c_0 c_2 - c_1^2}{c_0^2} \qquad \tilde{\beta}_2^2 = c_0 \frac{2 c_1 c_2 c_3 + c_0 c_2 c_4 - c_0 c_3^2 - c_1^2 c_4 - c_2^3}{(c_0 c_2 - c_1^2)^2} \tag{60.8}$$

from which the following relationships are readily established

$$\tilde{\alpha}_0 + \tilde{\alpha}_1 = \frac{c_0 c_3 - c_1 c_2}{c_0 c_2 - c_1^2} \qquad \tilde{\alpha}_0 \tilde{\alpha}_1 - \tilde{\beta}_1^2 = \frac{c_1 c_3 - c_2^2}{c_0 c_2 - c_1^2} \tag{60.9}$$

$$\tilde{\mu}_{0,1} = \alpha_0 \qquad \tilde{\mu}_{0,2} = \frac{c_2}{c_0} \qquad \tilde{\mu}_{1,2} = \frac{c_0 c_3 - c_1 c_2}{c_0^2 \beta_1}. \tag{60.10}$$

The new forms of (60.4) and (60.7) via the parameters from (60.8) are

$$|\tilde{\Psi}_0\rangle = |\Phi_0\rangle \qquad |\tilde{\Psi}_1\rangle = |\Phi_1\rangle - \tilde{\alpha}_0|\Phi_0\rangle$$
$$|\tilde{\Psi}_2\rangle = |\Phi_2\rangle - (\tilde{\alpha}_0 + \tilde{\alpha}_1)|\Phi_1\rangle + (\tilde{\alpha}_0 \tilde{\alpha}_1 - \tilde{\beta}_1^2)|\Phi_0\rangle \tag{60.11}$$
$$\tilde{N}_1^{-2} = \frac{c_2}{c_0} - \tilde{\mu}_{0,1}^2 = \tilde{\beta}_1^2 \qquad \tilde{N}_2^{-2} = \frac{c_4}{c_0} - \tilde{\mu}_{0,0}^2 - \tilde{\mu}_{1,2}^2 = \tilde{\beta}_1^2 \tilde{\beta}_2^2. \tag{60.12}$$

We see from (60.12) that the parameters $\tilde{\beta}_1$ and $\tilde{\beta}_2$ are the norms of orbitals $|\tilde{\Psi}_1\rangle$ and $|\tilde{\Psi}_2\rangle$, respectively. The Gram–Schmidt states $\{|\tilde{\Psi}_n\rangle\}$ are orthogonal, but unnormalized, in contrast to the corresponding orthonormalized states $\{|\Psi_n\rangle\}$. Inserting (60.7) and (60.12) into (60.4) we have

$$|\Psi_0\rangle = |\Phi_0\rangle \qquad |\Psi_1\rangle = \frac{1}{\tilde{\beta}_1}\{|\Phi_1\rangle - \tilde{\alpha}_0|\Phi_0\rangle\}$$

$$|\Psi_2\rangle = \frac{1}{\tilde{\beta}_1 \tilde{\beta}_2}\left\{|\Phi_2\rangle - \frac{c_0 c_3 - c_1 c_2}{c_0 c_2 - c_1^2}|\Phi_1\rangle - \frac{c_1 c_3 - c_2^2}{c_0 c_2 - c_1^2}|\Phi_0\rangle\right\}$$

$$= \frac{1}{\tilde{\beta}_1 \tilde{\beta}_2}\{|\Phi_2\rangle - (\tilde{\alpha}_0 + \tilde{\alpha}_1)|\Phi_1\rangle + (\tilde{\alpha}_0 \tilde{\alpha}_1 - \tilde{\beta}_1^2)|\Phi_0\rangle\}. \tag{60.13}$$

It is seen from (60.3) that the general Gram–Schmidt state $|\Psi_n)$ is computed recursively. Moreover, each of these states $|\Psi_n)$ is expressed as a linear combination of all the previously computed vectors $\{|\Psi_k)\}$ ($0 \le k \le n-1$). In other words, the whole string of the computed Gram–Schmidt states have to be kept in the computer memory throughout the recursion. Therefore, a more economical recursion is sought. Recursion (60.3) is not realistic from a physics viewpoint either, since it dictates that each given state $|\Psi_n)$ should be coupled with all its predecessors $\{|\Psi_k)\}$ ($0 \le k \le n-1$). However, in reality, any physical state $|\Psi_n)$ is anticipated to interact significantly only with a narrow band of the surounding vectors $\{|\Psi_k)\}$ ($k \in [k_{min}, k_{max}]$), whereas couplings with more distant states $\{|\Psi_k)\}$ ($k \notin [k_{min}, k_{max}]$) should be relatively less important. Yet, the goal is to achieve this without any approximation. In other words, we are looking for a way to implement a notion of local interaction of orbitals. Through such interactions only a few nearest orbitals would appear explicitly as being active, whereas the others would be disguised as dormant, i.e. latent or pending for potential interactions. In other words, the idea is to somehow condense the distant orbitals into a kind of a generating orbital. This latter device, in its 'undeveloped' form, would display explicitly only a very few orbitals at a time. However, when 'developed', the generating orbital would show that it implicitly also contains the distant orbitals.

One of the ways to obtain the simultaneous appearance of a few active orbitals at a time and still hold all the distant orbitals through a generating orbital is to use the evolution operator $\hat{U}(\tau) \equiv \hat{U}$ from the definition of the Schrödinger states $|\Phi_n) = \hat{U}^n|\Phi_0) \equiv \hat{U}^n(\tau)|\Phi_0)$, as per (8.2). Then the interaction of $|\Psi_0)$ with distant orbitals $|\Psi_n)$ for higher values of the subscript n would be implicit in the repeated action of the 'generator' $\hat{U}$ onto $|\Psi_m)$ with $m < n$. To show how this concept of local orbitals is implemented in the GSO, we take the developed form of the original GSOs from (60.7) with all the Schrödinger (Krylov) orbitals present explicitly and try to 'hide' the orbitals that are more distant from $|\Psi_0)$, as follows

$$|\tilde{\Psi}_1) = \tilde{\beta}_1|\Psi_1) = |\Phi_1) - \tilde{\alpha}_0|\Phi_0) = \hat{U}|\Phi_0) - \tilde{\alpha}_0|\Phi_0) = (\hat{U} - \tilde{\alpha}_0\hat{1})|\Phi_0)$$
$$\therefore \quad \tilde{\beta}_1|\Psi_1) = (\hat{U} - \tilde{\alpha}_0\hat{1})|\Psi_0) \quad |\Psi_0) = |\Phi_0). \quad (60.14)$$
$$\tilde{\beta}_1\tilde{\beta}_2|\Psi_2) = |\Phi_2) - (\tilde{\alpha}_0 + \tilde{\alpha}_1)|\Phi_1) + (\tilde{\alpha}_0\tilde{\alpha}_1 - \tilde{\beta}_1^2)|\Phi_0)$$
$$= \{|\Phi_2) - \tilde{\alpha}_0|\Phi_1)\} - \tilde{\alpha}_1\{|\Phi_1) - \tilde{\alpha}_0|\Phi_0)\} - \tilde{\beta}_1^2|\Phi_0)$$
$$= \hat{U}\{|\Phi_1) - \tilde{\alpha}_0|\Phi_0)\} - \tilde{\alpha}_1\tilde{\beta}_1|\Psi_1) - \tilde{\beta}_1^2|\Phi_0)$$
$$= \tilde{\beta}_1\hat{U}|\Psi_1) - \tilde{\beta}_1\{\tilde{\alpha}_1|\Psi_1) + \tilde{\beta}_1|\Phi_0)\}$$
$$\therefore \quad \tilde{\beta}_2|\Psi_2) = (\hat{U} - \tilde{\alpha}_1\hat{1})|\Psi_1) - \tilde{\beta}_1|\Psi_0) \quad |\Psi_0) = |\Phi_0). \quad (60.15)$$

With these relations, it is readily verified that the parameters $\tilde{\alpha}_0, \tilde{\alpha}_1, \tilde{\beta}_1$ and $\tilde{\beta}_2$ introduced provisionally by (60.8) can now be cast into their equivalent, but more structured and, hence, more useful forms

$$\tilde{\alpha}_0 = \frac{(\Psi_0|\hat{U}|\Psi_0)}{(\Psi_0|\Psi_0)} \qquad \tilde{\alpha}_1 = \frac{(\Psi_1|\hat{U}|\Psi_1)}{(\Psi_1|\Psi_1)}$$
$$\tilde{\beta}_1 = \frac{(\Psi_0|\hat{U}|\Psi_1)}{(\Psi_0|\Psi_0)} \qquad \tilde{\beta}_2 = \frac{(\Psi_1|\hat{U}|\Psi_2)}{(\Psi_1|\Psi_1)}. \qquad (60.16)$$

Therefore, from (60.12), (60.15) and (60.16), we can infer the general results

$$\tilde{\beta}_{n+1}|\Psi_{n+1}) = (\hat{U} - \tilde{\alpha}_n \hat{1})|\Psi_n) - \tilde{\beta}_n|\Psi_{n-1}) \qquad |\Psi_0) = |\Phi_0) \qquad (60.17)$$

$$\tilde{\alpha}_n = \frac{(\Psi_n|\hat{U}|\psi_n)}{(\Psi_n|\Psi_n)} \qquad \tilde{\beta}_n = \frac{(\Psi_{n-1}|\hat{U}|\Psi_n)}{(\Psi_{n-1}|\Psi_{n-1})} \qquad \tilde{\beta}_0 = 0 \qquad (60.18)$$

$$\tilde{N}_n^{-2} = \frac{c_{2n}}{c_0} - \sum_{k=0}^{n-1} \mu_{n,k}^2 = \tilde{\beta}_1^2 \tilde{\beta}_2^2 \cdots \tilde{\beta}_n^2. \qquad (60.19)$$

We see that 'hiding' the higher-order orbitals is achieved through the introduction of the generating orbital $\hat{U}|\Psi_n)$ which implicitly holds all the $n-1$ Schrödinger orbitals from the orignal GSO. Moreover, at any time only three orbitals $|\Psi_{n\pm 1})$ and $|\Psi_n)$ are strongly coupled in (60.17), while all the remaining ones are clustered into the generator of interactions, $\hat{U}|\Psi_n)$. This translates directly into a huge saving of the computer memory. Namely, instead of holding all the vectors $\{|\Phi_n)\}$ in the memory as in the original GSO from (60.13), the corresponding economised GSO from (60.17) needs to keep only three orbitals at a time.

A comparison of the particular results from (60.8), (60.10) and (60.11) with (22.5), (22.6) and (55.15) yields the identities

$$|\tilde{\Psi}_0) = |\tilde{\psi}_0) \qquad |\tilde{\Psi}_1) = |\tilde{\psi}_1) \qquad |\tilde{\Psi}_2) = |\tilde{\psi}_2)$$
$$|\Psi_0) = |\psi_0) \qquad |\Psi_1) = |\psi_1) \qquad |\Psi_2) = |\psi_2) \qquad (60.20)$$
$$\tilde{\alpha}_1 = \alpha_1 \qquad \tilde{\alpha}_2 = \alpha_2 \qquad \tilde{\beta}_1 = \beta_1 \qquad \tilde{\beta}_2 = \beta_2 \qquad (60.21)$$
$$\tilde{\mu}_{0,1} = \mu_{0,1} \qquad \tilde{\mu}_{0,2} = \mu_{0,2} \qquad \tilde{\mu}_{1,2} = \mu_{1,2}. \qquad (60.22)$$

Likewise, matching the expressions (60.6), (60.17) and (60.18) with the appropriate formulae (22.2), (22.4) and (55.15) gives the general relationships

$$|\tilde{\Psi}_n) = |\tilde{\psi}_n) \qquad |\Psi_n) = |\psi_n) \qquad (60.23)$$
$$\tilde{\alpha}_n = \alpha_n \qquad \tilde{\beta}_n = \beta_n \qquad (60.24)$$
$$\tilde{\mu}_{n,k} = \mu_{n,k}. \qquad (60.25)$$

Here, α_n and β_n are the Lanczos coupling parameters (22.2), whereas $\mu_{n,k}$ is the modified moment (59.7). This shows that the original Gram–Schmidt $|\Psi_n)$ and the Lanczos states $|\psi_n)$ are identical to each other in their explicitly developed respective forms. The same conclusion applies also to the corresponding unnormalized states $|\tilde{\Psi}_n)$ and $|\tilde{\psi}_n)$. Of course, such a conclusion is expected,

The Gram–Schmidt versus the Lanczos orthogonalization

since no matter which orthogonalization is used (Gram–Schmidt, Lanczos, etc), the final result must be the same. This implies that the *explicit* Lanczos states could be now written as

$$|\psi_n) = N_n |\tilde{\psi}_n) \qquad (60.26)$$

$$|\tilde{\psi}_n) = |\Phi_n) - \sum_{k=0}^{n-1} \mu_{n,k} |\tilde{\psi}_k) \qquad (60.27)$$

$$N_n^{-2} \equiv \|\psi_n\| = (\psi_n|\psi_n) = \frac{c_{2n}}{c_0} - \sum_{k=0}^{n-1} \mu_{n,k}^2 = \beta_1^2 \beta_2^2 \cdots \beta_n^2. \qquad (60.28)$$

Of course, this is only of a theoretical interest, not a practical one, for the same reason for which the GSO from (60.3) is not used in extensive computations. Observe specially that the recursions (22.1) and (60.17) are identical to each other. This complete equivalence indicates clearly in which sense the Lanczos recursive algorithm can be conceived as the economised Gram–Schmidt orthonormalization.

Our non-standard derivation of the Lanczos recurive algorithm directly from the Gram–Schmidt orthogonalization is pedagogically instructive, since the former method is usually not taught in regular physics courses. At such courses, the orthogonalization concept is invariably presented as the original Gram–Schmidt process. Moreover, no mention is customarily made to physics students that this standard GSO is of an exclusively theoretical interest with no real practical value, due to exhaustive computer memory demands. On the other hand, theoretical physicists consider the Lanczos algorithm of nearest orbitals as a purely research subject, which is unavailable in mathematical physics courses. One of the ways to bridge this gap is suggested in the present chapter from which the Lanczos algorithm is identified to be the corresponding economised version of the usual Gram–Schmidt orthogonalization.

Chapter 61

Analytical inversion of the Schrödinger overlap matrix

Here, we shall illustrate the usefulness of the methodology of the mappings from chapters 58 and 59. To this end, we take the mth derivative of the power function u^n from (58.1) to write

$$m!u^{n-m} = \sum_{r=0}^{n} \mu_{r,n} Q_{r,m}(u) \qquad (61.1)$$

where $Q_{r,m}(u)$ is the mth derivative of the Lanczos polynomial $Q_r(u)$, as per (24.2). Next we evaluate both sides of (61.1) at $u = 0$. It then follows

$$\lim_{u \to 0} u^{n-m} = \delta_{n,m}. \qquad (61.2)$$

This result reduces (61.1) to

$$\delta_{n,m} = \frac{1}{m!} \sum_{r=0}^{n} \mu_{r,n} Q_{r,m}(0) \qquad (61.3)$$

where $Q_{r,m}(0)$ is defined in (24.3). Using the relationship (26.10) between $Q_{r,m}(0)$ and $q_{r,r-m}$ the following equivalent expression for (61.3) is deduced

$$\delta_{n,m} = \sum_{r=0}^{n} \mu_{r,n} q_{r,r-m}. \qquad (61.4)$$

The matrices $\mathbf{q}$ and $\boldsymbol{\mu}$ from (26.4) and (55.3), respectively, are of the dimension $(n+1) \times (n+1)$. Thus setting $\mathbf{q}_{n+1} \equiv \mathbf{q}$, $\boldsymbol{\mu}_{n+1} \equiv \boldsymbol{\mu}$ and $\mathbf{1}_{n+1} \equiv \mathbf{1}$, we can write the matrix form of (61.4) as

$$\boldsymbol{\mu}\mathbf{q} = \mathbf{1} \qquad \mathbf{1} = \{\delta_{i,j}\} \qquad \boldsymbol{\mu} = \{\mu_{i,j}\} \qquad \mathbf{q} = \{q_{i,j}\} \qquad i,j \in [0,n]. \quad (61.5)$$

Analytical inversion of the Schrödinger overlap matrix

Therefore, the matrix **q** from (26.4) represents the inverse μ^{-1} of the matrix μ from (55.3)

$$\mu^{-1} = \mathbf{q}. \tag{61.6}$$

This is expected from (26.4) and (55.3) where **q** and μ are seen to be the lower and the upper triangular matrices, respectively. Eliminating $\mu_{r,n}$ from (61.4) with the help of (58.6), we have

$$\begin{aligned}
\delta_{n,m} &= \sum_{r=0}^{n} \mu_{r,n} q_{r,r-m} = \sum_{r=0}^{n} \left\{ \frac{1}{c_0} \sum_{k=0}^{r} c_{n+k} q_{r,r-k} \right\} q_{r,r-m} \\
&= \sum_{r=0}^{n} \left\{ \frac{1}{c_0} \left[\sum_{k=0}^{r} c_{n+k} q_{r,r-k} + \sum_{k=r+1}^{n} c_{n+k} q_{r,r-k} \right] \right\} q_{r,r-m} \\
&= \sum_{r=0}^{n} \left\{ \frac{1}{c_0} \sum_{k=0}^{n} c_{n+k} q_{r,r-k} \right\} q_{r,r-m} \\
&= \sum_{k=0}^{n} c_{n+k} \left\{ \frac{1}{c_0} \sum_{r=0}^{n} q_{r,r-k} q_{r,r-m} \right\}. \tag{61.7}
\end{aligned}$$

In the second line of (61.7), the superficially added sum $\sum_{k=r+1}^{n} c_{n+k} q_{r,r-k}$ is equal to zero, due to the property $q_{r,r-k} = 0$ for $k > r$, as stated in (26.3). This permits the extension of the sum over k to n, so that the upper limit becomes independent of r. In such a way, the sums over r and k can exchange their places, so that finally the end of (61.7) can be written as

$$\sum_{k=0}^{n} S_{n,k}^{(n+1)} T_{k,m}^{(n+1)} = \delta_{n,m}. \tag{61.8}$$

Here $\{S_{i,j}^{(n+1)}\}$ are the elements of the Schrödinger overlap matrix from (5.13), but with the dimension $(n+1) \times (n+1)$, i.e. $\{S_{i,j}^{(n+1)}\} = \{c_{i+j}\}_{i,j=0}^{n}$. Moreover, the element $T_{i,j}^{(n+1)}$ is identified from (61.7) as

$$T_{k,m}^{(n+1)} \equiv \frac{1}{c_0} \sum_{r=0}^{n} q_{r,r-k} q_{r,r-m}. \tag{61.9}$$

We shall now return to the original $n \times n$ dimension of the Schrödinger overlap matrix (5.13), and write simply $\{S_{i,j}^{(n)}\} \equiv \{S_{i,j}\} = \{c_{i+j}\}_{i,j=0}^{n-1}$ and $\{T_{i,j}^{(n)}\} \equiv \{T_{i,j}\}_{i,j=0}^{n-1}$. Then the matrix form of (61.8) can be given in a way which invokes only $n \times n$ matrices $\mathbf{S}_n \equiv \mathbf{S}$, $\mathbf{T}_n \equiv \mathbf{T}$ and $\mathbf{1}_n \equiv \mathbf{1}$

$$\mathbf{ST} = \mathbf{1} \Longrightarrow \mathbf{S}^{-1} = \mathbf{T} \tag{61.10}$$

$$\mathbf{S} = \{S_{\nu,k}\}_{\nu,k=0}^{n-1} \qquad \mathbf{T} = \{T_{k,m}\}_{k,m=0}^{n-1}$$

$$S_{\nu,k} = c_{\nu+k} \qquad T_{k,m} = \frac{1}{c_0} \sum_{r=0}^{n-1} q_{r,r-k} q_{r,r-m}. \tag{61.11}$$

Hence, the matrix $\mathbf{T}_n$ represents the inverse $\mathbf{S}_n^{-1}$ of the Schrödinger overlap matrix $\mathbf{S}_n$. The general matrix element $T_{k,m}$ from (61.11) is given in terms of the expansion coefficients $\{q_{n,n-r}\}$ of the power series representation (26.1) of the characteristic Lanczos polynomial of the first kind $Q_n(u)$. Finally, we shall list several explicit results for the matrix elements $\{T_{k,m}\}$ of the inverse matrix $\mathbf{S}^{-1}$ by employing (61.11) along with the conventional calculation based upon the corresponding standard formulae of linear algebra [228] (see also Appendix A). For example, using (26.6), (26.12) and (61.11), we obtain the explicit results for $T_{j,k} \equiv T_{j,k}^{(n)}$ with $n \leq 4$. A few of these findings are

$$T_{0,0}^{(1)} = \frac{1}{c_0} \qquad T_{0,1}^{(2)} = -\frac{c_1}{c_0 c_2 - c_1^2} \qquad T_{1,1}^{(2)} = \frac{c_0}{c_0 c_2 - c_1^2}$$

$$T_{0,0}^{(3)} = \frac{c_2 c_4 - c_3^2}{2 c_1 c_2 c_3 + c_0 c_2 c_4 - c_0 c_3^2 - c_1^2 c_4 - c_2^3}$$

$$T_{0,1}^{(3)} = \frac{-(c_1 c_4 - c_2 c_3)}{2 c_1 c_2 c_3 + c_0 c_2 c_4 - c_0 c_3^2 - c_1^2 c_4 - c_2^3}$$

$$T_{1,1}^{(3)} = \frac{(c_0 c_4 - c_2^2)}{2 c_1 c_2 c_3 + c_0 c_2 c_4 - c_0 c_3^2 - c_1^2 c_4 - c_2^3}$$

$$T_{0,2}^{(3)} = \frac{(c_1 c_3 - c_2^2)}{2 c_1 c_2 c_3 + c_0 c_2 c_4 - c_0 c_3^2 - c_1^2 c_4 - c_2^3}$$

$$T_{1,2}^{(3)} = \frac{-(c_0 c_3 - c_1 c_2)}{2 c_1 c_2 c_3 + c_0 c_2 c_4 - c_0 c_3^2 - c_1^2 c_4 - c_2^3}$$

$$T_{2,2}^{(3)} = \frac{(c_0 c_2 - c_1^2)}{2 c_1 c_2 c_3 + c_0 c_2 c_4 - c_0 c_3^2 - c_1^2 c_4 - c_2^3}. \tag{61.12}$$

These results can be rewritten in their alternative forms that involve the quotients of the determinants with the time signal points $\{c_n\}$

$$T_{0,0}^{(3)} = \frac{\begin{vmatrix} c_2 & c_3 \\ c_3 & c_4 \end{vmatrix}}{\begin{vmatrix} c_0 & c_1 & c_2 \\ c_1 & c_2 & c_3 \\ c_2 & c_3 & c_4 \end{vmatrix}} \qquad T_{0,1}^{(3)} = -\frac{\begin{vmatrix} c_1 & c_3 \\ c_2 & c_4 \end{vmatrix}}{\begin{vmatrix} c_0 & c_1 & c_2 \\ c_1 & c_2 & c_3 \\ c_2 & c_3 & c_4 \end{vmatrix}} \qquad T_{1,1}^{(3)} = \frac{\begin{vmatrix} c_0 & c_2 \\ c_2 & c_4 \end{vmatrix}}{\begin{vmatrix} c_0 & c_1 & c_2 \\ c_1 & c_2 & c_3 \\ c_2 & c_3 & c_4 \end{vmatrix}}$$

Analytical inversion of the Schrödinger overlap matrix 331

$$T_{0,2}^{(3)} = \frac{\begin{vmatrix} c_1 & c_2 \\ c_2 & c_3 \end{vmatrix}}{\begin{vmatrix} c_0 & c_1 & c_2 \\ c_1 & c_2 & c_3 \\ c_2 & c_3 & c_4 \end{vmatrix}} \qquad T_{1,2}^{(3)} = -\frac{\begin{vmatrix} c_0 & c_1 \\ c_2 & c_3 \end{vmatrix}}{\begin{vmatrix} c_0 & c_1 & c_2 \\ c_1 & c_2 & c_3 \\ c_2 & c_3 & c_4 \end{vmatrix}} \qquad T_{2,2}^{(3)} = \frac{\begin{vmatrix} c_0 & c_1 \\ c_1 & c_2 \end{vmatrix}}{\begin{vmatrix} c_0 & c_1 & c_2 \\ c_1 & c_2 & c_3 \\ c_2 & c_3 & c_4 \end{vmatrix}}$$

and so forth. The above expressions for $T_{j,k}^{(n)}$ can be put into another equivalent form of the type

$$T_{0,0}^{(3)} = \frac{\det S_{0,0}^{(3)}}{\det S_3} = \frac{\operatorname{cof} S_{0,0}^{(3)}}{\det S_3} \qquad T_{0,1}^{(3)} = -\frac{\det S_{0,1}^{(3)}}{\det S_3} = \frac{\operatorname{cof} S_{0,1}^{(3)}}{\det S_3}$$

$$T_{1,1}^{(3)} = \frac{\det S_{1,1}^{(3)}}{\det S_3} = \frac{\operatorname{cof} S_{1,1}^{(3)}}{\det S_3} \qquad T_{0,2}^{(3)} = \frac{\det S_{0,2}^{(3)}}{\det S_3} = \frac{\operatorname{cof} S_{0,2}^{(3)}}{\det S_3}$$

$$T_{1,2}^{(3)} = -\frac{\det S_{1,2}^{(3)}}{\det S_3} = \frac{\operatorname{cof} S_{1,2}^{(3)}}{\det S_3} \qquad T_{2,2}^{(3)} = \frac{\det S_{2,2}^{(3)}}{\det S_3} = \frac{\operatorname{cof} S_{2,2}^{(3)}}{\det S_3} \qquad (61.13)$$

where $\det S_{j,k}^{(n)}$ is the (j,k)th minor determinant at the position of the element $S_{j,k} = S_{j,k}^{(n)}$ of matrix $\mathbf{S}_n = \{S_{j,k}\}_{j,k=0}^{n-1}$. By definition, the (j,k)th minor determinant of $\det \mathbf{S}_n$ is obtained by deleting the row j and the column k from $\det \mathbf{S}_n$ (see Appendix A)

$$\det S_{j,k}^{(n)} = \begin{vmatrix} c_{0,0} & c_{0,1} & \cdots & c_{0,j-1} & c_{0,j+1} & \cdots & c_{0,n-1} \\ c_{1,0} & c_{1,1} & \cdots & c_{1,j-1} & c_{1,j+1} & \cdots & c_{1,n-1} \\ \vdots & \vdots & \vdots & \vdots & \vdots & \vdots & \cdots \\ c_{i-1,0} & c_{i-1,1} & \cdots & c_{i-1,j-1} & c_{i-1,j+1} & \cdots & c_{i-1,n-1} \\ c_{i+1,0} & c_{i+1,1} & \cdots & c_{i+1,j-1} & c_{i+1,j+1} & \cdots & c_{i+1,n-1} \\ \vdots & \vdots & \vdots & \vdots & \vdots & \ddots & \cdots \\ c_{n-1,0} & c_{n-1,1} & \cdots & c_{n-1,i-1} & c_{n-1,j+1} & \cdots & c_{n-1,n-1} \end{vmatrix}$$

(61.14)

where $c_{j,k}$ is the $(j+k)$th signal point $c_{j,k} \equiv c_{j+k}$. In the above expressions for $T_{j,k}^{(n)}$, the label $\operatorname{cof} S_{j,k}^{(n)}$ represents the (j,k)th cofactor at the position of the element $S_{j,k} = c_{j+k}$

$$\operatorname{cof} \mathbf{S}_n = \{\operatorname{cof} S_{j,k}^{(n)}\}_{j,k=0}^{n-1} \qquad \operatorname{cof} S_{j,k}^{(n)} = (-1)^{j+k} \det S_{j;k}^{(n)}. \qquad (61.15)$$

Thus, the cofactor determinant $\operatorname{cof} S_{j,k}^{(n)}$ is the minor determinant $\det S_{j,k}^{(n)}$ multiplied by the sign factor $(-1)^{j+k}$. We extended inductively this procedure beyond $n = 4$ to arrive at the following general result for the matrix elements $T_{j,k}^{(n)}$ from (61.11)

$$\frac{1}{c_0} \sum_{r=0}^{n-1} q_{r,r-j} q_{r,r-k} = \frac{\operatorname{cof} S_{j,k}^{(n)}}{\det \mathbf{S}_n}. \qquad (61.16)$$

The rhs of (61.16) suggests a direct link to S_n^{-1} due to the definition of the inverse matrix [228]

$$S_n^{-1} = \frac{(\text{cof } S_n)^T}{\det S_n}. \tag{61.17}$$

In (61.17) the term $(\text{cof } S_n)^T$ is the transpose matrix of the matrix $\text{cof } S_n$ (see Appendix A). The transpose matrix of the cofactor matrix of the matrix S_n is the *adjugate* of S_n. Of course, the transpose matrix of the matrix $S_n = \{S_{j,k}\}_{j,k=0}^{n-1}$ is obtained by exchanging the positions of rows and columns. But the determinants of the transpose matrix S_n^T and that of the matrix S_n are equal

$$S_n^T = \{S_{k,j}\}_{k,j=1}^{n-1} \qquad (\det S_n)^T = \det S_n$$

$$\therefore \quad S_n^{-1} = \left(\frac{\text{cof } S_n}{\det S_n}\right)^T. \tag{61.18}$$

However, the rhs of (61.16) is invariant under the exchange of rows and columns $j \longleftrightarrow k$

$$(\text{cof } S_n)^T = \text{cof } S_n \Longrightarrow S_n^{-1} = \frac{\text{cof } S_n}{\det S_n}. \tag{61.19}$$

Here, the elements of the matrix $\text{cof } S_n/\det S_n$ are identified using (61.15) through the expression

$$\frac{\text{cof } S_n}{\det S_n} = \left\{\frac{\text{cof } S_{j,k}^{(n)}}{\det S_n}\right\}. \tag{61.20}$$

The general element $\text{cof } S_{j,k}^{(n)}/\det S_n$ from the set $\{\text{cof } S_{j,k}^{(n)}/\det S_n\}$ on the rhs of (61.20) coincides with the rhs of (61.16), which is equal to the matrix element $T_{j,k}^{(n)}$. Therefore, given the overlap matrix S_n, or equivalently, the Hankel matrix $H_n(c_0)$ defined in (5.16), the inverse $S_n^{-1} = H_n^{-1}(c_0)$ is

$$S_n^{-1} = H_n^{-1}(c_0) = \begin{pmatrix} c_0 & c_1 & c_2 & \cdots & c_{n-1} \\ c_1 & c_2 & c_3 & \cdots & c_n \\ c_2 & c_3 & c_4 & \cdots & c_{n+1} \\ \vdots & \vdots & \vdots & \ddots & \vdots \\ c_{n-1} & c_n & c_{n+1} & \cdots & c_{2n-2} \end{pmatrix}^{-1}$$

$$= \{T_{j,k}\}_{j,k=0}^{n-1} \tag{61.21}$$

where the elements $T_{j,k} \equiv T_{j,k}^{(n)}$ are given in (61.11). The Lanczos expansion coefficients $\{q_{n,n-r}\}$ that are needed in $T_{j,k}$, as stated in (61.11), can be computed efficiently from their own recursion (26.3) or, equivalently, via the recursion (24.4) for the derivatives $Q_{n,m}(u)$. In the latter case, it follows at once that (61.9) takes the Geronimus form

$$T_{j,k}^{(n)} = \frac{1}{c_0} \frac{1}{j!k!} \sum_{r=0}^{n-1} Q_{r,j}(0) Q_{r,k}(0) \tag{61.22}$$

where use is made of the relationship $q_{n,n-m} = (1/m!) Q_{n,m}(0)$ from (26.10). The elements $\{T_{j,k}^{(n)}\}$ of the inverse Schrödinger or Krylov matrix $\mathbf{S}^{-1}$ in the form (61.22) have been first reported in 1929 by Geronimus [224] who used a variational principle for a quadratic Hankel form as analysed in the next chapter.

Chapter 62

A variational principle for a quadratic Hankel form

We shall now study another analytical method for inversion of the Schrödinger or Krylov overlap matrix $\mathbf{S}_n$ using a functional $\mathcal{B}$ defined by the following scalar product in the Lanczos space $\mathcal{L}$

$$\mathcal{B} = (A_{n-1}(u)|A_{n-1}(u)) \tag{62.1}$$

with $A_n(u)$ being any nth order polynomial

$$A_{n-1}(u) = \sum_{j=0}^{n-1} a_j u^j \tag{62.2}$$

where the expansion coefficients $\{a_j\}_{j=0}^{n-1}$ are unknown. Inserting (62.2) into (62.1) and using (58.3) leads to the following quadratic form

$$\mathcal{B} = \sum_{j=0}^{n-1}\sum_{k=0}^{n-1} c_{j+k} a_j a_k. \tag{62.3}$$

We shall determine the coefficients $\{a_j\}$ by setting up the extremum principle of Geronimus [224] and searching for the minimum of $\mathcal{B}$ under the constraint

$$a_k = 1. \tag{62.4}$$

In this variational principle, equating the derivatives of $\mathcal{B}$ with respect to the polynomial expansion coefficients, will immediately yield the result

$$a_j = \left[(-1)^{j+k}\frac{\det \mathbf{S}_{n;jk}}{\det \mathbf{S}_n}\right]\mathcal{B} \tag{62.5}$$

or equivalently

$$a_j = \left[\frac{\operatorname{cof} \mathbf{S}_{n;jk}}{\det \mathbf{S}_n}\right]\mathcal{B} \tag{62.6}$$

where $\det\mathbf{S}_n$ is the Schrödinger determinant (5.16) and $\det\mathbf{S}_{n;jk}$ is the (j,k)th minor (61.14) of $\det\mathbf{S}_n$. The relationship (61.15) between the cofactor $\mathrm{cof}\,\mathbf{S}_{n;jk}$ and $\det\mathbf{S}_{n;jk}$ is also used in (62.5). This procedure might be considered as being completed if we could eliminate $\mathcal{B}$ from the rhs of (62.5). To this end, we use the Lanczos orthonormal polynomials $\{Q_r(u)\}$ to expand $A_{n-1}(u)$

$$A_{n-1}(u) = \sum_{r=0}^{n-1} b_r Q_r(u) \tag{62.7}$$

where the new coefficients b_r are unknown. The rhs of (62.7) can be written explicitly as a polynomial in powers of u using the Maclaurin expansion

$$A_{n-1}(u) = \sum_{r=0}^{n-1} b_r \sum_{m=0}^{r} Q_{r,m}(0)\frac{u^m}{m!} \tag{62.8}$$

where $Q_{r,m}(u) = (d/du)^m Q_r(u)$, as in (24.2). The uniqueness theorem of a polynomial expansion, asserts that the coefficients multiplying the like powers of u in (62.2) and (62.8) must be equal. This establishes the following simple relationship between the sets of the expansion coefficients in (62.2) and (62.8)

$$a_j = \frac{1}{j!}\sum_{r=j}^{n-1} b_r Q_{r,j}(0) \tag{62.9}$$

where $Q_{n,m}(0)$ is from (24.3). Inserting the expansion (62.7) into (62.1) and using the orthonormality condition (31.20), we have

$$\mathcal{B} = \sum_{r=0}^{n-1} b_r^2. \tag{62.10}$$

We can determine the rhs of (62.7) by minimizing $\mathcal{B}$ from (62.10) with respect to variations of the coefficients $\{b_r\}$. In so doing, the uniqueness of $\mathcal{B}$, must be preserved, so that the new minimization ought to be accomplished with the same constraint (62.4), which combined with (62.9), gives

$$\frac{1}{k!}\sum_{r=0}^{n-1} b_r Q_{r,k}(0) = 1. \tag{62.11}$$

Therefore, imposing the extremum condition on $\mathcal{B}$ under the constraint (62.11) will determine the general coefficient b_r as

$$b_r = \frac{\mathcal{B}}{k!} Q_{r,k}(0). \tag{62.12}$$

It is apparent that the rhs of this equation depends explicitly upon the index k, and the k-dependence is the consequence of the constraint (62.4). Therefore, the lhs of (62.12) must depend implicitly upon the same index k via, e.g. $b_r \equiv b_r^{(k)}$, but omitting the superscript in b_r from (62.12) should not cause any confusion. Substituting the result (62.12) into (62.7) yields

$$A_{n-1}(u) = \frac{B}{k!} \sum_{r=0}^{n-1} Q_{r,k}(0) Q_r(u). \tag{62.13}$$

Thus, using (62.2) and (62.13), we can identify the coefficient a_j as

$$a_j = \frac{B}{j!k!} \sum_{r=0}^{n-1} Q_{r,j}(0) Q_{r,k}(0). \tag{62.14}$$

On comparing (62.5) and (62.14), it follows

$$\left[\frac{\operatorname{cof} \mathbf{S}_{n;jk}}{\det \mathbf{S}_n}\right] B = \frac{B}{j!k!} \sum_{r=0}^{n-1} Q_{r,j}(0) Q_{r,k}(0). \tag{62.15}$$

Here, we see that the functional B cancels out, as it was set to prove, and this permits the extraction of the sought quotient $\operatorname{cof} \mathbf{S}_{n;jk}/\det \mathbf{S}_n$ in the form

$$\frac{\operatorname{cof} \mathbf{S}_{n;jk}}{\det \mathbf{S}_n} = \frac{1}{j!k!} \sum_{r=0}^{n-1} Q_{r,j}(0) Q_{r,k}(0). \tag{62.16}$$

According to (61.20), the lhs of (62.16) is recognized as the general (j,k)th element of the inverse matrix $\mathbf{S}_n^{-1}$. Therefore, the current and the preceding chapters, working with the two different methods, one from the present work and the other from that of Geronimus [224], are found to agree with each other. For the given $n \times n$ determinant $\det \mathbf{S}_n$, these methods perform only $\sim n^2$ multiplications as opposed to $n!$ multiplications in direct evaluations of the same determinant. Both variants of the method can be used for determinants other than the one associated with the Schrödinger overlap determinant $\det \mathbf{S}_n$. Moreover, any determinant whose elements are signal points (5.11) or power moments (53.5) can be computed with analytical results. This has been demonstrated by Geronimus [225] for a number of determinants that are otherwise difficult to compute numerically.

Chapter 63

Eigenvalues without eigenproblems or characteristic equations

In this chapter, we shall show how the eigenvalues $\{u_k\}$ ($1 \le k \le K$) can be obtained without actually solving the eigenproblem $\hat{U}|\Upsilon_k) = u_k|\Upsilon_k)$ or the characteristic equation $Q_K(u) = 0$. Instead of these two latter classical procedures for arriving at the required eigenset $\{u_k\}$, we shall use the Shanks transform [72] and the Rutishauser [169] quotient difference algorithm, the QD. The form of the Shanks transform which will be used here is $e_m(c_n) = H_{m+1}(c_n)/H_m(\Delta^2 c_n)$ from (7.52). Employing (7.49) we can rewrite (7.52) as

$$e_k(c_n) = \frac{1}{W_k^2} \frac{H_{k+1}(c_n)}{H_k(c_n)} \tag{63.1}$$

where $W_k = \prod_{j=1}^{k}(u_j - 1) \neq 0$, as in (7.50). Moreover, by means of (7.26), it follows

$$\frac{H_{k+1}(c_n)}{H_k(c_n)} = R_k^2 d_{k+1} u_{k+1}^n \qquad R_k = \frac{V_{k+1}}{V_k}. \tag{63.2}$$

The quotient V_{k+1}/V_k is computed via the Vandermonde determinant (7.14) with the result

$$R_k = \prod_{j=1}^{k}(u_{k+1} - u_j). \tag{63.3}$$

Thus, the Shanks transform (7.52) becomes

$$e_k(c_n) = \rho_k^2 d_{k+1} u_{k+1}^n \tag{63.4}$$

$$\rho_k = \frac{R_k}{W_k}. \tag{63.5}$$

Next, we take the following ratio of two Shanks transforms

$$\frac{e_{k-1}(c_{n+1})}{e_{k-1}(c_n)} = \frac{\rho_{k-1}^2 d_k u_k^{n+1}}{\rho_{k-1}^2 d_k u_k^n} = u_k$$

$$\therefore \quad u_k = \lim_{n \to \infty} \frac{e_{k-1}(c_{n+1})}{e_{k-1}(c_n)}. \tag{63.6}$$

The limit $n \to \infty$ is taken on the rhs of this equation, since the lhs of the same equation is independent of n. The relation (63.6) can equivalently be written in terms of ε-vector from the Wynn recursive algorithm (7.54)

$$u_k = \lim_{n \to \infty} \frac{\varepsilon_{2k-2}^{(n+1)}}{\varepsilon_{2k-2}^{(n)}} \tag{63.7}$$

as announced in (19.71). According to (63.1), the transform $e_k(c_n)$ is proportional to the quotient $H_{k+1}(c_n)/H_k(c_n)$. Moreover, the constant of proportionality $1/W_k^2$ is independent on n. Since u_k is obtained from the ratio $e_{k-1}(c_{n+1})/e_{k-1}(c_n)$, as per (63.6), the constant $1/W_k^2$ can be ignored altogether. Therefore, for computation of u_k it is natural to use only the ratio $H_k(c_n)/H_{k-1}(c_n)$ which will be denoted as $H_k^{(n)}$

$$H_k^{(n)} \equiv \frac{H_k(c_n)}{H_{k-1}(c_n)}. \tag{63.8}$$

Employing (63.2) we can rewrite (63.8) via

$$H_k^{(n)} = R_{k-1}^2 d_k u_k^n. \tag{63.9}$$

Hence, to arrive at u_k, we only need to consider the quotient $H_k^{(n+1)}/H_k^{(n)}$, just like in $e_{k-1}(c_{n+1})/e_{k-1}(c_n)$ via (63.6). It is convenient to label the quotient $H_k^{(n+1)}/H_k^{(n)}$ by $q_k^{(n)}$

$$q_k^{(n)} \equiv \frac{H_k^{(n+1)}}{H_k^{(n)}}. \tag{63.10}$$

Thus, combining (63.9) and (63.10), it follows

$$q_k^{(n)} = \frac{R_{k-1}^2 d_k u_k^{n+1}}{R_{k-1}^2 d_k u_k^n} = u_k$$

$$\therefore \quad u_k = \lim_{n \to \infty} q_k^{(n)}. \tag{63.11}$$

In this way, the eigenvalue u_k is obtained directly from the determinantal quotient $q_k^{(n)} = H_k^{(n+1)}/H_k^{(n)}$ in the limit $n \to \infty$ via (63.11). Using (63.8) and (63.10) we can calculate $q_k^{(n)}$ explicitly as

$$q_k^{(n)} = \frac{H_k^{(n+1)}}{H_k^{(n)}} = \frac{H_k(c_{n+1})}{H_{k-1}(c_{n+1})} \frac{1}{\frac{H_k(c_n)}{H_{k-1}(c_n)}} = \frac{H_k(c_{n+1})}{H_{k-1}(c_{n+1})} \frac{H_{k-1}(c_n)}{H_k(c_n)}$$

$$\therefore \quad q_k^{(n)} = \frac{H_{k-1}(c_n)H_k(c_{n+1})}{H_{k-1}(c_{n+1})H_k(c_n)}. \tag{63.12}$$

It is also important to introduce another determinantal ratio obtained from (63.8) by the simultaneous changes $H_k^{(n+1)} \to H_{k+1}^{(n)}$ in the numerator and $H_k^{(n)} \to H_k^{(n+1)}$ in the denominator. The new quotient $H_{k+1}^{(n)}/H_k^{(n+1)}$ will be denoted by $e_k^{(n)}$

$$e_k^{(n)} \equiv \frac{H_{k+1}^{(n)}}{H_k^{(n+1)}}. \tag{63.13}$$

Inserting (63.9) into (63.13) gives

$$e_k^{(n)} = \left(\frac{R_k}{R_{k-1}}\right)^2 \frac{d_{k+1}}{d_k u_k} \left(\frac{u_{k+1}}{u_k}\right)^n. \tag{63.14}$$

If the eigenvalues $\{u_k\}$ are ordered according to

$$0 < |u_1| < |u_2| < \cdots < |u_{k-1}| < |u_k| < |u_{k+1}| < \cdots < |u_K| < 1 \tag{63.15}$$

then it follows from (63.14) that

$$\lim_{n\to\infty} e_k^{(n)} = 0. \tag{63.16}$$

Thus, the determinantal quotient $e_k^{(n)} = H_{k+1}^{(n)}/H_k^{(n+1)}$ tends to zero when $n \to \infty$ provided that the eigenvalues are ordered as in (63.15). Similarly as in (63.12), we can use (63.8) to calculate $e_k^{(n)}$ explicitly viz

$$e_k^{(n)} = \frac{H_{k+1}(c_n)}{H_k(c_n)} \frac{1}{\frac{H_k(c_{n+1})}{H_{k-1}(c_{n+1})}} = \frac{H_{k+1}(c_n)H_{k-1}(c_{n+1})}{H_k(c_n)H_k(c_{n+1})}$$

$$\therefore \quad e_k^{(n)} = \frac{H_{k-1}(c_{n+1})H_{k+1}(c_n)}{H_k(c_n)H_k(c_{n+1})}. \tag{63.17}$$

The vectors $\{q_k^{(n)}, e_k^{(n)}\}$ in their respective forms (63.12) and (63.17) are recognized to be the same as the corresponding ones from the QD algorithm [169].

The derivation from this chapter shows that the Shanks transform, the Wynn ε-algorithm and Rutishauser QD algorithm can all give the eigenvalues

$\{u_k\}$ ($1 \leq k \leq K$) without resorting to the more conventional procedures, e.g. the eigenproblem $\hat{U}|\Upsilon_k) = u_k|\Upsilon_k)$ or the characteristic equation $Q_K(u) = 0$. Of course, in practice, the vectors $e_k(c_n)$, $\varepsilon_k(c_n)$, $q_k^{(n)}$ and $e_k^{(n)}$ needed for obtaining the eigenvalues $\{u_k\}$ are not computed via the Hankel determinants at all, but rather through their respective recursions. Usually, the Shanks and the Rutishauser algorithms are presented and analysed separately in the literature. However, in the present chapter, we show how the need for the introduction of the two vectors $q_k^{(n)}$ and $e_k^{(n)}$ from the QD algorithm can be naturally motivated within the analysis of the Shanks transform. (See chapter 67 for further details on the QD algorithm.)

Chapter 64

Tridiagonal inhomogeneous systems of linear equations

Here, we shall develop a recursive algorithm for solving a tridiagonal inhomogeneous system of n linear equations of the typical type

$$B_1 x_1 + C_1 x_2 + 0 + 0 + 0 + \cdots + 0 + 0 + 0 + 0 = D_1$$
$$A_2 x_1 + B_2 x_2 + C_2 x_3 + 0 + \cdots + 0 + 0 + 0 + 0 = D_2$$
$$0 + A_3 x_2 + B_3 x_3 + C_3 x_4 + 0 + \cdots + 0 + 0 + 0 + 0 = D_3$$
$$\vdots \qquad (64.1)$$
$$0 + 0 + 0 + 0 + 0 + \cdots + A_{n-2} x_{n-3} + B_{n-2} x_{n-2} + C_{n-2} x_{n-1} + 0 = D_{n-2}$$
$$0 + 0 + 0 + 0 + 0 + \cdots + 0 + A_{n-1} x_{n-2} + B_{n-1} x_{n-1} + C_{n-1} x_n = D_{n-1}$$
$$0 + 0 + 0 + 0 + 0 + \cdots + 0 + 0 + A_n x_{n-1} + B_n x_n = D_n$$

where A_r, B_r, C_r and D_r are the known constants. The equivalent matrix form of (64.1) reads as

$$\begin{pmatrix} B_1 & C_1 & 0 & 0 & 0 & \cdots & 0 & 0 & 0 \\ A_2 & B_2 & C_3 & 0 & 0 & \cdots & 0 & 0 & 0 \\ 0 & A_3 & B_3 & C_3 & 0 & \cdots & 0 & 0 & 0 \\ \vdots & \vdots & \vdots & \vdots & \vdots & \ddots & \vdots & \vdots & \vdots \\ 0 & 0 & 0 & 0 & 0 & \cdots & A_{n-1} & B_{n-1} & C_{n-1} \\ 0 & 0 & 0 & 0 & 0 & \cdots & 0 & A_n & B_n \end{pmatrix} \begin{pmatrix} x_1 \\ x_2 \\ x_3 \\ \vdots \\ x_{n-1} \\ x_n \end{pmatrix}$$

$$= \begin{pmatrix} D_1 \\ D_2 \\ D_3 \\ \vdots \\ D_{n-1} \\ D_n \end{pmatrix} \qquad (64.2)$$

$$\mathbf{Ex} = \mathbf{D} \qquad (64.3)$$

341

where $\mathbf{E} = \{E_{r,s}\}$ is the $n \times n$ matrix with the elements $E_{r,s} = A_r \delta_{r,s+1} + B_s \delta_{r,s} + C_r \delta_{r+1,s}$ and $\delta_{n,m}$ is the Kronecker δ-symbol. In (64.1), $\mathbf{x} = \{x_r\}$ is an unknown column vector, whereas $\mathbf{D} = \{D_r\}$ is an inhomogeneous column vector with the given elements D_r ($1 \le r \le n$).

We solve the system (64.1) by the Gaussian elimination method. To this end, we multiply the last line $A_n x_{n-1} + B_n x_n = D_n$ in (64.1) by C_{n-1} and write

$$C_{n-1} x_n = \frac{C_{n-1} D_n}{B_n} - \frac{A_n C_{n-1}}{B_n} x_{n-1}. \tag{64.4}$$

Similarly, when the first to the last line in (64.1), i.e. $A_{n-1} x_{n-2} + B_{n-1} x_{n-1} + C_{n-1} x_n = D_{n-1}$, is multiplied by C_{n-2}, it follows

$$C_{n-2} x_{n-1} = C_{n-2} \frac{D_{n-1} - \dfrac{C_{n-1} D_n}{B_n}}{B_{n-1} - \dfrac{C_{n-1} A_n}{B_n}} - \frac{A_{n-1} C_{n-2}}{B_{n-1} - \dfrac{C_{n-1} A_n}{B_n}} x_{n-2}. \tag{64.5}$$

Here, we also used the last line in (64.1). Likewise, we multiply the second to the last line in (64.1), i.e. $A_{n-2} x_{n-3} + B_{n-2} x_{n-2} + C_{n-2} x_{n-1} = D_{n-2}$, by C_{n-3} and arrive at

$$C_{n-3} x_{n-2} = C_{n-3} \frac{D_{n-2} - C_{n-2} \dfrac{D_{n-1} - \dfrac{C_{n-1} D_n}{B_n}}{B_{n-1} - \dfrac{C_{n-1} A_n}{B_n}}}{B_{n-2} - \dfrac{A_{n-1} C_{n-2}}{B_{n-1} - \dfrac{C_{n-1} A_n}{B_n}}} - \frac{A_{n-2} C_{n-3}}{B_{n-2} - \dfrac{A_{n-1} C_{n-2}}{B_{n-1} - \dfrac{C_{n-1} A_n}{B_n}}} x_{n-3}. \tag{64.6}$$

From the particular expressions (64.4)–(64.6), we can immediately write down the general recursion for the solution x_r of the system (64.1). With this aim, let us introduce the two auxiliary vectors V_n and W_n, as follows

$$\left. \begin{aligned} V_n &= \frac{C_n}{B_n - A_n V_{n-1}} & V_0 &= 0 \\ W_n &= \frac{D_n - A_n W_{n-1}}{B_n - A_n V_{n-1}} & W_0 &= 0 \end{aligned} \right\}. \tag{64.7}$$

Then, the general element x_r of the column vector $\mathbf{x} = \{x_r\}$ from (64.1) or (64.2) is given by the following simple recursion

$$\left. \begin{aligned} x_n &= W_n \\ x_m &= W_{m-1} - V_{m-1} x_{m+1} \quad (1 \le m < n) \end{aligned} \right\} \tag{64.8}$$

Tridiagonal inhomogeneous systems of linear equations

where n is number of equations in (64.1). The relationship in (64.8) recurs downwards from the maximal value n of the subscript to the minimum which is 1. The recursion from (64.7) and (64.8) has previously been reported in [272] without any derivation.

To illustrate, the nested recursions (64.7) and (64.8), we take as an example the case with $n = 4$ in (64.1), i.e.

$$B_1 x_1 + C_1 x_2 + 0 \cdot x_3 + 0 \cdot x_4 = D_1$$
$$A_2 x_1 + B_2 x_2 + C_2 \cdot x_3 + 0 \cdot x_4 = D_2$$
$$0 \cdot x_1 + A_3 x_2 + B_3 x_3 + C_3 x_4 = D_3$$
$$0 \cdot x_1 + 0 \cdot x_2 + A_4 x_3 + B_4 \cdot x_4 = D_4. \tag{64.9}$$

Then, using (64.7) and (64.8) we calculate

$$x_4 = W_4 \qquad x_3 = W_3 - V_3 x_4 \tag{64.10}$$

$$x_2 = W_2 - V_2 x_3 \qquad x_1 = W_1 - V_1 x_2 \tag{64.11}$$

$$W_4 = \frac{D_4 - A_4 W_3}{B_4 - A_4 V_3} \qquad W_3 = \frac{D_3 - A_3 W_2}{B_3 - A_3 V_2} \tag{64.12}$$

$$W_2 = \frac{D_2 - A_2 W_1}{B_2 - A_2 V_1} \qquad W_1 = \frac{D_1 - A_1 W_0}{B_1 - A_1 V_0} \tag{64.13}$$

$$V_3 = \frac{C_3}{B_3 - A_3 V_2} \qquad V_2 = \frac{C_2}{B_2 - A_2 V_1} \qquad V_1 = \frac{C_1}{B_1 - A_1 V_0} = \frac{C_1}{B_1}. \tag{64.14}$$

This gives the following final results for W_r in terms of the constants A_r, B_r, C_r and D_r alone

$$W_1 = \frac{D_1}{B_1} \qquad W_2 = \frac{B_1 D_2 - A_2 D_1}{B_1 B_2 - A_2 C_1} \tag{64.15}$$

$$W_3 = \frac{D_3(B_1 B_2 - A_2 C_1) - A_3(B_1 D_2 - A_2 D_1)}{B_3(B_1 B_2 - A_2 C_1) - A_3 B_1 C_2} \tag{64.16}$$

$$W_4 = \frac{D_4[B_3(B_1 B_2 - A_2 C_1) - A_3 B_1 C_2]}{B_4[B_3(B_1 B_2 - A_2 C_1) - A_3 B_1 C_2] - A_4 C_3(B_1 B_2 - A_2 C_1)}$$
$$- \frac{A_4[D_3(B_1 B_2 - A_2 C_1) - A_3(B_1 D_2 - A_2 D_1)]}{B_4[B_3(B_1 B_2 - A_2 C_1) - A_3 B_1 C_2] - A_4 C_3(B_1 B_2 - A_2 C_1)}. \tag{64.17}$$

For convenience, in (64.17), W_4 is written as the difference of two quotients, but they have a common denominator. With these findings for V_m ($1 \leq m \leq 3$) and W_m ($1 \leq m \leq 4$) at hand, we can use (64.10) and (64.11) to arrive at the final solution for x_r

$$x_r = \frac{N_r}{D_4} \qquad (1 \leq r \leq 4) \tag{64.18}$$

$$N_1 = D_1[B_2(B_3B_4 - A_4C_3) - A_3B_4C_2]$$
$$- C_1[D_2(B_3B_4 - A_4C_3) - C_2(D_3B_4 - C_3D_4)] \quad (64.19)$$
$$N_2 = B_1[D_2(B_3B_4 - A_4C_3) - C_2(D_3B_4 - C_3D_4)]$$
$$- A_2D_1(B_3B_4 - A_4C_3) \quad (64.20)$$
$$N_3 = B_1[B_2(D_3B_4 - D_4C_3) - D_2A_3B_4]$$
$$- C_1A_2(D_3B_4 - C_3D_4) + A_2D_1A_3B_4 \quad (64.21)$$
$$N_4 = B_1[B_2(B_3D_4 - A_4D_3) - C_2A_3D_4 + D_2A_3A_4]$$
$$- C_1A_2(B_3D_4 - A_4D_3) - A_2A_3A_4D_1 \quad (64.22)$$
$$D_4 = B_1[B_2(B_3B_4 - A_4C_3) - A_3B_4C_2] - C_1A_2(B_3B_4 - A_4C_3). \quad (64.23)$$

For a check, we also solved the system (64.9) directly by using the Cramer rule and obtained exactly the same solutions $x_r = N_r/D_4$ ($1 \leq r \leq 4$) as in (64.18) and (64.19)–(64.23).

Next we shall consider two special cases of the general system (64.1). Both of these special cases are often encountered in physics and chemistry as well as in linear algebra. In both of these applications, there is only one non-zero element of the inhomogeneous column vector $\mathbf{D} = \{D_r\}$ from (64.1), i.e. $D_n \propto \delta_{n,1}$ where $\delta_{n,m}$ is the Kronecker δ-symbol from (3.7). The first such special system of equations is the one from the theory of the power moment [170]

$$B_n = \xi_n \qquad A_n = -\zeta_n^{1/2} \qquad C_n = -\zeta_{n+1}^{1/2} \qquad D_n = -\zeta_1 \delta_{n,1} \quad (64.24)$$

$$\xi_1 x_1 - \zeta_2^{1/2} x_2 + 0 \cdot x_3 + 0 \cdot x_4 + 0 \cdot x_5 + \cdots + 0 = -\zeta_1$$
$$- \zeta_2^{1/2} x_1 + \xi_2 x_2 - \zeta_3^{1/2} x_3 + 0 \cdot x_4 + 0 \cdot x_5 + \cdots + 0 = 0$$
$$0 \cdot x_1 - \zeta_3^{1/2} x_2 + \xi_3 x_3 - \zeta_4^{1/2} x_4 + 0 \cdot x_5 + \cdots + 0 = 0$$
$$\vdots \quad (64.25)$$
$$0 + 0 + 0 + 0 + 0 + \cdots + 0 - \zeta_{n-1}^{1/2} x_{n-2} + \xi_n x_{n-1} - \zeta_n^{1/2} x_n = 0$$
$$0 + 0 + 0 + 0 + 0 + \cdots + 0 - \zeta_n^{1/2} x_{n-1} + \xi_n x_n = 0$$

or in the matrix representation

$$\begin{pmatrix} \xi_1 & -\zeta_2^{1/2} & 0 & 0 & 0 & \cdots & 0 & 0 & 0 \\ -\zeta_2^{1/2} & \xi_2 & -\zeta_3^{1/2} & 0 & 0 & \cdots & 0 & 0 & 0 \\ 0 & -\zeta_3^{1/2} & \xi_3 & -\zeta_4^{1/2} & 0 & \cdots & 0 & 0 & 0 \\ \vdots & \vdots & \vdots & \vdots & \vdots & \ddots & \vdots & \vdots & \vdots \\ 0 & 0 & 0 & 0 & 0 & \cdots & -\zeta_{n-1}^{1/2} & \xi_n & -\zeta_n^{1/2} \\ 0 & 0 & 0 & 0 & 0 & \cdots & 0 & -\zeta_n^{1/2} & \xi_n \end{pmatrix}$$

Tridiagonal inhomogeneous systems of linear equations

$$\mathbf{x} \begin{pmatrix} x_1 \\ x_2 \\ x_3 \\ \vdots \\ x_{n-1} \\ x_n \end{pmatrix} = - \begin{pmatrix} \zeta_1 \\ 0 \\ 0 \\ \vdots \\ 0 \\ 0 \end{pmatrix}. \tag{64.26}$$

It then follows from (64.5) that

$$-\zeta_n^{1/2} x_n = -\frac{\zeta_n}{\xi_n} x_{n-1} \qquad -\zeta_n^{1/2} x_{n-1} = -\frac{\zeta_{n-1}}{\xi_{n-1} - \frac{\zeta_n}{\xi_n}} x_{n-2} \tag{64.27}$$

$$-\zeta_{n-2}^{1/2} x_{n-2} = -\frac{\zeta_{n-2}}{\xi_{n-2} - \dfrac{\zeta_{n-1}}{\xi_{n-1} - \dfrac{\zeta_n}{\xi_n}}} x_{n-3}. \tag{64.28}$$

The particular recursions (64.27) and (64.28) permit generalization to the general result for x_2 after n steps viz

$$\zeta_2^{1/2} x_2 = \left(\frac{\zeta_2}{\xi_2} - \frac{\zeta_3}{\xi_3} - \frac{\zeta_4}{\xi_4} - \frac{\zeta_5}{\xi_5} - \cdots - \frac{\zeta_{n-1}}{\xi_{n-1}} - \frac{\zeta_n}{\xi_n} \right) x_1. \tag{64.29}$$

The first equation from the system (64.25) allows elimination of x_2 from (64.29)

$$\xi_1 x_1 - \zeta_2^{1/2} x_2 = -\zeta_1 \tag{64.30}$$

$$-\zeta_1 = \xi_1 x_1 - \zeta_2^{1/2} x_2$$

$$= \xi_1 x_1 - \left(\frac{\zeta_2}{\xi_2} - \frac{\zeta_3}{\xi_3} - \frac{\zeta_4}{\xi_4} - \frac{\zeta_5}{\xi_5} - \cdots - \frac{\zeta_{n-1}}{\xi_{n-1}} - \frac{\zeta_n}{\xi_n} \right) x_1$$

$$\therefore \quad \zeta_1 = - \left(\xi_1 - \frac{\zeta_2}{\xi_2} - \frac{\zeta_3}{\xi_3} - \frac{\zeta_4}{\xi_4} - \frac{\zeta_5}{\xi_5} - \cdots - \frac{\zeta_{n-1}}{\xi_{n-1}} - \frac{\zeta_n}{\xi_n} \right) x_1.$$

Thus, the solution for the first element x_1 of the column vector $\mathbf{x} = \{x_n\}$ ($n = 1, 2, \ldots$) is given by

$$x_1 = -\frac{\zeta_1}{\xi_1} - \frac{\zeta_2}{\xi_2} - \frac{\zeta_3}{\xi_3} - \frac{\zeta_4}{\xi_4} - \frac{\zeta_5}{\xi_5} - \cdots - \frac{\zeta_{n-1}}{\xi_{n-1}} - \frac{\zeta_n}{\xi_n}. \tag{64.31}$$

The solution (64.31) is recognized as the continued fraction of the order n. The system (64.25) is encountered in statistical mechanics [170] when applying the method of power moments to obtain a sequence of approximations to the partition function $Q(\beta)$ defined as the Stieltjes integral $\int_0^\infty \exp(-\beta E) d\varphi(E)$ where β is a parameter proportional to the reciprocal temperature of the investigated system, $d\varphi(E)$ is a density of states and $\varphi(E)$ is a positive, non-decreasing function of energy E.

The second example from the general system (64.1) is of a direct relevance to the Lanczos continued fraction, the LCF, from chapter 41. More precisely, the Lanczos inhomogeneous tridiagonal system of linear equations is identified from (64.1) by the specification

$$B_n = u - \alpha_{n-1} \qquad A_n = -\beta_{n-1} \qquad C_n = -\beta_n \qquad D_n = c_0 \delta_{n,1} \qquad (64.32)$$

where α_n and β_n are Lanczos coupling parameters from (22.2) or (32.5). The system (64.1) now reads as

$$(u - \alpha_0)x_1 - \beta_1 x_2 + 0 + 0 + 0 + \cdots + 0 + 0 + 0 + 0 = c_0$$
$$- \beta_1 x_1 + (u - \alpha_1)x_2 - \beta_2 x_3 + 0 + 0 + \cdots + 0 + 0 + 0 + 0 = 0$$
$$0 - \beta_2 x_2 + (u - \alpha_2)x_3 - \beta_3 x_4 + 0 + \cdots + 0 + 0 + 0 + 0 = 0$$
$$\vdots \qquad (64.33)$$
$$0 + 0 + 0 + 0 + 0 + \cdots + 0 - \beta_{n-2} x_{n-2} + (u - \alpha_{n-2})x_{n-1} - \beta_{n-1} x_n = 0$$
$$0 + 0 + 0 + 0 + 0 + \cdots + 0 + 0 - \beta_{n-1} x_{n-1} + (u - \alpha_{n-1})x_n = 0$$

or in the equivalent matrix form

$$\begin{pmatrix} u - \alpha_0 & -\beta_1 & 0 & 0 & 0 & \cdots & 0 & 0 & 0 \\ -\beta_1 & u - \alpha_1 & -\beta_2 & 0 & 0 & \cdots & 0 & 0 & 0 \\ 0 & -\beta_2 & u - \alpha_2 & -\beta_3 & 0 & \cdots & 0 & 0 & 0 \\ \vdots & \vdots & \vdots & \vdots & \vdots & \ddots & \vdots & \vdots & \vdots \\ 0 & 0 & 0 & 0 & 0 & \cdots & -\beta_{n-2} & u - \alpha_{n-2} & -\beta_{n-1} \\ 0 & 0 & 0 & 0 & 0 & \cdots & 0 & -\beta_{n-1} & u - \alpha_{n-1} \end{pmatrix} \times \begin{pmatrix} x_1 \\ x_2 \\ x_3 \\ \vdots \\ x_{n-1} \\ x_n \end{pmatrix} = \begin{pmatrix} c_0 \\ 0 \\ 0 \\ \vdots \\ 0 \\ 0 \end{pmatrix}. \qquad (64.34)$$

In this case by using (64.4)–(64.6), we obtain

$$-\beta_{n-1} x_n = -\frac{\beta_{n-1}^2}{u - \alpha_{n-1}} x_{n-1} \qquad (64.35)$$

$$-\beta_{n-2} x_{n-1} = -\frac{\beta_{n-2}^2}{u - \alpha_{n-2} - \dfrac{\beta_{n-1}^2}{u - \alpha_{n-1}}} x_{n-2} \qquad (64.36)$$

$$-\beta_{n-3} x_{n-2} = -\frac{\beta_{n-3}^2}{u - \alpha_{n-3} - \dfrac{\beta_{n-2}^2}{u - \alpha_{n-2} - \dfrac{\beta_{n-1}^2}{u - \alpha_{n-1}}}} x_{n-3}. \qquad (64.37)$$

Tridiagonal inhomogeneous systems of linear equations

By continuing this recursion further for the next $n-2$ steps, it follows

$$\beta_1 x_2 = \left(\frac{\beta_1^2}{u-\alpha_1} - \frac{\beta_2^2}{u-\alpha_2} - \frac{\beta_3^2}{u-\alpha_3} - \cdots - \frac{\beta_{n-1}^2}{u-\alpha_{n-1}}\right) x_1 \quad (64.38)$$

where the symbolic notation for the continued fractions from chapter 41 is used. The first equation $c_0 = (u-\alpha_0)x_1 - \beta_1 x_2$ from the system (64.33) combined with (64.38) permits elimination of x_2 from (64.38) and this gives the final solution for x_1

$$c_0 = (u-\alpha_0)x_1 - \beta_1 x_2 = (u-\alpha_0)x_1$$
$$- \left(\frac{\beta_1^2}{u-\alpha_1} - \frac{\beta_2^2}{u-\alpha_2} - \frac{\beta_3^2}{u-\alpha_3} - \cdots - \frac{\beta_{n-1}^2}{u-\alpha_{n-1}}\right) x_1$$
$$= \left[u - \alpha_0 - \frac{\beta_1^2}{u-\alpha_1} - \frac{\beta_2^2}{u-\alpha_2} - \frac{\beta_3^2}{u-\alpha_3} - \cdots - \frac{\beta_{n-1}^2}{u-\alpha_{n-1}}\right] x_1$$

$$\therefore \quad x_1 = \frac{c_0}{u-\alpha_0} - \frac{\beta_1^2}{u-\alpha_1} - \frac{\beta_2^2}{u-\alpha_2} - \frac{\beta_3^2}{u-\alpha_3} - \cdots - \frac{\beta_{n-1}^2}{u-\alpha_{n-1}}. \quad (64.39)$$

Hence, the first component x_1 of the n-dimensional column vector $\mathbf{x} = \{x_r\}$ ($1 \leq r \leq n$) from (64.33) coincides with the Lanczos continued fraction $\mathcal{R}_n^{\mathrm{LCF}}(u)$ to the Green function $\mathcal{R}(u)$ from (3.1)

$$\mathcal{R}_n^{\mathrm{LCF}}(u) = x_1$$
$$= \frac{c_0}{u-\alpha_0} - \frac{\beta_1^2}{u-\alpha_1} - \frac{\beta_2^2}{u-\alpha_2} - \frac{\beta_3^2}{u-\alpha_3} - \cdots - \frac{\beta_{n-1}^2}{u-\alpha_{n-1}}. \quad (64.40)$$

A recursive algorithm for $\mathcal{R}_n^{\mathrm{CF}}(u)$ is available from (64.8) as

$$\left.\begin{array}{c} \Gamma_m = \dfrac{1}{u - \alpha_m - \beta_{m+1}^2 \Gamma_{m+1}} \quad \Gamma_{N+1} = 0 \\ \therefore \quad \mathcal{R}_N^{\mathrm{LCF}}(u) = c_0 \Gamma_0 \end{array}\right\} \quad (64.41)$$

where $\Gamma_m \equiv \Gamma_m(u)$, in agreement with (41.4). We shall list below the first two illustrations of (64.41) for $N=1$ and $N=2$. Thus for $N=1$ we have $\Gamma_2 = 0$, so that

$$\Gamma_1 = \frac{1}{u-\alpha_1} \quad (64.42)$$

$$\therefore \quad \mathcal{R}_1^{\mathrm{LCF}}(u) = \frac{c_0}{u-\alpha_1} \quad (64.43)$$

is in agreement with (41.11). Next, let $N = 2$ for which we have $\Gamma_3 = 0$ and this gives

$$\Gamma_2 = \frac{1}{u - \alpha_2} \qquad \Gamma_1 = \frac{1}{u - \alpha_1 - \dfrac{\beta_2^2}{u - \alpha_2}} \qquad (64.44)$$

$$\Gamma_0 = \frac{1}{u - \alpha_0 - \dfrac{\beta_1^2}{u - \alpha_1 - \dfrac{\beta_2^2}{u - \alpha_2}}} \qquad (64.45)$$

$$\therefore \quad \mathcal{R}_2^{\text{LCF}}(u) = \frac{c_0}{u - \alpha_0 - \dfrac{\beta_1^2}{u - \alpha_1 - \dfrac{\beta_2^2}{u - \alpha_2}}} \qquad (64.46)$$

in agreement with (41.11). Recall that according to (47.5), the LCF of the nth order, $\mathcal{R}_n^{\text{LCF}}(u)$, is the same as $\mathcal{R}_{2n}^{\text{CF}}(u)$ which is an even-order general continued fraction, the CF, obtained from (43.3) when the suffix n is taken to be $2n$. In other words, $\mathcal{R}_n^{\text{LCF}}(u)$ is the contracted continued fraction, the CCF, relative to $\mathcal{R}_n^{\text{CF}}(u)$ as pointed out in chapter 47. To underline that we are dealing here with the even part of $\mathcal{R}_n^{\text{CF}}(u)$ we shall add the subscript 'e' (for even)

$$\mathcal{R}_{e,n}^{\text{CCF}}(u) = \mathcal{R}_{2n}^{\text{CF}}(u) = \mathcal{R}_{e,n}^{\text{LCF}}(u) \qquad (n = 1, 2, 3, \ldots). \qquad (64.47)$$

The relationship (64.40) indicates that the same result for the Lanczos continued fraction $\mathcal{R}_n^{\text{LCF}}(u)$ can also be obtained by finding the first element x_1 of the system (64.2) of linear equations. The major practical significance of this equivalence is in the fact that systems of linear equations represent the most robust chapter of linear algebra packages [176]–[179]. Even such powerful algorithms are unnecessary for a special class of inhomogeneous linear systems with tridiagonal structure (64.1), since the solution $\{x_r\}$ can be obtained with the help of a remarkably simple recursion (64.8).

There is more to the presented strategy than the equivalence (64.40). To illuminate an additional advantage of the presented formalism, we shall analyse its application to the quantification problem (harmonic inversion) with the goal of determining the key spectral parameters $\{u_k, d_k\}$. To this end, we rewrite the matrix equation (64.34) as follows

$$(u\mathbf{1} - \mathbf{J}) \cdot \mathbf{x} = c_0 \mathbf{e}_1 \qquad (64.48)$$

where $\mathbf{x} = \{x_r\}$ is an unknown column vector, $\mathbf{1}$ is the unit $n \times n$ matrix, $\mathbf{e}_1 = \{1_r\}$ is the unit column vector $1_r = \delta_{r,1}$ and $\mathbf{J} = \{J_{i,j}\}$ is the symmetric tridiagonal Jacobi evolution matrix in the Lanczos representation with the element

$J_{i,j} = \alpha_i \delta_{i,j} + \beta_j \delta_{i+1,j} + \beta_i \delta_{i,j+1}$ as in (22.8)

$$\mathbf{J} = \begin{pmatrix} \alpha_0 & \beta_1 & 0 & 0 & 0 & \cdots & 0 & 0 & 0 \\ \beta_1 & \alpha_1 & \beta_2 & 0 & 0 & \cdots & 0 & 0 & 0 \\ 0 & \beta_2 & \alpha_2 & \beta_3 & 0 & \cdots & 0 & 0 & 0 \\ \vdots & \vdots & \vdots & \vdots & \vdots & \ddots & \vdots & \vdots & \vdots \\ 0 & 0 & 0 & 0 & 0 & \cdots & \beta_{n-2} & \alpha_{n-2} & \beta_{n-1} \\ 0 & 0 & 0 & 0 & 0 & \cdots & 0 & \beta_{n-1} & \alpha_{n-1} \end{pmatrix}. \quad (64.49)$$

The matrix equation (64.48) can formally be solved by writing

$$\mathbf{x} = c_0 (u\mathbf{1} - \mathbf{J})^{-1} \mathbf{e}_1 \quad (64.50)$$

assuming that the inverse of $u\mathbf{1} - \mathbf{J}$ exists. As before, we only need the first element x_1 from (64.49) to obtain $\mathcal{R}_n^{\text{LCF}}(u)$, as per (64.40). We can obtain x_1 by diagonalizing the matrix $\mathbf{J}$ on a basis belonging to a vector space with the symmetric scalar product. This amounts to transforming our system to another basis in which the matrix $\mathbf{J}$ is diagonal with the eigenvalues $\{u_j\}$ given by

$$u_j = (\mathbf{T}^{-1} \mathbf{J} \mathbf{T})_{jj} \quad (64.51)$$

where $\mathbf{T} = \{T_{ij}\}$ is the transformation unitary matrix. In the new basis, the vector $\mathbf{x}$ becomes

$$\mathbf{x} = c_0 \mathbf{T} \mathbf{T}^{-1} (u\mathbf{1} - \mathbf{J})^{-1} \mathbf{T} \mathbf{T}^{-1} \mathbf{e}_1. \quad (64.52)$$

From here, the first element x_1 is at once extracted as [170]

$$x_1 = \sum_{j=1}^{n} \frac{d_j}{u - u_j} \quad (64.53)$$

$$d_j = c_0 T_{1j}^2. \quad (64.54)$$

We recall that the Lanczos continued fraction is equivalent to the Padé–Lanczos approximant, which is defined in (36.7) through its polynomial quotient, or equivalently, via the sum of the Heaviside partial fractions. The form (64.53) also coincides with the Heaviside partial fractions fractions (36.7) of $\mathcal{R}_n^{\text{LCF}}(u)$ as expected due to the equivalence (64.40). Hence, with the spectral pairs $\{u_j, d_j\}$ found via (64.51) and (64.54), the formalism from this chapter based upon [170] solves fully the spectral problem of quantification, or the harmonic inversion nonlinear problem by using purely linear algebra for which the most stable algorithms are available [176]–[179].

Chapter 65

Delayed time series

Thus far, we always counted the time evolution of the considered systems from $t_0 = 0$, which corresponds to $n = 0$. This means that the first element of the data (Hankel) matrix $\mathbf{H}_n(c_0) = \{c_{i+j}\}$ is the signal point $c_0 = (\Phi_0|\Phi_0) \neq 0$. Otherwise the general element $c_{i,j} = c_{i+j}$ of the matrix $\mathbf{H}_n(c_0)$ is the overlap between the two Schrödinger (Krylov) states $c_{i+j} = (\Phi_j|\Phi_i)$. Of course, the matrix which eventually provides the sought spectrum is not $\mathbf{H}_n(c_0)$, but rather $\mathbf{H}_n(c_1)$. The general element of the matrix $\mathbf{H}_n(c_1)$ is given by $c_{i+j+1} = (\Phi_j|\hat{U}|\Phi_i)$ where $\hat{U} = \hat{U}(\tau)$ is the time evolution operator $\hat{U}^r|\Phi_0) = |\Phi_r)$ from (5.6). In many situations of practical interest, it is important to consider a non-zero initial time. In this chapter, in addition to $t_0 = 0$, we shall also spectrally analyse the delayed signals with the non-zero initial time $t_s = s\tau (0 \leq s < N-1)$. An appropriate mathematical model for the delayed time signal $\{c_{n+s}\}$ ($0 \leq n \leq N-1, 0 \leq s < N-1$) is a natural extension of its counterpart from (5.11)

$$c_{n+s} = \sum_{k=1}^{K} d_k e^{-i(n+s)\omega_k \tau} = \sum_{k=1}^{K} d_k^{(s)} u_k^n \qquad d_k^{(s)} = d_k u_k^s. \qquad (65.1)$$

We see from here that the whole spectral analysis on regular, non-delayed time signals $\{c_n\}$ with $s = 0$ can be extended directly to delayed signals $\{c_{n+s}\}$ for $s > 0$ solely via a formal replacement of the old amplitudes $\{d_k\}$ by the new ones $\{d_k^{(s)}\}$. Likewise, in the corresponding new contour integral representation, the delayed generalized Stieltjes measure $\sigma_s(z)$ should be substituted instead of its non-delayed counterpart $\sigma_0(z)$ from (30.5) and (53.8)

$$c_{n+s} = \frac{1}{2\pi i} \oint_C z^n d\sigma_s(z) \qquad \sigma_s(z) = \sum_{k=1}^{K} d_k u_k^s \vartheta(z - u_k) \qquad (65.2)$$

$$d\sigma_s(z) = \rho_s(z) dz \qquad \rho_s(z) = \sum_{k=1}^{K} d_k u_k^s \delta(z - u_k). \qquad (65.3)$$

Here again the net effect of the time delay is manifested through the appearance of the power function u_k^s in a multiplicative manner which changes the old quadrature weight function (the Christoffel number) d_k into the delayed one, $d_k^{(s)}$. Therefore, the relationships among the delayed generalized Stieltjes integral and the corresponding series follow at once from their non-delayed variants (53.1) and (53.3)

$$\frac{1}{2\pi i}\oint_C \frac{d\sigma_s(z)}{u-z} = \sum_{n=0}^{\infty} c_{n+s} u^{-n-1}. \tag{65.4}$$

By inserting (65.1) into the rhs of (65.4) we can calculate

$$\sum_{n=0}^{\infty} c_{n+s} u^{-n-1} = \sum_{n=0}^{\infty} \left\{ \sum_{k=1}^{K} d_k^{(s)} u_k^n \right\} u^{-n-1}$$

$$= \sum_{k=1}^{K} d_k^{(s)} u^{-1} \left\{ \sum_{n=0}^{\infty} (u_k/u)^n \right\} = \sum_{k=1}^{K} d_k^{(s)} u^{-1} (1 - u_k/u)^{-1}$$

where the geometric series $\sum_{n=0}^{\infty} x^n = 1/(1-x)$ is employed

$$\therefore \quad \sum_{n=0}^{\infty} c_{n+s} u^{-n-1} = \sum_{k=1}^{K} \frac{d_k^{(s)}}{u - u_k}. \tag{65.5}$$

This is the most important feature of the mathematical model (65.1). The lhs of (65.5) represents the *exact* delayed spectrum associated with the delayed signal (65.1). This exactness is due to the presence of an infinite sum, i.e. a series over signal points. It is precisely here that the model (65.1) displays its distinct power by using no approximation to reduce the *infinite* sum over the signal points $\sum_{n=0}^{\infty} c_{n+s} u^{-n-1}$ to the finite sum over the reconstructed signal's parameters $\sum_{k=1}^{K} d_k^{(s)}/(u - u_k)$. Of course, in practice, only a finite number of signal points is customarily available, so that the exact spectrum $\sum_{n=0}^{\infty} c_{n+s} u^{-n-1}$ cannot be obtained. Nevertheless, in such a case the finite geometric sum $\sum_{n=0}^{N-1} x^n = (1-x^N)/(1-x)$ can be used to get the following exact result for the truncated spectrum

$$\sum_{n=0}^{N-1} c_{n+s} u^{-n-1} = \sum_{k=1}^{K} d_k^{(s)} \frac{1 - (u_k/u)^N}{u - u_k}. \tag{65.6}$$

This result tends to (65.5) as $N \longrightarrow \infty$ if $|u_k/u| < 1$. In both responses (65.5) and (65.6) corresponding respectively to the infinite ($N = \infty$) and finite ($N < \infty$) signal length, the exact delayed spectrum for the model (65.1) is given by rational functions. The two rational functions from the rhs of (65.5) and (65.6) are the polynomial quotients and, hence, the Padé approximants for the input sums

$\sum_{n=0}^{\infty} c_{n+s} u^{-n-1}$ and $\sum_{n=0}^{N-1} c_{n+s} u^{-n-1}$, respectively. Thus, for any processing method used to reconstruct the spectral parameters $\{u_k, d_k^{(s)}\}$ from the input signal of the form (65.1), the resulting spectrum will invariably be the Padé approximant. Hence, there could be hardly any doubt as to which signal processor is optimal for parametric estimations of spectra due to the time signal (65.1).

By definition, the Padé approximant becomes the exact theory if the input function is a rational function given as a quotient of two polynomials (a rational polynomial). Here, the exactness overrides the requested optimality and, therefore, the Padé approximant has no competitor for studying input functions that are themselves defined as rational polynomials. However, the input function in signal processing is *not* a ratio of two polynomials, but rather a single sum $\sum_n c_{n+s} u^{-n-1}$ for which no processor is exact. The only way to salvage the situation and eventually still arrive at an exact theory of this latter spectrum is to invoke the appropriate prior information, e.g. the existence of a harmonic structure of the signal as in (65.1). In such a case, without any further approximation, the single sum $\sum_n c_{n+s} u^{-n-1}$ from the original input becomes exactly equivalent to a rational polynomial (65.5) or (65.6), for which the Padé approximation is again the exact theory.

To elaborate the general issue of prior information, the minimal knowledge needed in advance would be an assumption that the time signal *has a structure*. For example, by solely plotting a given signal, an experienced practitioner with time sequences could qualitatively discern certain oscillatory patterns pointing at a harmonic-type structure. Such structures could often become more pronounced by viewing the corresponding derivative of the time signal. Of course, the Fourier shape spectrum in the frequency domain would give a more definitive indication of an underlying structure through a clearer display of peaks. In this way the otherwise generic structure would become more specific indicating the resonant nature of the spectrum.

The Fourier method does not capitalize on this critical finding from its own analysis and, therefore, in the end it is left with the envelope spectrum alone. The lack of exploiting a structure of the processed signal is reflected directly in the so-named 'Fourier uncertainty principle'. According to this principle, for a fixed total acquisition time T, or equivalently, a given full signal length N, the corresponding Fourier spectrum cannot offer an angular frequency resolution better than the Fourier bound $2\pi/T$. The first drawback of such a principle is in an immediate implication that, irrespective of their intrinsic nature, all the signals of the same acquisition time T have the same frequency resolution $2\pi/T$. Stated equivalently, all Fourier spectra due to all time signals of the same length N will have the same resolution. Specifically, prior to both measurements and processing, the Fourier method predefines the frequencies at which all the spectra exist and these are the frequencies from the Fourier grid $2\pi k/T$ ($k = 0, 1, \ldots, N-1$). This fact rules out a chance for interpolation or extrapolation. Without an interpolation feature, the Fourier method is restricted to the predetermined minimal separation

$\omega_{\min} = 2\pi/T$ between any two adjacent frequencies. Moreover, without an extrapolation property, the Fourier approach has no predictive power. Instead of extrapolation, the Fourier analysis usually resorts to zero filling (zero padding) beyond N or to using the signal's periodic extensions $c_{n+N} = c_n$. For genuinely non-periodic signals encountered in most circumstances, neither of these two latter recipes invoke any extrapolation, let alone new information.

The first step in an attempt to simultaneously circumvent all the listed basic limitations of the Fourier method is to merely acknowledge the fact that each signal does possess its own inner structure. The second step would follow afterwards with the aim of unfolding the hidden spectral characteristics by reconstructing the resonance parameters $\{u_k, d_k^{(s)}\}$ from the signal. This opens the door to parametric estimations of spectra as opposed to shape processing by Fourier. In particular, the fact that the signal has an internal structure via, e.g. a number of constituent harmonics $\{d_k^{(s)} u_k^n\}$, clearly offers the possibility for resolution improvement beyond the Fourier bound $2\pi/T$. For parametric processors, the total acquisition time T of the signal ceases to be a prerequisite for definition of the spectral resolution. Therefore, parametric estimations of spectra can indeed lower the Fourier bound $2\pi/T$ which automatically leads to an improved resolution beyond the prescription of the Fourier uncertainty principle. What made the Fourier uncertainty principle obsolete in the realm of parametric processing is a milder restriction imposed by 'the informational principle'. The informational principle states that no more information could possibly be obtained from a spectrum in the frequency domain than what has been encoded originally in the time domain. This indispensable conservation of information translates into an algebraic condition which requests a minimum of $2K$ signal points to reconstruct all the spectral parameters $\{u_k, d_k^s\} (1 \leq k \leq K)$. Such a condition is guided by the demand that the underlying system of linear equations must be at least determined (the number of equations equal to the number of the unknown parameters). In practice, the experimentally measured time signals are usually corrupted with noise whose presence could be partially mitigated by solving a corresponding overdetermined system in which the number of signal points exceeds the number $2K$ of the sought parameters $\{u_k, d_k^{(s)}\}$.

This new signal $\{c_{n+s}\}$ stems from a physical system which has already evolved from $t_0 = 0$ to $t_s \neq 0$ before starting to count the time. In this case, the spectral analysis is concerned with the delayed Hankel matrix $\mathbf{H}_n(c_s) = \{c_{i+j+s}\}$ from (5.14) whose first element is $c_s = (\Phi_0 | \hat{U}^s | \Phi_0) \neq 0$ which is not limited to $s = 0$ or $s = 1$ found in the overlap and evolution data matrix $\mathbf{H}_n(c_0) = \{c_{i+j}\}$ and $\mathbf{H}_n(c_1) = \{c_{i+j+1}\}$ from (5.16) and (5.17), respectively.

In the FFT, such a delay effect in the signal is conceived merely as skipping the first s points $\{c_r\}$ $(0 \leq r \leq s-1)$ from the whole signal $\{c_n\}$ $(0 \leq n \leq N-1)$. However, the ensuing FFT spectra are known to be of unacceptably poor quality which cannot be corrected for the skipped data. Therefore it is important to design signal processors that can handle data matrices $\{c_{i+j+s}\}$ which describe the evolution of the investigated system from a non-zero initial time $t_s = s\tau \neq 0$.

The problem of spectral analysis of data records with delayed time series is of a great importance in many applications which rely upon processing methods. Naturally, theoretical developments always favour the analysis of the general delayed Hankel matrix $\mathbf{H}_n(c_s)$, since the corresponding counterparts $\mathbf{H}_n(c_0)$ and $\mathbf{H}_n(c_1)$ from signal processing are merely two particular cases of the former data taken at $s = 0$ and $s = 1$, respectively. Of course, as always in theory, generality is not chosen for the reason of achieving *l'art pour l'art*. Rather, a general approach is likely to offer potential advantages over a particular one in the practical domain of creating more fruitful algorithms. This will be partially documented in the present chapter. It will be shown that the mere introduction of a non-zero initial time $t_s \neq 0$ has far reaching consequences for spectral analysis, even if we are actually interested in keeping the whole signal $\{c_n\}$ ($0 \leq n \leq N-1$) of total length N. For example, processing the evolution matrix $\mathbf{H}_n(c_1)$ requires ordinarily matrix diagonalizations, or equivalently, rooting the corresponding secular/characteristic equation to obtain the spectral parameters $\{u_k, d_k\}$ of the signal (5.11). In sharp contrast to this, both of these latter procedures can be bypassed altogether when $\mathbf{H}_n(c_s)$ is used with any fixed integer $s > 0$ and not just $s = 1$ as in the evolution matrix $\mathbf{H}_n(c_1)$. The sought spectral parameters $\{u_k, d_k\}$ will follow directly from convergence of the appropriate 'delayed' continued fraction coefficients as the values of number s are systematically increased.

As opposed to the FFT, the space methods from the Schrödinger picture of quantum mechanics (FD, DSD, PA, etc) can spectrally analyse data matrices with an arbitrary initial time t_s. These estimators rely upon the evolution operator $\hat{U}$ which generates the state $|\Phi_s\rangle$ at the delayed moment t_s via the prescription $|\Phi_s\rangle = \hat{U}^s |\Phi_0\rangle$ from (5.6). Here, the delay is achieved by starting the analysis from $|\Phi_s\rangle$ rather than from $|\Phi_0\rangle$ which refers to $s = 0$. Evolution of the system in the time interval $[t_0, t_s]$ of non-zero length must be properly taken into consideration. This correction is readily adjusted in the state space methods via multiplication of the vector $|\Phi_0\rangle$ by the required operator $\hat{U}^{-s} = \exp(is\hat{\Omega}\tau)$ which would cancel out the evolution effect accumulated in the state vector in the time interval $t \in [0, t_s]$. In other words, if the first s signal points are skipped, all the matrix elements from the FD or DSD should be modified by the counteracting operator $\exp(is\hat{\Omega}\tau)$. Therefore, in the FD, DSD or PA, taking the instant $t_s \neq 0$ instead of $t_0 = 0$ for the initial time will not cause any difficulty. These methods perform spectral analysis by diagonalizing[1] the data matrix $\mathbf{H}_n(c_s)$ for any integer s precisely in the same fashion as for $\mathbf{H}_n(c_0)$ and $\mathbf{H}_n(c_1)$. Moreover, there is a very important advantage in diagonalizing the delayed evolution matrix $\mathbf{H}_n(c_s) = \mathbf{U}_n^{(s)} = \{\mathbf{U}_{i,j}^{(s)}\}$ relative to $\mathbf{H}_n(c_1)$, where the general element $\mathbf{U}_{i,j}^{(s)}$ stems from the sth power of $\hat{U}$ via $\mathbf{U}_{i,j}^{(s)} = \langle \Phi_j | \hat{U}^s | \Phi_i \rangle$, as in (5.13). This advantage

---

[1] Instead of diagonalization of the data matrix, the standard PA or DPA resorts to rooting the corresponding characteristic/secular equations.

is in the possibility of identifying spurious roots. To achieve such a goal in the FD and DSD, we would diagonalize the matrix $\mathbf{U}_n^{(s)}$, not only for the primary case of interest ($s = 1$), but also for, e.g. $s = 2$ and compare the obtained eigenfrequencies. The matching frequencies for different values of s would be retained as physical/genuine, whereas those frequencies which change when going from $s = 1$ to $s = 2$ should be rejected as unphysical/spurious. Likewise, the PA also encourages the usage of different powers s of the evolution operator $\hat{U}$ that are, in fact, implicit in the paradiagonal elements $[(n + s - 1)/n]_{\mathcal{R}}(u)$ of the Padé table associated with the delayed counterpart $\mathcal{R}^{(s)}(u)$ of the Green function $\mathcal{R}(u)$ from (38.2). Here, we would select several values of s to distinguish between physical and spurious eigenroots of the denominator polynomials, that are the same as the mentioned characteristic/secular polynomials. Those roots that are stable/unstable for different s are conceived as physical/unphysical, respectively. These are some of the practical advantages alluded to earlier that give support to considering the initial times t_s different from the customary one, $t_0 = 0$.

Chapter 66

Delayed Green function

The usage of a non-zero initial time $t_s \neq 0$ in lieu of $t_0 = 0$ can also be handled easily by spectral methods that are based upon the Green function (PA, RRGM, etc). Here, the replacement $|\Phi_0\rangle \longrightarrow |\Phi_s\rangle$ from state space methods is equivalent to the substitution of the Green operator $(\hat{1}u - \hat{U})^{-1}$ from (38.1) by $(\hat{1}u - \hat{U})^{-1}\hat{U}^s$

$$\hat{R}^{(s)}(u) \equiv (\hat{1}u - \hat{U})^{-1}\hat{U}^s = \sum_{n=0}^{\infty} \hat{U}^{n+s}(\tau)u^{-n-1}. \tag{66.1}$$

In such a case, instead of using $\sum_{n=0}^{\infty} c_n u^{-n-1}$ for the Green function $\mathcal{R}(u)$ from (38.2), one would employ the delayed Green function $\mathcal{R}^{(s)}(u)$ for spectral analysis

$$\mathcal{R}^{(s)}(u) = (\Phi_0|\hat{R}^{(s)}(u)|\Phi_0) = \sum_{n=0}^{\infty} c_{n+s} u^{-n-1} \tag{66.2}$$

where (66.1) is used. We also have

$$\mathcal{R}^{(0)}(u) \equiv \mathcal{R}(u) = \sum_{n=0}^{\infty} c_n u^{-n-1}. \tag{66.3}$$

The delayed counterpart of the usual Green function emerges naturally in signal processing when the infinite time interval $[0, \infty]$ is split into two parts according to $[0, \infty] = [0, s-1] + [s, \infty]$. In such a case, it follows from (66.3)

$$\mathcal{R}(u) = \sum_{n=0}^{s-1} c_n u^{-n-1} + u^{-s} \sum_{n=0}^{\infty} c_{n+s} u^{-n-1}$$

$$= \sum_{n=0}^{s-1} c_n u^{-n-1} + u^{-s} \mathcal{R}^{(s)}(u) \tag{66.4}$$

where $\sum_{n=0}^{-1} \equiv 0$. This gives the relationship between the two exact Green functions $\mathcal{R}^{(s)}(u)$ and $\mathcal{R}^{(0)}(u)$ that correspond to the case with and without the delay, i.e. $s \neq 0$ and $s = 0$, respectively. We see from (66.3) that, just like in the mentioned space methods, the delayed spectrum $\mathcal{R}^{(s)}(u)$ is multiplied by the proper overall term u^{-s} to compensate for the time evolution of the system from $t_0 = 0$ to $t_s \neq 0$.

Chapter 67

The quotient-difference (QD) recursive algorithm

In analogy with (43.1), the infinite- and mth-order delayed continued fraction to the time series (66.2) are respectively defined as

$$\mathcal{R}^{\text{CF}(s)}(u) = \frac{a_1^{(s)}}{u} - \frac{a_2^{(s)}}{1} - \frac{a_3^{(s)}}{u} - \cdots - \frac{a_{2r}^{(s)}}{1} - \frac{a_{2r+1}^{(s)}}{u} - \cdots \qquad (67.1)$$

$$\mathcal{R}_m^{\text{CF}(s)}(u) = \frac{a_1^{(s)}}{u} - \frac{a_2^{(s)}}{1} - \frac{a_3^{(s)}}{u} - \cdots - \frac{a_{2m}^{(s)}}{1} - \frac{a_{2m+1}^{(s)}}{u}. \qquad (67.2)$$

Here, $a_n^{(0)} = a_n$ where a_n is the general coefficient from the non-delayed continued fraction (43.1). All the elements of the assembly $\{a_n^{(s)}\}$ can be found from the equality between the expansion coefficients of the series of the rhs of (67.1) developed in powers of u^{-1} and the signal points $\{c_{n+s}\}$ from (66.2). The analysis is entirely similar to the one performed earlier for non-delayed time signals and Green functions when $s = 0$. Therefore, it suffices to merely quote some of the main final results that will be needed in the sequel. The first such result is the delayed version of (44.62), (44.63) and (46.12)

$$a_{2n}^{(s)} = \frac{H_n(c_{s+1})H_{n-1}(c_s)}{H_{n-1}(c_{s+1})H_n(c_s)} \qquad a_{2n+1}^{(s)} = \frac{H_{n-1}(c_{s+1})H_{n+1}(c_s)}{H_n(c_s)H_n(c_{s+1})} \qquad (67.3)$$

$$\alpha_n^{(s)} = a_{2n+1}^{(s)} + a_{2n+2}^{(s)} \qquad [\beta_n^{(s)}]^2 = a_{2n}^{(s)} a_{2n+1}^{(s)} \qquad (n \geq 1) \qquad (67.4)$$

where $\alpha_n^{(0)} \equiv \alpha_n$ and $[\beta_n^{(0)}]^2 = \beta_n^2$. The first three coupling coefficients computed from (67.4) are

$$\alpha_0^{(s)} = \frac{c_{s+1}}{c_s} \qquad \alpha_1^{(s)} = \frac{c_s^2 c_{s+3} - 2c_s c_{s+1} c_{s+2} + c_{s+1}^3}{c_s(c_s c_{s+2} - c_{s+1}^2)} \qquad (67.5)$$

$$[\beta_1^{(s)}]^2 = \frac{c_s c_{s+2} - c_{s+1}^2}{c_s^2}. \qquad (67.6)$$

The quotient-difference (QD) recursive algorithm

By means of (67.3) and (67.4), a recursive algorithm can be derived for computations of all the coefficients $\{a_n^{(s)}\}$. To this end, we shall interchangeably use the following alternative notation

$$a_{2n}^{(s)} \equiv q_n^{(s)} \qquad a_{2n+1}^{(s)} \equiv e_n^{(s)} \qquad (67.7)$$

as in (63.12), (63.17) and (67.3). Then the product of $q_n^{(s)}$ with $e_n^{(s)}$ becomes

$$q_n^{(s)} e_n^{(s)} = \frac{H_{n-1}(c_s)H_n(c_{s+1})}{H_{n-1}(c_{s+1})H_n(c_s)} \frac{H_{n-1}(c_{s+1})H_{n+1}(c_s)}{H_n(c_s)H_n(c_{s+1})}$$

$$\therefore \quad q_n^{(s)} e_n^{(s)} = \frac{H_{n-1}(c_s)H_{n+1}(c_s)}{H_n^2(c_s)}. \qquad (67.8)$$

In a similar way, we can evaluate the product of $q_{n+1}^{(s)}$ and $e_n^{(s)}$ as

$$q_{n+1}^{(s)} e_n^{(s)} = \frac{H_n(c_s)H_{n+1}(c_{s+1})}{H_n(c_{s+1})H_{n+1}(c_s)} \frac{H_{n-1}(c_{s+1})H_{n+1}(c_s)}{H_n(c_s)H_n(c_{s+1})}$$

$$= \frac{H_{n-1}(c_{s+1})H_{n+1}(c_{s+1})}{H_n(c_{s+1})H_n(c_{s+1})}$$

$$= \frac{H_{n-1}(c_{s+1})H_n(c_{s+2})}{H_{n-1}(c_{s+2})H_n(c_{s+1})} \frac{H_{n-1}(c_{s+2})H_{n+1}(c_{s+1})}{H_n(c_{s+2})H_n(c_{s+1})}$$

$$= q_n^{(s+1)} e_n^{(s+1)}$$

$$q_{n+1}^{(s)} e_n^{(s)} = q_n^{(s+1)} e_n^{(s+1)} \qquad s \geq 0. \qquad (67.9)$$

Using the following well-known identity among Hankel determinants

$$[H_n(c_s)]^2 = H_n(c_{s-1})H_n(c_{s+1}) - H_{n+1}(c_{s-1})H_{n-1}(c_{s+1}) \qquad (67.10)$$

we can calculate the sum of $q_n^{(s+1)}$ and $e_{n-1}^{(s+1)}$

$$q_n^{(s+1)} + e_{n-1}^{(s+1)} = \frac{H_{n-1}(c_{s+1})H_n(c_{s+2})}{H_{n-1}(c_{s+2})H_n(c_{s+1})} + \frac{H_{n-2}(c_{s+2})H_n(c_{s+1})}{H_{n-1}(c_{s+1})H_{n-1}(c_{s+2})}$$

$$= [H_{n-1}(c_{s+2})H_{n-1}(c_{s+1})H_n(c_{s+1})]^{-1}$$

$$\times [H_{n-1}^2(c_{s+1})H_n(c_{s+2}) + H_{n-2}(c_{s+2})H_n^2(c_{s+1})]$$

$$= [H_{n-1}(c_{s+2})H_{n-1}(c_{s+1})H_n(c_{s+1})H_n(c_s)]^{-1}$$

$$\times \{H_{n-1}^2(c_{s+1})[H_n(c_s)H_n(c_{s+2})]$$

$$+ [H_n(c_s)H_{n-2}(c_{s+2})]H_n^2(c_{s+1})\}$$

$$= [H_{n-1}(c_{s+2})H_{n-1}(c_{s+1})H_n(c_{s+1})H_n(c_s)]^{-1}$$

$$\times \{H_{n-1}^2(c_{s+1})[H_n^2(c_{s+1}) + H_{n+1}(c_s)H_{n-1}(c_{s+2})]$$

$$+ H_n^2(c_{s+1})[H_{n-1}(c_s)H_{n-1}(c_{s+2}) - H_{n-1}^2(c_{s+1})]\}$$

$$= \frac{H_{n-1}^2(c_{s+1})H_{n+1}(c_s)H_{n-1}(c_{s+2})}{H_{n-1}(c_{s+2})H_{n-1}(c_{s+1})H_n(c_{s+1})H_n(c_s)}$$

$$+ \frac{H_n^2(c_{s+1})H_{n-1}(c_s)H_{n-1}(c_{s+2})}{H_{n-1}(c_{s+2})H_{n-1}(c_{s+1})H_n(c_{s+1})H_n(c_s)}$$

$$= \frac{H_{n-1}(c_s)H_n(c_{s+1})}{H_{n-1}(c_{s+1})H_n(c_s)} + \frac{H_{n-1}(c_{s+1})H_{n+1}(c_s)}{H_n(c_s)H_n(c_{s+1})} = q_n^{(s)} + e_n^{(s)}$$

$$\therefore \quad q_n^{(s)} + e_n^{(s)} = q_n^{(s+1)} + e_{n-1}^{(s+1)}. \qquad (67.11)$$

The derived relationships (67.9) and (67.11) for the delayed continued fraction coefficients are recognized as the Rutishauser quotient-difference (QD) algorithm [169]

$$\left.\begin{array}{l} e_n^{(s)} = e_{n-1}^{(s+1)} + q_n^{(s+1)} - q_n^{(s)} \\[4pt] q_{n+1}^{(s)} = q_n^{(s+1)} \dfrac{e_n^{(s+1)}}{e_n^{(s)}} \\[4pt] e_0^{(s)} = 0 \quad (s \geq 1) \qquad q_1^{(s)} = \dfrac{c_{s+1}}{c_s} \quad (s \geq 0) \end{array}\right\} . \qquad (67.12)$$

From here the first few coefficients are readily deduced in the form

$$q_1^{(s)} = \frac{c_{s+1}}{c_s}$$

$$e_1^{(s)} = \frac{c_s c_{s+2} - c_{s+1}^2}{c_s c_{s+1}}$$

$$q_2^{(s)} = \frac{c_s(c_{s+1}c_{s+3} - c_{s+2}^2)}{c_{s+1}(c_s c_{s+2} - c_{s+1}^2)}$$

$$e_2^{(s)} = \frac{c_{s+1}(c_s c_{s+2} c_{s+4} - c_{s+1}^2 c_{s+4} - c_{s+2}^3 - c_s c_{s+3}^2 + 2c_{s+1}c_{s+2}c_{s+3})}{(c_s c_{s+2} - c_{s+1}^2)(c_{s+1}c_{s+3} - c_{s+2}^2)}.$$

$$(67.13)$$

As seen from (67.12), the vectors $q_n^{(s)}$ and $e_n^{(s)}$ are generated by forming interchangeably their quotients and differences. Hence the name 'quotient-difference' for this algorithm. The QD algorithm is one of the most extensively used tools in the field of numerical analysis. The vectors $q_n^{(s)}$ and $e_n^{(s)}$ form a two-dimensional table as a double array of a lozenge form which can be depicted as

$$\begin{array}{cccccc}
& & q_1^{(0)} & & & \\
e_0^{(1)} & & & & e_1^{(0)} & \\
& q_1^{(1)} & & & & q_2^{(0)} \\
e_0^{(2)} & & & e_1^{(1)} & & & e_2^{(0)} \\
& q_1^{(2)} & & & q_2^{(1)} & & \\
e_0^{(3)} & & & e_1^{(2)} & & e_2^{(1)} & \quad\quad (67.14) \\
\vdots & q_1^{(3)} & & & q_2^{(2)} & & \\
& & e_1^{(3)} & & & e_2^{(2)} & \\
& \vdots & & \vdots & & \vdots & \ddots
\end{array}$$

where the first column is filled with zeros $e_0^{(0)} = 0$. We see from the table (67.14) that the subscript (n) and superscript (s) denote a column and a counterdiagonal, respectively. The columns of the arrays $q_n^{(s)}$ and $e_n^{(s)}$ are interleaved. The starting values are $e_0^{(s)} = 0$ ($s = 1, 2, \ldots$) and $q_1^{(s)} = c_{s+1}/c_s$ ($s = 0, 1, 2, \ldots$). Further arrays $q_n^{(s)}$ and $e_n^{(s)}$ are generated through two intertwined recursions of quantities that are located at the vertices (corners) of the lozenge in the table (67.14). The column containing only the vectors $e_n^{(s)}$ are derived via the differences $e_n^{(s)} = q_n^{(s+1)} - q_n^{(s)} + e_{n-1}^{(s+1)}$ ($n = 1, 2, \ldots$ and $s = 0, 1, 2, \ldots$). Likewise, the columns with the arrays $q_n^{(s)}$ are constructed by means of the quotients $q_n^{(s)} = q_{n-1}^{(s+1)} e_{n-1}^{(s+1)} / e_{n-1}^{(s)}$ ($n = 2, 3, \ldots$ and $s = 0, 1, 2, \ldots$). This is the whole procedure by which the QD algorithm generates one column at a time by alternatively forming the quotients and differences of the q- and e-quantities via the recursive relations from (67.12).

Chapter 68

The product-difference (PD) recursive algorithm

The Lanczos algorithm (22.1) is known to experience numerical difficulties such as loss of orthogonality among the elements of the basis set $\{|\psi_n\rangle\}$. This can, in turn, severely deteriorate the required accuracy of the coupling parameters $\{\alpha_n, \beta_n\}$ that are generated during the construction of the state vectors $\{|\psi_n\rangle\}$. Since the constants $\{\alpha_n, \beta_n\}$ are of paramount importance for spectral analysis, it is imperative to search for more stable algorithms than the Lanczos recursion (22.1) for state vectors $\{|\psi_n\rangle\}$, but still rely upon the signal points $\{c_n\}$ or equivalently, the power moments $\{\mu_n\}$ as the only input data. Since the main goal is to obtain the couplings $\{\alpha_n, \beta_n\}$, it is natural to try to alleviate any unnecessary calculations for this purpose, and especially the construction of state vectors $\{|\psi_n\rangle\}$ whose orthogonality could be destroyed during the recursion (22.1). Fortunately, there are at least two recursive algorithms that fulfil the two said requests by securing the reliance solely upon the signal points and by simultaneously avoiding the Lanczos state vectors altogether. One of them is Rutishauser's QD algorithm [169] and the other is Gordon's [170] product-difference (PD) algorithm. Both of them can compute the complete set $\{\alpha_n, \beta_n\}$ for arbitrarily large values of n. This is important especially in view of a statement from *Numerical Recipes* [176] claiming that computing the parameters $\{\alpha_n, \beta_n\}$ from the power moments must be considered as useless due to their ill-conditioning. However, such a claim does not apply to the power moments generated by the QD and PD algorithms [169, 170].

In order to write the general prescription for the PD algorithm, we first introduce an auxiliary matrix $\boldsymbol{\lambda}^{(s)} = \{\lambda_{n,m}^{(s)}\}$ with zero elements below the main

The product-difference (PD) recursive algorithm

counter-diagonal

$$\lambda^{(s)} = \begin{pmatrix} \lambda^{(s)}_{1,1} & \lambda^{(s)}_{1,2} & \lambda^{(s)}_{1,3} & \cdots & \lambda^{(s)}_{1,n-2} & \lambda^{(s)}_{1,n-1} & \lambda^{(s)}_{1,n} \\ \lambda^{(s)}_{2,1} & \lambda^{(s)}_{2,2} & \lambda^{(s)}_{2,3} & \cdots & \lambda^{(s)}_{2,n-2} & \lambda^{(s)}_{2,n-1} & 0 \\ \lambda^{(s)}_{3,1} & \lambda^{(s)}_{3,2} & \lambda^{(s)}_{3,3} & \cdots & \lambda^{(s)}_{3,n-2} & 0 & 0 \\ \vdots & \vdots & \vdots & \ddots & \vdots & \vdots & \vdots \\ \lambda^{(s)}_{n-2,1} & \lambda^{(s)}_{n-2,2} & \lambda^{(s)}_{n-2,3} & \cdots & 0 & 0 & 0 \\ \lambda^{(s)}_{n-1,1} & \lambda^{(s)}_{n-1,2} & 0 & \cdots & 0 & 0 & 0 \\ \lambda^{(s)}_{n,1} & 0 & 0 & \cdots & 0 & 0 & 0 \end{pmatrix}. \quad (68.1)$$

The first column of this matrix is initialized to zero except for the element $\lambda^{(s)}_{1,1}$ which is set to unity, whereas the second column is filled with the signal points with the alternating sign according to

$$\lambda^{(s)} = \begin{pmatrix} 1 & c_s & c_{s+1} & \cdots & \lambda^{(s)}_{1,n-2} & \lambda^{(s)}_{1,n-1} & \lambda^{(s)}_{1,n} \\ 0 & -c_{s+1} & -c_{s+2} & \cdots & \lambda^{(s)}_{2,n-2} & \lambda^{(s)}_{2,n-1} & 0 \\ 0 & c_{s+2} & c_{s+3} & \cdots & \lambda^{(s)}_{3,n-2} & 0 & 0 \\ \vdots & \vdots & \vdots & \ddots & \vdots & \vdots & \vdots \\ 0 & (-1)^{n-1}c_{n+s-3} & (-1)^{n-1}c_{n+s-2} & \cdots & 0 & 0 & 0 \\ 0 & (-1)^{n}c_{n+s-2} & 0 & \cdots & 0 & 0 & 0 \\ 0 & 0 & 0 & \cdots & 0 & 0 & 0 \end{pmatrix}. \quad (68.2)$$

Here, the general matrix element $\lambda^{(s)}_{n,m}$ is defined by

$$\lambda^{(s)}_{n,m} \equiv - \begin{vmatrix} \lambda^{(s)}_{1,m-2} & \lambda^{(s)}_{1,m-1} \\ \lambda^{(s)}_{n+1,m-2} & \lambda^{(s)}_{n+1,m-1} \end{vmatrix}. \quad (68.3)$$

This can be rewritten as a simple recursion

$$\lambda^{(s)}_{n,m} = \lambda^{(s)}_{1,m-1}\lambda^{(s)}_{n+1,m-2} - \lambda^{(s)}_{1,m-2}\lambda^{(s)}_{n+1,m-1} \quad (68.4)$$

with the initialization

$$\lambda^{(s)}_{n,1} = \delta_{n,1} \qquad \lambda^{(s)}_{n,2} = (-1)^{n+1}c_{n+s-1} \qquad \lambda^{(s)}_{n,3} = (-1)^{n+1}c_{n+s} \quad (68.5)$$

where $\delta_{n,1}$ is the Kronecker symbol (3.7). Once the arrays $\{\lambda^{(s)}_{i,j}\}$ are generated, we can compute all the coefficients $\{a^{(s)}_n\}$ of the delayed continued fractions (67.1) by using the following expression

$$a^{(s)}_n = \frac{\lambda^{(s)}_{1,n+1}}{\lambda^{(s)}_{1,n-1}\lambda^{(s)}_{1,n}} \qquad (n = 1, 2, 3, \ldots). \quad (68.6)$$

Substituting (68.6) into (67.7) it follows

$$q_n^{(s)} = \frac{\lambda_{1,2n+1}^{(s)}}{\lambda_{1,2n-1}^{(s)}\lambda_{1,2n}^{(s)}} \qquad e_n^{(s)} = \frac{\lambda_{1,2n+2}^{(s)}}{\lambda_{1,2n}^{(s)}\lambda_{1,2n+1}^{(s)}}. \qquad (68.7)$$

Finally, the Lanczos coupling parameters $\{\alpha_n^{(s)}, \beta_n^{(s)}\}$ are obtained by inserting the string $\{a_n^{(s)}\}$ into (67.4). The explicit dependence of the pair $\{\alpha_n^{(s)}, \beta_n^{(s)}\}$ upon the auxiliary elements $\{\lambda_{1,n}^{(s)}\}$ follows from (67.4) and (68.6) as

$$\alpha_n^{(s)} = \frac{[\lambda_{1,2n+2}^{(s)}]^2 + \lambda_{1,2n}^{(s)}\lambda_{1,2n+3}^{(s)}}{\lambda_{1,2n}^{(s)}\lambda_{1,2n+1}^{(s)}\lambda_{1,2n+2}^{(s)}} \qquad \beta_n^{(s)} = \frac{\lambda_{1,2n+2}^{(s)}}{\lambda_{1,2n-1}^{(s)}[\lambda_{1,2n}^{(s)}]^2}. \qquad (68.8)$$

We see that the recursion (68.3) on the vectors $\{\lambda_{n,m}^{(s)}\}$ involves only their products and differences, but no divisions. Hence the name 'product-difference' algorithm. The PD algorithm for non-delayed signals/moments ($s = 0$) has been introduced in [170]. The extension of the PD algorithm to delayed time signals/moments $\{c_{n+s}\} = \{\mu_{n+s}\}$ is supplied in this chapter. As seen from (68.6), the PD algorithm performs the division only once at the end of the computations to arrive straight at the delayed CF coefficients $\{a_n^{(s)}\}$. For this reason, the PD algorithm is error-free for signal points $\{c_{n+s}\}$ that are integers. Such integer data matrices $\{c_{n+s}\}$ are measured experimentally throughout magnetic resonance phenomena (NMR, MRS, etc). The same infinite-order precision (no round-off errors) is achievable within the PD algorithm for auto-correlation functions or power moments $\{C_{n+s}\} = \{\mu_{n+s}\}$ given as rational numbers. In many cases of physical interest (e.g. systems exposed to external fields), the role of signal points is played by expansion coefficients that are obtained exactly as rational numbers from the quantum-mechanical perturbation theory. Here, one would operate directly with rational numbers by means of symbolic language programings, such as MAPLE and the like [91]. The computational complexity of the PD algorithm for the CF coefficients $\{a_m^{(s)}\}$ ($1 \leq m \leq n$) is of the order of n^2 multiplications. By comparison, a direct computation of the Hankel determinant $H_n(c_s)$ of the dimension n entering the definition (67.3) for $\{a_n^{(s)}\}$ requires, within the Cramer rule, some formidable $n!$ multiplications that would preclude any meaningful application for large n.

We saw in chapter 67 that the QD algorithm (67.12) for the auxiliary double array $\{q_n^{(s)}, e_n^{(s)}\}$ carries out divisions in each iteration. In a finite-precision arithmetic, this could lead to round-off errors that might cause the QD algorithm to break down for non-integer signal points. However, if the input data $\{c_{n+s}\}$ are non-zero integers, then divisions would produce rational numbers during the QD recursion. This would be innocuous leading to error-free results, provided that the infinite-precision arithmetic with rational numbers is employed via, e.g. MAPLE [91].

It is also possible to show that the vectors $\{\lambda_{1,m}^{(s)}\}$ can advantageously be combined to produce Hankel determinants of arbitrary orders. To this end, using (68.3), we have generated the first several vectors $\{\lambda_{n,m}^{(s)}\}$ and their particular values yield the following relationships with the Hankel determinants $\{H_n(c_s)\}$ from (7.20), (7.22) and (7.24) [142]

$$H_2(c_s) = \frac{\lambda_{1,4}^{(s)}}{\prod_{m=1}^{2} \lambda_{1,m}^{(s)}} H_1(c_s)$$

$$H_3(c_s) = \frac{\lambda_{1,6}^{(s)}}{\prod_{m=1}^{4} \lambda_{1,m}^{(s)}} H_2(c_s)$$

$$H_4(c_s) = \frac{\lambda_{1,8}^{(s)}}{\prod_{m=1}^{6} \lambda_{1,m}^{(s)}} H_3(c_s) \quad \text{etc}$$

$$\therefore \quad H_n(c_s) = \frac{\lambda_{1,2n}^{(s)}}{\lambda_{2n-2}^{(s)}} H_{n-1}(c_s) \qquad \lambda_n^{(s)} = \prod_{m=1}^{n} \lambda_{1,m}^{(s)}. \qquad (68.9)$$

The recursion (68.9) can be trivially solved by iterations with the explicit result

$$H_n(c_s) = c_s \prod_{m=2}^{n} \frac{\lambda_{1,2m}^{(s)}}{\lambda_{2m-2}^{(s)}} \quad (n = 2, 3, \ldots) \qquad H_1(c_s) = c_s. \qquad (68.10)$$

This completes the demonstration that the general-order Hankel determinant $H_n(c_s)$ can be easily obtained from the recursively pre-computed string $\{\lambda_{1,m}^{(s)}\}$. We have verified explicitly that, e.g. the particular results (7.20), (7.22) and (7.24) for $H_2(c_s)$, $H_3(c_s)$ and $H_4(c_s)$, respectively, can be reproduced exactly from (68.10). A remarkable feature of the expression (68.10) is that it effectively carries out only the computation of the simplest 2×2 determinant from (68.3). For integer data $\{c_{n+s}\}$, the determinant $H_n(c_s)$ is also an integer number, say $N_n^{(s)}$. But in such a case, the expression (68.10) for $H_n(c_s)$ would evidently be a rational number. However, by construction (68.9), the numerator in (68.10) is, in fact, equal to $\prod_{m=2}^{n} \lambda_{1,2m}^{(s)} = N_n^{(s)} \prod_{m=2}^{n} \lambda_{2m-2}^{(s)}$ so that $H_n(c_s) = N_n^{(s)}$, as it should be. To achieve this in practice, integer algebra could be used in which the generated integers $\{\lambda_{i,j}^{(s)}\}$ should be kept in their composite intermediate forms without carrying out the final multiplications in (68.10). This would allow the exact cancellation of the denominator $\prod_{m=2}^{n} \lambda_{2m-2}^{(s)}$ by the corresponding part of the numerator in $H_n(c_s)$ from (68.10) to yield the exact result in the integer form $H_n(c_s) = N_n^{(s)}$.

The generalization of the PD algorithm from its original non-delayed variant of Gordon [170] to the present delayed version is particularly advantageous regarding the eigenvalues $\{u_k\}$. Namely, having only the non-delayed CF coefficients $\{a_n\} \equiv \{a_n^{(0)}\}$ as in [170], the eigenvalues $\{u_k\}$ can be obtained either by rooting the characteristic polynomial $Q_K(u) = 0$ or by solving the eigenproblem for the Jacobi matrix $\mathbf{J}$ from (64.49). However, the delayed CF coefficients $\{a_n^{(s)}\}$ can bypass altogether these two latter standard procedures and provide an alternative way of obtaining the eigenvalues of data matrices using (63.11)

$$u_k = \lim_{s\to\infty} a_{2k}^{(s)} = \lim_{s\to\infty} \frac{\lambda_{1,2k+1}^{(s)}}{\lambda_{1,2k-1}^{(s)} \lambda_{1,2k}^{(s)}}. \tag{68.11}$$

For the purpose of checking, it is also useful to consider the same limit $s \longrightarrow \infty$ in the string $\{a_{2n+1}^{(s)}\}$ by employing (63.16) which is reduced to

$$\lim_{s\to\infty} a_{2k+1}^{(s)} = \lim_{s\to\infty} \frac{\lambda_{1,2k+2}^{(s)}}{\lambda_{1,2k}^{(s)} \lambda_{1,2k+1}^{(s)}} = 0. \tag{68.12}$$

The accuracy of the results for the eigenvalues $\{u_k\}$ computed in this way using the delayed PD algorithm can be checked against the formula $u_k = \lim_{s\to\infty} a_{2k}^{(s)}$ by means of the analytical expression for $a_{2k}^{(s)}$. The exact closed formula for the general delayed CF coefficients $a_n^{(s)}$ can be obtained directly from $a_n^{(0)} \equiv a_n$ by replacing every c_n by c_{n+s}. Thus it follows from (44.49)–(44.52)

$$a_{n+1}^{(s)} = \frac{c_{n+s} - \sigma_n^{(s)} \pi_{n-1}^{(s)} - \lambda_n^{(s)} \pi_{n-4}^{(s)}}{\pi_n^{(s)}} \tag{68.13}$$

$$\pi_n^{(s)} = \prod_{i=1}^{n} a_i^{(s)} \qquad \sigma_n^{(s)} = \left[\sum_{j=2}^{n} a_j^{(s)}\right]^2 \tag{68.14}$$

$$\lambda_n^{(s)} = \sum_{j=\left[\frac{n-1}{2}\right]}^{n-3} a_j^{(s)} [\xi_j^{(s)}]^2 \qquad \xi_j^{(s)} = \sum_{k=2}^{j+1} a_k^{(s)} \sum_{\ell=2}^{k+1} a_\ell^{(s)} \tag{68.15}$$

$$\pi_n^{(s)} \equiv 0 \quad (n \leq 0) \qquad \sigma_n^{(s)} \equiv 0 \quad (n \leq 3). \tag{68.16}$$

Of course, just like in (44.50), if we have the CF coefficients $\{a_n^{(s)}\}$, then the input data $\{c_{n+s}\}$ could be retrieved exactly by means of the formula

$$c_{n+s} = \pi_{n+1}^{(s)} + \sigma_n^{(s)} \pi_{n-1}^{(s)} + \lambda_n^{(s)} \pi_{n-4}^{(s)}. \tag{68.17}$$

Clearly, with the availability of the exact delayed CF coefficients $\{a_n^{(s)}\}$ from (68.13), one can immediately obtain the exact delayed continued fraction of a fixed order as the explicit polynomial quotient, which is the corresponding Padé approximant. For example, the even-order delayed CF is the following Padé approximant as in (43.7)

$$\mathcal{R}_{2n}^{\text{CF}(s)}(u) = a_1^{(s)} \frac{\tilde{P}_n^{\text{CF}(s)}(u)}{\tilde{Q}_n^{\text{CF}(s)}(u)}. \tag{68.18}$$

Similarly to (43.8), the delayed odd-order CF, which is denoted by $\mathcal{R}_{2n-1}^{\text{CF}(s)}(u)$, is obtained from the even-order CF by setting $a_{2n}^{(s)} \equiv 0$

$$\mathcal{R}_{2n-1}^{\text{CF}(s)}(u) \equiv \{\mathcal{R}_{2n}^{\text{CF}(s)}(u)\}_{a_{2n}^{(s)}=0} \qquad (n=1,2,3,\ldots). \tag{68.19}$$

The polynomials $\tilde{P}_n^{\text{CF}(s)}(u)$ and $\tilde{Q}_n^{\text{CF}(s)}(u)$ from (68.18) can be defined through their general power series representations as in (43.12)

$$\tilde{P}_n^{\text{CF}(s)}(u) = \sum_{r=0}^{n-1} \tilde{p}_{n,n-r}^{(s)} u^r \qquad \tilde{Q}_n^{\text{CF}(s)}(u) = \sum_{r=0}^{n} \tilde{q}_{n,n-r}^{(s)} u^r. \tag{68.20}$$

The expansion coefficients $\tilde{p}_{n,n-r}^{(s)}$ and $\tilde{q}_{n,n-r}^{(s)}$ are the generalizations of (43.19) and (43.20), respectively

$$\tilde{p}_{n,m}^{(s)} = (-1)^{m-1} \underbrace{\sum_{r_1=3}^{2(n-m+2)} a_{r_1}^{(s)} \sum_{r_2=r_1+2}^{2(n-m+3)} a_{r_2}^{(s)} \cdots \sum_{r_{m-1}=r_{m-2}+2}^{2n} a_{r_{m-1}}^{(s)}}_{m-1 \text{ summations}} \tag{68.21}$$

$$\tilde{q}_{n,m}^{(s)} = (-1)^{m} \underbrace{\sum_{r_1=2}^{2(n-m+1)} a_{r_1}^{(s)} \sum_{r_2=r_1+2}^{2(n-m+2)} a_{r_2}^{(s)} \sum_{r_3=r_2+2}^{2(n-m+3)} a_{r_3}^{(s)} \cdots \sum_{r_m=r_{m-1}+2}^{2n} a_{r_m}^{(s)}}_{m \text{ summations}}$$

$$\tag{68.22}$$

where $n \geq m$.

Chapter 69

Delayed Lanczos continued fractions

Using (67.7), the infinite-order and the mth-order delayed continued fractions $\mathcal{R}^{\mathrm{CF}(s)}(u)$ and $\mathcal{R}_m^{\mathrm{CF}(s)}(u)$ from (67.1) and (67.2) can respectively be written as

$$\mathcal{R}^{\mathrm{CF}(s)}(u) = \frac{c_s}{u} - \frac{q_1^{(s)}}{1} - \frac{e_1^{(s)}}{u} - \cdots - \frac{q_r^{(s)}}{1} - \frac{e_r^{(s)}}{u} - \cdots \qquad (69.1)$$

$$\mathcal{R}_m^{\mathrm{CF}(s)}(u) = \frac{c_s}{u} - \frac{q_1^{(s)}}{1} - \frac{e_1^{(s)}}{u} - \cdots - \frac{q_m^{(s)}}{1} - \frac{e_m^{(s)}}{u}. \qquad (69.2)$$

Likewise, the infinite-order and the mth-order of the even part of the corresponding Lanczos continued fractions are respectively defined as

$$\begin{aligned}
\mathcal{R}_e^{\mathrm{LCF}(s)}(u) &= \frac{c_s}{u - q_1^{(s)} - e_0^{(s)}} - \frac{q_1^{(s)} e_1^{(s)}}{u - q_2^{(s)} - e_1^{(s)}} - \cdots - \frac{q_r^{(s)} e_r^{(s)}}{u - q_{r+1}^{(s)} - e_r^{(s)}} - \cdots \\
&= \frac{c_s}{u - \alpha_0^{(s)}} - \frac{[\beta_1^{(s)}]^2}{u - \alpha_1^{(s)}} - \cdots - \frac{[\beta_r^{(s)}]^2}{u - \alpha_r^{(s)}} - \cdots \qquad (69.3)
\end{aligned}$$

$$\begin{aligned}
\mathcal{R}_{e,m}^{\mathrm{LCF}(s)}(u) &= \frac{c_s}{u - q_1^{(s)} - e_0^{(s)}} - \frac{q_1^{(s)} e_1^{(s)}}{u - q_2^{(s)} - e_1^{(s)}} - \cdots - \frac{q_m^{(s)} e_m^{(s)}}{u - q_{m+1}^{(s)} - e_m^{(s)}} \\
&= \frac{c_s}{u - \alpha_0^{(s)}} - \frac{[\beta_1^{(s)}]^2}{u - \alpha_1^{(s)}} - \cdots - \frac{[\beta_m^{(s)}]^2}{u - \alpha_m^{(s)}}. \qquad (69.4)
\end{aligned}$$

To establish a general relationship between $\mathcal{R}_{e,m}^{\mathrm{LCF}(s)}(u)$ and $\mathcal{R}_m^{\mathrm{CF}(s)}(u)$ it is sufficient to extract explicitly a few of the first terms from (69.1)–(69.4). For example, setting $m = 2$ in (69.2) and $m = 1$ in (69.4) gives

$$\mathcal{R}_2^{\text{CF}(s)}(u) = \frac{c_s}{u} - \frac{q_1^{(s)}}{1} = \frac{c_s}{u - q_1^{(s)}} = \frac{c_s}{u - \alpha_0^{(s)}} \tag{69.5}$$

$$\mathcal{R}_{e,1}^{\text{LCF}(s)}(u) = \frac{c_s}{u - q_1^{(s)}} = \frac{c_s}{u - \alpha_0^{(s)}} \tag{69.6}$$

$$\therefore \quad \mathcal{R}_{e,1}^{\text{LCF}(s)}(u) = \mathcal{R}_2^{\text{CF}(s)}(u). \tag{69.7}$$

Similarly, letting $m = 4$ in (69.2) and $m = 2$ in (69.4) yields

$$\mathcal{R}_4^{\text{CF}(s)}(u) = \frac{c_s}{u} - \frac{q_1^{(s)}}{1} - \frac{e_1^{(s)}}{u} - \frac{q_2^{(s)}}{1}$$

$$= \frac{c_s}{u - q_1^{(s)} - \dfrac{q_1^{(s)} e_1^{(s)}}{u - q_2^{(s)} - e_1^{(s)}}} = \frac{c_s}{u - q_1^{(s)}} - \frac{q_1^{(s)} e_1^{(s)}}{u - q_2^{(s)} - e_1^{(s)}} \tag{69.8}$$

$$\mathcal{R}_{e,2}^{\text{LCF}(s)}(u) = \frac{c_s}{u - \alpha_0^{(s)}} - \frac{[\beta_1^{(s)}]^2}{u - \alpha_1^{(s)}} = \frac{c_s}{u - q_1^{(s)}} - \frac{q_1^{(s)} e_1^{(s)}}{u - q_2^{(s)} - e_1^{(s)}}$$

$$= \frac{c_s[u - \alpha_1^{(s)}]}{u^2 - [\alpha_0^{(s)} + \alpha_1^{(s)}]u + \{\alpha_0^{(s)} \alpha_1^{(s)} - [\beta_1^{(s)}]^2\}} \tag{69.9}$$

$$\therefore \quad \mathcal{R}_{e,2}^{\text{LCF}(s)}(u) = \mathcal{R}_4^{\text{CF}(s)}(u). \tag{69.10}$$

The polynomial coefficients in (69.9) can be expressed via signal points alone as follows

$$\alpha_1^{(s)} = \frac{c_s^2 c_{s+3} - 2 c_s c_{s+1} c_{s+2} + c_{s+1}^3}{c_s(c_s c_{s+2} - c_{s+1}^2)} \tag{69.11}$$

$$\alpha_0^{(s)} + \alpha_1^{(s)} = \frac{c_s c_{s+3} - c_{s+1} c_{s+2}}{c_s c_{s+2} - c_{s+1}^2} \tag{69.12}$$

$$\alpha_0^{(s)} \alpha_1^{(s)} - [\beta_1^{(s)}]^2 = \frac{c_{s+2}^2 - c_{s+1} c_{s+3}}{c_{s+1}^2 - c_s c_{s+2}}. \tag{69.13}$$

We carried out similar calculation for higher orders m and observed that all the particular results (69.7), (69.10), etc, satisfy the following general pattern

$$\mathcal{R}_{e,n}^{\text{LCF}(s)}(u) = \mathcal{R}_{2n}^{\text{CF}(s)}(u). \tag{69.14}$$

We see that $\mathcal{R}_{2n}^{\text{CF}(s)}(u)$ is matched by $\mathcal{R}_{e,n}^{\text{LCF}(s)}(u)$ which is called equivalently the contracted continued fraction $\mathcal{R}_{e,n}^{\text{CCF}(s)}(u) = \mathcal{R}_{2n}^{\text{CF}(s)}(u)$. The relation (69.14) for delayed time signals is an extension of the corresponding result (64.47) for non-delayed signals.

There is also the infinite-order and the mth-order odd part of (69.1) denoted respectively by $\mathcal{R}_o^{\text{LCF}(s)}(u)$ and $\mathcal{R}_{o,m}^{\text{LCF}(s)}(u)$. By definition

$$\mathcal{R}_o^{\text{LCF}(s)}(u) = \frac{c_s}{u}$$
$$\times \left[1 + \frac{q_1^{(s)}}{u - q_1^{(s)} - e_1^{(s)}} - \frac{q_2^{(s)} e_1^{(s)}}{u - q_2^{(s)} - e_2^{(s)}} - \cdots - \frac{q_r^{(s)} e_{r-1}^{(s)}}{u - q_r^{(s)} - e_r^{(s)}} - \cdots \right] \quad (69.15)$$

$$\mathcal{R}_{o,m}^{\text{LCF}(s)}(u) = \frac{1}{u}$$
$$\times \left[c_s + \frac{c_{s+1}}{u - q_1^{(s)} - e_1^{(s)}} - \frac{q_2^{(s)} e_1^{(s)}}{u - q_2^{(s)} - e_2^{(s)}} - \cdots - \frac{q_m^{(s)} e_{m-1}^{(s)}}{u - q_m^{(s)} - e_m^{(s)}}\right]$$
$$= \frac{1}{u} \left\{ c_s + \frac{c_{s+1}}{u - \alpha_0^{(s+1)}} - \frac{[\beta_1^{(s+1)}]^2}{u - \alpha_1^{(s+1)}} - \cdots - \frac{[\beta_{m-1}^{(s+1)}]^2}{u - \alpha_{m-1}^{(s+1)}} \right\} \quad (69.16)$$

where $m = 1, 2, 3, \ldots$. Comparing (69.15) with the identity

$$\sum_{n=0}^{\infty} c_{n+s} u^{-n-1} = \frac{c_s}{u} + \frac{1}{u} \sum_{n=0}^{\infty} c_{n+s+1} u^{-n-1} \quad (69.17)$$

it follows

$$\sum_{n=0}^{\infty} c_{n+s+1} u^{-n-1}$$
$$= c_s \left[\frac{q_1^{(s)}}{u - q_1^{(s)} - e_1^{(s)}} - \frac{q_2^{(s)} e_1^{(s)}}{u - q_2^{(s)} - e_2^{(s)}} - \cdots - \frac{q_{r+1}^{(s)} e_r^{(s)}}{u - q_{r+1}^{(s)} - e_{r+1}^{(s)}} - \cdots \right]$$
$$= c_s \left\{ \frac{q_1^{(s)}}{u - \gamma_1^{(s)}} - \frac{[\delta_1^{(s)}]^2}{u - \gamma_2^{(s)}} - \cdots - \frac{[\delta_r^{(s)}]^2}{u - \gamma_{r+1}^{(s)}} - \cdots \right\} \quad (69.18)$$

$$\gamma_n^{(s)} = q_n^{(s)} + e_n^{(s)} \qquad [\delta_n^{(s)}]^2 = q_{n+1}^{(s)} e_n^{(s)}. \quad (69.19)$$

However, using (67.4) and (67.12) we have

$$q_n^{(s)} + e_n^{(s)} = q_n^{(s+1)} + e_{n-1}^{(s+1)} = \alpha_{n-1}^{(s+1)} \quad (69.20)$$
$$q_{n+1}^{(s)} e_n^{(s)} = q_n^{(s+1)} e_n^{(s+1)} = [\beta_n^{(s+1)}]^2 \quad (69.21)$$
$$\gamma_n^{(s)} = \alpha_{n-1}^{(s+1)} \qquad [\delta_n^{(s)}]^2 = [\beta_n^{(s+1)}]^2 \quad (69.22)$$

so that

$$\sum_{n=0}^{\infty} c_{n+s+1} u^{-n-1}$$

$$= \frac{c_{s+1}}{u - q_1^{(s)} - e_1^{(s)}} - \frac{q_2^{(s)} e_1^{(s)}}{u - q_2^{(s)} - e_2^{(s)}} - \cdots - \frac{q_{r+1}^{(s)} e_r^{(s)}}{u - q_{r+1}^{(s)} - e_{r+1}^{(s)}} - \cdots$$

$$= \frac{c_{s+1}}{u - \alpha_0^{(s+1)}} - \frac{[\beta_1^{(s+1)}]^2}{u - \alpha_1^{(s+1)}} - \cdots - \frac{[\beta_r^{(s+1)}]^2}{u - \alpha_r^{(s+1)}} - \cdots . \qquad (69.23)$$

It is seen now that the second line of (69.23) coincides with the second line of (69.3), when s is replaced by $s+1$ as it should be. Therefore (69.23) is a retrospective proof that (69.15) is correct. Then, returning to (69.16) for the odd part of $\mathcal{R}_n^{\text{CCF}(s)}(u)$, which is $\mathcal{R}_{e,n}^{\text{CCF}(s)}(u) = \mathcal{R}_{e,n}^{\text{LCF}(s)}(u)$, we have for $m = 1$

$$\mathcal{R}_{o,1}^{\text{LCF}(s)}(u) = \frac{1}{u}\left[c_s + \frac{c_{s+1}}{u - q_1^{(s)} - e_1^{(s)}}\right]. \qquad (69.24)$$

Inserting $m = 3$ in (69.2), it follows

$$\mathcal{R}_3^{\text{CF}(s)}(u) = \frac{c_s}{u} - \frac{q_1^{(s)}}{1} - \frac{e_1^{(s)}}{u} = \frac{c_s}{u - \dfrac{q_1^{(s)}}{1 - \dfrac{e_1^{(s)}}{u}}} = \frac{c_s}{u - \dfrac{q_1^{(s)} u}{u - e_1^{(s)}}}$$

$$= \frac{c_s}{u} + \frac{c_s}{u - \dfrac{q_1^{(s)} u}{u - e_1^{(s)}}} - \frac{c_s}{u}$$

$$= \frac{c_s}{u} + \left\{\frac{c_s[u - e_1^{(s)}]}{u[u - e_1^{(s)}] - q_1^{(s)} u} - \frac{c_s}{u}\right\}$$

$$= \frac{c_s}{u} + \frac{c_s}{u} \frac{[u - e_1^{(s)}] - [u - q_1^{(s)} - e_1^{(s)}]}{u - q_1^{(s)} - e_1^{(s)}}$$

$$= \frac{c_s}{u} + \frac{c_s}{u} \frac{q_1^{(s)}}{u - q_1^{(s)} - e_1^{(s)}}$$

$$\therefore \quad \mathcal{R}_3^{\text{CF}(s)}(u) = \frac{1}{u}\left[c_s + \frac{c_{s+1}}{u - q_1^{(s)} - e_1^{(s)}}\right] = \frac{1}{u}\left[c_s + \frac{c_{s+1}}{u - \alpha_0^{(s+1)}}\right]. \qquad (69.25)$$

Comparison of (69.24) and (69.25) yields the equality

$$\mathcal{R}_{o,1}^{\text{LCF}(s)}(u) = \mathcal{R}_3^{\text{CF}(s)}(u). \qquad (69.26)$$

We also have the following result for $\mathcal{R}_3^{CF(s)}(u)$ deduced from its non-delayed counterpart (43.10)

$$\mathcal{R}_3^{CF(s)}(u) = \frac{a_1^{(s)}}{u} \frac{u - a_3^{(s)}}{u - a_2^{(s)} - a_3^{(s)}}. \tag{69.27}$$

Employing (67.7) and (69.20), we can rewrite (69.27) as

$$\mathcal{R}_3^{CF(s)}(u) = \frac{c_s}{u} \frac{u - e_1^{(s)}}{u - q_1^{(s)} - e_1^{(s)}} = \frac{c_s}{u} \frac{u - e_1^{(s)}}{u - \alpha_0^{(s+1)}}. \tag{69.28}$$

Furthermore, we have

$$e_1^{(s)} = \alpha_0^{(s+1)} - \alpha_0^{(s)} \tag{69.29}$$

so that (69.28) becomes

$$\mathcal{R}_3^{CF(s)}(u) = \frac{c_s}{u} \frac{u - \alpha_0^{(s+1)} + \alpha_0^{(s)}}{u - \alpha_0^{(s+1)}}. \tag{69.30}$$

The results (69.25) and (69.28) or (69.30) must be identical to each other, and to check this we calculate

$$\frac{1}{u}\left[c_s + \frac{c_{s+1}}{u - q_1^{(s)} - e_1^{(s)}}\right] = \frac{c_s}{u}\left[1 + \frac{q_1^{(s)}}{u - q_1^{(s)} - e_1^{(s)}}\right]$$

$$= \frac{c_s}{u} \frac{u - q_1^{(s)} - e_1^{(s)} + q_1^{(s)}}{u - q_1^{(s)} - e_1^{(s)}} = \frac{c_s}{u} \frac{u - e_1^{(s)}}{u - q_1^{(s)} - e_1^{(s)}}$$

$$\therefore \quad [\mathcal{R}_3^{CF(s)}(u)]_{\text{equation (69.25)}} = [\mathcal{R}_3^{CF(s)}(u)]_{\text{equation (69.28)}} \quad \text{(QED).} \tag{69.31}$$

Due to (69.26) and (69.30) we can write (69.24) as

$$\mathcal{R}_{o,1}^{LCF(s)}(u) = \frac{c_s}{u} \frac{u - \alpha_0^{(s+1)} + \alpha_0^{(s)}}{u - \alpha_0^{(s+1)}} \tag{69.32}$$

$$\alpha_0^{(s+1)} = \frac{c_{s+2}}{c_{s+1}} \qquad \alpha_0^{(s+1)} - \alpha_0^{(s)} = \frac{c_s c_{s+2} - c_{s+1}^2}{c_s c_{s+1}}. \tag{69.33}$$

Next, we set $m = 2$ in (69.16) and extract the term

$$\mathcal{R}_{o,2}^{LCF(s)}(u) = \frac{1}{u}\left\{c_s + \frac{c_{s+1}}{u - \alpha_0^{(s+1)}} - \frac{[\beta_1^{(s+1)}]^2}{u - \alpha_1^{(s+1)}}\right\}. \tag{69.34}$$

This with the help of (69.15) can also be written explicitly as

$$\mathcal{R}_{0,2}^{LCF(s)}(u) = \frac{c_s}{u} + \frac{1}{u} \frac{c_{s+1}}{u - \alpha_0^{(s+1)} - \dfrac{[\beta_1^{(s+1)}]^2}{u - \alpha_1^{(s+1)}}}$$

$$= \frac{c_s}{u} + \frac{1}{u} \frac{q_1^{(s)}}{u - q_1^{(s)} - e_1^{(s)} - \dfrac{q_2^{(s)} e_1^{(s)}}{u - q_2^{(s)} - e_2^{(s)}}}. \quad (69.35)$$

For $m = 5$ we have from (67.2) and (69.2)

$$\mathcal{R}_5^{CF(s)}(u) = \frac{c_s}{u} - \frac{q_1^{(s)}}{1} - \frac{e_1^{(s)}}{u} - \frac{q_2^{(s)}}{1} - \frac{e_2^{(s)}}{u}$$

$$= \frac{a_1^{(s)}}{u} - \frac{a_2^{(s)}}{1} - \frac{a_3^{(s)}}{u} - \frac{a_4^{(s)}}{1} - \frac{a_5^{(s)}}{u}. \quad (69.36)$$

The corresponding form of the explicit polynomial quotient from (69.36) is

$$\mathcal{R}_5^{CF(s)}(u) = \frac{c_s}{u} \frac{u^2 - [a_3^{(s)} + a_4^{(s)} + a_5^{(s)}]u + a_3^{(s)} a_5^{(s)}}{u^2 - [a_2^{(s)} + a_3^{(s)} + a_4^{(s)} + a_5^{(s)}]u + \{a_2^{(s)}[a_4^{(s)} + a_5^{(s)}] + a_3^{(s)} a_5^{(s)}\}}. \quad (69.37)$$

Using (67.4) and (67.7) it can be shown that the following relations hold

$$a_2^{(s)} + a_3^{(s)} + a_4^{(s)} + a_5^{(s)} = \alpha_0^{(s+1)} + \alpha_1^{(s+1)} \quad (69.38)$$

$$a_2^{(s)}[a_4^{(s)} + a_5^{(s)}] + a_3^{(s)} a_5^{(s)} = \alpha_0^{(s+1)} \alpha_1^{(s+1)} - [\beta_1^{(s+1)}]^2 \quad (69.39)$$

$$a_3^{(s)} a_5^{(s)} = e_1^{(s)} e_2^{(s)} = [\alpha_0^{(s+1)} - \alpha_0^{(s)}]\alpha_1^{(s+1)} - [\beta_1^{(s+1)}]^2 \quad (69.40)$$

$$a_3^{(s)} = e_1^{(s)} = \alpha_0^{(s+1)} - \alpha_0^{(s)} \quad (69.41)$$

$$a_3^{(s)} + a_4^{(s)} + a_5^{(s)} = a_3^{(s)} + \alpha_1^{(s+1)} = \alpha_0^{(s+1)} - \alpha_0^{(s)} + \alpha_1^{(s+1)}. \quad (69.42)$$

We can also express (69.38)–(69.42) in terms of signal points only

$$\alpha_0^{(s+1)} + \alpha_1^{(s+1)} = \frac{c_{s+2} c_{s+3} - c_{s+1} c_{s+4}}{c_{s+2}^2 - c_{s+1} c_{s+3}} \quad (69.43)$$

$$\alpha_0^{(s+1)} \alpha_1^{(s+1)} - [\beta_1^{(s+1)}]^2 = \frac{c_{s+3}^2 - c_{s+2} c_{s+4}}{c_{s+2}^2 - c_{s+1} c_{s+3}}. \quad (69.44)$$

$$[\alpha_0^{(s+1)} - \alpha_0^{(s)}]\alpha_1^{(s+1)} - [\beta_1^{(s+1)}]^2$$

$$= \frac{c_s c_{s+2} c_{s+4} - c_{s+1}^2 c_{s+4} + 2c_{s+1} c_{s+2} c_{s+3} - c_{s+2}^3 - c_s c_{s+3}^2}{c_s (c_{s+1} c_{s+3} - c_{s+2}^2)} \quad (69.45)$$

$$\alpha_0^{(s+1)} - \alpha_0^{(s)} + \alpha_1^{(s+1)}$$

$$= \frac{c_{s+1} c_{s+2}^2 - c_{s+1}^2 c_{s+3} - c_s c_{s+2} c_{s+3} + c_s c_{s+1} c_{s+4}}{c_s (c_{s+1} c_{s+3} - c_{s+2}^2)}. \quad (69.46)$$

Inserting (69.38)–(69.42) into (69.37) finally gives

$$\mathcal{R}_5^{\text{CF}(s)}(u) = \frac{c_s}{u}$$
$$\times \frac{u^2 - [\alpha_0^{(s+1)} - \alpha_0^{(s)} + \alpha_1^{(s+1)}]u + \{[\alpha_0^{(s+1)} - \alpha_0^{(s)}]\alpha_1^{(s+1)} - [\beta_1^{(s+1)}]^2\}}{u^2 - [\alpha_0^{(s+1)} + \alpha_1^{(s+1)}]u + \{\alpha_0^{(s+1)}\alpha_1^{(s+1)} - [\beta_1^{(s+1)}]^2\}}.$$

(69.47)

On the other hand, we can transform (69.35) as follows

$$\frac{u}{c_s}\mathcal{R}_{o,2}^{\text{LCF}(s)}(u)$$

$$= 1 + \cfrac{q_1^{(s)}}{u - \alpha_0^{(s+1)} - \cfrac{[\beta_1^{(s+1)}]^2}{u - \alpha_1^{(s+1)}}} = 1 + \cfrac{\alpha_0^{(s)}}{u - \alpha_0^{(s+1)} - \cfrac{[\beta_1^{(s+1)}]^2}{u - \alpha_1^{(s+1)}}}$$

$$= \frac{[u - \alpha_0^{(s+1)}][u - \alpha_1^{(s+1)}] - [\beta_1^{(s+1)}]^2 + \alpha_0^{(s)}[u - \alpha_1^{(s+1)}]}{[u - \alpha_0^{(s+1)}][u - \alpha_1^{(s+1)}] - [\beta_1^{(s+1)}]^2}$$

$$= \frac{u^2 - [\alpha_0^{(s+1)} - \alpha_0^{(s)} + \alpha_1^{(s+1)}]u + \{[\alpha_0^{(s+1)} - \alpha_0^{(s)}]\alpha_1^{(s+1)} - [\beta_1^{(s+1)}]^2\}}{u^2 - [\alpha_0^{(s+1)} + \alpha_1^{(s+1)}]u + \{\alpha_0^{(s+1)}\alpha_1^{(s+1)} - [\beta_1^{(s+1)}]^2\}}$$

so that

$$\mathcal{R}_{o,2}^{\text{LCF}(s)}(u) = \frac{c_s}{u}$$
$$\times \frac{u^2 - [\alpha_0^{(s+1)} - \alpha_0^{(s)} + \alpha_1^{(s+1)}]u + \{[\alpha_0^{(s+1)} - \alpha_0^{(s)}]\alpha_1^{(s+1)} - [\beta_1^{(s+1)}]^2\}}{u^2 - [\alpha_0^{(s+1)} + \alpha_1^{(s+1)}]u + \{\alpha_0^{(s+1)}\alpha_1^{(s+1)} - [\beta_1^{(s+1)}]^2\}}.$$

(69.48)

An inspection of (69.47) and (69.48) gives the identity

$$\mathcal{R}_{o,2}^{\text{LCF}(s)}(u) = \mathcal{R}_5^{\text{CF}(s)}(u). \tag{69.49}$$

We continued this type of calculation for higher orders and verified that all these particular results, such as (69.26), (69.49), etc, conform with the general relationship

$$\mathcal{R}_{o,n}^{\text{LCF}(s)}(u) = \mathcal{R}_{2n+1}^{\text{CF}(s)}(u). \tag{69.50}$$

Thus, we see from (69.14) and (69.50) that the even and odd parts of the delayed Lanczos approximants $\mathcal{R}_{e,n}^{\text{LCF}(s)}(u)$ and $\mathcal{R}_{o,n}^{\text{LCF}(s)}(u)$ of order n ($n = 1, 2, 3, \ldots$) are equal to the delayed continued fractions $\mathcal{R}_{2n}^{\text{CF}(s)}(u)$ and $\mathcal{R}_{2n+1}^{\text{CF}(s)}(u)$ of orders $2n$ and $2n + 1$, respectively.

Chapter 70

Delayed Padé–Lanczos approximant

The delayed Padé–Lanczos approximant is defined in analogy with (36.6) as

$$\mathcal{R}_{L,K}^{\text{PLA}(s)}(u) = \frac{c_s}{\beta_1^{(s)}} \frac{P_L^{(s)}(u)}{Q_K^{(s)}(u)}. \tag{70.1}$$

The paradiagonal case $L = K$ of (70.1) will hereafter be denoted by

$$\mathcal{R}_{n,n}^{\text{PLA}(s)}(u) \equiv \mathcal{R}_n^{\text{PLA}(s)}(u) \qquad \mathcal{R}_n^{\text{PLA}(s)}(u) = \frac{c_s}{\beta_1^{(s)}} \frac{P_n^{(s)}(u)}{Q_n^{(s)}(u)}. \tag{70.2}$$

Here, $Q_n^{(s)}(u)$ and $P_n^{(s)}(u)$ are delayed Lanczos polynomials of the first and second kind, respectively. They can be defined via their recursions similar to their non-delayed counterparts (23.16) and (23.20)

$$\left. \begin{array}{c} \beta_{n+1}^{(s)} P_{n+1}^{(s)}(u) = [u - \alpha_n^{(s)}] P_n^{(s)}(u) - \beta_n^{(s)} P_{n-1}^{(s)}(u) \\ P_0^{(s)}(u) = 0 \qquad P_1^{(s)}(u) = 1 \end{array} \right\} \tag{70.3}$$

$$\left. \begin{array}{c} \beta_{n+1}^{(s)} Q_{n+1}^{(s)}(u) = [u - \alpha_n^{(s)}] Q_n^{(s)}(u) - \beta_n^{(s)} Q_{n-1}^{(s)}(u) \\ Q_{-1}^{(s)}(u) = 0 \qquad Q_0^{(s)}(u) = 1 \end{array} \right\}. \tag{70.4}$$

Equivalently, the polynomials $P_n^{(s)}(u)$ and $Q_n^{(s)}(u)$ can be introduced by power series as in (26.1)

$$P_n^{(s)}(u) = \sum_{r=0}^{n-1} p_{n,n-r}^{(s)} u^r \qquad Q_n^{(s)}(u) = \sum_{r=0}^{n} q_{n,n-r}^{(s)} u^r. \tag{70.5}$$

The expansion coefficients $p_{n,n-r}^{(s)}$ and $q_{n,n-r}^{(s)}$ can be generated recursively just as in (26.2) and (26.3)

$$\left.\begin{array}{c} \beta_{n+1}^{(s)} p_{n+1,n+1-r}^{(s)} = p_{n,n+1-r}^{(s)} - \alpha_n^{(s)} p_{n,n-r}^{(s)} - \beta_n^{(s)} p_{n-1,n-1-r}^{(s)} \\ p_{0,0}^{(s)} = 0 \qquad p_{1,1}^{(s)} = 1 \end{array}\right\} \quad (70.6)$$

$$\left.\begin{array}{c} \beta_{n+1}^{(s)} q_{n+1,n+1-r}^{(s)} = q_{n,n+1-r}^{(s)} - \alpha_n^{(s)} q_{n,n-r}^{(s)} - \beta_n^{(s)} q_{n-1,n-1-r}^{(s)} \\ q_{0,0}^{(s)} = 1 \qquad q_{1,1}^{(s)} = -\dfrac{\alpha_0^{(s)}}{\beta_1^{(s)}} \end{array}\right\} \quad (70.7)$$

$$p_{n,-1}^{(s)} = 0 \qquad q_{n,-1}^{(s)} = 0 \qquad p_{n,m}^{(s)} = 0 \qquad q_{n,m}^{(s)} = 0 \qquad (m > n). \quad (70.8)$$

The degrees of the polynomials $P_n^{(s)}(u)$ and $Q_n^{(s)}(u)$ are $n-1$ and n, respectively. The recursion (70.3) and (70.4) for these two latter polynomials are exactly the same, except for the different initializations.

Next, we want to establish the connections of $\mathcal{R}_{e,n}^{\text{LCF}(s)}(u)$ and $\mathcal{R}_{o,n}^{\text{LCF}(s)}(u)$ with the delayed Padé–Lanczos approximant (70.1). For this purpose, we need the first few explicit delayed Lanczos polynomials from (70.3) and (70.4)

$$\left.\begin{array}{c} P_0^{(s)}(u) = 0 \qquad P_1^{(s)}(u) = 1 \qquad \beta_2^{(s)} P_2^{(s)}(u) = u - \alpha_1^{(s)} \\ \beta_2^{(s)} \beta_3^{(s)} P_3^{(s)}(u) = u^2 - [\alpha_1^{(s)} + \alpha_2^{(s)}] u + \{\alpha_1^{(s)} \alpha_2^{(s)} - [\beta_2^{(s)}]^2\} \end{array}\right\} \quad (70.9)$$

$$\left.\begin{array}{c} Q_0^{(s)}(u) = 1 \qquad \beta_1^{(s)} Q_1^{(s)}(u) = u - \alpha_0^{(s)} \\ \beta_1^{(s)} \beta_2^{(s)} Q_2^{(s)}(u) = u^2 - [\alpha_0^{(s)} + \alpha_1^{(s)}] u + \{\alpha_0^{(s)} \alpha_1^{(s)} - [\beta_1^{(s)}]^2\} \end{array}\right\}. \quad (70.10)$$

Therefore, using (70.9) and (70.10), it follows from (70.1) for, e.g. $n = 1$

$$\mathcal{R}_1^{\text{PLA}(s)}(u) = \dfrac{c_s}{u - \alpha_0^{(s)}}. \quad (70.11)$$

The results (69.6) and (70.11) are seen to coincide with each other

$$\therefore \qquad \mathcal{R}_1^{\text{PLA}(s)}(u) = \mathcal{R}_{e,1}^{\text{LCF}(s)}(u). \quad (70.12)$$

Similarly, for $n = 2$ the usage of (70.1), (70.9) and (70.10) gives

$$\mathcal{R}_2^{\text{PLA}(s)}(u) = c_s \dfrac{u - \alpha_1^{(s)}}{u^2 - [\alpha_0^{(s)} + \alpha_1^{(s)}] u + \{\alpha_0^{(s)} \alpha_1^{(s)} - [\beta_1^{(s)}]^2\}}. \quad (70.13)$$

On the other hand, by means of (69.4), we have

$$\mathcal{R}_{e,2}^{\text{LCF}(s)}(u) = \cfrac{c_s}{u - \alpha_0^{(s)} - \cfrac{[\beta_1^{(s)}]^2}{u - \alpha_1^{(s)}}} = c_s \frac{u - \alpha_1^{(s)}}{[u - \alpha_0^{(s)}][u - \alpha_1^{(s)}] - [\beta_1^{(s)}]^2}$$

$$\therefore \quad \mathcal{R}_{e,2}^{\text{LCF}(s)}(u) = c_s \frac{u - \alpha_1^{(s)}}{u^2 - [\alpha_0^{(s)} + \alpha_1^{(s)}]u + \{\alpha_0^{(s)}\alpha_1^{(s)} - [\beta_1^{(s)}]^2\}}. \tag{70.14}$$

Hence, it follows from (70.13) and (70.14) that

$$\mathcal{R}_2^{\text{PLA}(s)}(u) = \mathcal{R}_{e,2}^{\text{LCF}(s)}(u). \tag{70.15}$$

We continued further with similar calculations for $n \geq 3$ and recorded that all the particular cases (70.12), (70.15), etc, are in accord with the general rule

$$\mathcal{R}_n^{\text{PLA}(s)}(u) = \mathcal{R}_{e,n}^{\text{LCF}(s)}(u) \qquad (n = 1, 2, 3, \ldots). \tag{70.16}$$

Hence, the delayed Padé–Lanczos approximant $\mathcal{R}_n^{\text{PLA}(s)}(u)$ and the even part of the delayed Lanczos continued fraction $\mathcal{R}_{e,n}^{\text{LCF}(s)}(u)$ give exactly the same results for any order n. This generalizes the analogous conclusion (47.7) or (64.47) reached earlier for the non-delayed versions of these two methods.

Obviously, it will be important to see whether the odd part of the delayed LCF, namely $\mathcal{R}_{o,n}^{\text{LCF}(s)}(u)$, could also be found among the elements of the Padé–Lanczos general table for $\mathcal{R}_{n,m}^{\text{PLA}(s)}(u)$. For instance, let us consider the diagonal case ($L = K + 1$) in (70.1) and write

$$\mathcal{R}_{n+1,n}^{\text{PLA}(s)}(u) \equiv \tilde{\mathcal{R}}_n^{\text{PLA}(s)}(u) = \frac{c_s}{\beta_1^{(s)}} \frac{P_{n+1}^{(s)}(u)}{Q_n^{(s)}(u)}. \tag{70.17}$$

In this case, using (70.9) and (70.10), we have for $n = 1$

$$\tilde{\mathcal{R}}_1^{\text{PLA}(s)}(u) = \frac{c_s}{\beta_1^{(s)}} \frac{u - \alpha_1^{(s)}}{u - \alpha_0^{(s)}}. \tag{70.18}$$

Thus, it follows from (69.24) and (70.18) that

$$\tilde{\mathcal{R}}_1^{\text{PLA}(s)}(u) \neq \mathcal{R}_{o,1}^{\text{LCF}(s)}(u). \tag{70.19}$$

We have calculated explicitly $\tilde{\mathcal{R}}_n^{\text{PLA}(s)}(u)$ for the next few higher orders n and always confirmed the inequality

$$\tilde{\mathcal{R}}_n^{\text{PLA}(s)}(u) \neq \mathcal{R}_{o,n}^{\text{LCF}(s)}(u) \qquad (n = 1, 2, \ldots). \tag{70.20}$$

More generally, there are no integers n and m for which $\mathcal{R}_{n,m}^{\text{PLA}(s)}(u)$ will match $\mathcal{R}_{o,n}^{\text{LCF}(s)}(u)$. This is because the denominator in $\mathcal{R}_{o,n}^{\text{LCF}(s)}(u)$ is a polynomial with no free term ($\propto u^0$), namely $\gamma_0 u + \gamma_1 u^2 + \cdots + \gamma_n u^{n+1}$. This extra u in the denominator of $\mathcal{R}_{o,n}^{\text{LCF}(s)}(u)$ relative to $\mathcal{R}_{e,n}^{\text{LCF}(s)}(u)$ suggests that $\mathcal{R}_{o,n}^{\text{LCF}(s)}(u)$ could stem from the Padé approximant in the variable u^{-1} rather than u. In the next chapter, we shall see that this is indeed the case.

Chapter 71

Delayed Padé approximant convergent outside the unit circle

Here, our starting point of the analysis is the exact delayed Green function (66.2). The series (66.2) is the Maclaurin expansion in powers of $u^{-1} \equiv z = \exp(i\omega\tau)$ and, therefore, convergent for $|u| > 1$, i.e. outside the unit circle. Let us first introduce an auxiliary function $\mathcal{G}^{(s)}(u^{-1})$ as in the non-delayed counterpart (38.3)

$$\mathcal{R}^{(s)}(u) = \sum_{n=0}^{\infty} c_{n+s} u^{-n-1} = u^{-1} \mathcal{G}^{(s)}(u^{-1}) \tag{71.1}$$

$$\mathcal{G}^{(s)}(u^{-1}) = \sum_{n=0}^{\infty} c_{n+s} u^{-n} = \lim_{N \to \infty} \mathcal{G}_N^{(s)}(u^{-1}) \tag{71.2}$$

$$\mathcal{G}_N^{(s)}(u^{-1}) = \sum_{n=0}^{N-1} c_{n+s} u^{-n} \tag{71.3}$$

$$\mathcal{R}_N^{(s)}(u) \equiv \sum_{n=0}^{N-1} c_{n+s} u^{-n-1} = u^{-1} \mathcal{G}_N^{(s)}(u^{-1}). \tag{71.4}$$

Then we define the diagonal delayed Padé approximant $\mathcal{G}_K^{\text{PA}(s)-}(u^{-1})$ to $\mathcal{G}_N^{(s)}(u^{-1})$ by

$$\mathcal{G}_K^{\text{PA}(s)-}(u^{-1}) = \frac{A_K^{(s)-}(u^{-1})}{B_K^{(s)-}(u^{-1})}. \tag{71.5}$$

The corresponding diagonal delayed Padé approximant to $\mathcal{R}_N^{(s)}(u)$ is

$$\mathcal{R}_K^{\text{PA}(s)-}(u^{-1}) \equiv u^{-1} \mathcal{G}_K^{\text{PA}(s)-}(u^{-1}) = u^{-1} \frac{A_K^{(s)-}(u^{-1})}{B_K^{(s)-}(u^{-1})}. \tag{71.6}$$

Here, the numerator and denominator polynomials $A_K^{(s)-}(u^{-1})$ and $B_K^{(s)-}(u^{-1})$ are in the same variable u^{-1} as the function $\mathcal{G}_K^{(s)}(u^{-1})$ itself. Both polynomials $A_K^{(s)-}(u^{-1})$ and $B_K^{(s)-}(u^{-1})$ are of the same degree K

$$A_K^{(s)-}(u^{-1}) = \sum_{r=0}^{K} a_r^{(s)-} u^{-r} \qquad B_K^{(s)-}(u^{-1}) = \sum_{r=0}^{K} b_r^{(s)-} u^{-r}. \qquad (71.7)$$

We shall determine the unknown expansion coefficients $a_r^{(s)-}$ and $b_r^{(s)-}$ from (71.7) by imposing the equality $\mathcal{G}_N^{(s)}(u^{-1}) = \mathcal{G}_K^{\text{PA}(s)-}(u^{-1})$, i.e.

$$\mathcal{R}_N^{(s)}(u) \equiv \sum_{n=0}^{N-1} c_{n+s} u^{-n} = \frac{A_K^{(s)-}(u^{-1})}{B_K^{(s)-}(u^{-1})}. \qquad (71.8)$$

Then, we multiply (71.8) by $B_K^{(s)-}(u^{-1})$ to write

$$\left. \begin{array}{c} B_K^{(s)-}(u^{-1}) \sum_{n=0}^{N-1} c_{n+s} u^{-n} = A_K^{(s)-}(u^{-1}) \\[2mm] \left[\sum_{r=0}^{K} b_r^{(s)-} u^{-r}\right] \left[\sum_{n=0}^{N-1} c_{n+s} u^{-n}\right] = \sum_{r=0}^{K} a_r^{(s)-} u^{-r} \end{array} \right\}. \qquad (71.9)$$

When the two sums on the lhs of (71.9) are multiplied out as indicated and the ensuing coefficients of the same powers of u^{-1} are equated with their counterparts from the rhs of (71.9), the following results emerge

$$a_\nu^{(s)-} = \sum_{r=0}^{\nu} b_r^{(s)-} c_{\nu-r+s} \qquad (0 \le \nu \le K) \qquad (71.10)$$

$$c_\nu = -\sum_{r=1}^{K} b_r^{(s)-} c_{\nu-r}. \qquad (71.11)$$

Let us set

$$M = N - 1 - K - s \qquad (71.12)$$

so that we can rewrite (71.11) as

$$c_{K+s+m} + \sum_{r=1}^{K} b_r^{(s)-} c_{K+s+m-r} = 0 \qquad 0 \le m \le M. \qquad (71.13)$$

This is an implicit system of linear equations for the unknown coefficients $b_r^{(s)-}$. The system (71.13) can be made explicit by varying the integer m from 1 to M as follows

$$\begin{rcases}c_{K+s+1}b_0^{(s)-} + c_{K+s}b_1^{(s)-} + \cdots + c_{s+1}b_K^{(s)-} = 0 \\ c_{K+s+2}b_0^{(s)-} + c_{K+s+1}b_1^{(s)-} + \cdots + c_{s+2}b_K^{(s)-} = 0 \\ \vdots \\ c_{M+K+s}b_0^{(s)-} + c_{M+K+s-1}b_1^{(s)-} + \cdots + c_{M+s}b_K^{(s)-} = 0\end{rcases}. \quad (71.14)$$

It is also clear that (71.10) represents a system of linear equations when the suffix v is varied from 0 to K and thus

$$\begin{rcases}a_0^{(s)-} = c_s b_0^{(s)-} \\ a_1^{(s)-} = c_{s+1}b_0^{(s)-} + c_s b_1^{(s)-} \\ \vdots \\ a_K^{(s)-} = c_{K+s}b_0^{(s)-} + c_{K+s-1}b_1^{(s)-} + \cdots + c_s b_K^{(s)-}\end{rcases}. \quad (71.15)$$

Both systems (71.14) and (71.15) can equivalently be cast into their respective matrix forms viz

$$\begin{pmatrix} c_{K+s} & c_{K-1+s} & \cdots & c_{s+1} \\ c_{K+s+1} & c_{K+s} & \cdots & c_{s+2} \\ \vdots & \vdots & \ddots & \vdots \\ c_{K+s+M-1} & c_{K+s+M-2} & \cdots & c_{s+K} \end{pmatrix} \begin{pmatrix} b_1^{(s)-} \\ b_2^{(s)-} \\ \vdots \\ b_K^{(s)-} \end{pmatrix} = -b_0^{(s)-} \begin{pmatrix} c_{K+s+1} \\ c_{K+s+2} \\ \vdots \\ c_{K+s+M} \end{pmatrix} \quad (71.16)$$

$$\begin{pmatrix} a_0^{(s)-} \\ a_1^{(s)-} \\ \vdots \\ a_K^{(s)-} \end{pmatrix} = \begin{pmatrix} c_s & 0 & \cdots & 0 \\ c_{s+1} & c_s & \cdots & 0 \\ \vdots & \vdots & \ddots & \vdots \\ c_{K+s} & c_{K+s-1} & \cdots & c_s \end{pmatrix} \begin{pmatrix} b_0^{(s)-} \\ b_1^{(s)-} \\ \vdots \\ b_K^{(s)-} \end{pmatrix}. \quad (71.17)$$

It will prove convenient to write

$$\tilde{a}_{r,K}^{(s)-} \equiv \frac{a_r^{(s)-}}{c_s b_0^{(s)-}} \qquad \tilde{b}_{r,K}^{(s)-} \equiv \frac{b_r^{(s)-}}{b_0^{(s)-}} \quad (71.18)$$

where it is understood that the coefficients $a_r^{(s)-}$ and $b_r^{(s)-}$ are implicitly dependent upon K.

For an illustration, we set $K = 1$ and obtain

$$\tilde{b}_{1,1}^{(s)-} = -\frac{c_{s+2}}{c_{s+1}} \quad (71.19)$$

$$\tilde{a}_{0,1}^{(s)-} = 1 \qquad \tilde{a}_{1,1}^{(s)-} = \frac{c_{s+1}^2 - c_s c_{s+2}}{c_s c_{s+1}}. \quad (71.20)$$

Inserting (71.18)–(71.20) into (71.7) yields

$$A_1^{(s)-}(u^{-1}) = c_s b_0^{(s)-}[1 + \tilde{a}_{1,1}^{(s)-} u^{-1}] \tag{71.21}$$

$$B_1^{(s)-}(u^{-1}) = b_0^{(s)-}[1 + \tilde{b}_{1,1}^{(s)-} u^{-1}]. \tag{71.22}$$

Substitution (71.21) and (71.22) into (71.6) gives

$$\mathcal{R}_1^{PA(s)-}(u^{-1}) = \frac{c_s}{u} \frac{1 + \tilde{a}_{1,1}^{(s)-} u^{-1}}{1 + \tilde{b}_{1,1}^{(s)-} u^{-1}}. \tag{71.23}$$

It follows from (69.33), (71.19) and (71.20) that

$$\alpha_0^{(s+1)} = -\tilde{b}_{1,1}^{(s)-} \qquad \alpha_0^{(s+1)} - \alpha_0^{(s)} = -\tilde{a}_{1,1}^{(s)-}. \tag{71.24}$$

This permits recasting (69.32) in the form

$$\mathcal{R}_{o,1}^{LCF(s)}(u) = \frac{c_s}{u} \frac{1 + \tilde{a}_{1,1}^{(s)-} u^{-1}}{1 + \tilde{b}_{1,1}^{(s)-} u^{-1}} \tag{71.25}$$

which coincides with (71.23)

$$\therefore \qquad \mathcal{R}_1^{PA(s)-}(u^{-1}) = \mathcal{R}_{o,1}^{LCF(s)}(u). \tag{71.26}$$

In the same way, we consider the case with $K = 2$ for which (71.14) and (71.15) yield

$$\tilde{b}_{1,2}^{(s)-} = \frac{c_{s+1} c_{s+4} - c_{s+3} c_{s+2}}{c_{s+2}^2 - c_{s+1} c_{s+3}} \qquad \tilde{b}_{2,2}^{(s)-} = \frac{c_{s+3}^2 - c_{s+2} c_{s+4}}{c_{s+2}^2 - c_{s+1} c_{s+3}} \tag{71.27}$$

$$\tilde{a}_{0,2}^{(s)-} = 1 \tag{71.28}$$

$$\tilde{a}_{1,2}^{(s)-} = \frac{c_{s+1}(c_{s+2}^2 - c_{s+1} c_{s+3}) + c_s(c_{s+1} c_{s+4} - c_{s+2} c_{s+3})}{c_s(c_{s+2}^2 - c_{s+1} c_{s+3})} \tag{71.29}$$

$$\tilde{a}_{2,2}^{(s)-} = \frac{c_{s+2}^3 - 2c_{s+1} c_{s+2} c_{s+3} + c_{s+1}^2 c_{s+4} + c_s c_{s+3}^2 - c_s c_{s+2} c_{s+4}}{c_s(c_{s+2}^2 - c_{s+1} c_{s+3})}. \tag{71.30}$$

Placing (71.27)–(71.30) into (71.7) gives

$$A_2^{(s)-}(u^{-1}) = b_0^{(s)-} c_s [1 + \tilde{a}_{1,2}^{(s)-} u^{-1} + \tilde{a}_{2,2}^{(s)-} u^{-2}] \tag{71.31}$$

$$B_2^{(s)-}(u^{-1}) = b_0^{(s)-} [1 + \tilde{b}_{1,2}^{(s)-} u^{-1} + \tilde{b}_{2,2}^{(s)-} u^{-2}]. \tag{71.32}$$

Inserting (71.31) and (71.32) into (71.6) yields

$$\mathcal{R}_2^{PA(s)-}(u^{-1}) = \frac{c_s}{u} \frac{1 + \tilde{a}_{1,2}^{(s)-} u^{-1} + \tilde{a}_{2,2}^{(s)-} u^{-2}}{1 + \tilde{b}_{1,2}^{(s)-} u^{-1} + \tilde{b}_{2,2}^{(s)-} u^{-2}}. \tag{71.33}$$

Comparison of (69.43)–(69.46) with (71.27)–(71.30) leads to

$$\alpha_0^{(s+1)} + \alpha_1^{(s+1)} = -\tilde{b}_{1,2}^{(s)-} = a_2^{(s)} + a_3^{(s)} + a_4^{(s)} + a_5^{(s)} \qquad (71.34)$$

$$\alpha_0^{(s+1)}\alpha_1^{(s+1)} - [\beta_1^{(s+1)}]^2 = \tilde{b}_{2,2}^{(s)-} = a_2^{(s)}[a_4^{(s)} + a_5^{(s)}] + a_3^{(s)}a_5^{(s)} \qquad (71.35)$$

$$\alpha_0^{(s+1)} - \alpha_0^{(s)} + \alpha_1^{(s+1)} = -\tilde{a}_{1,2}^{(s)-} = a_3^{(s)} + a_4^{(s)} + a_5^{(s)} \qquad (71.36)$$

$$[\alpha_0^{(s+1)} - \alpha_0^{(s)}]\alpha_1^{(s+1)} - [\beta_1^{(s+1)}]^2 = \tilde{a}_{2,2}^{(s)-} = a_3^{(s)}a_5^{(s)}. \qquad (71.37)$$

Here, we have

$$\left.\begin{array}{l}[\alpha_0^{(s+1)} - \alpha_0^{(s)}]\alpha_1^{(s+1)} - [\beta_1^{(s+1)}]^2 = \{\alpha_0^{(s+1)}\alpha_1^{(s+1)} - [\beta_1^{(s+1)}]^2\} \\ -\alpha_0^{(s)}\alpha_1^{(s+1)} = a_2^{(s)}[a_4^{(s)} + a_5^{(s)}] + a_3^{(s)}a_5^{(s)} = a_3^{(s)}a_5^{(s)} \iff a_2^{(s)} = 0\end{array}\right\} \qquad (71.38)$$

$$\therefore \quad a_r^{(s)-} = \{b_r^{(s)-}\}_{a_2^{(s)}=0} \quad (0 \leq r \leq 2) \qquad (71.39)$$

$$A_2^{(s)-}(u^{-1}) = \{B_2^{(s)-}(u^{-1})\}_{a_2^{(s)}=0}. \qquad (71.40)$$

Moreover, it follows, in general

$$a_r^{(s)-} = \{b_r^{(s)-}\}_{a_2^{(s)}=0} \quad (0 \leq r \leq K) \qquad (71.41)$$

$$A_n^{(s)-}(u^{-1}) = \{B_n^{(s)-}(u^{-1})\}_{a_2^{(s)}=0}. \qquad (71.42)$$

Using (71.34)–(71.37), we can rewrite (69.48) viz

$$\mathcal{R}_{o,2}^{LCF(s)}(u) = \frac{c_s}{u} \frac{1 + \tilde{a}_{1,2}^{(s)-}u^{-1} + \tilde{a}_{2,2}^{(s)-}u^{-2}}{1 + \tilde{b}_{1,2}^{(s)-}u^{-1} + \tilde{b}_{2,2}^{(s)-}u^{-2}} \qquad (71.43)$$

which agrees exactly with (71.33)

$$\therefore \quad \mathcal{R}_2^{PA(s)-}(u^{-1}) = \mathcal{R}_{o,2}^{LCF(s)}(u). \qquad (71.44)$$

This type of derivation has been pursued further for $K \geq 3$ and all these particular results for (71.26), (71.44) etc, are found to invariably obey the following general relation

$$\mathcal{R}_n^{PA(s)-}(u^{-1}) = \mathcal{R}_{o,n}^{LCF(s)}(u) \quad (n = 1, 2, 3 \ldots). \qquad (71.45)$$

Hence, we can conclude that the delayed Padé approximant with the convergence region outside the unit circle ($|u| > 1$) is identical to the odd part of the delayed Lanczos continued fraction to any order n, as per (71.45). Moreover, both $\mathcal{R}_n^{PA(s)-}(u^{-1})$ and the original truncated Green function (71.3) are convergent for $|u| > 1$ as $N \to \infty$. Therefore, outside the unit circle, the delayed Padé approximant $\mathcal{R}_n^{PA(s)-}(u^{-1})$ plays the role of an accelerator of an already convergent series which is the Green function (71.1).

Chapter 72

Delayed Padé approximant convergent inside the unit circle

There is another variant of the delayed diagonal Padé approximant for the same function $\mathcal{G}_N^{(s)}(u^{-1})$ from (71.3). This variant can be deduced from (65.5) which we rewrite as

$$\mathcal{G}^{(s)}(u) = \sum_{n=0}^{\infty} c_{n+s} u^{-n} = \sum_{k=1}^{K} \frac{d_k^{(s)} u}{u - u_k}. \qquad (72.1)$$

Here, the sum over k is an implicit quotient of two polynomials in the variable u and, hence, it is the Padé approximant. A special feature of (72.1) is that the numerator polynomial has no free term independent of u. Thus, it is natural that the needed version of the delayed diagonal Padé approximant, hereafter denoted by $\mathcal{R}_K^{PA(s)+}(u)$, should be a polynomial quotient in u. Thus, the Padé approximant to $\mathcal{G}_N^{(s)}(u^{-1})$ will be introduced by

$$\mathcal{G}_K^{PA(s)+}(u) = \frac{A_K^{(s)+}(u)}{B_K^{(s)+}(u)}. \qquad (72.2)$$

The corresponding delayed Padé approximant to the truncated Green function $\mathcal{G}_N^{(s)}(u^{-1})$ from (71.3) is

$$\mathcal{R}_K^{PA(s)+}(u) = u^{-1} \mathcal{G}_K^{PA(s)+}(u) = u^{-1} \frac{A_K^{(s)+}(u)}{B_K^{(s)+}(u)}. \qquad (72.3)$$

Both polynomials $A_K^{(s)+}(u)$ and $B_K^{(s)+}(u)$ are of the same degree K. Following (72.1), the variable of the numerator and denominator polynomials in (72.3) is set to be u as opposed to u^{-1} in the original sum (71.4)

$$A_K^{(s)+}(u) = \sum_{r=1}^{K} a_r^{(s)+} u^r \qquad B_K^{(s)+}(u) = \sum_{r=0}^{K} b_r^{(s)+} u^r. \qquad (72.4)$$

Here, as per (72.1), the numerator polynomial $A_K^{(s)+}(u)$ does not have the free term, i.e. $a_0^{(s)+} = 0$, so that the sum starts from $r = 1$ with the first term $a_1^{(s)+}u$. The convergence range of $\mathcal{R}_K^{PA(s)+}(u)$ is inside the unit circle ($|u| < 1$) where the original sum $\mathcal{G}^{(s)}(u^{-1})$ from (71.2) is divergent. The polynomials $A_K^{(s)+}(u)$ and $B_K^{(s)+}(u)$ are readily identified from the condition

$$\mathcal{G}_N^{(s)}(u) \equiv \sum_{n=0}^{N-1} c_{n+s} u^{-n} = \frac{A_K^{(s)+}(u)}{B_K^{(s)+}(u)}. \tag{72.5}$$

We multiply (72.5) by $B_K^{(s)+}(u)$ so that

$$\left. \begin{array}{c} B_K^{(s)+}(u) \sum_{n=0}^{N-1} c_{n+s} u^{-n} = A_K^{(s)+}(u) \\[6pt] \left[\sum_{r=0}^{K} b_r^{(s)+} u^r \right] \left[\sum_{n=0}^{N-1} c_{n+s} u^{-n} \right] = \sum_{r=1}^{K} a_r^{(s)+} u^r \end{array} \right\}. \tag{72.6}$$

The same procedure as in (71.9) followed by equating the coefficients of the like powers of the expansion variable gives

$$b_0^{(s)+} c_{n+s} + \sum_{r=1}^{K} b_r^{(s)+} c_{n+s+r} = 0 \qquad (n = 1, 2, \ldots, M). \tag{72.7}$$

Using (71.12), we can write (72.7) explicitly as

$$\begin{aligned} c_s b_0^{(s)+} &+ c_{s+1} b_1^{(s)+} &+ c_{s+2} b_2^{(s)+} &+ \cdots &+ c_{s+K} b_K^{(s)+} &= 0 \\ c_{s+1} b_0^{(s)+} &+ c_{s+2} b_1^{(s)+} &+ c_{s+3} b_2^{(s)+} &+ \cdots &+ c_{s+K+1} b_K^{(s)+} &= 0 \\ &&\vdots && \\ c_{M+s} b_0^{(s)+} &+ c_{M+s+1} b_1^{(s)+} &+ c_{M+s+2} b_2^{(s)+} &+ \cdots &+ c_{M+s+K} b_K^{(s)+} &= 0 \end{aligned} \tag{72.8}$$

or in the equivalent matrix form

$$\begin{pmatrix} c_{s+1} & c_{s+2} & c_{s+3} & \cdots & c_{s+K} \\ c_{s+2} & c_{s+3} & c_{s+4} & \cdots & c_{s+K+1} \\ c_{s+3} & c_{s+4} & c_{s+5} & \cdots & c_{s+K+2} \\ \vdots & \vdots & \vdots & \ddots & \vdots \\ c_{s+M+1} & c_{s+M+2} & c_{s+M+3} & \cdots & c_{s+K+M} \end{pmatrix} \begin{pmatrix} b_1^{(s)+} \\ b_2^{(s)+} \\ b_3^{(s)+} \\ \vdots \\ b_K^{(s)+} \end{pmatrix}$$

$$= - \begin{pmatrix} c_s b_0^{(s)+} \\ c_{s+1} b_0^{(s)+} \\ c_{s+2} b_0^{(s)+} \\ \vdots \\ c_{s+M} b_0^{(s)+} \end{pmatrix}. \tag{72.9}$$

The coefficients $\{a_r^{(s)+}\}$ of the numerator polynomial $A_K^{(s)+}(u)$ follow from the inhomogeneous part of the positive powers of the expansion variable from (72.6)

$$\begin{aligned} c_s b_1^{(s)+} + c_{s+1} b_2^{(s)+} + c_{s+3} b_3^{(s)+} + \cdots + c_{s+K-1} b_K^{(s)+} &= a_1^{(s)+} \\ c_s b_2^{(s)+} + c_{s+1} b_3^{(s)+} + \cdots + c_{s+K-2} b_K^{(s)+} &= a_2^{(s)+} \\ &\ddots \qquad \vdots \qquad \vdots \\ c_s b_K^{(s)+} &= a_K^{(s)+} \end{aligned} \tag{72.10}$$

or via the matrix representation

$$\begin{pmatrix} a_1^{(s)+} \\ a_2^{(s)+} \\ a_3^{(s)+} \\ \vdots \\ a_K^{(s)+} \end{pmatrix} = \begin{pmatrix} c_s & c_{s+1} & c_{s+2} & \cdots & c_{s+K-1} \\ 0 & c_s & c_{s+1} & \cdots & c_{s+K-2} \\ 0 & 0 & c_s & \cdots & c_{s+K-3} \\ \vdots & \vdots & \vdots & \ddots & \vdots \\ 0 & 0 & 0 & \cdots & c_s \end{pmatrix} \begin{pmatrix} b_1^{(s)+} \\ b_2^{(s)+} \\ b_3^{(s)+} \\ \vdots \\ b_K^{(s)+} \end{pmatrix}. \tag{72.11}$$

For convenience, let us write

$$\tilde{a}_{r,K}^{(s)-} \equiv \frac{a_r^{(s)+}}{b_0^{(s)+}} \qquad \tilde{b}_{r,K}^{(s)+} = \frac{b_s^{(s)+}}{b_0^{(s)+}} \tag{72.12}$$

where the K-dependence of $a_r^{(s)+}$ and $b_r^{(s)+}$ is implicit. To illustrate this variant of the delayed Padé approximant, we shall again consider a few examples. For $K = 1$, it follows from (72.8) and (72.10)

$$\tilde{b}_{1,1}^{(s)+} = -\frac{c_s}{c_{s+1}} \tag{72.13}$$

$$\tilde{a}_{1,1}^{(s)+} = -\frac{c_s^2}{c_{s+1}} = c_s \tilde{b}_{1,1}^{(s)+}. \tag{72.14}$$

With this, the polynomials from (72.4) become

$$B_1^{(s)+}(u) = b_0^{(s)+}[1 + \tilde{b}_{1,1}^{(s)+}u] \tag{72.15}$$

$$A_1^{(s)+}(u) = b_0^{(s)+}\tilde{a}_{1,1}^{(s)+}. \tag{72.16}$$

Inserting (72.15) and (72.16) into (72.3) gives

$$R_1^{PA(s)+}(u) = \frac{\tilde{a}_{1,1}^{(s)+}}{1 + \tilde{b}_{1,1}^{(s)+}u}. \tag{72.17}$$

Using (67.5), (72.13) and (72.14), it follows

$$\alpha_0^{(s)} = -\frac{1}{\tilde{b}_{1,1}^{(s)+}} = -\frac{c_s}{\tilde{a}_{1,1}^{(s)+}}. \tag{72.18}$$

This maps (69.6) into the form

$$R_{e,1}^{LCF(s)+}(u) = \frac{\tilde{a}_{1,1}^{(s)+}}{1 + \tilde{b}_{1,1}^{(s)+}u} \tag{72.19}$$

which agrees with (72.17), so that

$$R_1^{PA(s)+}(u) = R_{e,1}^{LCF(s)}(u). \tag{72.20}$$

Likewise, for $K = 2$ it follows from (72.8) and (72.10)

$$\tilde{b}_{1,2}^{(s)+} = \frac{c_{s+1}c_{s+2} - c_s c_{s+3}}{c_{s+1}c_{s+3} - c_{s+2}^2} \tag{72.21}$$

$$\tilde{b}_{2,2}^{(s)+} = \frac{c_s c_{s+2} - c_{s+1}^2}{c_{s+1}c_{s+3} - c_{s+2}^2} \tag{72.22}$$

$$\tilde{a}_{1,2}^{(s)+} = \frac{2c_s c_{s+1} c_{s+2} - c_s^2 c_{s+3} - c_{s+1}^3}{c_{s+1}c_{s+3} - c_{s+2}^2} \tag{72.23}$$

$$\tilde{a}_{2,2}^{(s)+} = c_s \frac{c_s c_{s+2} - c_{s+1}^2}{c_{s+1}c_{s+3} - c_{s+2}^2} = c_s \tilde{b}_{2,2}^{(s)+}. \tag{72.24}$$

By inserting (72.21)–(72.24) in (72.4) we obtain

$$\therefore \quad B_2^{(s)+}(u) = b_0^{(s)+}[1 + \tilde{b}_{1,2}^{(s)+}u + \tilde{b}_{2,2}^{(s)+}u^2] \tag{72.25}$$

$$A_2^{(s)+}(u) = b_0^{(s)+}u[\tilde{a}_{1,2}^{(s)+} + \tilde{a}_{2,2}^{(s)+}u]. \tag{72.26}$$

Substituting (72.25) and (72.26) into (72.3) leads to

$$R_2^{PA(s)+}(u) = \frac{\tilde{a}_{1,2}^{(s)+} + \tilde{a}_{2,2}^{(s)+}u}{1 + \tilde{b}_{1,2}^{(s)+}u + \tilde{b}_{2,2}^{(s)+}u^2}. \tag{72.27}$$

Comparing (69.11)–(69.13) with (72.21)–(72.24) yields

$$\alpha_1^{(s)} = -\frac{\tilde{a}_{1,2}^{(s)+}}{c_s \tilde{b}_{2,2}^{(s)+}} \tag{72.28}$$

$$\alpha_0^{(s)} + \alpha_1^{(s)} = -\frac{\tilde{b}_{1,2}^{(s)+}}{\tilde{b}_{2,2}^{(s)+}} \tag{72.29}$$

$$\alpha_0^{(s)} \alpha_1^{(s)} - [\beta_1^{(s)}]^2 = \frac{1}{\tilde{b}_{2,2}^{(s)+}}. \tag{72.30}$$

This converts (69.9) into the following form

$$\mathcal{R}_{e,2}^{\text{LCF}(s)}(u) = \frac{\tilde{a}_{1,2}^{(s)+} + \tilde{a}_{2,2}^{(s)+} u}{1 + \tilde{b}_{1,2}^{(s)+} u + \tilde{b}_{2,2}^{(s)+} u^2} \tag{72.31}$$

which is identical to (72.27)

$$\therefore \quad \mathcal{R}_2^{\text{PA}(s)+}(u) = \mathcal{R}_{e,2}^{\text{LCF}(s)}(u). \tag{72.32}$$

Our further explicit calculation for $K \geq 3$ revealed that all the particular cases (72.20), (72.32), etc, satisfy the general relationship

$$\mathcal{R}_n^{\text{PA}(s)+}(u) = \mathcal{R}_{e,n}^{\text{LCF}(s)}(u) \quad (n = 1, 2, 3, \ldots). \tag{72.33}$$

Thus, it follows that the delayed Padé approximant with the convergence region inside the unit circle ($|u| < 1$) is identical to the even part of the delayed Lanczos continued fraction to any order n, as given by (72.33). Here, $\mathcal{R}_n^{\text{PA}(s)+}(u)$ is convergent for $|u| < 1$, whereas the original truncated Green function (71.3) is divergent in the same region, i.e. inside the unit circle. Hence, inside the unit circle, the delayed Padé approximant $\mathcal{R}_n^{\text{PA}(s)+}(u)$ uses the Cauchy concept of analytical continuation to induce/force convergence into the initially divergent series, i.e. the Green function (71.1).

Overall, we see that the introduction of $\mathcal{R}_n^{\text{PA}(s)\pm}(u^{\pm 1})$ helped to prove that the same LCF, i.e. $\mathcal{R}_n^{\text{LCF}(s)}(u)$ contains both $\mathcal{R}_n^{\text{PA}(s)-}(u^{-1})$ (as an accelerator of monotonically converging series/sequences) as well as $\mathcal{R}_n^{\text{PA}(s)+}(u)$ (as an analytical continuator of divergent series/sequences), where $\mathcal{R}_n^{\text{PA}(s)-}(u^{-1})$ and $\mathcal{R}_n^{\text{PA}(s)+}(u)$ are equal to the odd and even part of $\mathcal{R}_n^{\text{CF}(s)}(u)$, i.e. $\mathcal{R}_{o,n}^{\text{LCF}(s)}(u) = \mathcal{R}_{2n+1}^{\text{CF}(s)}(u)$ and $\mathcal{R}_{e,n}^{\text{LCF}(s)}(u) = \mathcal{R}_{2n}^{\text{CF}(s)}(u)$, respectively. Once these equivalences/correspondences have been established, clearly it is optimal to use only the quantities $\mathcal{R}_{e,n}^{\text{LCF}(s)}(u)$ and $\mathcal{R}_{o,n}^{\text{LCF}(s)}(u)$, for a fixed s, to extract the LCF in the forms $\mathcal{R}_{2n}^{\text{LCF}(s)}(u)$ and $\mathcal{R}_{2n+1}^{\text{LCF}(s)}(u)$ for obtaining the two sets of observables that converge inside $|u| < 1$ and outside $|u| > 1$ the unit circle. These results represent respectively the lower and upper bounds of the computed observables (spectra, eigenfrequencies, density of states, etc). For example $|\mathcal{R}_{e,n}^{\text{LCF}(s)}(u)|$ and $|\mathcal{R}_{o,n}^{\text{LCF}(s)}(u)|$ are respectively the lower and the upper limits of the envelope of the magnitude shape spectrum for a given signal $\{c_{n+s}\}$ ($0 \leq n \leq N-1$). Similarly, the eigenfrequencies and residues $\{\omega_k^{(s)+}, d_k^{(s)+}\}$ and $\{\omega_k^{(s)-}, d_k^{(s)-}\}$ that emanate from $\mathcal{R}_K^{\text{PA}(s)+}(u) = \mathcal{R}_{e,K}^{\text{LCF}(s)}(u)$ and $\mathcal{R}_K^{\text{PA}(s)-}(u^{-1}) = \mathcal{R}_{o,K}^{\text{LCF}(s)}(u)$ represent respectively the lower and upper limits of the true (exact) values $\{\omega_k, d_k\}$.

Chapter 73

Illustrations

We use the name 'fast Padé transform' (FPT) for the Padé approximant (PA) applied to signal processing. This is done to highlight the so-called 'transform' feature of the PA by which the shape spectrum is obtained non-parametrically as reminiscent of the fast Fourier transform (FFT). In contrast to the FFT, as discussed, the FPT can perform parametric analysis. In this chapter, we study resolution improvement of the FPT relative to the FFT for signal processing. Numerical examples are given to illustrate the two variants of the *diagonal* FPT denoted by $\text{FPT}^+ \equiv \text{FPT}(z)$ and $\text{FPT}^- \equiv \text{FPT}(z^{-1})$ whose convergence regions are inside ($|z| < 1$) and outside ($|z| > 1$) the unit circle (see figure 73.1). Here, z is the complex variable $z = \exp{(i\tau\omega)}$, whereas τ is the sampling time and ω is any angular frequency (real or complex). Both forms $\text{FPT}(z^{\pm 1})$ are used in this chapter only in the parametric variants for the Padé spectra $A^{\pm}/B^{\pm}$. The ansatz frequency spectrum is defined by

$$S(z^{-1}) = \frac{1}{N} \sum_{n=0}^{N-1} c_n z^{-n} \qquad z = e^{i\tau\omega}. \tag{73.1}$$

This coincides with the FFT only at the Fourier grid $\omega = \tilde{\omega}_k \equiv 2\pi k/N$ ($k = 0, 1, \ldots, N-1$). The usual diagonal PA for $S(z^{-1})$ is denoted by $\text{FPT}(z^{-1})$ and introduced *from the onset* as the quotient of two polynomials both of degree K in the same variable z^{-1}, as in the input sum from the rhs of (73.1), i.e. $A^-/B^- \equiv \sum_{n=0}^{K} a_n^- z^{-n} / \sum_{m=0}^{K} b_m^- z^{-m}$. In the $\text{FPT}(z^{-1})$ the signal $\{c_n\}$ is generic, i.e. *not* assumed to be a sum of complex damped exponentials. By contrast, the variant $\text{FPT}(z)$ is *not set up* as a ratio of two polynomials from the very beginning. Rather, it is introduced with the explicit reliance upon the damped exponential model $c_n = \sum_{k=1}^{K} d_k \exp{(in\tau\omega_k)}$ and together with the causal z-

transform ends up again as a quotient of two polynomials of the type of the PA

$$\sum_{n=0}^{\infty} c_n z^{-n} = \sum_{k=1}^{K} d_k \sum_{n=0}^{\infty} [z^{-1} \exp(i\tau\omega_k)]^n = \sum_{k=1}^{K} \frac{d_k z}{z - \exp(i\tau\omega_k)}$$

$$\equiv \frac{\sum_{n=1}^{K} a_n^+ z^n}{\sum_{m=0}^{K} b_m^+ z^m} \equiv \frac{A^+}{B^+}.$$

The coefficients $\{a_n^\pm, b_m^\pm\} \equiv \{a_n^{(0)\pm}, b_m^{(0)\pm}\}$ as well as the complex spectra $A^\pm/B^\pm \equiv A_K^{(0)\pm}(z^{\pm 1})/B_K^{(0)\pm}(z^{\pm 1})$ are obtained from the following definition and the related procedure outlined in chapters 71 and 72 with no delay ($s = 0$) in the signal data

$$\text{FPT}(z) \equiv \text{FPT}^+ : \quad \frac{1}{N}\sum_{n=0}^{N-1} c_n z^{-n} = \frac{\sum_{n=1}^{K} a_n^+ z^n}{\sum_{m=0}^{K} b_m^+ z^m} = \frac{A^+}{B^+} \quad (73.2)$$

$$\text{FPT}(z^{-1}) \equiv \text{FPT}^- : \quad \frac{1}{N}\sum_{n=0}^{N-1} c_n z^{-n} = \frac{\sum_{n=0}^{K} a_n^- z^{-n}}{\sum_{m=0}^{K} b_m^- z^{-m}} = \frac{A^-}{B^-}. \quad (73.3)$$

The same input function $S(z^{-1})$ is employed in FPT(z) and FPT(z^{-1}) from (73.2) and (73.3), respectively. Therefore, the two converged results $A^\pm/B^\pm$ from FPT($z^{\pm 1}$) must coincide (up to the random noise) due to the uniqueness of the PA, as is indeed the case (see figure 73.1). We set $K = N/2$ and use the SVD to solve the systems of linear equations encountered in (73.2) and (73.3) for an experimentally measured time signal $\{c_n\}$, i.e. the free induction decay (FID) from [13]. This FID has been encoded from the brain of a healthy volunteer using Magnetic Resonance Spectroscopy (MRS) at the magnetic field strength of 7T. The data $\{c_n\}$ are of a very good signal-to-noise ratio and this, together with a relatively long signal length ($N = 2048$), yields the optimal absorption spectrum in the FFT itself (see figure 73.1)[1]. In figure 73.1 perfect agreement is seen among the FFT, FPT(z) and FPT(z^{-1}) that are all computed with $N = 2048$. The convergence rates of the FFT and FPT(z^{-1}) are compared in figures 73.2 and 73.3. Here, both methods use either the truncated or the full signal ($N/M = 2048/M$, $M = 1 - 32$). At $M > 1$ the FFT uses $N - N/M$ zeros for completion to $N = 2048$, but FPT(z^{-1}) does not. The resolution seen in figures 73.2 and 73.3 is distinctly better in the FPT(z^{-1}) than in the FFT at any truncation $M > 1$ (see, e.g. $N/16 = 128$). The convergence pattern of the FPT without any spikes or other artefacts is summarized in figure 73.4. Finally, figure 73.5 displays the residual absorption spectra as the difference between the real parts of the complex spectra from the FFT and FPT(z^{-1}). Throughout

---

[1] The symbols $\Re(z)$ and $\Im(z)$ in figures 73.1–73.5 stand for the real and imaginary part Re(z) and Im(z) of a complex number z, respectively. The abcissas are relative dimensionless frequencies in ppm (parts per million) and the ordinates are in arbitrary units (a.u.).

figure 73.5, the FFT is taken at $N = 2048$, whereas the FPT(z^{-1}) is computed at N/M ($M = 1 - 32$). In figure 73.5 a strikingly robust convergence is obtained in the FPT(z^{-1}) with the decrease of the truncation level M of the full signal length $N = 2048$ [27]–[29].

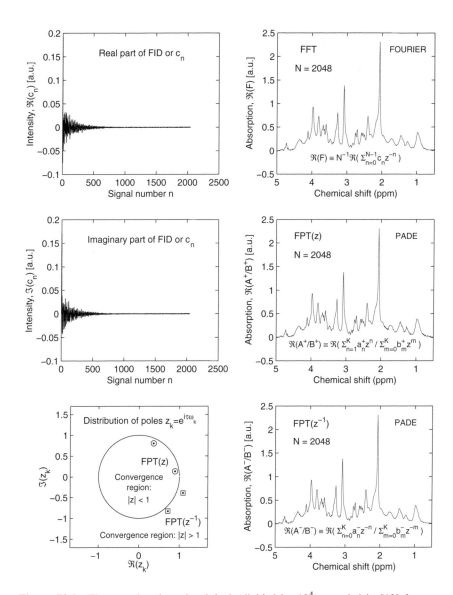

Figure 73.1. The complex time signal $\{c_n\}$ (divided by 10^4) encoded in [13] from a healthy human brain (occipital gray matter) and the corresponding absorption spectra of metabolites computed presently using the fast Fourier transform (FFT) and the fast Padé transform (FPT) both with the full signal length $N = 2048$. Two complementary, but equivalent variants of the FPT are employed being symbolized as FPT(z) and FPT(z^{-1}) with different convergence regions $|z| < 1$ and $|z| > 1$ having their poles located inside and outside the unit circle, respectively.

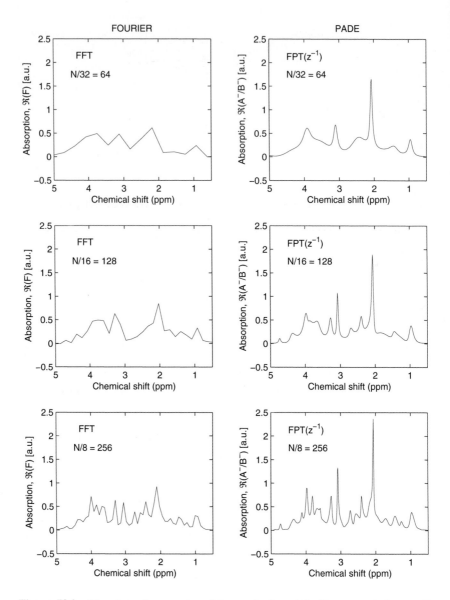

Figure 73.2. The absorption spectra of human brain metabolites computed presently by applying the fast Fourier transform (FFT) and the fast Padé transform (FPT) to the complex valued time signal $\{c_n\}$ from figure 73.1 with the truncated signal length N/M ($M = 8 - 32$) where $N = 2048$. The FPT is employed with the variant $\text{FPT}(z^{-1})$ which converges for $|z| > 1$ and has its poles located outside the unit circle.

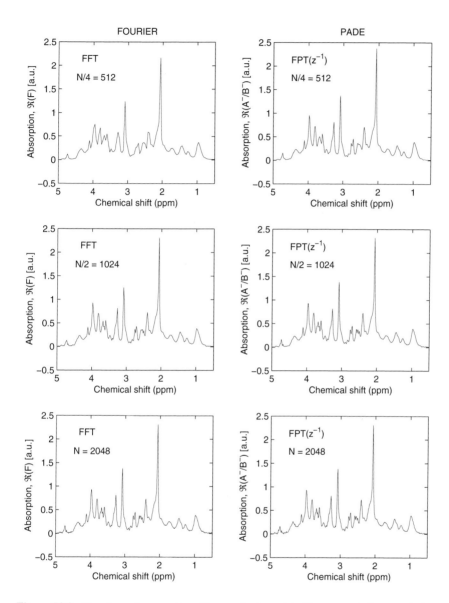

Figure 73.3. The absorption spectra of human brain metabolites computed presently by applying the fast Fourier transform (FFT) and the fast Padé transform (FPT) to the complex valued time signal $\{c_n\}$ from figure 73.1 with the truncated or full signal length N/M ($M = 1 - 4$) where $N = 2048$. The FPT is employed with the variant FPT(z^{-1}) which converges for $|z| > 1$ and has its poles located outside the unit circle.

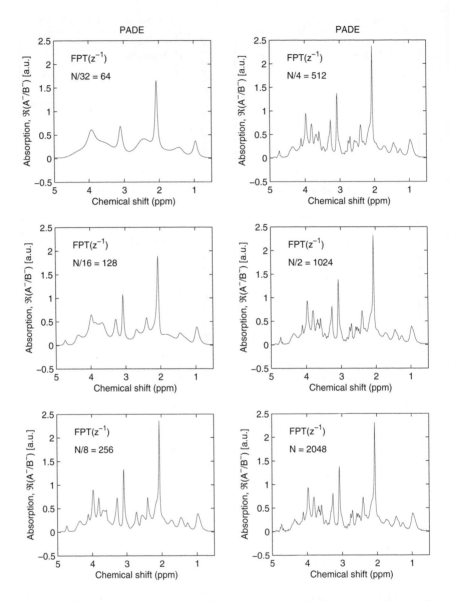

Figure 73.4. The absorption spectra of human brain metabolites computed presently by applying the fast Padé transform (FPT) to the complex valued time signal $\{c_n\}$ from figure 73.1 with the truncated or full signal length N/M ($M = 1 - 32$) where $N = 2048$. The FPT is employed with the variant FPT(z^{-1}) which converges for $|z| > 1$ and has its poles located outside the unit circle.

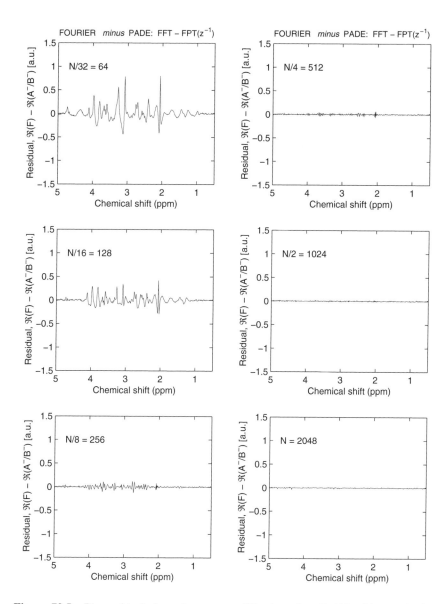

Figure 73.5. The residual absorption spectra ('Fourier *minus* Padé') of human brain metabolites computed presently by applying the fast Fourier transform (FFT) and the fast Padé transform (FPT) to the complex valued time signal $\{c_n\}$ from figure 73.1. The FFT is evaluated only with the full signal length $N = 2048$ whereas the FPT is computed with the truncated or full signal length N/M ($M = 1 - 32$). The variant of the FPT employed is FPT(z^{-1}) which converges for $|z| > 1$ and has its poles located outside the unit circle.

Chapter 74

The uncertainty principle and resolution improvement

Quantum mechanics can answer the critical question in general signal processing: *does any improvement beyond the Fourier resolution contradict the Heisenberg uncertainty principle* $\Delta\omega\Delta t \geq \hbar$? The correct answer requires a sharp distinction between the informational content *and* the resolving power (see the discussion regarding figure 73.2). Reliability of the FPT and the FFT will be guaranteed only if they preserve information from the original time signal. They actually both do, the FFT in a linear and the FPT in a nonlinear manner. The resolution of the FFT is determined from the onset by the preselected time-bandwidth. This resolution is given by the smallest possible distance called the Rayleigh limit or bound $\Delta\omega_{min} = 2\pi/T$ between any two adjacent equidistant Fourier frequencies for the chosen total acquisition time T. It is this Rayleigh bound $\Delta\omega_{min}$ which inherently limits the resolution in the FFT. Across interdisciplinary research on spectral analysis, the Heisenberg uncertainty principle for the conjugate pair, time-frequency, from quantum mechanics has been repeatedly evoked to claim that no theoretical processing method can surpass the Rayleigh limit $\Delta\omega_{min}$. However, this claim is confusing and misleading. The Heisenberg principle demands that any *two* conjugate observables $\mathcal{A}$ and $\mathcal{B}$, described by the operators A and B that do not commute ($AB \neq BA$), cannot be *measured* simultaneously. This means that a measurement of, e.g. $\mathcal{A}$ precludes the knowledge of $\mathcal{B}$. Any attempt, to determine $\mathcal{B}$ experimentally would *destroy* the knowledge of $\mathcal{A}$. The Heisenberg principle is one of the cornerstones of quantum mechanics, which is the only theory to date determining what actually *cannot* be measured experimentally, irrespective of any conceivable advance in instrumentation. However, this uncertainty principle has nothing to do with the resolution limit $\Delta\omega_{min}$ on the frequency within the FFT. This is simply because, in the considered phenomenon, there is only *one* physical quantity which is subject to measurement, and that is the time signal $c(t)$. By contrast, the corresponding spectrum $F(\omega)$ is *not* measured experimentally, but rather *calculated* theoretically. Since signal processing which is, at any rate, used after completion of the experiment, cannot possibly *destroy* the measurement of $c(t)$, it follows that the Rayleigh bound $\Delta\omega_{min}$ could only be a limitation of theoretical tools of linear analysis, such as the one used in the FFT. The FPT as a nonlinear processor takes full advantage of that opportunity and achieves the frequency resolution, which could be better than $\Delta\omega_{min}$. The resolution in the FPT is given by the average separation $\Delta\omega_{ave}$ which is usually smaller than $\Delta\omega_{min}$ in a chosen window. Therefore, the FPT can surpass the Rayleigh limit $\Delta\omega_{min} = 2\pi/T$. The resolution $2\pi/T$ of the FFT is limited by a sharp cut-off of $c(t)$ at $t = T$. By contrast, the FPT extrapolates the time signal $c(t)$ to $t > T$ and this can yield a resolution better than $2\pi/T$.

Finally, we ask the key question: *why should quantum mechanics be used in processing time signals from NMR and MRS?* It definitely should, because such time signals $c(t)$ stem from purely quantum-mechanical phenomena. The function $c(t)$ itself is the overall result of the induced transitions between two *quantum-mechanically allowed* states of a sample, perturbed by static, as well as varying magnetic fields and radio-waves tuned to a resonant frequency of the investigated spin-active nuclei with non-zero values of the magnetic moment. Moreover, quantum physics is the only complete theory which can provide both data driven and purely mechanistic descriptions of spectra *and* time signals $c(t)$. For example, in mechanistic modelling by quantum physics, one would insert the known spin–spin and spin–lattice interactions into the Hamiltonian of the studied system and predict a functional dependence of $c(t)$ which turns out to fully agree with experimentally encoded time signals [273]. The same analysis also leads to a quantum-mechanical spectrum which could be reduced to the Fourier spectrum only under certain restrictive approximations. For instance, in the setting of the nearest neighbour method, while applying the strategy of the well-known inverse scattering problem to the measured $c(t)$, quantum mechanics can retrieve interactions (e.g. protein–protein interactions in tissue [8]), as the most important part of the dynamics which produces the encoded time signals. Furthermore, quantum mechanics is used to design modern experiments in NMR chemistry and biology through practically infinite possibilities of manipulations with spins of nuclei.

Chapter 75

Prediction/extrapolation for resolution improvement

The solution c_n of the Kth-order direct problem of an ordinary difference equation is the geometric sequence $c_n = \sum_{k=1}^{K} d_k u_k^n$ with $u_k = \exp(-i\omega_k \tau)$ where τ is the sampling time. The frequencies $\{\omega_k\}$ and the amplitudes $\{d_k\}$ are complex with $\operatorname{Im}(\omega_k) < 0$ and $u_k \neq 0$ as well as $u_k \neq 1$. Here c_n with exactly K transients is unique if the data/Hankel matrices satisfy $H_K(c_n) \neq 0$ and $H_{K+1}(c_n) = 0$. In the harmonic inversion problem, the solution c_n is given as the above geometric sequence, but the triple $\{K, \omega_k, d_k\}$ is unknown. Order K is surmised prior to analysis, but this is incorrect due to either missing certain physical/genuine ω_k or yielding some unphysical/spurious ω_k. But K is actually built into c_n. Therefore K is extractable from the data matrix $\{c_n\}$ without any guessing. This is unequivocally done by the Shanks transform $e_K(c_n)$ which can filter out exactly all the K transients from $\{c_n\}$ via $e_K(c_n) = X_K H_{K+1}(c_n)/H_K(c_n)$ with $H_K(c_n) = (d_1 d_2 \cdots d_K) Y_K$, where X_m and Y_m are non-zero for any m. Since K is unknown, a set $\{H_m(c_n)\}$ is computed for varying m. If $\{c_n\}$ has K transients, then $d_k \neq 0$ $(1 \leq k \leq K)$ and $d_k = 0$ $(k \geq K+1)$. Thus $H_K(c_n) = (d_1 d_2 \cdots d_K) X_K \neq 0$ and $H_{K+1}(c_n) = (d_1 d_2 \cdots d_K) d_{K+1} Y_{K+1} = 0$, as in the above direct problem. Then $e_K(c_n) = 0$ is the signature for the exact filtering of all K transients from $\sum_{k=1}^{K} d_k u_k^n$. The transform $e_K(c_n)$ is nonlinear, but it obeys a linear scaling $e_K(\gamma + c_n) = \gamma + e_K(c_n)$, where γ is a constant independent of n. The result $e_K(c_n) = 0$ also emerges from a more general geometric sequence $\{\bar{c}_n\}$ with $\bar{c}_n = \bar{c}_\infty + \sum_{k=1}^{K} d_k u_k^n$, where $\bar{c}_\infty$ is the baseline constant. If no diverging/exploding transients are present, i.e. $|u_k| < 1$ $(1 \leq k \leq K)$, the limiting/equilibrium value $\bar{c}_\infty$ is reached at $n \longrightarrow \infty$, when all the transients are extinguished by decaying to zero. But the above scaling property for $e_K(\bar{c}_n)$ gives $e_K(\bar{c}_n) = \bar{c}_\infty$ for any n, not just $n \longrightarrow \infty$, including convergence $|u_k| < 1$ $(1 \leq k \leq K)$ and divergence with some or all exploding transients $|u_k| > 1$ $(1 \leq k \leq m, m \leq K)$. The result $e_K(c_n) = 0$ is obtained from $e_K(\bar{c}_n) = \bar{c}_\infty$ for $\bar{c}_\infty = 0$, in which case $\bar{c}_n = c_n$. Hence

$e_K(\bar{c}_n)$ is a regular transform for $\{\bar{c}_n\}$. This means that if $\{\bar{c}_n\}$ has no exploding transients and, thus, converges to its limit $\bar{c}_\infty$ as $n \longrightarrow \infty$, then $e_K(c_n)$ generates another sequence which converges to the same limit $\bar{c}_\infty$. The new convergence is faster, so that the Shanks transform is an accelerator of slowly converging sequences/series. The sequence $\{\bar{c}_n\}$ diverges if at least one transient is exploding. The sequence from the Shanks transform converges to its anti-limit per analytical continuation even if all the transients are exploding. Then the Shanks transform induces/forces convergence of divergent sequences. For spectral sequences of partial sums $\{\sum_{n=0}^{m-1} c_n u^{-n-1}\}$ $(1 \leq m \leq N)$ where $u = \exp(-i\omega\tau)$, the Shanks transform coincides with the Padé approximant P/Q. This result is a complex spectrum as the unique quotient of two polynomials P and Q of degrees $K-1$ and K, respectively. Crucially, $e_K(c_n)$ can extract the vital information from $\{c_n\}$ without the need for all the transients to decay as $n \longrightarrow \infty$. Therefore, short signals $\{c_n\}$ suffice for retrieving the key information by the Shanks transform, in sharp contrast to the Fourier transform. This is because the Shanks transform is a predictor, whereas the Fourier transform is not. Beyond the total duration T of the signal of length $N = T/\tau$, the Fourier transform replaces every c_n by zero or by the periodic extension $c_{n+N} = c_n$ without new information, so that in either case prediction is impossible. The Shanks transform can interpolate $(n < N)$ and extrapolate $(n > N)$ via a polynomial quotient P/Q. The inverse Q^{-1} of Q is not a terminating sum, but rather an infinite power series and so is $P/Q = PQ^{-1}$, although the sum of the degrees of P and Q is equal to $N-1$ from the input $\sum_{n=0}^{N-1} c_n u^{-n-1}$. This leads to the extrapolation feature of P/Q. Hence, the predictive power of the Shanks transform is in extrapolating $\{c_n\}$ beyond N, i.e. in drawing inferences on the hypothetical infinite signal $\{c_n\}$ $(0 \leq n \leq \infty)$ from an available finite subset $\{c_n\}$ $(0 \leq n \leq N-1)$ for $N < \infty$. It is this extrapolation/prediction which enables the Shanks transform to reach a resolution higher than that of the Fourier transform. Specifically, the frequency resolution in the Shanks transform/extrapolation can be better than the Fourier resolution, which is the Rayleigh limit $2\pi/T$ pre-assigned by a fixed $T = N\tau$ for any signal $\{c_n\}$.

Chapter 76

Main advantages of the Padé-based spectral analysis

Although the primary focus of the present book is on the harmonic inverse problem, we saw that it was very instructive to begin the analysis with a direct problem of the Kth ordinary differential equations (ODE) and the ordinary difference equations (OΔE) with constant coefficients. This is simply because the most general solutions of the ODE and OΔE are continuous and discrete functions $c(t)$ and c_n, respectively, given by their linear combinations of K complex damped exponentials/transients $c(t) = \sum_{k=1}^{K} d_k \exp(-i\omega_k t)$ and $c_n = \sum_{k=1}^{K} d_k \exp(-in\omega_k \tau)$. Here, the running index n counts the time with $t = t_n \equiv n\tau$ ($n = 0, 1, 2, \ldots, N-1$), where τ is the sampling rate. The frequencies $\{\omega_k\}$ are the normal mode complex frequencies and $\{d_k\}$ are the corresponding complex amplitudes/residues. The solutions $\{c_n\}$ are unique for the given K initial conditions, i.e. there will be exactly K such transients in every c_n, if the data matrices or the Hankel matrices of the dimensions $K \times K$ and $(K+1) \times (K+1)$ fulfil simultaneously the two conditions $H_K(c_n) \neq 0$ and $H_{K+1}(c_n) = 0$.

In the harmonic inversion problem, the same two conditions $H_K(c_n) \neq 0$ and $H_{K+1}(c_n) = 0$, as above, are also the prerequisites for the existence of a local spectrum which is a truncated version of the Maclaurin expansion or the Green function $\mathcal{R}(u) = (\Phi_0|(u\hat{1} - \hat{U})^{-1}|\Phi_0) = \sum_{n=0}^{\infty} c_n u^{-n-1}$, where $u = \exp(-i\omega\tau)$. The symbol $(\zeta|\xi) = (\xi|\zeta)$ is the symmetric inner or scalar product without conjugation of any vector. The object $|\Phi_0)$ is the initial state of the studied system. The quantity $\hat{U}(\tau)$ is the evolution/relaxation operator $\hat{U}(\tau) \equiv \hat{U} = \exp(-i\hat{\Omega}\tau)$, where $\hat{\Omega}$ is the dynamical system operator or the 'Hamiltonian'. Given the time signal $\{c_n\}$ ($0 \leq n \leq N-1$), the N rank Green function $\mathcal{R}_N(u) = \sum_{n=0}^{N-1} c_n u^{-n-1}$ becomes exactly the Padé–Lanczos approximant (PLA) defined by the unique ratio $\mathcal{R}_K^{\text{PLA}}(u) = (c_0/\beta_K) P_K(u)/Q_K(u)$ of two polynomials $P_K(u)$ and $Q_K(u)$ of degrees $K - 1$ and K, respectively, with $K = N/2$ and where $\{\alpha_n, \beta_n\}$ are the Lanczos coupling parameters. The above confluence of the existence condition for the PLA and the uniqueness condition for the solutions of the OΔE is not a coincidence at all. This is because, the expansion coefficients $\{q_{K,K-n}\}$ of the power series representation $Q_K(u) = \sum_{r=0}^{K} q_{K,K-r} u^r$ of the characteristic/secular polynomial $Q_K(u)$ in the PLA and $\{c_n\}$ satisfy the OΔE of order K. In the OΔE of the direct problem for $\{c_n\}$, the building components $\{u_k, d_k\} = \{\exp(-i\omega_k\tau), d_k\}$ of the damped complex exponentials $c_n = \sum_{k=1}^{K} d_k \exp(-in\omega_k\tau)$ are respectively determined by the zeros of $Q_K(u)$ and the initial conditions. By contrast, in the harmonic inversion problem, the signal $\{c_n\}$ as a sum of attenuated exponentials $c_n = \sum_{k=1}^{K} d_k u_k^n$ is known, i.e. given as a whole, but not its harmonic components. The task is to find the triple of the spectral parameters $\{K, u_k, d_k\}$.

The set $\{q_{K,K-r}\}$ also represents the predicting coefficients of the signal processing method called the linear predictor (LP). In order to solve the system of linear equations for the predicting coefficients $\{q_{K,K-n}\}$, the LP must guess the number K of such equations prior to the analysis. This inevitably leads to under-fitting by missing some of the genuine ω_k or over-fitting by yielding some spurious ω_k. After determining $\{q_{K,K-n}\}$ the LP solves the characteristic equation $Q_K(u_k) = 0$ to arrive at the eigenfrequencies $\{\omega_k\} = \{i\tau^{-1} \ln(u_k)\}$. Inserting the found roots $\{u_k\}$ into the damped exponential model for the signal, yet another system of linear equations $c_n = \sum_{k=1}^{K} d_k u_k^n$ for a set of values of n is solved to give the amplitudes $\{d_k\}$. This system uses all the K eigenroots $\{u_k\}$ to arrive at the set $\{d_k\}$. Any spurious solution from the set $\{u_k\}$ would undermine the accuracy of the computed amplitudes from $\{d_k\}$. Thus the LP carries out spectral analysis by first guessing the order K and then solving two successive systems of linear equations for $\{q_{K,K-r}\}$ and $\{d_k\}$. These systems of linear equations are usually solved employing the SVD which can be optimized with least square fittings. Exactly the same numerical procedure from the LP for determining d_k is also encountered in another parametric estimator from MRS called the Hankel–Lanczos singular value decomposition (HLSVD). The only difference between the LP and the HSLVD is in that the latter method determines the quantities $\{u_k\}$ as the eigenvalues of the data (Hankel) matrix.

On top of being exact for the same signals $c_n = \sum_{k=1}^{K} d_k u_k^n$, the PLA is by far more efficient than either the LP or the HLSVD. This is achieved first by the unequivocal determination of the order/rank K directly from the available data matrix comprised of the signal points $\{c_n\}$ ($0 \le n \le N-1$) by verifying the simultaneous conditions $H_K(c_n) \ne 0$ and $H_{K+1}(c_n) = 0$. Remarkably, the PLA avoids altogether solving the system of equations for $\{d_k\}$, by computing the residues directly from either of the two explicit formulae: $d_k = c_0 P_{K,k}/(\beta_1 Q'_{K,k})$ or $d_k = c_0/[\beta_K Q_{K-1,k} Q'_{K,k}]$, where $Q'_K(u) = (d/du) Q_K(u)$ and $P_{K,k} \equiv P_K(u_k)$ as well as $Q_{K,k} \equiv Q_K(u_k)$. The second mentioned formula for d_k is possible, since the polynomials $Q_K(u)$ and $Q_{K-1}(u)$ have no common roots and, therefore, $Q_{K-1,k} \ne 0$. The mentioned Maclaurin expansion for the Green function $\mathcal{R}(u)$ stems from the equivalence $C_n = c_n$ between the quantum-mechanical auto-correlation function $C_n = (\Phi_0|\Phi_n)$ and the time signal $c_n = \sum_{k=1}^{K} d_k u_k^n$. The vector $|\Phi_n) = \hat{U}^n|\Phi_0)$ is the Schrödinger time-dependent state whose continuous counterpart satisfies the Schrödinger non-stationary equation $i(\partial/\partial t)|\Phi(t)\rangle = \hat{\Omega}|\Phi(t)\rangle$, such that $|\Phi_n\rangle \equiv |\Phi(t_n)\rangle$. The stationary total eigenstate $|\Upsilon_k\rangle$ of the system satisfies the stationary Schrödinger equation $\hat{U}|\Upsilon_k\rangle = u_k|\Upsilon_k\rangle$. This state is locally complete, i.e. $\sum_{k=1}^{K}|\Upsilon_k\rangle\langle\Upsilon_k| = \hat{1}$. When the operator binomial expansion $(u\hat{1} - \hat{U})^{-1} = \sum_{n=0}^{\infty}\hat{U}^n u^{-n-1}$ is taken between two copies of the same state $|\Phi_0\rangle$, it then follows $\mathcal{R}(u) = \sum_{n=0}^{\infty} c_n u^{-n-1}$ provided that $C_n = c_n$. This latter equality stems from using the simple rule $\hat{U}(t_n) = \hat{U}(n\tau) = \hat{U}^n(\tau)$ and the local completeness of $\{|\Upsilon_k\rangle\}(1 \leq k \leq K)$, so that $\hat{U}^n = \hat{U}^n\hat{1} = \sum_{k=1}^{K}u_k^n|\Upsilon_k\rangle\langle\Upsilon_k|$. Thus $C_n = \langle\Phi_0|\hat{U}^n|\Phi_0\rangle = \sum_{k=1}^{K}u_k^n\langle\Phi_0|\Upsilon_k\rangle\langle\Upsilon_k|\Phi_0\rangle = \sum_{k=1}^{K}d_k u_k^n = c_n$, where the complex amplitude d_k is identified as the matrix element $d_k \equiv \langle\Phi_0|\Upsilon_k\rangle\langle\Upsilon_k|\Phi_0\rangle = \langle\Phi_0|\Upsilon_k\rangle^2$. This latter equivalence $(C_n = c_n)$ is the first motivation for the introduction of the novel concept of quantum-mechanical signal processing.

Quantum-mechanical signal processing is carried out using, e.g. the stationary Schrödinger eigenvalue problem $\hat{U}|\Upsilon(u)\rangle = u|\Upsilon(u)\rangle$ for $\hat{U}$. The eigenvalue $\{u_k\}$ and eigenstates $\{|\Upsilon_k\rangle\} = \{|\Upsilon(u_k)\rangle\}$ can be obtained by means of a diagonalization in a preselected basis $\{|\psi_n\rangle\}$. This basis might be either non-orthogonal $(\mathbf{S} \neq \mathbf{1})$ or orthogonal $(\mathbf{S} = \mathbf{1})$, where $\mathbf{S}$ is the overlap matrix for two different basis functions $|\psi_n\rangle$ and $|\psi_m\rangle$. The former and the latter choices yield the ordinary and generalized Schrödinger eigenvalue problem, respectively. In particular, if the total eigenvector $|\Upsilon(u)\rangle$ of the examined physical system is developed in terms of $\{|\psi_n\rangle\}$ as $|\Upsilon(u)\rangle = \sum_n A_n(u)|\psi_n\rangle$, then the original ordinary Schrödinger eigenproblem $\hat{U}|\Upsilon_k\rangle = u_k|\Upsilon_k\rangle$ becomes a system of linear homogeneous equations $(\mathbf{U} - u_k\mathbf{S})\mathbf{A}_k = \mathbf{0}$ for the unknown pair, i.e. the eigenvalues $\{u_k\}$ and the expansion coefficients $\mathbf{A}_k = \{A_{n,k}\} \equiv \{A_n(u_k)\}$. The unique solution $\{u_k, A_{n,k}\}$ of this latter system exists if the corresponding secular equation $\det[\mathbf{U} - u_k\mathbf{S}] = 0$ is satisfied, which is the characteristic equation $Q_K(u_k) = 0$, i.e the equation for the zeros of the characteristic polynomial. There are K roots $\{u_k\}$ of this latter equation and they are also the eigenvalues of the evolution matrix $\mathbf{U} = \{U_{n,m}\} = \{\langle\psi_m|\hat{U}|\psi_n\rangle\}$. The eigenvalues $\{u_k\}$ obtained from the secular equation are afterwards inserted into the system of linear equations $(\mathbf{U} - u_k\mathbf{S})\mathbf{A}_k = \mathbf{0}$ to obtain the corresponding eigenvectors $\{\mathbf{A}_k\}$. This step permits deduction of the total wavefunction $|\Upsilon_k\rangle$. Then by projecting $|\Upsilon_k\rangle$ onto $|\Phi_0\rangle$ and squaring the resulting matrix element gives the associated residue $d_k = \langle\Upsilon_k|\Phi_0\rangle^2$.

The PLA can also be conceived as a state space method. This is because the PLA can obtain the total eigenstates $\{|\Upsilon_k\rangle\}$ of the system and thus become a complete solver of quantum-mechanical ordinary and generalized Schrödinger eigenproblems. To accomplish this goal, the polynomials $\{Q_n(u)\}$ are again observed to play the pivotal role. It appears multiply advantageous to choose the expansion basis $\{|\psi_n\rangle\}$ to be the orthonormalized Lanczos states. In such a case, each vector $|\psi_n\rangle$ is an explicit sum of the Schrödinger, i.e. Krylov states $\{|\Phi_n\rangle\}$ via $|\psi_n\rangle = \sum_{r=0}^{n}q_{n,n-r}|\Phi_r\rangle$. Moreover $|\Upsilon(u)\rangle$ is also obtainable directly

Main advantages of the Padé-based spectral analysis 405

from the explicit sum $|\Upsilon(u)\rangle = \sum_n Q_n(u)|\psi_n\rangle$. In other words, the above expansion coefficients $\{A_{n,k}\}$ will be automatically available via $A_{n,k} = Q_{n,k}$, as soon as the pair $\{q_{n,n-m}, u_k\}$ has been generated. Therefore, the entire quantum-mechanical spectral analysis would be possible to carry out by purely algebraic means from the above closed analytical expressions of the PLA, if only the predicting/secular coefficients $\{q_{n,n-s}\}$ of the characteristic polynomials $\{Q_n(u)\}$ could be extracted via an analytical formula directly from the experimentally measured time signals $\{c_n\}$ or theoretically generated auto-correlation functions $\{C_n\}$. We achieve such a final step by obtaining analytically the whole needed set $\{q_{n,n-s}\}$ as an explicit algebraic expression without the usual implicit Cramer quotients of two general Hankel determinants. This explicit expression for $\{q_{n,n-s}\}$ follows from another general analytical formula for any coefficient $\{a_n\}$ of the contracted continued fractions for the total Green function or the spectrum or the system function or the response function $\mathcal{R}(u)$. Hence, the PLA obtains all the eigenvalues and the corresponding total eigenstates $\{u_k, |\Upsilon_k\rangle\}$, as well as the total Green function $\mathcal{R}(u)$, via the explicit analytical expressions that are practical for versatile applications. The collection of these analytical expressions valid for any exponentially damped time signal can be used to generate the needed exact numerical results. Such a data base for signal processing can also be exploited for testing the accuracy of all the existing numerical procedures for spectral and parametric estimations. The results from the PLA represent the full information which can be extracted from the experimentally measured time signal $\{c_n\}$ in accordance with quantum mechanics, whose main hypothesis is that everything which could possibly be learned about *any* examined system is contained in the total eigenstates $\{|\Upsilon(u)\rangle\}$, or equivalently, in the total Green function $\mathcal{R}(u)$.

From the above list of our analytical results within the PLA, the particular transformation $\{c_n\} \longleftrightarrow \{a_n\}$ deserves to be specially highlighted. Here, a single set $\{a_n\}$ from the frequency domain is linked to the input set $\{c_n\}$ from the time domain and *vice versa*. The experimentally measured data are entirely contained in the time signal $\{c_n\}$, whereas the theoretically computed spectral quantities for the shape spectrum in the PLA are fully confined to the continued fraction coefficients $\{a_n\}$. The Lanczos continued fractions (LCF) are an equivalent form of the PLA used for the shape spectrum in non-parametric estimations. The two sets $\{a_n\}$ and $\{c_n\}$ can be exactly retrieved from each other according to their compact explicit expressions $a_{n+1} = (c_n - \sigma_n \pi_{n-1} - \lambda_n \pi_{n-4})/\pi_n$ and $c_n = \pi_{n+1} + \sigma_n \pi_{n-1} + \lambda_n \pi_{n-4}$. Here, we employ the notation: $\pi_n = \prod_{i=1}^n a_i$, $\sigma_n = (\sum_{j=2}^n a_j)^2$, $\lambda_n = \sum_{j=j'}^{n-3} a_j \xi_j^2$, $\xi_j = \sum_{k=2}^{j+1} a_k \sum_{\ell=2}^{k+1} a_\ell$ with the symbol $\pi_n \equiv 0$, $(n \leq 0)$, $\sigma_n \equiv 0$ $(n \leq 3)$ and $j' = [(n-1)/2]$, where $[n/2]$ is the largest integer in $n/2$ and $a_1 = c_0 \neq 0$. The exact transformation $\{c_n\} \longleftrightarrow \{a_n\}$ is of paramount importance since it enables preservation of the full informational content when passing from the time to the frequency domain. The PLA is both a non-parametric and parametric estimator of spectra. For the shape spectrum the non-parametric variant is used by simply skipping the rooting $Q_K(u_k) = 0$

and evaluating directly the quotient $\mathcal{R}_K^{\text{PLA}}(u) = (c_0/\beta_K)P_K(u)/Q_K(u)$ using $u = \exp(-i\omega\tau)$ at any real or complex frequency ω. Moreover, for the shape spectrum there is no need to generate the polynomials $P_K(u)$ and $Q_K(u)$ explicitly. Rather, the PLA can be obtained analytically from the explicit form of the LCF truncated at the Kth order, i.e. $\mathcal{R}^{\text{LCF}}(u) = c_0/\mathcal{D}(u)$ with $\mathcal{D}(u) = u - \alpha_0 - \beta_1^2/\{u - \alpha_1 - \beta_2^2/[u - \alpha_2 - \cdots - \beta_{K-1}^2/(u - \alpha_K - \cdots - \cdots)]\}$ where $\alpha_n = a_{2n+1} + a_{2n+2}$ and $\beta_n^2 = a_{2n}a_{2n+1}$. Due to our mentioned explicit closed formula for $\{a_n\}$, it is seen that the present results for the Lanczos couplings $\{\alpha_n, \beta_n\}$ are also available from the simple analytical expressions.

Signal processing can be advantageously examined also from two other powerful theoretical strategies, i.e. the concept of transients and the method of moments. Transients are presently reviewed using the Shanks transform which stems from sequence-to-sequence mappings performed either in the time or in the frequency domain. The Shanks transform $e_K(c_n)$ is the exact filter for geometric sequences in the time domain. Such sequences $\{c_n\}$ are sums of exponentially attenuated time signals. One can apply the Shanks transform also to sequences comprised of the elements that are partial sums from the Maclaurin series of the total Green function $\mathcal{R}(u)$, which is the complex Lorentzian spectrum. For these sequences, the Shanks transform coincides with the Padé approximant. Further refinements of the results are expected when the iterated Shanks transforms are employed. These iterations are obtained by treating the results from the first Shanks sequence as a new entry to the Shanks transform. In this way, the second Shanks sequence is generated which could also be subjected to the Shanks transform and so forth. The iterated Shanks transforms exhibit enhanced convergence rates. Furthermore, parametric estimations can also be performed by the Shanks transform to obtain the spectral constants $\{K, u_k, d_k\}$. Here the Shanks transform has a unique virtue, since it can extract the order K exactly from the geometric sequence of time signals. This is due to the definition $e_K(c_n) = X_K H_{K+1}(c_n)/H_K(c_n) = 0$ for $c_n = \sum_{k=1}^{K} d_k u_k^n$ where $X_K > 0$. To obtain K unequivocally, the sequence of the auxiliary Hankel determinants $\{H_M(c_n)\}$ is computed for different values of the intermediate orders M. The first value of M (say M') for which one detects simultaneously $H_{M'}(c_n) \neq 0$ and $H_{M'+1}(c_n) = 0$ will give the sought order $K = M'$. The determinants $\{H_M(c_n)\}$ are computed recursively by the product-difference (PD) recursive algorithm of Gordon [170] with only M^2 multiplications, relative to the formidable $M!$ multiplications in the corresponding direct evaluation by the Cramer rule. The same algorithm can also be used for computation of the continued fraction coefficients $\{a_n\}$ and the expansion coefficients $\{q_{v,v-s}\}$ of the characteristic polynomials $\{Q_v(u)\}$. The Shanks transform itself can be computed by, e.g. the Wynn epsilon recursion yielding the vectors $\{\varepsilon_k^{(m)}(u)\}$ at any fixed u. This powerful and versatile algorithm originates from interconnecting four neighbouring elements of the Padé table. The Wynn recursion can find the eigenroots $\{u_k\}$ of $Q_K(u)$ from $u_k = \lim_{m \to \infty} \{\varepsilon_{2k-2}^{(m+1)}/\varepsilon_{2k-2}^{(m)}\}$. The associated residue d_k is the inverse of the first derivative of the reciprocal of the corresponding ε-vector evaluated at the found u_k.

We shall also recapitulate the issue of spurious resonances that could impose serious limitations on successful completion of most parametric estimators of spectra from, e.g. NMR, MRS, MRSI, etc. Several numerical techniques exist in the literature for dealing with spurious resonances. As an alternative avenue, we have presently elaborated an analytical tool for unambiguous identification of all the existing spurious roots. The resulting stable method is called the Padé–Schur approximant (PSA). Within the general PA, we use the constrained root reflection for regularization of any spurious roots. This procedure automatically assures that both the numerator and denominator polynomials of the PSA are stable or Schur polynomials. Such polynomials possess all their roots exclusively on the same physical side of the unit circle, leading to exponentials that tend to zero as time is increased infinitely. The physical and mathematical realm which sets up the framework for introducing and implementing the PSA for application to generic time signals can be summarized through the following five consecutive steps:

(i) As an approximation to the exact finite Fourier integral $F_{\text{ex}}(\omega)$, a complex Riemann spectrum $F(\omega)$ is introduced at the given frequencies ω that need not coincide with the Fourier grid $\tilde{\omega}_k = 2\pi k/N$ ($k = 0, 1, 2, 3, \ldots$) via: $F_{\text{ex}}(\omega) \equiv (1/T) \int_0^T dt\, c(t) e^{i\omega t} \approx (1/N) \sum_{n=0}^{N-1} c_n e^{in\omega\tau} \equiv F(\omega)$. Here, N is the total length of the signal c_n which is sampled at the rate $\tau = T/N$ where T is the total acquisition time or epoch. The function $F(\omega)$ is the Riemann sum defined as the modified trapezoidal quadrature rule for the exact Fourier integral $F_{\text{ex}}(\omega)$. Since ω is arbitrary, $F(\omega)$ is a more general function than the DFT, i.e. $F_k^{\text{DFT}} \equiv F(2\pi k/N)$, which is obtained from $F(\omega)$ by restricting ω to the Fourier grid $\{\tilde{\omega}_k\}$ only.

The power spectrum $|F(\omega)|^2$, as a real and non-negative function, evaluated at real frequencies $\omega \in [\omega_{\min}, \omega_{\max}]$, represents the local spectral density of the signal. The associated local energy of the signal is given by the definite integral of the function $|F(\omega)|^2$ from $\omega_{\min}$ to $\omega_{\max}$. In the special case where $\omega_{\min,\max} = \pm\pi/\tau$, the local energy becomes the total energy of the signal for the whole Nyquist interval $[-\pi/\tau, +\pi/\tau]$. To avoid any dependence of the abscissa upon the strength of the static external magnetic field, spectra in MRS are customarily plotted as a function of the so-called chemical shift, which is a dimensionless quantity expressed in parts per million (ppm) of hertz (Hz). In this case the surface under the given magnitude spectrum $|F(\text{ppm})|$ from $\text{ppm}_{\min}$ to $\text{ppm}_{\max}$ represents the concentration of the corresponding resonance, i.e. metabolite located at the centre $(\text{ppm}_{\min} + \text{ppm}_{\max})/2$ of the interval $[\text{ppm}_{\min}, \text{ppm}_{\max}]$. The associated full width at the half maximum (FWHM) is the inverse of the relaxation time T_2 of the given metabolite.

(ii) The Padé complex spectrum $F_{L,K}(u^{-1})$ is set up by expressing $F(\omega)$ *exactly* via the rational polynomial $F_{L,K}(u^{-1}) = A_L(u^{-1})/B_K(u^{-1})$ where u is a complex exponential $u = \exp(-i\omega\tau)$. Here, the numerator $A_L(u^{-1})$ and denominator $B_K(u^{-1})$ are complex polynomials of degrees L and K, respectively, such that $K + L = N$. These two latter polynomials are determined by requiring that the Padé complex spectrum coincides with the original spectrum $F_{L,K}(u^{-1}) = F(\omega)$ at any real and/or complex ω : $(1/N)\sum_{n=0}^{N-1} c_n u^{-n} = A_L(u^{-1})/B_K(u^{-1})$. Any of the pertinent algorithms existing in the literature can be used to generate the polynomial pair $\{A_L(u^{-1}), B_K(u^{-1})\}$. For example, the algorithm due to Wheeler [174], as a relatively simple, robust and efficient procedure, established in the research field of the moment problem, yields the tight binding coupling parameters $\{\alpha_n, \beta_n\}$ within machine accuracy. Using these parameters one can immediately obtain the polynomials $A_L(u^{-1})$ and $B_K(u^{-1})$. Crucially, the parameters $\{\alpha_n, \beta_n\}$ hold the entire information about the studied system and they can be used to instantaneously construct the Green function, i.e. the spectrum either as the PLA, LCF, etc. Moreover, we have presently shown that the auto-correlation functions, i.e. time signals can also be reconstructed exactly by using only the coupling constants $\{\alpha_n, \beta_n\}$ without necessarily taking recourse to the spectral parameters $\{u_k, d_k\}$. This is advantageous relative to the standard Lanczos algorithm, as well as the filter diagonalization (FD) or decimated signal diagonalization (DSD) where the set $\{C_n\}$ cannot be obtained without prior computation of the eigenvalues $\{u_k\}$ and residues $\{d_k\}$.

For the given input data $F(\omega)$ there is one and only one rational polynomial $F_{L,K}(u^{-1})$. In the case when the input spectrum $F(\omega)$ is of a purely Lorentzian nature, its mathematical model in a form of the rational polynomial $A_L(u^{-1})/B_K(u^{-1})$ will have K simple poles. Such poles are the K zeros of the denominator polynomial $B_K(u^{-1})$. This is because the only singularities of the PA in $A_L(u^{-1})/B_K(u^{-1})$ are its poles. In other words, the PA is a meromorphic function. If the poles of the PA are simple, the corresponding spectrum is non-degenerate and, therefore, one-to-one correspondence exists between the eigenvalues $\{u_k\}$ and the eigenfunctions $|\Upsilon_k)$ of the quantum-mechanical evolution matrix $\mathbf{U}$. In a non-degenerate spectrum, all the peaks are separated, since only one d_k is associated with the given u_k. A non-degenerate spectrum $F(\omega)$ is represented exactly by the PA, i.e. $A_L(u^{-1})/B_K(u^{-1})$ whose time signal c_n is given by a linear combination of damped complex exponentials with constant residues $\{d_k\}$, i.e. $c_n = \sum_{k=1}^{K} d_k u_k^n$. If the spectrum $F(\omega)$ is degenerate, then its PA, i.e. $A_L(u^{-1})/B_K(u^{-1})$ will find the proper multiplicity, say M_k, of the general kth degenerate eigenfrequency. This leads to a non-Lorentzian Padé spectrum $A_L(u^{-1})/B_K(u^{-1})$. Such a degenerate Padé spectrum will retrieve the time signal c_n which is given again by a linear combination of attenuated complex exponentials, but with the time-varying residues via a polynomial of degree M_k in the variable t such as: $c(t) = \sum_{k=1}^{J} \sum_{m_k=1}^{M_k} d_{k,m_k} t^{m_k-1} u_k^n$, where $\sum_{k=1}^{J} M_k = K$, $d_{k,m_k} = P_K(u_k)/Q_{K,m_k}(u_k)$ and $Q_{K,m}(u) = (d/du)^m Q_K(u)$.

(iii) The Padé power spectrum $|F_{L,K}(u^{-1})|^2$ is written exactly as the real and non-negative PA, i.e. $R_{L,K}(\cos \omega \tau)$ in the variable $\cos \omega \tau$ via $|F_{L,K}(u^{-1})|^2 = R_{L,K}(\cos \omega \tau)$. Here, $R_{L,K}(\cos \omega \tau)$ is a real positive definite rational polynomial, $R_{L,K}(\cos \omega \tau) = P_L(\cos \omega \tau)/Q_K(\cos \omega \tau)$ and, moreover, $|F_{L,K}(u^{-1})|^2 =$

$|A_L(u^{-1})/B_K(u^{-1})|^2 = P_L(\cos\omega\tau)/Q_K(\cos\omega\tau) \equiv R_{L,K}(\cos\omega\tau)$. The numerator $P_L(\cos\omega\tau)$ and denominator $Q_K(\cos\omega\tau)$ are given by the closed analytical expressions via the expansion coefficients of the polynomials $A_L(u^{-1})$ and $B_K(u^{-1})$ that are available from step (ii).

(iv) Spurious roots of $\{P_L(\cos\omega\tau), Q_K(\cos\omega\tau)\}$ cannot be identified by non-computational, i.e. analytical/algebraic means. However, such an identification becomes possible by returning to the original complex variable u^{-1}. This is done by inserting $\cos\omega\tau = (u + u^{-1})/2$ into $R_{L,K}(\cos\omega\tau)$ and renaming the ensuing PA, i.e. the transformed rational polynomial as $T_{L,K}(u^{-1})$: $R_{L,K}(\cos\omega\tau) \longrightarrow T_{L,K}(u^{-1})$. The quantity $T_{L,K}(u^{-1})$ is a complex function whose absolute value $|T_{L,K}(u^{-1})|$ represents a non-negative power spectrum. Both the original functions $R_{L,K}(\cos\omega\tau)$ and $|T_{L,K}(u^{-1})|$ represent the power spectra. This is because, the definite integrals of functions $R_{L,K}(\cos\omega\tau)$ and $|T_{L,K}(u^{-1})|$ have their customary physical meaning which is the energy of the signal, since both functions are real and non-negative. The critical part of the spectrum $|T_{L,K}(u^{-1})|$ to be regularized is: $|T_{L,K}(u^{-1})| \propto \{|\prod_{\ell=1}^{L}(u^{-1} - \rho_\ell e^{i\phi_\ell})(u^{-1} - \rho_\ell^{-1}e^{-i\phi_\ell})|\}\{|\prod_{k=1}^{K}(u^{-1} - \sigma_k e^{i\varphi_k})(u^{-1} - \sigma_k^{-1}e^{-i\varphi_k})|\}^{-1}$, where $\rho_\ell > 1$ and $\sigma_k > 1$. Here, the genuine and spurious roots of the numerator polynomial are $\rho_\ell e^{i\phi_\ell}$ and $\rho_\ell^{-1}e^{-i\phi_\ell}$, respectively, whereas the genuine and spurious roots of the denominator polynomial are $\sigma_k e^{i\varphi_k}$ and $\sigma_k^{-1}e^{-i\varphi_k}$, respectively. Therefore, the genuine and spurious roots are identified analytically from both polynomials in $|T_{L,K}(u^{-1})|$ as those roots that lie outside ($\rho_\ell e^{i\phi_\ell}, \sigma_k e^{i\varphi_k}$) and inside ($\rho_\ell^{-1}e^{-i\phi_\ell}, \sigma_k^{-1}e^{-i\varphi_k}$) the unit circle, respectively. The spurious roots in the denominator of $|T_{L,K}(u^{-1})|$ lead to unphysical spikes in the spectrum. The extraneous roots of numerator of $|T_{L,K}(u^{-1})|$ fill in the valleys in the spectrum with the unphysical content and increase the background, thus obscuring some weak genuine resonances. Such weak resonances can be found in, e.g. MRS spectra of healthy human brain gray matter at the locations of taurine, lactate, etc. Additionally, these latter roots destroy the uniqueness of the PA, i.e. $|T_{L,K}(u^{-1})|$. Most critically, due to the presence of all the spurious roots, the energy conservation of the signal is not preserved, since the two power spectra satisfy the inequality

$$|T_{L,K}(u^{-1})|_{\text{all the spurious roots unregularized}} \neq R_{L,K}(\cos\omega\tau).$$

(v) Finally, all the spurious roots from both the numerator and denominator of the power spectrum $|T_{L,K}(u^{-1})|$ are reflected from inside to outside the unit circle via complex conjugation of the whole term $(u^{-1} - \rho_\ell^{-1}e^{-i\phi_\ell}) \longrightarrow (u^{-1} - \rho_\ell^{-1}e^{-i\phi_\ell})^*$ and *not just* $(\rho_\ell^{-1}e^{-i\phi_\ell}) \longrightarrow (\rho_\ell^{-1}e^{-i\phi_\ell})^*$, which is done in the usual root reflection. Simultaneously, all the genuine roots are left intact according to the prescription: $(u^{-1} - \rho_\ell e^{i\phi_\ell})(u^{-1} - \rho_\ell^{-1}e^{-i\phi_\ell}) \longrightarrow (u^{-1} - \rho_\ell e^{i\phi_\ell})(u^{-1} - \rho_\ell^{-1}e^{-i\phi_\ell})^*$ and $(u^{-1} - \sigma_k e^{i\varphi_k})(u^{-1} - \sigma_k^{-1}e^{-i\varphi_k}) \longrightarrow (u^{-1} - \sigma_k e^{i\varphi_k})(u^{-1} - \sigma_k^{-1}e^{-i\varphi_k})^*$. Such a regularization establishes the equality between the original and the transformed power spectra leading to the preservation of the energy of the signal in any chosen frequency window

$$|T_{L,K}(u^{-1})|_{\text{all the spurious roots regularized}} = R_{L,K}(\cos\omega\tau).$$

This is the constrained root reflection which regularizes all the encountered extraneous roots by relocating them to the proper physical site of the genuine poles. The resulting stable and energy-preserving rational polynomial given by $|T_{L,K}(u^{-1})|_{\text{all the spurious roots regularized}}$ is the PSA. Within the PSA, we deliberately *avoid eliminating* any of the encountered spurious roots. This is because, in practice, some of the spurious roots are often associated with the genuine, physical transients that were supposed to decay, but nevertheless remained in the signal for the given total acquisition time T of the FID. The presence of such 'spurious' resonances might influence the overall spectral information in the signal, in which case elimination of the extraneous roots would be unjustified. The constrained root reflection, as a concentration-preserving (or signal-energy-preserving) regularization of spurious resonances, should be contrasted to the usual root reflection, which simply replaces the frequencies with the incorrect (unphysical) sign of their imaginary parts $\text{Im}(\omega_k) > 0$ by $-\text{Im}(\omega_k)$, but without the key achievement of the equality between the resulting magnitude and/or power spectra before and after this replacement. In other words, the usual root reflection does not conserve the total energy of the signal.

Chapter 77

Conclusions

We study non-classical orthogonal Lanczos polynomials and their role as complete sets of expansion functions. The auto-correlation functions, the Green function, the Padé–Lanzos approximant (PLA), the Stieltjes contracted continued fractions (CCF) and the Lanczos polynomials are intertwined in a very useful manner within the nearest neighbour method or tight binding model which is one of the most frequently used methods in quantum physics and quantum chemistry. Specifically, when applied to signal processing and spectral analysis, the standard Padé approximant (PA) is called the fast Padé transform (FPT). This is done to emphasize both the parametric and non-parametric features of the PA. Specifically, the word 'transform' in the acronym FPT highlights the transform property of the PA for shape spectra, in the spirit of the fast Fourier transform (FFT). The spectrum in the FPT can be computed in many different, but mathematically equivalent algorithms. One such algorithm elaborated in the present book is the Padé–Lanczos approximant (PLA), which is given by the rational polynomial $P_L(u)/Q_K(u)$ to the Green function, expressed as a power series with theoretically generated auto-correlation functions $\{C_n\}$ or experimentally measured time signals $\{c_n\}$ in the role of the expansion coefficients. Here, the variable u is $u = \exp(-i\omega\tau)$ with τ being the sampling time and ω is any frequency (real or complex). The response function $P_L(u)/Q_K(u)$ is the unique quotient of two polynomials $P_L(u)$ and $Q_K(u)$ of degrees $L-1$ and K. The Maclaurin series of this latter ratio coincides exactly with the first $L+K-1$ terms of the series development for the Green function. We show that the FPT is a versatile eigenproblem solver. The FPT yields complete spectra with the eigenvalues, eigenfunctions and residues of any ordinary or generalized eigenproblems, including the evolution/relaxation matrix **U** whose elements are either C_n or c_n. The necessary dimensionality reduction of the original large matrix **U** is presently achieved by the Lanczos tridiagonalization. The ensuing sparse Jacobi matrix is subsequently subjected to the usual procedure within the PA. This two-step hybrid method represents the parametric or non-parametric PLA, which is a low-storage and high-resolution signal processor. The non-parametric version of the PLA does not need to generate the polynomials $P_L(u)$ and $Q_K(u)$ explicitly, since their quotient $P_L(u)/Q_K(u)$ can be computed recursively. We obtain the general exact analytical expressions for the following key quantities: (i) expansion coefficients $\{a_n\}$ of the Stieltjes CCF, (ii) the Lanczos coupling parameters $\{\alpha_n, \beta_n\}$, (iii) the characteristic or secular Lanczos polynomials or eigenpolynomials of the first and second kind $\{Q_n(u), P_n(u)\}$, (iv) the fast Padé transform $P_K(u)/Q_K(u)$, (v) the Lanczos state vectors ψ_n, (vi) the total wavefunction $\Upsilon(u)$ of the studied generic system at any value of the complex frequency variable u and (vii) the 'Hamiltonian' operator $\hat{\Omega}$. All the closed formulae (i)–(vii) use C_n or c_n as the only input data stored in the so-called data matrix or the Hankel matrix. In the FPT, obtaining, e.g. the shape spectrum $P_K(u)/Q_K(u)$ at any u, amounts merely to inserting the time domain data $\{C_n\}$ or $\{c_n\}$ into an explicit analytical formula. This bypasses altogether mathematical ill-conditioning in the usual numerical generations of the Padé numerator and denominator polynomials. Conversely, for a given rational response function in the frequency domain via, e.g. the expansion coefficients $\{a_n\}$ of the CCF or the quotient $P_K(u)/Q_K(u)$, generating the auto-correlation functions $\{C_n\}$ or the time signals $\{c_n\}$ is presently carried out also from an explicit analytical formula. This establishes the symmetry between the main quantities $\{c_n\}$ and $\{a_n\}$ in the time and frequency domain, respectively. Moreover, the FPT unambiguously extracts the true number K of resonances from the input data $\{c_n\}$. This follows readily from the uniqueness condition for the existence of the FPT, as operationalized through accurate and fast recursive computations of two Hankel determinants $H_K(c_n)$ and $H_{K+1}(c_n)$ via the product-difference (PD) algorithm.

For the given generic time signal $\{c_n\}$ $(0 \le n \le N-1)$, one can immediately construct the associated Riemann sum/spectrum $R(z^{-1}) = N^{-1} \sum_{n=0}^{N-1} c_n z^{-n}$, where $z = u^{-1} = \exp(i\omega\tau)$. Then the diagonal FPT for $R(z^{-1})$ is given by the ratio A^-/B^- of two polynomials both of degree K in variable z^{-1}. The expansion coefficients $\{a_n^-, b_m^-\}$ of $\{A^-, B^-\}$ are determined from the requirement $(1/N)\sum_{n=0}^{N-1} c_n z^{-n} = (\sum_{n=0}^{K} a_n^- z^{-n})/(\sum_{m=0}^{K} b_m^- z^{-m}) \equiv A^-/B^-$. Pre-multiplication by B^- yields two systems of linear equations for $\{a_n^-, b_m^-\}$. The system for $\{b_m^-\}$ is solved numerically, whereas the set $\{a_n^-\}$ emerges as an explicit formula. This PA with A^-/B^- is referred to as FPT(z^{-1}). Its convergence region with exponentially decaying harmonics $\{z_k = \exp(i\omega_k \tau)\}$ for the physical/genuine poles Im$(\omega_k) > 0$ is located outside the unit circle $|z| > 1$. The variant FPT(z^{-1}) considers $\{c_n\}$ as generic time signal points that do not need to be linear combinations of damped complex exponentials. This implies that the quotient A^-/B^- can describe both Lorentzian and non-Lorentzian spectra. The former and the latter stem from the zeros of B^- without and with multiplicities, respectively. Riemann sum $R(z^{-1})$ as an expansion in z^{-1} converges for $|z| > 1$ with the increased values of the signal length N. Therefore, the role of the FPT(z^{-1}) is to accelerate this convergence by means of interpolation/extrapolation via a rational function.

Conclusions   413

The spectrum $R(z^{-1})$ diverges for $|z| < 1$. In the same region inside the unit circle, there exists another PA which is convergent and suitable for signal processing. This analytical continuation of $R(z^{-1})$ into the domain $|z| < 1$ is referred to as FPT(z). It is obtained considering each c_n as a sum of damped complex exponentials $c_n = \sum_{k=1}^{K} d_k z_k^n$. Then $R(z^{-1}) = \sum_{k=1}^{K} d_k [1 - (z_k/z)^N]/[1 - (z_k/z)] \approx \sum_{k=1}^{K} d_k z/(z - z_k) \equiv A^+/B^+$ ($N \gg 1, |z_k/z| < 1$) where polynomials A^+ and B^+ are of degree K in variable z. The expansion coefficients $\{a_n^+, b_m^+\}$ of $\{A^+, B^+\}$ are found from the condition $(1/N) \sum_{n=0}^{N-1} c_n z^{-n} \equiv A^+/B^+ = (\sum_{n=1}^{K} a_n^+ z^n)/(\sum_{m=0}^{K} b_m^+ z^m)$. Unlike A^-/B^- where $a_0^- \neq 0$, the free term in A^+/B^+ is equal to zero ($a_0^+ \equiv 0$) and the sum in A^+ starts from $a_1^+ z$. Multiplication by B^+ yields two systems of linear equations for $\{a_n^+, b_m^+\}$. The system for $\{b_m^+\}$ is solved numerically and the set $\{a_m^+\}$ follows from an analytical expression. This PA with A^+/B^+ represents FPT(z). Its convergence region is inside the unit circle $|z| < 1$. For $|z| < 1$ the ansatz $R(z^{-1})$ diverges as N increases and the task of the FPT(z) is to induce/force convergence by the Cauchy concept of analytical continuation. Thus, although $R(z^{-1})$ was originally undefined inside the unit circle, it emerged as being well defined also in this region by means of A^+/B^+, i.e. the FPT(z). Due to the above-assumed form for the c_n, the FPT(z) is apparently limited to Lorentzian spectra only. However, the roots of B^+ can have their multiplicities just like those from B^-. Indeed, supposing that $c_n = \sum_{k=1}^{J} \sum_{m_k=1}^{M_k} d_{k,m_k}(n\tau)^{m_k-1} z_k^{-n} (M_1 + M_2 + \cdots + M_J = K)$, which leads to a non-Lorentzian spectrum, a new $R(z^{-1})$ in the limit $N \gg 1$ becomes a ratio $\tilde{A}^+/\tilde{B}^+$ of two polynomials each of degree K in variable z. But both $\tilde{A}^+/\tilde{B}^+$ and A^+/B^+ stem from $R(z^{-1})$. Then the uniqueness of the PA imposes $\tilde{A}^+ = A^+$ and $\tilde{B}^+ = B^+$. Hence, the FPT(z) is also valid for both Lorentzian and non-Lorentzian spectra.

The parametric versions of the FPT(z) and FPT(z^{-1}) are presently illustrated on an experimentally measured time signal of length $N = 2048$. Barely a few minutes were needed at $K \leq 1024$ in our fully automated user-friendly software with the standard, least square optimized, Singular Value Decomposition (SVD) to compute highly-resolved absorption spectra with minimal effort for the whole Nyquist interval without windowing. Another factor of four can be gained in computer time if the information-preserving band-limited decimation is used for the main window from 0 ppm to 5 ppm of the chemical shift, which is a relative dimensionless frequency. Extremely stable and rapid convergence has been reached particularly in the FPT(z^{-1}) with the varying fractions N/M of the full signal length N where $M = 1 - 32$. This is unprecedented among all the existing nonlinear parameter estimators.

In all parametric estimators in signal processing, some of the complex eigenvalues of data matrices or eigenroots of characteristic polynomials are unphysical, i.e. spurious. In fact, adequacy and utility of all the existing parametric methods ultimately depends upon their ability to unambiguously identify and cope with spurious resonances. These extraneous eigenvalues of, e.g. the evolution matrix $\mathbf{U}$ lead to exponentially increasing harmonics of the corresponding time signal with the increase of time. The main difficulty with spurious roots is not so much in their appearance in parametric estimators, but rather in the implementation of theoretically devised procedures securing that the spectral density (power or magnitude spectrum) is invariant under the selected regularization. Recently, the filter diagonalization (FD) and decimated signal diagonalization (DSD) have been applied to a number of problems in signal processing, but it was found that these methods generate spurious eigenfrequencies leading to exponentially diverging fundamental harmonics in time signals. However, both the FD and the DSD have a joint and reliable device for handling spurious eigenenergies. In the FD and DSD, one compares the eigenvalues of the powers of the matrix $\mathbf{U}^{(s)}$ (comprised of the matrix elements of the sth power of the evolution operator $\hat{\mathbf{U}}$) for two values of the integer s, e.g. $s = 1$ and $s = 2$. From these two separate diagonalizations, only those eigenvalues are retained as genuine that are within a prescribed level of accuracy, whereas the other unstable solutions are discarded as spurious. Similarly, the recursive Lanczos algorithm for solving large eigenproblems produces extraneous eigenvalues, but they are also managed by an adequate procedure. In the decimated Padé approximant (DPA), spurious roots are recognized by comparing the results between the main diagonal and several successive paradiagonals of the two-dimensional Padé table for varying orders of the numerator and denominator polynomials. Within a pre-assigned threshold of accuracy, some of the roots are virtually unaltered when passing from the main diagonal to several neighbouring paradiagonals. These stable roots are classified as physical, i.e. genuine, and the corresponding peaks are retained in the final spectrum. However, the same comparison also finds some peaks whose parameters change considerably beyond the prescribed accuracy level. These resonances are considered as unphysical, i.e. spurious, and they are rejected from the spectrum. Diagonalization of the matrix $\mathbf{U}^{(s)}$ in the DSD for an integer s is mathematically equivalent to the DPA for the main diagonal ($s = 1$) and the paradiagonals ($s \neq 1$). Therefore, the techniques for recognizing spurious resonances in the FD, DSD and DPA are the same. However, two diagonalizations of data matrices for, e.g. $s = 1$ and $s = 2$ might find acceptable roots regarding the accuracy threshold, but the sign of $\text{Im}(\omega_k)$ could still be wrong, in which case $\text{Im}(\omega_k)$ is replaced by $-\text{Im}(\omega_k)$ which is the standard root reflection procedure.

As an alternative to these numerical techniques for dealing with extraneous resonances, we presently elaborate an analytical tool for unambiguous identification of all the existing spurious roots. To this end, within the generic PA, we use the so-called constrained root reflection for regularization of any spurious root. This procedure guarantees that both the numerator and denominator polynomials of the PA are the Schur polynomials, i.e. stable. A variant of the root reflection is designed by subjecting all the spurious roots of both Padé polynomials to complex conjugation, but with the important constraint that the total energy of the signal is conserved. This is the mentioned constrained root reflection, which regularizes all the encountered extraneous poles by relocating them to the side of the genuine poles. Such a method deliberately avoids eliminating any of the encountered spurious roots. This is because some of the spurious roots might well be associated with the genuine, i.e. physical transients that were supposed to decay, but nevertheless remained in the signal for the given total acquisition time. The presence of such 'spurious' resonances influences the spectral information in the signal and, therefore, they should not be eliminated from the analysis. The constrained root reflection as a power-spectrum-preserving regularization of spurious resonances is different from the usual root reflection, which replaces $\text{Im}(\omega_k)$ by $-|\text{Im}(\omega_k)|$ whenever the sign of $\text{Im}(\omega_k)$ leads to diverging exponentials, but without a simultaneous request that the ensuing magnitude and/or power spectra are identical for the parametric and non-parametric estimations in the same PA.

In conclusion, the shape spectrum of the FPT is resolved better than in the FFT for the same signal length. Also, by using only half signal length, the FPT can achieve the same resolution as in the FFT which is computed with the full signal length. Crucially, the FPT is the most natural parametric method for solving the quantification problem, since the best established quantum-mechanical non-degenerate (Lorentzians) $\sum_{k=1}^{K} d_k/(u-u_k)$ as well as degenerate (non-Lorentzians) $\sum_{k=1}^{J} \sum_{m_k}^{M_k} d_{k,m_k}/(u-u_k)^{m_k}$ $(M_1 + M_2 + \cdots + M_J = K)$ spectra are, by definition, the Heaviside partial fraction expansions of the Padé polynomial quotient $P_K(u)/Q_K(u)$ for the distinct and coincident zeros $\{u_k\}$ of the characteristic equation $Q_K(u) = 0$, respectively. Both of these spectra are, in fact, among the major quantum-mechanical cornerstones recognized as the resolvent or the Green function. The Fourier integral of the Green operator is the evolution operator whose matrix element between any two Schrödinger states represents the auto-correlation function, or equivalently, the time signal. Hence, the data matrix or Hankel matrix from MRS and NMR is the quantum-mechanical evolution/relaxation matrix. From the mathematical viewpoint, the quantum-mechanical auto-correlation functions are the linear combination of damped complex exponentials with constant or time dependent coefficients (amplitudes, residues) precisely as those given by the fast Padé transform. As is well known, no more information could be extracted from any studied system than what is provided by the corresponding quantum-mechanical wavefunction, or equivalently, the Green function. Since this Green function is the Padé approximant *per se*, it comes as no surprise that this latter method provides the optimal spectral analysis of generic time signals. Optimality is rooted in the the fact that the fast Padé transform is a method from the category of variational principles which have no errors of the first-order such as the Schwinger variational principle from quantum-mechanical scattering theory.

Appendix A

Linear mappings among vector spaces

Here, we shall outline a number basic features of vector spaces that are needed in the main text. In particular, we shall consider three linear vector spaces $\mathcal{X}, \mathcal{Y}$ and $\mathcal{Z}$ of finite dimensions $\dim \mathcal{X} = n$, $\dim \mathcal{Y} = \ell$ and $\dim \mathcal{Z} = m$, respectively. They are subspaces of the Hilbert space $\mathcal{H}$. Let the sets $\{|x_i\rangle\}_{i=1}^{n}$, $\{|y_i\rangle\}_{i=1}^{\ell}$ and $\{|z_i\rangle\}_{i=1}^{m}$ represent our fixed choices for the basis in the spaces $\mathcal{X}, \mathcal{Y}$ and $\mathcal{Z}$ respectively. We shall introduce the linear mappings A and B between any two of these vector spaces [228]

$$B : \mathcal{X} \longrightarrow \mathcal{Y} \qquad B(x_j) = \sum_{k=1}^{\ell} b_{kj} y_k \qquad (j = 1, 2, 3, \ldots, n) \qquad (A.1)$$

$$A : \mathcal{Y} \longrightarrow \mathcal{Z} \qquad A(y_k) = \sum_{i=1}^{m} a_{ik} z_i \qquad (k = 1, 2, 3, \ldots, \ell) \qquad (A.2)$$

where $a_{i,j} = a_{ij}$ and $b_{i,j} = b_{ij}$. The matrices corresponding to the mappings A and B will be denoted by $\mathbf{M}(A)$ and $\mathbf{M}(B)$

$$\mathbf{M}(A) = \{a_{ij}\}_{i,j=1}^{m,\ell} \qquad \mathbf{M}(B) = \{b_{ij}\}_{i,j=1}^{\ell,n}. \qquad (A.3)$$

To find a relationship between the matrices $\mathbf{M}(A), \mathbf{M}(B)$ and $\mathbf{M}(AB)$, we compute the vector $AB(x_j)$

$$AB(x_j) = A[B(x_j)] = A\sum_{k=1}^{\ell} b_{kj} y_k = \sum_{k=1}^{\ell} b_{kj} A(y_k)$$

$$= \sum_{k=1}^{\ell} b_{kj} \sum_{i=1}^{m} a_{ik} z_i = \sum_{i=1}^{m} \left(\sum_{k=1}^{\ell} a_{ik} b_{kj} \right) z_i \qquad (A.4)$$

which is recognized as the jth element of the product matrix $\mathbf{M}(AB)$. Namely, according to the standard rule, the product of the two matrices $\mathbf{M}(A)$ and $\mathbf{M}(B)$

is the matrix $\mathbf{M}(C)$ with the elements $\{c_{ij}\}$ defined by

$$\mathbf{M}(C) = \mathbf{M}(AB) \qquad \mathbf{M}(C) = \{c_{ij}\}_{i,j=1}^{m,n} \qquad c_{ij} = \sum_{k=1}^{\ell} a_{ik} b_{kj} \qquad (A.5)$$

$$\therefore \qquad \mathbf{M}(AB) = \mathbf{M}(A)\mathbf{M}(B). \qquad (A.6)$$

In the special case $B : \mathcal{Y} \longrightarrow \mathcal{Y}$ where the transformation B maps the $m \times m$ vector space $\mathcal{Y}$ onto the unit coordinate vectors, the general element c_{ij} is the Kronecker δ-symbol $\delta_{i,j}$ from (3.7)

$$c_{ij} = \delta_{i,j} \qquad (A.7)$$

so that the product $\mathbf{M}(AB)$ is the unit matrix $\mathbf{1}$. This circumstance permits the identification of the $m \times m$ matrix $[\mathbf{M}(A)]^{-1}$ as the inverse matrix of $\mathbf{M}(B)$

$$\mathbf{1} = \mathbf{M}(A)\mathbf{M}(B) \Longrightarrow \mathbf{M}(B) = [\mathbf{M}(A)]^{-1}. \qquad (A.8)$$

Finding the inverse $\mathbf{A}^{-1}$ of matrix $\mathbf{A}$ is mathematically equivalent to the problem of solving the following system of n linear inhomogeneous equations

$$\sum_{k=1}^{n} a_{ik} b_{kj} = \delta_{i,j}. \qquad (A.9)$$

In a general system of linear equations, the two sets of constants $\{a_{ij}\}_{i,j=1}^{n}$ and $\{b_i\}_{i=1}^{n}$ are given and the unknown vectors $\{x_j\}_{j=1}^{n}$ are obtained from

$$\mathbf{AX} = \mathbf{B} \qquad \sum_{j=1}^{n} a_{ij} x_i = b_i \qquad (i = 1, 2, 3, \ldots, n) \qquad (A.10)$$

$$\mathbf{A} = \{a_{ij}\}_{i,j=1}^{n} \qquad \mathbf{B} = \{b_i\}_{i=1}^{n} \qquad \mathbf{X} = \{x_i\}_{i=1}^{n} \qquad (A.11)$$

with $\mathbf{A}$ being the square $n \times n$ matrix

$$\mathbf{A} = \begin{pmatrix} a_{11} & a_{12} & \cdots & a_{1,j-1} & a_{1j} & a_{1,j+1} & \cdots & a_{1n} \\ a_{21} & a_{22} & \cdots & a_{2,j-1} & a_{2j} & a_{2,j+1} & \cdots & a_{2n} \\ \vdots & \vdots & \vdots & \vdots & \vdots & \vdots & & \vdots \\ a_{i-1,1} & a_{i-1,2} & \cdots & a_{i-1,j-1} & a_{i-1,j} & a_{i-1,j+1} & \cdots & a_{i-1,n} \\ a_{i1} & a_{i2} & \cdots & a_{i,j-1} & a_{i,j} & a_{i,j+1} & \cdots & a_{in} \\ a_{i+1,1} & a_{i+1,2} & \cdots & a_{i+1,j-1} & a_{i+1,j} & a_{i+1,j+1} & \cdots & a_{i+1,n} \\ \vdots & \vdots & \vdots & \vdots & \vdots & \vdots & \ddots & \vdots \\ a_{n1} & a_{n2} & \cdots & a_{n,j-1} & a_{nj} & a_{n,j+1} & \cdots & a_{nn} \end{pmatrix} \qquad (A.12)$$

whereas **B** and **X** are the $n \times 1$ matrices, i.e. column vectors of length n

$$\mathbf{B} = \begin{pmatrix} b_1 \\ b_2 \\ b_3 \\ \vdots \\ b_n \end{pmatrix} \qquad \mathbf{X} = \begin{pmatrix} x_1 \\ x_2 \\ x_3 \\ \vdots \\ x_n \end{pmatrix}. \tag{A.13}$$

Associated with the $n \times n$ matrix **A** is the $(n-1) \times (n-1)$ matrix called the (i, j)th minor matrix $\mathbf{A}_{ij} = \{a_{i,j}\}_{i,j=1}^{n-1}$ which is obtained by deleting all the elements from the ith row and jth column of $\mathbf{A}_n$

$$\mathbf{A}_{ij} = \begin{pmatrix} a_{11} & a_{12} & \cdots & a_{1,j-1} & a_{1,j+1} & \cdots & a_{1n} \\ a_{21} & a_{22} & \cdots & a_{2,j-1} & a_{2,j+1} & \cdots & a_{2n} \\ \vdots & \vdots & \vdots & \vdots & \vdots & \cdots & \vdots \\ a_{i-1,1} & a_{i-1,2} & \cdots & a_{i-1,j-1} & a_{i-1,j+1} & \cdots & a_{i-1,n} \\ a_{i+1,1} & a_{i+1,2} & \cdots & a_{i+1,j-1} & a_{i+1,j+1} & \cdots & a_{i+1,n} \\ \vdots & \vdots & \vdots & \vdots & \vdots & \ddots & \vdots \\ a_{n1} & a_{n2} & \cdots & a_{n,i-1} & a_{n,j+1} & \cdots & a_{nn} \end{pmatrix}. \tag{A.14}$$

We shall interchangeably say that $\mathbf{A}_{ij}$ is the minor of **A** corresponding to the element a_{ij}. The explicit solution of the system (A.10) is obtained by using the Cramer rule in the matrix form

$$\mathbf{X} = \mathbf{A}^{-1}\mathbf{B} = \frac{1}{\det \mathbf{A}}(\operatorname{cof} \mathbf{A})^T \mathbf{B}. \tag{A.15}$$

Here, the matrix cof **A** is the cofactor matrix associated with matrix **A**

$$\operatorname{cof} \mathbf{A} = \{\operatorname{cof} A_{ij}\}_{i,j=1}^{n} \qquad \operatorname{cof} A_{ij} = (-1)^{i+j} \det \mathbf{A}_{ij}. \tag{A.16}$$

This relation shows that the cofactor cof A_{ij} is, except for a plus or a minus sign, equal to the corresponding determinant of the minor $\det \mathbf{A}_{ij}$. The superscript T on the matrix **A** denotes the transpose matrix $\mathbf{A}^T = \{a_{ji}\}_{j,i=1}^{n}$, which is obtained by exchanging the places of the rows and columns in the original matrix $\mathbf{A} = \{a_{ij}\}_{i,j=1}^{n}$

$$\operatorname{cof} \mathbf{A}^T = \{\operatorname{cof} A_{ji}\}_{i,j=1}^{n} \qquad \operatorname{cof} A_{ji} = (-1)^{j+i} \det \mathbf{A}_{ji}. \tag{A.17}$$

The determinant of the transpose and the original matrices are the same

$$\det \mathbf{A}^T = \det \mathbf{A}. \tag{A.18}$$

The transpose of a cofactor matrix cof **A** is the *adjugate* of **A** denoted by adj **A**

$$\operatorname{adj} \mathbf{A} = (\operatorname{cof} \mathbf{A})^T. \tag{A.19}$$

Linear mappings among vector spaces

Adjugate adj $\mathbf{A}$ of matrix $\mathbf{A} = \{a_{ij}\}$ should not be confused with adjoint $\mathbf{A}^\dagger$ which some authors denote by adj $\mathbf{A}$. An adjoint matrix $\mathbf{A}^\dagger$ of $\mathbf{A} = \{a_{ij}\}$ is defined by $\mathbf{A}^\dagger = \{a^*_{ji}\}$. In the special case $\mathbf{A}^\dagger = \mathbf{A}$, which is of particular relevance for quantum mechanics, the matrix $\mathbf{A}$ is said to be Hermitean. The elements $\{x_j\}_{j=1}^n$ of the matrix $\mathbf{X}$ from (A.15) are

$$x_j = \frac{1}{\det \mathbf{A}} \sum_{k=1}^n b_k \operatorname{cof} A_{kj} \qquad (j = 1, 2, \ldots, n). \tag{A.20}$$

The solution $\mathbf{X}$ exists and it is unique if and only if the determinant $\det \mathbf{A}$ of the system (A.10) is non-zero

$$\det \mathbf{A} \neq 0 \tag{A.21}$$

$$\det A = \begin{vmatrix} a_{11} & a_{12} & \cdots & a_{1,j-1} & a_{1j} & a_{1,j+1} & \cdots & a_{1n} \\ a_{21} & a_{22} & \cdots & a_{2,j-1} & a_{2j} & a_{2,j+1} & \cdots & a_{2n} \\ \vdots & \vdots & \vdots & \vdots & \vdots & \vdots & \cdots & \vdots \\ a_{i-1,1} & a_{i-1,2} & \cdots & a_{i-1,j-1} & a_{i-1,j} & a_{i-1,j+1} & \cdots & a_{i-1,n} \\ a_{i1} & a_{i2} & \cdots & a_{i,j-1} & a_{i,j} & a_{i,j+1} & \cdots & a_{in} \\ a_{i+1,1} & a_{i+1,2} & \cdots & a_{i+1,j-1} & a_{i+1,j} & a_{i+1,j+1} & \cdots & a_{i+1,n} \\ \vdots & \vdots & \vdots & \vdots & \vdots & \vdots & \ddots & \vdots \\ a_{n1} & a_{n2} & \cdots & a_{n,j-1} & a_{nj} & a_{n,j+1} & \cdots & a_{nn} \end{vmatrix}.$$

$$\tag{A.22}$$

The Cramer solution can equivalently be stated as the following ratio of two determinants

$$x_j = \frac{\det \mathbf{C}_j}{\det \mathbf{A}} \tag{A.23}$$

where the dimensions of both matrices $\mathbf{A}$ and $\mathbf{C}_j$ are equal ($\dim \mathbf{A} = n = \dim \mathbf{C}_j$). The matrix $\mathbf{C}_j$ is obtained by replacing the jth column of $\mathbf{A}$ by the column matrix $\mathbf{B}$

$$\mathbf{C}_j = \begin{pmatrix} a_{11} & a_{12} & \cdots & a_{1,j-1} & b_1 & a_{1,j+1} & \cdots & a_{1n} \\ a_{21} & a_{22} & \cdots & a_{2,j-1} & b_2 & a_{2,j+1} & \cdots & a_{2n} \\ \vdots & \vdots & \vdots & \vdots & \vdots & \vdots & \cdots & \vdots \\ a_{i-1,1} & a_{i-1,2} & \cdots & a_{i-1,j-1} & b_{i-1} & a_{i-1,j+1} & \cdots & a_{i-1,n} \\ a_{i1} & a_{i2} & \cdots & a_{i,j-1} & b_i & a_{i,j+1} & \cdots & a_{in} \\ a_{i+1,1} & a_{i+1,2} & \cdots & a_{i+1,j-1} & b_{i+1} & a_{i+1,j+1} & \cdots & a_{i+1,n} \\ \vdots & \vdots & \vdots & \vdots & \vdots & \vdots & \ddots & \vdots \\ a_{n1} & a_{n2} & \cdots & a_{n,j-1} & b_n & a_{n,j+1} & \cdots & a_{nn} \end{pmatrix}.$$

$$\tag{A.24}$$

The square matrix $\mathbf{A}$ is an *orthogonal* matrix if we have

$$\mathbf{A}\mathbf{A}^T = \mathbf{1}. \tag{A.25}$$

It follows from the above outlines that if (A.21) is satisfied, then the inverse $\mathbf{A}^{-1}$ exists and is given by the Cramer solution of the system (A.9)

$$\mathbf{A}^{-1} = \frac{1}{\det \mathbf{A}} (\operatorname{cof} \mathbf{A})^T \qquad (A.26)$$

or, on account of (A.18)

$$\mathbf{A}^{-1} = \left(\frac{\operatorname{cof} \mathbf{A}}{\det \mathbf{A}} \right)^T. \qquad (A.27)$$

In the special case $\mathbf{B} = \{b_{if}\} = \{0\}_{i,j=1}^n = \mathbf{0}$ where $\mathbf{0}$ is a zero matrix, the expression (A.10) becomes a system of n homogeneous linear equations

$$\mathbf{AX} = \mathbf{0} \qquad \sum_{j=1}^n a_{ij} x_i = 0 \quad (i = 1, 2, 3, \ldots, n). \qquad (A.28)$$

Inserting $\mathbf{B}=\mathbf{0}$ into (A.10), it follows that the system (A.28) will have a non-trivial, i.e. non-zero solution $\mathbf{X}^{-1}$ if and only if the following condition is satisfied [212]

$$\det \mathbf{A} = 0. \qquad (A.29)$$

This special case is important in, e.g. eigenvalue problems. Suppose we are given a linear transformation $\Lambda : \mathcal{X} \longrightarrow \mathcal{X}$ (dim $\mathcal{X} = n$) such that the corresponding matrix $\mathbf{\Lambda}$ possesses a diagonal representation

$$\mathbf{\Lambda} = \{x_{i,j}\}_{i,j=1}^n \qquad x_{i,j} = x_i \delta_{i,j} \qquad (A.30)$$

where $\delta_{i,j}$ is the Kronecker δ-symbol from (3.7) and

$$\mathbf{\Lambda} = \operatorname{diag}(x_1, x_2, \ldots, x_n) = \begin{pmatrix} x_1 & 0 & 0 & \cdots & 0 \\ 0 & x_2 & 0 & \cdots & 0 \\ 0 & 0 & x_3 & \cdots & 0 \\ \vdots & \vdots & \vdots & \ddots & \vdots \\ 0 & 0 & 0 & \cdots & x_n \end{pmatrix}_{n \times n} \qquad (A.31)$$

with $\{x_k\}_{k=1}^n$ being a set of scalars. Then, there exists a linearly independent sequence of state vectors $\{|x_k\rangle\}_{k=1}^n$ satisfying the eigenvalue problem

$$\Lambda |x_k\rangle = x_k |x_k\rangle \qquad \mathbf{\Lambda X}_k = x_k \mathbf{X}_k \qquad (k = 1, 2, \ldots, n) \qquad (A.32)$$

where $\{x_1, x_2, \ldots, x_n\}$ and $\{|x_1\rangle, |x_2\rangle, \ldots, |x_n\rangle\}$ are the eigenvalues, or equivalently, the characteristic values/numbers, and the eigenstates vectors, respectively. Referring to (A.28), we rewrite (A.32) as a system of homogeneous linear equations

$$(x_k - \Lambda)|x_k\rangle = 0 \qquad (x_k \mathbf{1} - \mathbf{\Lambda})\mathbf{X}_k = \mathbf{0} \qquad (k = 1, 2, \ldots, n) \qquad (A.33)$$

where **1** is the unity or the identity matrix. According to (A.29) the system (A.28) has non-trivial solutions $\{|x_k\rangle\} \neq \{|0\rangle\}$ where $|0\rangle$ is zero state vector if and only if

$$\det[x_k\mathbf{1} - \mathbf{\Lambda}] = 0 \qquad (k = 1, 2, \ldots, n) \tag{A.34}$$

which is the *secular equation*. Hence the eigenvalues $\{\lambda_k\}$ of Λ are also obtainable from the secular equation (A.34) which can be rewritten as

$$Q_k(x_k) = 0 \qquad Q_k(x) = \det[x\mathbf{1} - \mathbf{\Lambda}] \tag{A.35}$$

where $Q_k(x)$ is the characteristic polynomial of order k. Thus the zeros $\{x_k\}$ of the characteristic polynomial $Q_k(x)$ are the eigenvalues of the corresponding matrix $\mathbf{\Lambda}$. Of course, the matrix elements $\{x_{ij}\}$ of $\mathbf{\Lambda}$ depend upon a concrete choice of the basis $\{|x_k\rangle\}$. By contrast, both the eigenvalues $\{x_k\}$ and the characteristic polynomial $Q_k(x)$ are defined with no reference to any basis and, therefore, they are independent of the choice of basis [228]. In general, the roots $\{x_k\}$ of the characteristic polynomial $Q_k(x)$ are complex numbers, but some of them could be purely real. Even if all the coefficients of the power series representation of $Q_k(x)$ are real, some of the roots x_k may be complex, in which case they come in pairs with the corresponding complex conjugate numbers. The name 'secular' equation comes from celestial mechanics in studies on certain motions of planets as a function of time $t \geq 0$. There it was observed that the set of the eigenroots $\{x_k\}$ of (A.35) contains mainly the exponentially decaying harmonics $\{\exp(-x_k t)\}$ with $\text{Re}(x_k) > 0$ that have a plausible meaning, since the probability of occurrence of all temporal physical phenomena must decay as the time is increased infinitely $t \longrightarrow \infty$. However, occasionally there were some elements $\exp(-x_k t) = \exp\{-\text{Re}(x_k)t - i\,\text{Im}(x_k)t\}$ of the set $\{\exp(-x_k t)\}$ with $\text{Re}(x_k) < 0$ that diverge exponentially as $t \longrightarrow \infty$. These complex roots x_k with the wrong sign $\text{Re}(x_k) < 0$ form the so-called secular solutions of (A.34). They are meaningless, since they cannot have any realistic interpretation. The ansatz (A.34) itself is known as the secular equation to point at the possible existence of some secular roots among the elements of the set $\{x_k\}$. It is sufficient that there exists only one such secular element x_k to cause the whole assembly $\{x_k\}$ to totally lose its physical meaning if a general solution of the problem is the sum of all the harmonics $\{\exp(-x_k t)\}_{k=1}^n$. This latter superposition of exponentials follows from solving, e.g. an nth-order differential equation $\sum_{k=1}^n \gamma_k (d/dt)^k x(t) = 0$ with the given constant coefficients $\{\gamma_k\}$ as discussed in chapter 7. This differential equation has the unique general solution expressed as $x(t) = \sum_{k=1}^n a_k \exp(-x_k t)$ where the amplitudes $\{a_k\}_{k=1}^n$ can be determined from the *a priori* pre-assigned n boundary conditions of the differential equation. Here, $\{x_k\}_{k=1}^n$ are the roots of the corresponding characteristic polynomial. Clearly, $x(t)$ diverges $x(t) \longrightarrow \infty$ as $t \longrightarrow \infty$ if there is only one secular root, say x_1 with $\text{Re}(x_1) < 0$ because it yields $\exp(-x_1 t) \longrightarrow \infty$ in the limit of infinitely large times. This causes the whole sum $x(t) = \sum_{k=1}^n a_k \exp(-x_k t)$ to diverge. In the literature across sciences and engineering, the term 'secular roots' is almost never in use. Instead, whenever referring to those roots x_k having the wrong sign $\text{Re}(x_k) < 0$ in the damping factor $\exp\{-\text{Re}(x_k)t\}$ of the harmonic $\exp\{-\text{Re}(x_k)t - i\,\text{Im}(x_k)t\}$, it is customary to employ the term 'spurious' or 'extraneous' roots or 'ghosts'. From this intriguing etymology, the only nomenclature which has been retained in the literature is the term 'secular equation' for (A.34) [212].

Appendix B

Non-classical polynomials in the expansion methods

Here, we shall use real orthogonal polynomials $\{Q_r(x)\}$ as basis functions to develop an expansion method for computation of the most important physical quantities, such as Green functions, density of states, integrated density of states, etc. These polynomials are defined on the real segment $[a, b]$ which can be finite or infinite. The polynomial coefficients are also real, but since $Q_r(x)$ is an analytic function, the quantity $Q_r(z)$ exists for a complex z in the limit $\text{Im}(z) \longrightarrow 0$. For a real x and a complex z we have

$$Q_n^*(x) = Q_n(x) \qquad Q_n^*(z) = Q_n(z^*) \qquad \text{(B.1)}$$

where the star superscript denote the usual complex conjugation. Let $d\sigma(x)$ be a non-negative Riemann measure on $[a, b]$

$$d\sigma(x) = W(x)dx \qquad \text{(B.2)}$$

where the weight function $W(x) > 0$ has at least $n + 1$ points of increase. Then, in the space of polynomials $\{Q_r(x)\}$, we can define the asymmetric scalar/inner product with respect to the measure $d\sigma(x)$ as

$$\langle Q_m(x) | Q_n(x) \rangle \equiv \int_a^b Q_m^*(x) Q_n(x) d\sigma(x). \qquad \text{(B.3)}$$

Asymmetric scalar products are customary in the conventional formulation of quantum mechanics defined on the field of Hermitean operators. Of course, the star in $Q_m^*(x)$ is superfluous in (B.3), since $Q_m^*(x) = Q_m(x)$, as per (B.1). Nevertheless, we shall keep the star superscript on all the 'bra' vectors $\langle f(x)|$ even for a real function $f(x)$ simply to indicate the use of the asymmetric inner product. However, if the value $Q_n(z)$ is encountered with a complex z then obviously the star superscript on such a polynomial cannot be ignored any longer. We shall see that such circumstances become relevant for resolvents of Hermitean operators (Hamiltonians) from quantum mechanics. The polynomials $\{Q_n(x)\}$ are assumed to be orthonormalized with respect to the measure $d\sigma(x)$ viz

$$\int_a^b Q_n^*(x) Q_m(x) d\sigma(x) = w_n \delta_{nm} \tag{B.4}$$

where the coefficients $\{w_n\}$ are called the weights ($w_n > 0$). These weights should not be confused with the Christoffel numbers $\{w_n\}$ from numerical integrations that we shall encounter later on. There exists also the completeness relation given by

$$\delta(x - x_0) = \sum_{n=0}^{\infty} W_n(x) Q_n^*(x) Q_m(x_0) \qquad W_n(x) = \frac{W(x)}{w_n}. \tag{B.5}$$

The same relationship is also formally valid for operator-valued functions

$$\delta(E\hat{I} - \hat{H}) = \sum_{n=0}^{\infty} W_n(E) Q_n^*(E) Q_m(\hat{H}) \tag{B.6}$$

where E is the real energy and $\hat{H}$ is the Hermitean Hamiltonian operator ($\hat{H}^\dagger = \hat{H}$) of the studied physical system. The Dirac δ-operator (B.6) is called the density operator for the total energy E and Hamiltonian $\hat{H}$. In the limit $\eta \longrightarrow 0^+$, the following operator relationship exists

$$\frac{1}{(E + i\eta)\hat{I} - \hat{H}} = \mathcal{P}\left(\frac{1}{E\hat{I} - \hat{H}}\right) - i\pi \delta(E\hat{I} - \hat{H}) \tag{B.7}$$

where the symbol $\mathcal{P}$ denotes the principle value. This permits writing the density operator in the form

$$\delta(E\hat{I} - \hat{H}) = -\frac{1}{\pi} \operatorname{Im} \hat{G}(z) \tag{B.8}$$

where $\hat{G}(z)$ is the Green operator

$$\hat{G}(z) = \frac{1}{z\hat{I} - \hat{H}}. \tag{B.9}$$

Here, the usage of a Hermitean Hamiltonian $\hat{H}$ implies

$$\hat{G}^*(z) = \hat{G}(z^*). \tag{B.10}$$

The inverse operator $\hat{G}(z)$ exists if z is not an eigenenergy E_k from the spectrum of $\hat{H}$. For a Hermitean Hamiltonian $\hat{H}$, this is assured by setting

$$z = E + i\eta \qquad z^+ = E + i0^+ \tag{B.11}$$

where η is an infinitesimally small positive number. Equivalently, the resolvent $\hat{G}^+(z)$ can be conceived as the usual Fourier integral of the evolution operator

$$\hat{G}(E) = -i \int_{-\infty}^{\infty} dt\, \theta(t) \hat{U}(t) e^{iEt} \tag{B.12}$$

$$\hat{U}(t) = e^{-i\hat{H}t} \tag{B.13}$$

where $\theta(t)$ is the Heaviside step function (9.20) and t is the real time variable.

By definition, $\hat{G}(z)$ is the operator Padé approximant (OPA), since it is the product of the unity operator and $(z\hat{1} - \hat{H})^{-1}$, i.e. $\hat{G}(z) = \hat{1}(z\hat{1} - \hat{H})^{-1} = (z\hat{1} - \hat{H})^{-1}\hat{1}$. Therefore, diagonal or off-diagonal matrix elements of $\hat{G}$, i.e. the Green functions, will necessarily be the usual (scalar) Padé approximant. This can be easily shown by using the corresponding Schrödinger eigenvalue problem

$$\hat{H}|\Upsilon_k\rangle = E_k|\Upsilon_k\rangle \tag{B.14}$$

where $|\Upsilon_k\rangle$ is the eigenstate corresponding to the eigenenergy E_k. Using completeness of the set $\{|\Upsilon_k\rangle\}$, in the form of the spectral decomposition of the unity operator, i.e. $\hat{1} = \sum_{k=1}^{K} |\Upsilon_k\rangle\langle\Upsilon_k|$, it follows from (B.9)

$$\hat{G}(z) = \sum_{k=1}^{K} \frac{|\Upsilon_k\rangle\langle\Upsilon_k|}{z - E_k}. \tag{B.15}$$

Here, K is finite or infinite, depending of the number of eigenstates E_k in the spectrum of $\hat{H}$. The corresponding diagonal ($r = s$) and off-diagonal ($r \neq s$) matrix elements of $\hat{G}(z)$ taken over the states $|\Phi_s\rangle$ and $\langle\Phi_r|$ are obtained from (B.15) via

$$G_{r,s}(z) \equiv \langle\Phi_r|\hat{G}|\Phi_s\rangle = \sum_{k=1}^{K} \frac{d_k^{(r,s)}}{z - E_k} \tag{B.16}$$

$$d_k^{(r,s)} = \langle\Phi_r|\Upsilon_k\rangle\langle\Upsilon_k|\Phi_s\rangle. \tag{B.17}$$

In the diagonal case, the residue (B.17) simplifies as $d_k^{(s)} = |\langle\Phi_s|\Upsilon_k\rangle|^2$ where $d_k^{(s,s)} \equiv d_k^{(s)}$. It is obvious from (B.16) that $G_{r,s}(z)$ is the quotient of two polynomials $A_{K-1}(z)/B_K(z)$, which is the Padé approximant and that was set to prove (QED). This shows that the Padé approximant is the most natural method for computation of the Green function. This is because, by definition, the Padé approximant becomes exact if the function to be modelled happens to be a ratio of two polynomials. This is precisely the case with the Green function (B.16) which is a polynomial quotient, since it stems from the matrix element of the operator Padé approximant in the form of the resolvent operator (B.9).

Being orthonormal, the polynomials $\{Q_r(x)\}$ satisfy the standard three-term recursion

$$\left.\begin{array}{c} \beta_{n+1} Q_{n+1}(x) = (x - \alpha_n) Q_n(x) - \beta_n Q_{n-1}(x) \\ Q_{-1}(x) = 0 \qquad Q_0(x) = 1 \end{array}\right\}. \tag{B.18}$$

If we multiply (B.18) by $Q_m^*(x)$ and integrate over x from a to b using (B.2) and orthogonality (B.4), it follows

$$\alpha_n = \frac{1}{w_n} \int_a^b Q_n^*(x) x Q_n(x) d\sigma(x). \tag{B.19}$$

By an analogous argument, multiplication of (B.18) by $Q_{n-1}(x)$ leads to

$$\beta_n = \frac{1}{w_{n-1}} \int_a^b Q_{n-1}^*(x) x Q_n(x) d\sigma(x). \tag{B.20}$$

The expressions (B.19) and (B.20) are considered as the definitions of α_n and β_n. As a check, we can eliminate $xQ_n(x)$ from (B.19) and (B.20) to show via (B.4) that (B.19) and (B.20) lead to the identities $\alpha_n \equiv \alpha_n$ and $\beta_n \equiv \beta_n$. As such, α_n from (B.19) appears as the diagonal element of the position operator $\hat{x}$ in the coordinate representation. Likewise, in this representation, the same operator $\hat{x}$ has the parameter β_n from (B.20) as the coupling constant between the nth and $(n-1)$st polynomials. However, there is an alternative interpretation for β_n. It follows by eliminating $xQ_{n-1}^*(x)$ from (B.20) by means of (B.18) viz $xQ_{n-1}^*(x) = \beta_n Q_n^*(x) + \alpha_{n-1} Q_{n-1}^*(x) + \beta_{n-1} Q_{n-2}^*(x)$ and subsequently employing the orthonormality condition (B.4)

$$\beta_n = \frac{1}{w_{n-1}} \int_a^b |Q_n(x)|^2 d\sigma(x) = \frac{1}{w_{n-1}} \int_a^b Q_n^2(x) d\sigma(x)$$

$$= \frac{1}{w_{n-1}} \|Q_n\|^2 \implies \beta_n > 0. \tag{B.21}$$

Thus, we see that β_n is proportional to the squared norm of $Q_n(x)$ and, therefore, all the parameters $\{\beta_n\}$ are strictly positive. Moreover, since the diagonal case of the matrix element (B.4) is the norm $\|Q_n\|^2 = \int_a^b |Q_n(x)|^2 d\sigma(x) = w_n$, we can obtain a very simple expression for β_n from (B.21) as

$$\beta_n = \frac{w_n}{w_{n-1}}. \tag{B.22}$$

It then clearly follows that the whole procedure of numerically constructing the orthonormalized polynomials $\{Q_r(x)\}$ with the self-generating coefficients $\{\alpha_r, \beta_r\}$ is entirely reminiscent of the Lanczos algorithm when applied to the quantum-mechanical position operator $\hat{x}$. In fact, precisely the Lanczos algorithm for physical state vectors will be derived in this appendix by simply replacing operator $\hat{x}$ by the Hamiltonian $\hat{H}$ in the scalar recursion (B.18). The polynomials $\{Q_n(x)\}$ are called polynomials of the first kind. Each polynomial $Q_n(x)$ of the degree n has n zeros $\{x_k\}$ ($1 \leq k \leq n$). Because of the said assumptions on $W(x)$, all the quantities $\{x_k\}$ are simple zeros ($x_k \neq x_{k'}$ for $k' \neq k$), real and confined to the interval (a, b).

Non-classical polynomials in the expansion methods 427

Closely related to basis functions $\{Q_r(x)\}$ are the polynomials of the second kind denoted by $\{P_r(x)\}$. They satisfy the same recursive relation as in (B.18), except for two different initial conditions

$$\left.\begin{array}{c}\beta_{n+1}P_{n+1}(x) = (x-\alpha_n)P_n(x) - \beta_n P_{n-1}(x) \\ P_0(x) = 0 \qquad P_1(x) = 1\end{array}\right\}. \qquad \text{(B.23)}$$

Both recursions (B.18) and (B.23) are known to be numerically stable. We can define $Q_n(x)$ and $P_n(x)$ through their respective power series representations

$$Q_n(x) = \sum_{r=0}^{n} q_{r,n-r} x^r \qquad P_n(x) = \sum_{r=0}^{n-1} p_{r,n-r} x^r. \qquad \text{(B.24)}$$

As in chapter 26, it can be shown that the expansion coefficients $p_{r,n-r}$ and $q_{r,n-r}$ can be computed recursively by the same type of relations encountered in (B.18) and (B.23), i.e.

$$\left.\begin{array}{c}\beta_{n+1} p_{n+1,n+1-r} = p_{n,n+1-r} - \alpha_n p_{n,n-r} - \beta_n p_{n-1,n-1-r} \\ p_{n,-1} = 0 \qquad p_{n,m} = 0 \quad (m>n) \qquad p_{0,0} = 1 \qquad p_{1,1} = 1\end{array}\right\} \qquad \text{(B.25)}$$

$$\left.\begin{array}{c}\beta_{n+1} q_{n+1,n+1-r} = q_{n,n+1-r} - \alpha_n q_{n,n-r} - \beta_n q_{n-1,n-1-r} \\ q_{n,-1} = 0 \qquad q_{n,m} = 0 \quad (m>n) \qquad q_{0,0} = 1\end{array}\right\}. \qquad \text{(B.26)}$$

The zeros $\{x'_k\}$ of $P_n(x)$ interleave with the zeros x_k of $Q_n(x)$

$$x_1 < x'_1 < x_2 < x'_2 < \cdots < x_n < x'_n. \qquad \text{(B.27)}$$

In mathematics, this is called the Cauchy–Poincaré interlacing theorem, which has been reinvented in quantum chemistry where it is known as the Hylleraas–Undheim theorem [274].

Due to the scaled initial conditions relative to (B.18), the degree of the polynomial $P_n(x)$ is by 1 less than that of $Q_n(x)$. In other words, the degrees of the polynomials $Q_n(x)$ and $P_n(x)$ are n and $n-1$, respectively, as per (B.24). The polynomials $Q_n(x)$ and $P_n(x)$ are related to each other by the integral

$$P_n(z) = \frac{\beta_1}{\mu_0} \int_a^b \frac{Q_n(z) - Q_n(x)}{z - x} d\sigma(x) \qquad \text{(B.28)}$$

$$P_n(z^*) = P_n^*(z) \qquad \text{(B.29)}$$

where μ_0 is the simplest case $n=0$ of the power moment

$$\mu_n = \int_a^b x^n d\sigma(x). \qquad \text{(B.30)}$$

In order to connect these polynomials naturally to the Gaussian quadratures, it is convenient to introduce the Stieltjes integral $S(z)$ by

$$S(z) = \int_a^b \frac{d\sigma(x)}{z-x} \qquad S(z^*) = S^*(z). \tag{B.31}$$

The Stieltjes integral can be modelled by the Padé approximant $\mathcal{R}_n(u)$ via

$$S(z) \approx \mathcal{R}_n(z) \tag{B.32}$$

$$\mathcal{R}_n(x) \equiv \frac{\mu_0}{\beta_1} \frac{P_n(x)}{Q_n(x)}. \tag{B.33}$$

With the purpose of estimating the remainder $S(z) - \mathcal{R}_n(z)$ in (B.32), we introduce the error term $\mathcal{E}_n(z)$ which is invoked by the approximation $S(z) \approx \mathcal{R}_n(z)$

$$S(z) = \mathcal{R}_n(z) + \mathcal{E}_n(z). \tag{B.34}$$

We shall now carry out a derivation of an expression for the error or remainder $\mathcal{E}_n(z)$ by inserting (B.34) into (B.28) and using (B.33)

$$P_n(z) = \frac{\beta_1}{\mu_0} \left\{ Q_n(z) \int_a^b \frac{d\sigma(x)}{z-x} - \int_a^b \frac{d\sigma(x)}{z-x} Q_n(x) \right\}$$

$$= \frac{\beta_1}{\mu_0} \left[Q_n(z)[\mathcal{R}_n(z) + \mathcal{E}_n(z)] - \int_a^b \frac{d\sigma(x)}{z-x} Q_n(x) \right]$$

$$= \frac{\beta_1}{\mu_0} \left\{ Q_n(z) \left[\frac{\mu_0}{\beta_1} \frac{P_n(z)}{Q_n(z)} + \mathcal{E}_n(z) \right] - \int_a^b \frac{d\sigma(x)}{z-x} Q_n(x) \right\}$$

$$P_n(z) = P_n(z) + \frac{\beta_1}{\mu_0} \left[Q_n(z) \mathcal{E}_n(z) - \int_a^b \frac{d\sigma(x)}{z-x} Q_n(x) \right]$$

$$\therefore \quad \mathcal{E}_n(z) = \frac{\pi_n(z)}{Q_n(z)} \tag{B.35}$$

$$\pi_n(z) \equiv \int_a^b \frac{d\sigma(x)}{z-x} Q_n(x). \tag{B.36}$$

In deriving the result (B.35), we made a division by the constant quotient β_1/μ_0 and this is permissible, since $\beta_1 \neq 0$ and $\mu_0 \neq 0$. Using (B.1) and (B.36), it follows

$$\pi_n^*(z) = \pi_n(z^*) \tag{B.37}$$

and in this way the relation (B.10) is reflected onto $\pi_n(z)$. This is because (B.37) can also be obtained from the complex conjugate counterpart of (B.36) if we account for (B.1) and (B.10). The obtained formula (B.35) for the error $\mathcal{E}_n(z)$ is indeed remarkable, since the function $Q_n(z)\mathcal{E}_n(z)$ coincides exactly with the Hilbert transform of the polynomial $Q_n(x)$ in the vector space defined by the measure $d\sigma(x)$ and the asymmetric scalar product for $x \in [a, b]$.

Non-classical polynomials in the expansion methods

The rational polynomial $\mathcal{R}_n(x)$ from (B.33) is a meromorphic function and, as such, can be expressed through its spectral representation which is a linear combination of the Heaviside partial fractions

$$\mathcal{R}_n(x) = \sum_{k=1}^{n} \frac{w_k}{x - x_k}. \tag{B.38}$$

Here, x_k are the zeros of the polynomials $Q_n(x)$, whereas w_k are corresponding Christoffel numbers. The quantities $\{w_k\}$ are obtained by applying the Cauchy residue theorem to $\mathcal{R}_n(x)$ viz

$$\begin{aligned} w_k &= \lim_{x \to x_k} (x - x_k)\mathcal{R}_n(x) \\ &= \frac{\mu_0}{\beta_1} \frac{P_n(x_k)}{Q'_n(x_k)} = \frac{\mu_0}{\beta_n} \frac{1}{Q_{n-1}(x_k)Q'_n(x_k)} \end{aligned} \tag{B.39}$$

$$P_n(x_k) = \frac{\beta_1}{\beta_n Q_{n-1}(x_k)} \qquad Q_n(x_k) = 0 \tag{B.40}$$

where $Q'_n(x_k) = (d/dx)Q_n(x)$. Similarly to (31.13), the same result (B.39) can likewise be obtained by using the Christoffel–Darboux formula

$$\sum_{m=0}^{n-1} Q_m(x)Q_m(y) = \beta_n \frac{Q_n(x)Q_{n-1}(y) - Q_{n-1}(x)Q_n(y)}{x - y} \tag{B.41}$$

as in (31.9). If we sum up explicitly each term $w_k/(x - x_k)$ on the rhs of (B.38), the result will be the quotient of the two polynomials $(\mu_0/\beta_1)P_n(x)/Q_n(x)$ and this procedure retrieves faithfully the Padé approximant (B.33) from which we started.

The expressions (B.34) and (B.35) demonstrate that the error $\mathcal{E}_n(z)$ can be computed with absolute certainty while representing the Stieltjes integral (B.31) by the rational polynomial $\mathcal{R}_n(z)$ from (B.33). This is accomplished by the exact Gauss-type numerical quadrature rule

$$\int_a^b \frac{d\sigma(x)}{z - x} = \lim_{n \to \infty} \sum_{k=1}^{n} \frac{w_k}{z - x_k}. \tag{B.42}$$

This formula is of the same type as the standard one used for the corresponding Gauss numerical integrations with the classical polynomials. The only difference is in using non-classical Lanczos polynomials in (B.42). Of course, in all realistic computations, the infinite upper limit in the sum over n in (B.42) is replaced by a cut-off number K which represents the order or rank of the quadrature rule. In other words, if the expansion in (B.42) is truncated to the first K terms, we shall obtain the following approximation

$$\int_a^b \frac{d\sigma(x)}{z - x} \approx \sum_{k=1}^{K} \frac{w_k}{z - x_k}. \tag{B.43}$$

Nevertheless, even with only K terms retained in the Heaviside partial fractions, the final result for $S(z)$ can still be exact provided that the error $\mathcal{E}_n(z)$ is included

$$S(z) \equiv \int_a^b \frac{d\sigma(x)}{z-x} = \sum_{k=1}^K \frac{w_k}{z-x_k} + \mathcal{E}_n(z)$$

$$= \frac{\mu_0}{\beta_1} \frac{P_n(z)}{Q_n(z)} + \mathcal{E}_n(z). \tag{B.44}$$

This analysis extends at once to the corresponding exact Gauss-type quadrature for the case of a general function $f(x)$ used instead of $(z-x)^{-1}$

$$\int_a^b f(x) d\sigma(x) = \lim_{n\to\infty} \sum_{k=1}^n w_k f(x_k) + \mathcal{E}_n \tag{B.45}$$

$$\mathcal{E}_n = \frac{1}{Q_n(z)} \int_a^b d\sigma(x) f(x) Q_n(x). \tag{B.46}$$

If we use the binomial expansion for $(z-x)^{-1} = \sum_{n=0}^\infty x^n z^{-n-1}$ in (B.36), we will obtain

$$\pi_n(z) = \sum_{m=0}^\infty \mu_{n,m} z^{-m-1} \tag{B.47}$$

where $\mu_{n,m}$ is the modified moment

$$\mu_{n,m} = \langle Q_n(x) | x^m \rangle = \int_a^b Q_n^*(x) x^m d\sigma(x) \qquad \mu_{n,0} = \mu_n. \tag{B.48}$$

An important feature of $\mu_{n,m}$ follows from the orthogonality (B.3) as

$$\langle Q_n(x) | x^m \rangle = 0 \qquad m = 0, 1, \ldots, n-1$$
$$\therefore \quad \mu_{n,m} = 0 \qquad m \leq n-1. \tag{B.49}$$

This annihilates the first n terms of series in (B.47) so that the final expression for $\pi_n(z)$ becomes

$$\pi_n(z) = \sum_{m=n}^\infty \mu_{n,m} z^{-m-1} = \sum_{m=0}^\infty \mu_{n,n+m} z^{-n-m-1}. \tag{B.50}$$

The coefficients $\mu_{n,m}$ from (B.50) can be computed recursively, as in (55.22)

$$\beta_{n+1} \mu_{n+1,m} = \mu_{n,m+1} - \alpha_n \mu_{n,m} - \beta_n \mu_{n-1,m} \qquad \mu_{0,0} = 1 \tag{B.51}$$

where the parameters $\{\alpha_n\}$ and $\{\beta_n\}$ are the same Lanczos coupling constants as in (B.18). Using (B.24) and (B.30) we can rewrite (B.49) as

$$\sum_{r=0}^n q_{r,n-r} \mu_{r+m} = 0 \qquad (0 \leq m < n) \tag{B.52}$$

Non-classical polynomials in the expansion methods 431

or in the equivalent matrix form

$$\begin{pmatrix} \mu_0 & \mu_1 & \mu_2 & \cdots & \mu_{n-1} \\ \mu_1 & \mu_2 & \mu_3 & \cdots & \mu_n \\ \mu_2 & \mu_3 & \mu_4 & \cdots & \mu_{n+1} \\ \vdots & \vdots & \vdots & \ddots & \vdots \\ \mu_{n-1} & \mu_n & \mu_{n+1} & \cdots & \mu_{2n-2} \end{pmatrix} \begin{pmatrix} q_{0,n} \\ q_{1,n-1} \\ q_{2,n-2} \\ \vdots \\ q_{n-1,1} \end{pmatrix} = - \begin{pmatrix} \mu_n \\ \mu_{n+1} \\ \mu_{n+2} \\ \vdots \\ \mu_{2n-1} \end{pmatrix}. \tag{B.53}$$

The $n \times n$ matrix on the lhs of (B.53) is recognized as the Hankel matrix $\mathbf{H}_n(\mu_0)$ which is obtained from (5.16) via replacement of c_r by μ_r. For the given $2n$ moments $\{\mu_r\}$, the system (B.53) of n linear inhomogeneous equations can be solved by considering $q_{n,n-r}$ as the unknown. The solution of the system (B.53) will exist and shall be unique, provided that the Hankel determinant is non-zero

$$\det \mathbf{H}_n(\mu_0) \neq 0. \tag{B.54}$$

We shall prove (B.54) by assuming that the opposite is valid, $\det \mathbf{H}_n(\mu_0) = 0$, and subsequently show the contradiction. The equation $\det \mathbf{H}_n(\mu_0) = 0$ implies that there is a linear dependence among the rows of $\det \mathbf{H}_n(\mu_0)$. By definition, linear dependence means that there exists a non-zero column vector $\mathbf{y} = \{y_r\}(0 \leq r \leq n-1)$ such that $\mathbf{H}_n(\mu_0)\mathbf{y} = \mathbf{0}$

$$\begin{pmatrix} \mu_0 & \mu_1 & \mu_2 & \cdots & \mu_{n-1} \\ \mu_1 & \mu_2 & \mu_3 & \cdots & \mu_n \\ \mu_2 & \mu_3 & \mu_4 & \cdots & \mu_{n+1} \\ \vdots & \vdots & \vdots & \ddots & \vdots \\ \mu_{n-1} & \mu_n & \mu_{n+1} & \cdots & \mu_{2n-2} \end{pmatrix} \begin{pmatrix} y_0 \\ y_1 \\ y_2 \\ \vdots \\ y_{n-1} \end{pmatrix} = \begin{pmatrix} 0 \\ 0 \\ 0 \\ \vdots \\ 0 \end{pmatrix} \tag{B.55}$$

or equivalently

$$\mathbf{y}^T \mathbf{H}_n(\mu_0) \qquad \mathbf{y} \neq \mathbf{0} \tag{B.56}$$

where $\mathbf{y}^T$ is the transpose matrix of $\mathbf{y}$. Now a simple calculation by means of (B.56) shows that

$$0 = \mathbf{y}^T \mathbf{H}_n(\mu_0)\mathbf{y} = \sum_{r=0}^{n-1} y_r \sum_{s=0}^{n-1} y_s \mu_{r+s}$$

$$= \sum_{r=0}^{n-1} y_r \left\langle \sum_{s=0}^{n-1} y_s x^s \middle| x^r \right\rangle = \left\langle \sum_{s=0}^{n-1} y_s x^s \middle| \sum_{r=0}^{n-1} y_r x^r \right\rangle = \left\| \sum_{s=0}^{n-1} y_s x^s \right\|^2$$

$$\therefore \qquad \left\| \sum_{s=0}^{n-1} y_s x^s \right\|^2 = 0. \tag{B.57}$$

The zero norm of the polynomial $\sum_{s=0}^{n-1} y_s x^s$ is inadmissible, contradicting the assumption $\det \mathbf{H}_n(\mu_0) = 0$ and, therefore, the opposite is true which proves (B.54) (QED).

If we replace n by $n+1$ in (B.36), then multiply the result with β_{n+1} and finally use the recursion (B.18), it will follow

$$\beta_{n+1}\pi_{n+1}(z) = (z - \alpha_n)\pi_n(z) - \beta_n\pi_{n-1}(z). \tag{B.58}$$

This shows that the denominator $\{\pi_n(z)\}$ of the error $\mathcal{E}_n(z)$ from (B.35) can be generated by means of the same recursion as in (B.18). Thus, while the Lanczos process continues, the error at each order can be computed in parallel with the same type of recursion as seen from (B.18) and (B.58). Needless to say, such an efficient error analysis is of utmost importance in practical applications of the Lanczos algorithm.

Since the set $\{Q_r(x)\}$ represents a basis, we can expand, e.g. the Green operator $\hat{G}(z)$ from (B.9) in terms of $Q_n(x)$ via

$$\hat{G}(z) = \sum_{n=0}^{\infty} \gamma_n(z) Q_n(\hat{H}) \qquad \gamma_n(z) \equiv \frac{C_n(z)}{w_n} \tag{B.59}$$

where $\{C_n(z)\}$ are the expansion coefficients. These coefficients $C_n(z)$ can be obtained by the usual projection procedure which reveals them as the Hilbert transform of $Q_n^*(x)$

$$C_n(z) = \int_a^b \frac{d\sigma(x)}{z-x} Q_n^*(x). \tag{B.60}$$

Comparison of (B.36) and (B.60) with the help of (B.1) gives

$$C_n(z) = \pi_n(z). \tag{B.61}$$

In (B.59), the coefficients $\{C_n(z)\}$ are independent of the Hamiltonian $\hat{H}$, which appears in the expansion only through $Q_n(\hat{H})$. In other words, once the basis $\{Q_r\}$ has been constructed numerically or selected from classical polynomials, the set of the coefficients $\{C_n(E)\}$ will be the same for different systems provided that they are all considered at the same energy E [275]–[277].

It should be realized that the integral representation (B.28) of the polynomial $P_n(z)$ also permits extraction of the coefficient $C_n(z)$. This can be done by replacing z by z^* in (B.28) to obtain $P_n(z^*)$ which is $P_n^*(z)$, as per (B.29)

$$P_n(z^*) = P_n^*(z) = \frac{\beta_1}{\mu_0}\left\{Q_n(z^*)\int_a^b \frac{d\sigma(x)}{z^*-x} - \int_a^b \frac{d\sigma(x)}{z^*-x}Q_n(x)\right\}$$

$$= \frac{\beta_1}{\mu_0}\left\{Q_n(z)\int_a^b \frac{d\sigma(x)}{z-x} - \int_a^b \frac{d\sigma(x)}{z-x}Q_n^*(x)\right\}^*$$

$$= \frac{\beta_1}{\mu_0}\left\{Q_n(z)\int_a^b \frac{d\sigma(x)}{z-x} - C_n(z)\right\}^*$$

$$\therefore \quad C_n(z) = S(z)Q_n(z) - \frac{\mu_0}{\beta_1}P_n(z) \tag{B.62}$$

where $S(z)$ is the Stieltjes integral (B.31). Using (B.1) while comparing directly (B.35) with (B.60) or substituting (B.60) into (B.62), it follows

$$C_n(z) = Q_n(z)\mathcal{E}_n(z). \tag{B.63}$$

We see that the expansion coefficient $C_n(z)$ in the development of the Green operator (B.60) is equal to the product of the polynomial $Q_n(z)$ and the error term $\mathcal{E}_n(z)$ from the representation of the Stieltjes integral (B.31) by the Padé approximant $\mathcal{R}_n(z)$ or its Heaviside partial fractions (B.44).

In the method of propagation of wave packets in the Schrödinger picture of quantum mechanics, the evolution operator $\hat{U}(t)$ is used. Similarly to $\hat{H}$, the evolution operator $\hat{U}(t)$ can also be expanded in the polynomial basis $\{Q_n(x)\}$ as

$$\hat{U}(t) = \sum_{n=0}^{\infty} \tilde{C}_n(t) Q_n(\hat{H}) \tag{B.64}$$

where the expansion coefficients $\tilde{C}_n(t)$ are related to $C_n(E)$ from (B.59) by the Fourier integral

$$C_n(E) = -i \int_{-\infty}^{\infty} dt\, \theta(t) \tilde{C}_n(t) e^{iEt}. \tag{B.65}$$

If the set $\{Q_n(x)\}$ is taken to be among the classical polynomials [176, 180], then the Stieltjes integral (B.31) with a known weight function $W(x)$ can be obtained in analytical forms.

The relations (B.18) and (B.23) can also be used in their operator forms

$$\left.\begin{array}{l} \beta_{n+1} Q_{n+1}(\hat{H}) = (\hat{H} - \alpha_n \hat{1}) Q_n(\hat{H}) - \beta_n Q_{n-1}(\hat{H}) \\ Q_0(\hat{H}) = \hat{1} \qquad Q_1(\hat{H}) = \hat{H} - \alpha_1 \hat{1}. \end{array}\right\} \tag{B.66}$$

For a given physical states $|\Phi_n\rangle$ which could be an eigenvector of a part of $\hat{H}$, or an orbital from a basis set, we can introduce two state vectors $|\Psi_n\rangle$ and $|\psi_n^{(s)}\rangle$ by

$$|\Psi_n\rangle = \hat{G}(z)|\Phi_n\rangle \qquad |\psi_n^{(s)}\rangle = Q_n(\hat{H})|\Phi_s\rangle. \tag{B.67}$$

For a fixed s, the vector $|\psi_n^{(s)}\rangle$ can be generated by repeatedly applying (B.66), leading to the Lanczos algorithm for physical states

$$\beta_{n+1}|\psi_{n+1}^{(s)}\rangle = (\hat{H} - \alpha_n \hat{1})|\psi_n^{(s)}\rangle - \beta_n |\psi_{n-1}^{(s)}\rangle. \tag{B.68}$$

In this way, we can calculate both diagonal and off-diagonal matrix elements of the Green operator via

$$\langle \Phi_r | \hat{G}(z) | \Phi_s \rangle = \sum_{n=0}^{\infty} \gamma_n(z) I_{r,n,s} \qquad I_{r,n,s} = \langle \Phi_r | \psi_n^{(s)} \rangle. \tag{B.69}$$

It is also possible to calculate matrix elements of any number of Green operators multiplied by other operators by setting

$$|\Psi_s\rangle = \hat{G}(z)|\Phi_s\rangle = \sum_{n=0}^{\infty} \gamma_n(z)|\psi_n^{(s)}\rangle \tag{B.70}$$

so that, for instance

$$\langle\Phi_r|\hat{G}(z)\hat{V}\hat{G}(z)|\Phi_s\rangle = \langle\Psi_r|\hat{V}|\Psi_s\rangle. \tag{B.71}$$

The density of states (DOS) is usually introduced via the quantum-mechanical trace (Tr) of the Green function

$$\rho(E) = -\frac{1}{\pi} \operatorname{Im} \operatorname{Tr}\{\hat{G}(z^+)\}. \tag{B.72}$$

The trace is defined as the sum of the diagonal elements of $\delta(E\hat{1} - \hat{H})$

$$\rho(E) = \sum_{s=0}^{L} \langle\Phi_s|\delta(E\hat{1} - \hat{H})|\Phi_s\rangle.$$

Inserting (B.6) into this expression for $\rho(E)$, we have

$$\rho(E) = \sum_{n=0}^{\infty} W_n(E) Q_n(E) I_n \qquad I_n = \sum_{s=0}^{L} I_{s,n,s} \tag{B.73}$$

where L is the number of the retained vectors from the collection $\{|\Phi_s\rangle\}$. Integration of $\rho(x)$ gives the integrated density of state (IDOS) denoted by $N(E)$

$$N(E) = \int_{-\infty}^{E} dx \rho(x) = \sum_{n=0}^{\infty} I_n \int_{-\infty}^{E} W_n(x) Q_n(x) d\sigma(x). \tag{B.74}$$

Given that the set $\{|\psi_r\rangle\}$ is a basis, we can use it to expand the complete Schrödinger state vector $|\Upsilon_k\rangle$ of $\hat{H}$ from (B.14). More generally, the complete state vector can be written at any energy E, and not only E_k, via the following expansion

$$|\Upsilon(E)\rangle = \sum_{n=0}^{\infty} B_n(E)|\psi_n^{(s)}\rangle \tag{B.75}$$

for any vector $|\Phi_s\rangle$ which is inherently present in $|\psi_n^{(s)}\rangle$ via the state $|\psi_n^{(s)}\rangle = Q_n(\hat{H})|\Phi_s\rangle$. In other words, $|\Upsilon(E)\rangle$ is independent of $|\Phi_s\rangle$ at any energy E. Similarly to (23.14), we can show that the expansion coefficients $B_n(E)$ are given by

$$B_n(E) = W_n(E) Q_n(E). \tag{B.76}$$

Thus, the exact Schrödinger eigenvector $|\Upsilon(E)\rangle$ of $\hat{H}$ can be calculated by propagating the state $|\Phi_s\rangle$ as

$$|\Upsilon(E)\rangle = \mathcal{P}(E, \hat{H})|\Phi_s\rangle \qquad (B.77)$$

$$\mathcal{P}(E, \hat{H}) = \sum_{n=0}^{\infty} W_n(E) Q_n(E) Q_n(\hat{H}). \qquad (B.78)$$

The vector $|\Phi_s\rangle$ can be the initial state of the investigated system, but this is not indispensable. An arbitrary state can be used for $|\Phi_s\rangle$ including a random state. Comparing the completeness relation (B.6) with (B.78), we deduce

$$|\Upsilon(E)\rangle = \delta(E\hat{1} - \hat{H})|\Phi_s\rangle. \qquad (B.79)$$

The eigenstates $|\Upsilon_k\rangle \equiv |\Upsilon(E_k)\rangle$, corresponding to the eigenvalue E_k, are extracted from the expansion (B.75) taken at $E = E_k$

$$|\Upsilon_k\rangle = \sum_{n=0}^{\infty} B_n(E_k)|\psi_n^{(s)}\rangle = \sum_{n=0}^{\infty} W_n(E_k) Q_n(E_k) Q_n(\hat{H})|\Phi_s\rangle. \qquad (B.80)$$

To summarize, we shall enumerate several important features of the expansion method built from non-classical polynomials, that are intertwined with rational polynomials as presented in this appendix (see also [276, 277]):

- both diagonal and off-diagonal elements are calculable, not only for the Green function itself, but also for the product of any number of such functions with other quantum-mechanical operators,
- the energy resolution in the calculated spectrum is controlled by the expansion order, which is the number of the retained polynomials,
- it can be used for systems with discrete energies as well as for resonances and scattering,
- it provides the integrated density of states (IDOS), eigenvalues E_k and corresponding eigenvectors $|\Upsilon_k\rangle$ of the Schrödinger eigenvalue problem without ever solving this problem explicitly,
- the computational and storage costs scale linearly with the size of the considered system,
- it is suitable for parallel processing.

Appendix C

Classical polynomials in the expansion methods

For generality, the expansion method from Appendix B is introduced in the space of non-classical Lanczos polynomials $\{P_n(x), Q_n(x)\}$. However, the entire analysis and the pertinent conclusions also apply automatically to classical polynomials. This is because all the classical polynomials belong to their families of orthogonal polynomials and, consequently, they satisfy the same three-term recursion (B.18) with the unaltered definition of the coupling constants α_n and β_n from (B.19) and (B.20). The only difference is that α_n and β_n can be calculated in the analytical forms for all the classical polynomials. The Chistofel numbers $\{w_n\}$ for classical polynomials are computed from the same formula and, moreover, the relation (B.22) can be verified analytically to hold true.

What is extraordinary about the error or remainder $\mathcal{E}_n(z)$ from (B.34) is that it can be calculated analytically for classical polynomials. For example, if we take $Q_n(x)$ to be the Chebyshev polynomial $T_n(x)$ of the first kind, then the polynomial $P_n(x)$ could readily be identified as the Chebyshev polynomial $U_n(x)$ of the second kind

$$T_n(z) = \cos(n\varphi) \qquad U_n(z) = \frac{\sin([n+1]\varphi)}{\sin\varphi} \qquad \varphi = \cos^{-1}(z). \qquad (C.1)$$

In this case, it follows from (B.35) that $\mathcal{E}_n(z) = \pi_n(z)/T_n(z)$ and

$$\text{Chebyshev:} \qquad \pi_n(z) = \frac{1+\delta_{n,0}}{2}\pi\left[U_{n-1}(z) - \frac{i}{\sqrt{1-z^2}}T_n(z)\right] \qquad (C.2)$$

where $\delta_{n,0}$ is the Kronecker δ-symbol (3.7). We have also derived the simple analytical expressions for $\mathcal{E}_n(z)$ by replacing $Q_n(x)$ with the Hermite $H_n(x)$, Jacobi $P_n^{(\alpha,\beta)}(x)$, Legendre, Laguerre polynomials, etc, with some of the results given by $\mathcal{E}_n(E) = \pi_n(E)/P_n^{(\alpha,\beta)}(E)$ and (see also [277])

Hermite: $\quad \pi_n(E) = -\dfrac{1}{n!\sqrt{2\pi}}[i\pi w(E/\sqrt{2})\mathcal{H}_n(E) - \mathcal{H}_n(E)] \quad$ (C.3)

Jacobi: $\quad \pi_n(E) = -\dfrac{2^{\alpha+\beta+1}B(\alpha+1,\beta+1)}{1-E}$

$$\times {}_2F_1\left(1, \beta+1; \alpha+\beta+2; \dfrac{2}{1-E}\right) P_n^{(\alpha,\beta)}(E)$$

$$- Q_n^{(\alpha,\beta)}(E) \quad \text{(C.4)}$$

$$w(z^+) = \dfrac{i}{\pi}\int_0^\infty dx \dfrac{e^{-x^2}}{z^+ - x} = e^{-x^2/2}[\mathrm{erf}(iE/\sqrt{2}) - 1] \quad \text{(C.5)}$$

where z^+ is given in (B.11). Here, $w(z)$ is the w-function closely related to the error-function (erf), $\{a,b\}$ are fixed parameters from the Jacobi polynomial, $B(\alpha,\beta)$ is the beta-function and ${}_2F_1(a,b;c;z)$ is Gauss hypergeometric function [240]. For accurate and fast computations of the w-function of a complex variable via continued fractions, the algorithm of Gautschi [172] is recommended. In (C.3) and (C.4), $\mathcal{H}_n(E)$ and $Q_n^{(\alpha,\beta)}(E)$ are the Hermite and Jacobi polynomials of the second kind. They are the appropriate specifications of (B.23) where the constants α_n and β_n are the same as in the standard recursions for $H_n(E)$ and $P_n^{(\alpha,\beta)}(E)$ except for the changed initial conditions $\mathcal{H}_0(E) = 0$, $\mathcal{H}_1(E) = 1$ and $Q_0^{(\alpha,\beta)}(E) = 0$, $Q_1^{(\alpha,\beta)}(E) = 1$. The choice of the Legendre polynomials is a special case obtained from $P_n^{(\alpha,\beta)}(x)$ by setting $\alpha = 0 = \beta$. Also, $\pi_n(E)$ given in (C.2) can be deduced from (C.4) for $\alpha = -1/2 = \beta$ which maps the Jacobi into the Chebyshev polynomials.

Moreover, the integral in (B.74) can be carried out analytically for classical polynomials. For instance, using the Hermite $H_n(x)$ and Jacobi $P_n^{(\alpha,\beta)}(x)$ polynomials, it follows [277]

Hermite: $\quad N(E) = M\left[1 + \dfrac{1}{\sqrt{\pi}}\mathrm{erf}(E/\sqrt{2\pi})\right]$

$$- \dfrac{1}{\sqrt{2\pi}}\sum_{n=1}^\infty \dfrac{I_n}{n!}e^{-E^2/2}H_{n-1}(E) \quad \text{(C.6)}$$

Jacobi: $\quad N(E) = M\dfrac{B_{E'}(\alpha+1,\beta+1)}{B(\alpha+1,\beta+1)} - \dfrac{1}{2}\sum_{n=1}^\infty \dfrac{I_n}{n}\mathcal{W}_n(E)P_{n-1}^{(\alpha+1,\beta+1)}(E)$

(C.7)

$$\mathcal{W}_n(x) = \dfrac{W(x)}{w_n} \qquad W(x) = (1-x)^\alpha(1+x)^\beta \quad \text{(C.8)}$$

$$w_n = \dfrac{2^{\alpha+\beta+1}}{2n+1}\dfrac{\Gamma(n+\alpha+1)\Gamma(n+\beta+1)}{n!\Gamma(n+\alpha+\beta+1)} \quad \text{(C.9)}$$

where $E' = (1+E)/2$, $B_\gamma(\alpha, \beta)$ is the incomplete beta function and $\Gamma(a)$ is gamma function [240].

The expansion methods in terms of classical orthogonal polynomials, and most notably the Chebyshev polynomials, have been used extensively over the years in solving quantum-mechanical problems that require comprehensive computations of Green functions and the related observables [275, 276]. This has recently been revived under the name 'recursive orthogonal polynomial expansion method' (ROPEM) [277], which was explicitly implemented for the Chebyshev, Jacobi and Legendre polynomials. All the main features enumerated at end of Appendix B for non-classical polynomials have originally been found to hold true in applications of classical polynomials via the ROEPM [277]. Moreover, in the ROEPM with the Legendre polynomials (and presumably with the other classical polynomials), the truncation error, caused by using a finite number M of the terms in the expansion, has been found to be proportional to M^{-1} for IDOS and to $M^{-3/2}$ for certain tight-binding chain models with continuous spectra [277]. In several applications [277] aimed primarily at developing expansion methods for large physical systems, the ROPEM has been proven accurate, stable as well as efficient, and this should motivate further explorations of this method based on classical polynomials.

Relative to the ROEPM, there is a very important advantage of a more general expansion method constructed from non-classical Lanczos polynomials from chapter B. In physics and chemistry, the functions to be expanded on a basis often satisfy certain prescribed boundary conditions. For example, continuum wave functions are required to behave at large distances as a sinusoidal function which should include a phase shift for a short and/or a long-range Coulombic potential. Similarly, discrete wave functions should possess exponentially declining terms in the asymptotic region ($x \longrightarrow \infty$). Also, the boundary conditions at the origin are frequently imposed on functions that describe physical states as $f(x) \longrightarrow 0$ as $x \longrightarrow 0$. Functions used in physics often have essential singularities within the interval of interest $[a, b]$. It is, by far, more advantageous that the prescribed asymptotic behaviour(s) of the considered function is included in the employed basis set from the onset. This can be easily accomplished with non-classical polynomials, but *not* with the classical ones. For instance, if a solution of second-order differential equations is desired with a singularity at $x \approx a$, classical polynomials would necessarily undergo large cancellations (leading to considerable round-off errors) to fullfil the imposed boundary condition. This is because, in order to be complete for expansions of all functions, any classical polynomial of degree n must contain a free term, which is a constant, i.e. a factor independent on x. Even a small admixture of an irregular solution of the second-order differential equation from this example could produce a dramatic loss of accuracy when classical polynomials are used to expand the regular solution. However, this kind of problem is readily solved by employing the non-classical Lanczos polynomials $\{Q_n(x)\}$ whose recursion should be modified to start with x rather than with a constant [261].

References

[1] Rabi I I 1937 *Phys. Rev.* **51** 652
   Rabi I I, Zacharias J R, Millmann S and Kusch P 1938 *Phys. Rev.* **53** 318
   Rabi I I, Zacharias J R, Millmann S and Kusch P 1939 *Phys. Rev.* **55** 526
[2] Purcell E U, Torrey H C and Pound R V 1946 *Phys. Rev.* **69** 37
[3] Bloch F, Hansen W W and Packard M 1946 *Phys. Rev.* **69** 127
[4] Ernst R R and Anderson W A 1966 *Rev. Sci. Instrum.* **37** 93
   Ernst R R 1966 *Adv. Magnetic Reson.* **2** 1
   Ernst R R 1991 *Computational Aspects of the Study of Biological Macromolecules by NMR Spectroscopy* ed J C Hosh *et al* (New York: Plenum) p 1
[5] Hoch J C and Stern A S 1996 *NMR Data Processing* (New York: John Willey & Sons Inc.) p 34
[6] Deschampes M, Burghardt I, Derouet C, Bodenhausen G and Belkić Dž 2000 *J. Chem. Phys.* **113** 1630
[7] Comisarow M B and Marshall A G 1974 *Can. J. Chem.* **52** 1997
   Comisarow M B and Marshall A G 1976 *J. Chem. Phys.* **64** 110
   Guan S and Marshall A G 1997 *Anal. Chem.* **69** 1156
   Shi S D-H, Hendrikson C L and Marshall A G 1998 *Proc. Natl Acad. Sci., USA* **95** 11 532
[8] Figeys D, McBroom L D and Moran M F 2001 *Methods* **24** 230
[9] Belkić Dž, Dando P A, Taylor H S, Main J and Shin S̩-K 2000 *J. Phys. Chem. A* **104** 11 677
   Belkić Dž 2000 *Many-Particle Spectroscopy of Atoms, Molecules and Surfaces* ed J Berakdar (Halle: Max-Planck-Institut für Mikrostrukturphysik) p 122
   Belkić Dž 2000 *Many-Particle Spectroscopy of Atoms, Molecules and Surfaces* ed J Berakdar (Halle: Max-Planck-Institut für Mikrostrukturphysik) p 134
[10] Frahm J, Bruhn H, Gyngell M L, Merbolt K D, Hänicke V and Sauter R 1989 *Magn. Reson. Med.* **11** 47
   Leibfritz D 1996 *Anticancer Research* **16** 1317
   Hanefeld F J 1997 *Advances in Neurochemistry* vol 8, ed H Bachelard (New York: Plenum) p 380
   Cabanes E, Confirt-Gouny S, Le Fur Y, Simond G and Cozzone P J 2001 *J. Magn. Res.* **150** 116
[11] Provencher S W 1993 *Magn. Reson. Med.* **30** 672
   Provencher S W 2001 *NMR Biomed.* **14** 260
   Frahm J and Hanefeld F 1997 *Advances in Neurochemistry* vol 8, ed H Bachelard (New York: Plenum) p 329

[12] Govindaraju V, Young K and Maudsley A A 2000 *NMR Biomed.* **13** 129
[13] Tkáč I, Kim J, Uğurbil K and Gruetter R 2001 *Proc. Int. Soc. Mag. Reson. Med.* **9** 214
Tkáč I, Andersen P, Adriany G, Merkle H, Uğurbil K and Gruetter R 2001 *Magn. Reson. Med.* **46** 451
[14] Roelants-Van Rijn A M, Van Der Grond J, DeVries L S and Groenendaal F 2001 *Pediatric Research* **49** 396
[15] Zandt H J A, van der Graaf M and Heerschap A 2001 *NMR Biomed.* **14** 224
Venhamme L, Sundin T, van Hecke P and van Huffel S 2001 *NMR Biomed.* **14** 233
Miwerisová Š and Ala-Korpela M 2001 *NMR Biomed.* **14** 247
[16] Belkić Dž 2002 *Magn. Res. Mater. Phys. Biol. Med. (MAGMA) Suppl. no 1* **15** 36
[17] Belkić Dž 2003 *J. Comput. Meth. Sci. Eng.* **3** 109
[18] Belkić Dž 2003 *J. Comput. Meth. Sci. Eng.* **3** 299
[19] Belkić Dž 2003 *J. Comput. Meth. Sci. Eng.* **3** 563
[20] Belkić Dž 2003 *Computational Methods in Sciences and Engineering* ed T E Simos (Singapore: World Scientific) p 87
[21] Belkić K 2003 *Computational Methods in Sciences and Engineering* ed T E Simos (Singapore: World Scientific) p 659
[22] Belkić Dž and Belkić K 2003 *Computational Methods in Sciences and Engineering* ed T E Simos (Singapore: World Scientific) p 701
[23] Belkić Dž 2003 *Computational Methods in Sciences and Engineering* ed T E Simos (Singapore: World Scientific) p 727
[24] Belkić Dž 2003 High-resolution parametric estimation of two-dimensional magnetic resonance spectroscopy *20th Annual Meeting of European Soc. Magn. Res. Med. Biol. (ESMRMB), Abstract Number 365 (CD), Rotterdam (Netherlands)*
[25] Belkić Dž 2003 *Spectroscopic Positron Emission Tomography (sPET) for Monitoring Radiation Therapy, Nobel Symposium 'Imaging 2003': Int. Conf. on Imaging Technologies in Subatomic Physics, Astrophysics, Medicine, Biology and Industry Stockholm, Sweden, Book of Abstracts* p 93
[26] Belkić Dž 2003 *Fast Padé Transform (FPT) as Opposed to Conventional Fitting in Signal and Image Processing, Nobel Symposium 'Imaging 2003': Int. Conf. on Imaging Technologies in Subatomic Physics, Astrophysics, Medicine, Biology and Industry Stockholm, Sweden, Book of Abstracts* p 103
[27] Belkić Dž 2004 *Nucl. Instrum. Meth. Phys. Res.* A **525** 366
[28] Belkić Dž 2004 *Nucl. Instrum. Meth. Phys. Res.* A **525** 372
[29] Belkić Dž 2004 *Nucl. Instrum. Meth. Phys. Res.* A **525** 379
[30] Brown T R, Kincaid B M and Uğurbil K 1982 *Proc. Natl Acad. Sci., USA* **79** 3523
Brown T R, Kincaid B M and Uğurbil K 1982 *Proc. Soc. Phot. Opt. Instrum. Eng.* **347** 354
Tweig D B 1983 *Med. Phys.* **10** 610
Ljunrggren S 1983 *J. Magn. Res.* **54** 338
Haacke E M 1987 *Inverse Problems* **3** 421
Sebastini G and Barone P 1991 *Signal Processing* **25** 227
Barone P 1992 *IEEE Trans. Med. Imag.* **11** 250
March R and Barone P 1998 *SIAM J. Appl. Math.* **58** 324
March R and Barone P 2000 *SIAM J. Appl. Math.* **60** 1137
[31] Belkić Dž 2001 *Nucl. Instrum. Methods* A **471** 165

[32] Belkić Dž and Belkić K 2003 *High-Resolution Magnetic Resonance Imaging (MRI)*, IEEE Medical Imaging Conference (MIC), Abstract Number, 1971 (CD), Portland (Oregon)
[33] Brown T R, Kincaid B M and Uğurbil K 1982 *Proc. Natl Acad. Sci.* **79** 3523
Nelson S J 2001 *Magn. Reson. Med.* **46** 228
Maudslay A A, Hilal S K, Perman W H and Simon H E 1983 *J. Magn. Res.* **51** 147
Yu K K, Scheidler J, Hricak H, Vigneron D B, Zaloudek C J, Males R J, Nelson S J, Carroll P R and Kurhanewicz J 1999 *Radiology* **213** 481
[34] Belkić K and Belkić Dž 2003 *The Fast Padé transform (FPT) for Magnetic Resonance Spectroscopic Imaging (MRSI) in Oncology*, IEEE Medical Imaging Conference (MIC), Abstract Number 1918 (CD) Portland (Oregon)
[35] Belkić K 2003 *J. Comput. Meth. Sci. Eng.* **3** 505
[36] Belkić K 2003 *J. Comput. Meth. Sci. Eng.* **3** 535
[37] Belkić K 2004 *Nucl. Instrum. Meth. Phys. Res.* A **525** 313
[38] Belkić K and Belkić Dž 2004 *J. Comput. Meth. Sci. Eng.* **4** 157
[39] Belkić K 2004 *Isr. Med. Assoc. J. (IMAJ)* **6** 610
[40] Natarrer F 1986 *The Mathematics of Computerized Tomography* (New York: Wiley) ch 2
Kak A C and Slaney M 1987 *Principles of Computerized Tomography* (New York: IEEE)
[41] Tretiak O and Metz C 1980 *SIAM J. Appl. Math.* **39** 341
Aguilar V and Kuchment P 1995 *Inverse Problems* **11** 977
Palamodov V P 1996 *Inverse Problems* **12** 717
Arbuzov E V, Bukhgeim A L, Kuzentsev S G 1998 *Siberian Adv. Math.* **8** 1
[42] Novikov R 1999 *Ark. Math.* **37** 141
Novikov R 2001 *C. R. Acad. Sci., Paris* **332** 1059
Natarrer F 2001 *Inverse Problems* **17** 113
Kunyansky L A 2001 *Inverse Problems* **17** 293
[43] Gauss C F 1866 *Werke* [Collected works of Gauss C F] vol 3 (Göttingen: König. Gesellsch. der Wissensch.) p 265
[44] Runge C 1903 *Z. Math. Phys.* **48** 443
Runge C 1905 *Z. Math. Phys.* **52** 117
[45] Stumpff K 1939 *Tafeln und Aufgaben zur Harmonischen Analyse und Periodigrammrechnung* (Berlin: Springer)
[46] Danielson G C and Lanczos C 1942 *J. Franklin Inst.* **233** 365
Danielson G C and Lanczos C 1942 *J. Franklin Inst.* **233** 432
[47] Good I J 1958 *J. R. Statist. Soc.* B **20** 361
Good I J 1960 *J. R. Statist. Soc.* B (*Addendum*) **22** 372
Good I J 1979 *Applic. Anal.* **9** 205
[48] Thomas L H 1963 *Using Computers to Solve Problems in Physics; Application of Digital Computers* (Boston, MA: Ginn)
[49] Cooley J W and Tukey J W 1965 *Math. Comput.* **19** 297
[50] Rudnick P 1966 *Math. Comput.* **20** 429
[51] Cooley J W, Lewis P A W and Welch P D 1967 *IEEE Trans. Aud. Electroacoust.* **19** 76
Cooley J W 1987 *Microchimica Acta* **3** 33
[52] Huang T S 1971 *Computer* **4** 15
Heideman M T, Johnson D H and Burrus C S 1985 *Arch. History Exact Sci.* **34** 265

[53] Birgham E O 1988 *The Fast Fourier Transform and its Applications* (Englewood Cliffs, NJ: Prentice-Hall)
[54] Bracewell R N 1990 *Science* **248** 697
    Bracewell R N 2000 *The Fourier Transform and its Applications* 3rd edn (Boston, MA: McGraw-Hill)
[55] Rockmore D N 2000 *Comput. Sci. Eng.* **2** 60
[56] Schur I 1917 *J. Reine Angew. Math.* **147** 205
    Schur I 1918 *J. Reine Angew. Math.* **148** 122
    Ralston A 1965 *A First Course in Numerical Analysis* (New York: McGraw-Hill)
[57] Prony R 1795 *J. l'École Polytechnique* **1** 24
[58] Makhoul J 1975 *Proc. IEEE* **63** 561
[59] Holmes R B 1979 *SIAM Rev.* **21** 361
[60] Barkhuisen H, de Beer R, Brové W M M and van Ormondt D 1985 *J. Magn. Res.* **61** 465
    Barkhuisen H, de Beer R, Brové W M M and van Ormondt D 1987 *J. Magn. Res.* **73** 553
    Barkhuisen H, de Beer D and van Ormondt D 1986 *J. Magn. Res.* **67** 371
    Pijnappell W W F, van den Boogaart A, de Beer R and van Ormondt D 1992 *J. Magn. Res.* **97** 122
    Desluc M A, Ni F and Levy G C 1987 *J. Magn. Res.* **73** 548
    Tang J and Norris J R 1988 *J. Magn. Res.* **78** 23
    Tang J and Norris J R 1988 *J. Magn. Res.* **79** 190
    Zhu G and Bax A 1990 *J. Magn. Res.* **90** 405
[61] Marple S L Jr 1987 *Digital Spectral Analysis with Applications* (Englewood Cliffs, NJ: Prentice-Hall)
[62] Stephenson D S 1988 *Prog. Nucl. Magn. Reson. Spectrosc.* **20** 515
[63] Prabhu K M M and Bagan K B 1989 *IEE Proc.* **136** 135
[64] Porat B 1994 *Digital Processing of Random signals, Theory and Methods* (Englewood Cliffs, NJ: Prentice-Hall)
    Porat B 1997 *A Course in Digital Signal Processing* (New York: Wiley)
    Grenader U and Szegö G 1958 *Nonlinear methods in Spectral Analysis* (Berkeley, CA: University of California Press)
    Stoica P and Moses R 1997 *Introduction to Spectral Analysis* (Englewood Cliffs, NJ: Prentice-Hall)
[65] Prony R 1797 *J. de l'Ecole Polytechnique* **1** 459
    Jacobi C G J 1845 *J. Reine Angew. Math.* **30** 127
[66] Frobenius G 1879 *J. Für Math.* **90** 1
    Frobenius 1895 G *J. Reine Angew. Math.* **114** 187
[67] Padé H 1892 *Ann. Fac. Sci. de l'École Norm. Suppl.* **9** 1
[68] Cesàro E 1904 *Elementares Lehrbuch der Algebraischen Analysis und der Infinitesimalrechnung* (Leipzig: Teubner)
[69] Wall H S 1931 *Trans. Am. Math. Soc.* **33** 511
[70] Aitken A C 1926 *Proc. R. Soc. Edinburgh* **46** 289
    Aitken A C 1937 *Proc. R. Soc. Edinburgh* **57** 269
[71] Schmidt R J 1941 *Phil. Mag.* **32** 369
[72] Shanks D 1949 *Naval Ordnance Laboratory Memorandum* (NOLM 9994: Project no NOL-4-Re9d-21-2, White Oak, MD) p 1
    Shanks D 1955 *J. Math. Phys.* **34** 1

Lubkin S 1952 *J. Res. Nat. Bur. Stand.* **48** 228
[73] Wynn P 1956 *Math. Tabl. Aids. Comput.* **10** 91
Wynn P 1956 *Proc. Camb. Phil. Soc.* **52** 663
Wynn P 1960 *Math. Comput.* **14** 147
[74] Weiss L and McDonough R N 1963 *SIAM Rev.* **5** 145
[75] Schlessinger L 1967 *PhD Thesis* University of California, Berkeley, unpublished
Schlessinger L and Schwartz C 1966 *Phys. Rev. Lett.* **16** 1173
Schlessinger L 1968 *Phys. Rev.* **167** 1411
Miller K 1968 *SIAM J. Appl. Math.* **18** 346
[76] Parsons D H 1968 *Math. Biosci.* **2** 123
Parsons D H 1970 *Math. Biosci.* **9** 37
[77] Istratov A A and Vivenko O F 1999 *Rev. Sci. Instrum.* **70** 1233
[78] Basdevant J L, Bessis D an Zinn-Justin J 1969 *Nuovo Cimento* **60** 185
[79] Goscinski O and Brändas E 1968 *Chem. Phys. Lett.* **2** 299
Brändas E and Goscinski O 1970 *Phys. Rev. A* **1** 552
Goscinski O and Brändas E 1971 *Int. J. Quantum Chem.* **5** 131
Brändas E and Bartlett R J 1971 *Chem. Phys. Lett.* **8** 153
Bartlett R J and Brändas E 1971 *Int. J. Quantum Chem.* **5** 151
Micha D A and Brändas E 1971 *J. Chem. Phys.* **55** 4792
Brändas E and Micha D A 1972 *J. Math. Phys.* **13** 155
Bartlett R J and Brändas E 1972 *J. Chem. Phys.* **56** 5467
Brändas E and Goscinski O 1972 *Int. J. Quantum Chem.* **6** 56
Bartlett R J and Brändas E 1973 *J. Chem. Phys.* **59** 2032
[80] Longman I M 1971 *Int. J. Comput. Math. B* **3** 53
Chisholm J S R 1973 *Padé Approximant* (Lectures Notes, Summer School, University of Kent, Canterbury), ed P R Graves-Morris (Bristol: Institute of Physics Publishing) p 1
[81] Baker G A and Gammel J L 1970 *The Padé Approximant in Theoretical Physics* (New York: Academic)
Baker G A 1975 *Essentials of the Padé Approximants* (New York: Academic)
Baker G A and Graves-Morris P 1996 *Padé Approximants* 2nd edn (Cambridge: Cambridge University Press)
[82] Claessens G 1975 *J. Comput. Appl. Math.* **1** 141
Wuytack L 1976 in *Approximation Theory (Lecture Notes in Mathematics 556)* ed R Schaback and K Scherer (Berlin: Springer) p 453
Wuytack L 1979 in *Padé Approximation and its Applications (Lecture Notes in Mathematics 765)* ed L Wuytack (Berlin: Springer) p 373
[83] Hadamard J 1892 *J. Math. Pures Appl.* **8** 101
[84] Rutishauser H 1990 *Lectures on Numerical Mathematics* (Boston, MA: Birkhäuser)
[85] Brezinski C 1976 *J. Comput. Appl. Math.* **2** 113
Brezinski C 1980 *Padé-Type Approximants and General Orthogonal Polynomials* (Basel: Birkhäuser)
Brezinski C and Redivo Zaglia M 1991 *Extrapolation Methods* (Amsterdam: North-Holland)
Brezinski C and Redivo Zaglia M 1994 *Adv. Comput. Math.* **2** 461
Brezinski C and Redivo Zaglia M 1994 *Appl. Numer. Math.* **16** 239
[86] Sidi A 2003 *Practical Extrapolation Methods: Theory and Applications* (Cambridge: Cambridge University Press)

[87] Genz A 1973 *Padé Approximant* (Lectures Notes, Summer School, University of Kent, Canterbury), ed P R Graves-Morris (Bristol: Institute of Physics Publishing) p 112
Garibotti C R and Grinstein F F 1979 *J. Math. Phys.* **20** 141
[88] Belkić Dž 1989 *J. Phys. A: Math. Gen.* **22** 3003
[89] Weniger E J 1989 *Comput. Phys. Rep.* **10** 189
[90] Epele L N, Fanchiotti H, Garcia Canal C A and Ponciano J A 1999 *Phys. Rev.* A **60** 280
[91] Grotendorst J 1989 *Comput. Phys. Commun.* **55** 325
Grotendorst J 1990 *Comput. Phys. Commun.* **59** 289
Grotendorst J 1991 *Comput. Phys. Commun.* **67** 325
[92] Driscoll T A and Fornberg B 2001 *Num. Algorithms* **26** 77
[93] Lanczos C 1950 *J. Res. Nat. Bur. Stand.* **45** 255
Lanczos C 1952 *J. Res. Nat. Bur. Stand.* **49** 33
[94] Lanczos C 1956 *Applied Analysis* (Englewood Cliffs, NJ: Prentice-Hall)
[95] Karush W 1951 *Pacific J. Math.* **1** 233
Wilkinson J W 1958 *The Computer Journal* **1** 90
Wilkinson J W 1958 *The Computer Journal* **1** 148
Ericsson T and Ruhe A 1980 *Math. Comput.* **35** 1251
Parlett B N 1980 *The Symmetric Eigenvalue Problem* (Englewood Cliffs, NJ: Prentice-Hall) ch 13
Wyatt R E 1995 *J. Chem. Phys.* **103** 8433
Grosso G, Martinelli L and Parravincini G P 1995 *Phys. Rev.* B **51** 13033
Chen R and Gua H 1997 *J. Comput. Chem.* **136** 494
Yu H-G and Nyman G 1999 *J. Chem. Phys.* **110** 11 133
[96] Slater J C and Koster G F 1954 *Phys. Rev.* **94** 1498
Goringe C M, Bowler D R and Hernández E *Rep. Prog. Phys.* **60** 1447
[97] Haydock R, Heine V and Kelly P J 1972 *J. Phys. C: Solid State Phys.* **5** 2845
Haydock R, Heine V and Kelly P J 1975 *J. Phys. C: Solid State Phys.* **8** 2591
Heine V 1980 *Solid State Physics; Advances in Research and Application* vol 35, ed H Ehrenreich, F Seitz and D Turnbull (New York: Academic) p 1
Heine V 1980 *Solid State Physics; Advances in Research and Application* vol 35, ed H Ehrenreich, F Seitz and D Turnbull (New York: Academic) p 315
[98] Pettifor D G and Weaire D L 1985 *The Recursion Method and Its Applications* ed D G Pettifor and A Weaire (Berlin: Springer) p 104
[99] Cullum J K and Willoughby R A 1985 *Lanczos Algorithm for Large Symmetric Eigenvalue Computations* (Boston, MA: Birkhäuser)
Saad Y 1992 *Numerical Methods for Large Eigenvalue Problems* (New York: Halsted–Wiley)
[100] Nauts A and Wyatt R E 1983 *Phys. Rev. Lett.* **51** 2238
Nauts A and Wyatt R E 1984 *Phys. Rev.* A **30** 872
Wyatt R E 1989 *Adv. Chem. Phys.* **73** 231
Karlsson H O and Gorscinski O 1992 *J. Phys. B: At. Mol. Opt. Phys.* **25** 5015
Karlsson H O and Gorscinski 1994 *J. Phys. B: At. Mol. Opt. Phys.* **27** 1061
[101] Horn D and Weinstein M 1984 *Phys. Rev.* D **30** 1256
[102] Duncan A and Roskies R 1985 *Phys. Rev.* D **31** 364
Duncan A and Roskies R 1985 *Phys. Rev.* D **32** 3277
Duncan A and Roskies R 1986 *Phys. Rev.* D **33** 2500 (erratum)

References

[103] Duncan A and Roskies R 1986 *J. Symbolic Comput.* **2** 201
Choe J-W, Duncan A and Roskies R 1988 *Phys. Rev.* A **37** 472
[104] Hiller J R 1991 *Phys. Rev.* D **44** 2504
[105] Schmidt R O 1986 *IEEE Trans. Antennas Propagat.* AP **34** 276
Lee H B and Wengrovitz M S 1990 *IEEE Trans. Acoust. Speech and Signal Process.* ASSP **38** 1545
[106] Roy R, Paulraj A and Kailath T 1986 *IEEE Trans. Acoust. Speech and Signal Process.* ASSP **34** 1340
Roy R and Kailath T 1989 *IEEE Trans. Acoust. Speech and Signal Process.* ASSP **37** 984
Roy R, Sumpter B G, Pfeffer G A, Gray S K and Noid D W 1991 *Phys. Rep.* **205** 109
Rao B D and Arun K S 1992 *Proc. IEEE* **80** 283
[107] Chisholm J S R, Genz A C and Pusterla M, 1976 *J. Comput. Appl. Math.* **2** 73
Claverie P, Denis A and Yeramian E 1989 *Comput. Phys. Rep.* **9** 247
Yeramian E, Schaffer F, Caudron B, Claverie P and Buc H 1990 *Biopolymers* **30** 481
Yeramian E 1994 *Europhys. Lett.* **25** 49
Yeramian E 2000 *Gene* **255** 139
Yeramian E 2000 *Gene* **255** 151
[108] Neuhauser D 1990 *J. Chem. Phys.* **93** 2611
Wall M R and Neuhauser D 1995 *J. Chem. Phys.* **102** 8011
[109] Mandelshtam V A and Taylor H S 1997 *J. Chem. Phys.* **107** 6756
Mandelstham V A 1999 *Progr. Nucl. Magn. Reson. Spectrosc.* **38** 159 and references therein
[110] Roy P-N and Carrington T Jr 1995 *Chem. Phys.* **103** 5600
Huang S-W and Carrington Jr. T 1999 *Chem. Phys. Lett.* **312** 311
[111] Chen R and Guo H 1998 *Phys. Rev.* E **57** 7288
Beck M H and Meyer H-D 1998 *J. Chem. Phys.* **109** 3730
[112] Vijay A and Wyatt R E 2000 *Phys. Rev.* E **62** 4351
Mandelshtam V A and Carrington T Jr 2002 *Phys. Rev.* E **65** 028701
Vijay A 2002 *Phys. Rev.* E **65** 028702
[113] Belkić Dž, Dando P A, Taylor H S and Main J 1999 *Chem. Phys. Lett.* **315** 135
[114] Main J, Dando P A, Belkić Dž and Taylor H S 2000 *J. Phys. A: Math. Gen.* **33** 1247
[115] Belkić Dž, Dando P A, Main J and Taylor H S 2000 *J. Chem. Phys.* **133** 6542
[116] Kunikeev S D and Taylor H S 2004 *J. Chem. Phys.* **108** 743
Kunikeev S D, Abligan E, Taylor H S, Kaledin A L and Main J 2004 *J. Chem. Phys.* **120** 6478
[117] Holtz H and Leondes C T 1966 *J. Assoc. Comput. Mach.* **13** 262
[118] Weinberg L 1962 *Network Analysis and Synthesis* (New York: McGraw-Hill)
Scanlan J O and Levy R 1970 *Circuit Theory* (Edinburgh: Oliver and Boyd)
Scanlan J O 1973 *Padé Approximant* (Lectures Notes, Summer School, University of Kent, Canterbury), ed P R Graves-Morris (Bristol: Institute of Physics Publishing) p 101
Papageorgiou C D and Raptis A D 1987 *Comput. Phys. Commun.* **43** 325
[119] Gutknecht M H 1992 *SIAM J. Matrix Anall. Appl.* **13** 594 and references therein
Freund R W, Gutknecht M H and Nachtigal N 1993 *SIAM J. Sci. Comput.* **1** 137

[120] Feldman P and Freund R F 1995 *IEEE Trans. Computer-Aided Design Integr. Circuits Syst.* **14** 639
Freund R F and Feldman P 1996 *IEEE Trans. Circuits Syst.—Anal. Digit. Sign. Process.* **43** 577
[121] Carathéodory C 1907 *Math. Ann.* **64** 95
Carathéodory C 1911 *Rendic. Circ. Matemat. Palermo* **32** 193
Carathéodory C and Fejér C 1911 *Rendic. Circ. Matemat. Palermo* **32** 218
[122] Nevanlinna R 1922 *Ann. Acad. Sci. Fenn.* A **18** 1
Weyl H *Ann. Math.* **36** 230
Delsarte Ph, Genin Y and Kamp Y 1981 *Circuit Theory Appl.* **9** 177
Delsarte Ph, Genin Y, Kamp Y and van Dooren P 1982 *Philips J. Res.* **37** 277
Burns C I, Georgiou T T and Lindquist A 1999 *IEEE Trans. Automat. Control.* **44** 211
[123] Georgiou T T 1987 *SIAM J. Math. Anal.* **18** 1248
Georgiou T T 1987 *IEEE Trans. Acoust. Speech, Signal Process.* ASSP **35** 438
Georgiou T T 1999 *IEEE Trans. Automat. Control.* **44** 631
[124] Fant G 1960 *Acoustic Theory of Speech Production* (The Hague: Mouton) p 42
[125] Atal B S and Hanauer S L 1971 *J. Acoust. Soc. Am.* **50** 637
[126] Markel J D and Gray A H 1976 *Linear Prediction of Speech* (Berlin: Springer)
[127] Oppenheim A V, Kopec G E and Tribolet J M 1976 *IEEE Trans. Acoust. Speech and Signal Process.* ASSP **24** 327
Kopec G E, Oppenheim A V and Tribolet J M 1977 *IEEE Trans. Acoust. Speech and Signal Process.* ASSP **25** 40
Cadzov J A 1982 *Proc. IEEE* **70** 907
[128] Mammone R J and Zhang X 1998 *The Digital Signal Processing Handbook* ed V K Madisetti and D B Williams (New York: IEEE) ch 27
[129] Berger R D, Saul J P and Cohen R J 1989 *IEEE Trans. Biomed. Eng.* **36** 1061
Pagani M *et al* 1986 *Circulation Res.* **59** 178
Appel M L, Saul J P, Berger E D and Cohen R J 1992 *Blood Pressure and Heart Rate Variability*, ed M Di Rienzo *et al* (Amsterdam: IOS) p 68
Parati G, Saul J P, Di Rienzo M and Mancia G 1995 *Hypertension* **25** 1276
Litvack D A, Oberlander T F, Carney L H and Saul J P 1995 *Psychophysiology* **32** 492
Belkić K and Belkić Dž 2002 in *Computational and Mathematical Methods in Sciences and Engineering* vol 2, ed J Vigo-Aguiar and B A Wade (Salamanca: University of Salamanca) p 372
[130] Belkić Dž 2001 *Modern Spectral Methods for Direct and Inverse Problems (J. Comput. Meth. Sci. Eng., Topical Issue (no 2))* vol 1, ed Dž Belkić, p i
[131] Belkić Dž and Belkić K 2003 *Computational Methods in Magnetic Resonance Spectroscopy (MRS) (J. Comput. Meth. Sci. Eng., Topical Issue (no 4))* vol 3, ed Dž Belkić and K Belkić, p i
[132] Heaviside O 1899 *Electromagnetic Theory* (London: Macmillan)
[133] Bromwich T J I'A 1916 *Proc. London Soc.* **15** 401
Bromwich T J I'A 1920 *Proc. Camb. Phil. Soc.* **20** 423
[134] Carson J R 1917 *Phys. Rev.* **10** 217
Carson J R 1925 *Bell Technol. J.* **4** 685
[135] Wagner K W 1916 *Arch. Elektrotech.* **4** 159

References 447

Carson J R 1926 *Electric Circuit Theory and the Operational Calculus* (New York: McGraw-Hill)
Carson J R 1926 *Bell Technol. J.* **5** 50
Carson J R 1926 *Bell Technol. J.* **5** 336
Carson J R 1926 *Bull. Am. Math. Soc.* **22** 43
[136] Lévy M P 1926 *Bul. Sci. Math.* **50** 174
[137] March H W 1927 *Bull. Am. Math. Soc.* **33** 311
[138] van der Pol B 1929 *Phil. Mag.* **7** 1153
van der Pol B 1929 *Phil. Mag.* **8** 861
van der Pol B and Niessen K F 1932 *Phil. Mag.* **13** 537
van der Pol B and Weijers Th J 1933 *Physica* **1** 78
[139] Pipes L A 1937 *Phil. Mag.* **24** 502
[140] Condon E U and Greenwood R 1937 *Phil. Mag.* **24** 281
[141] Gabutti B and Lepora P 1987 *J. Comput. Appl. Math.* **19** 189
Oppenheim A V and Schafer R W 1975 *Digital Signal Processing* (Englewood Cliffs, NJ: Prentice-Hall)
[142] Belkić Dž 2004 *J. Comput. Meth. Sci. Eng.* **4** 355
[143] Dirac P A M 1947 *Quantum Mechanics* 3rd edn (New York: Oxford University Press)
[144] Landau L D and Lifshitz E M 1965 *Quantum Mechanics: Nonrelativistic Theory* (Reading, MA: Addison-Wesley-Longman)
[145] Brenig W and Haag R 1959 *Fortschr. Phys.* **7** 183 (in German) (Engl. transl. M H Ross (ed) 1963 *Quantum Scattering Theory* (Bloomington, IN: Indiana University Press) p 13)
[146] Jauch J M 1957 *Helv. Phys. Acta* **30** 143
[147] Jauch J M 1958 *Helv. Phys. Acta* **31** 127
[148] Jauch J M 1958 *Helv. Phys. Acta* **31** 661
[149] Jauch J M 1972 *Aspects of Quantum Theory* ed A Salam (Cambrudge: Cambridge University Press) p 137
[150] Newton R G 1966 *Scattering Theory of Waves and Particles* (New York: McGraw-Hill)
[151] Pearson D B 1988 *Quantum Scattering and Spectral Theory* (New York: Academic)
[152] Moiseyev N, Friesner R A and Wyatt R E 1986 *J. Chem. Phys.* **85** 331
[153] Tannor D J and Weeks D E 1993 *J. Chem. Phys.* **98** 3884
Kouri D J, Huang Y and Zhu W 1994 *J. Chem. Phys.* **100** 3662
[154] Mandelshtam V A 1998 *J. Chem. Phys.* **108** 9999
[155] Pike R and Sabatier P (ed) 2001 *Scattering and Inverse Scattering in Pure and Applied Science* (London: Academic)
[156] Belkić Dž 2001 *J. Comput. Meth. Sci. Eng.* **1** 353
[157] Belkić Dž 2004 *Principles of Quantum Scattering Theory* (Bristol: Institute of Physics Publishing)
[158] Belkić Dž 2001 *J. Comput. Meth. Sci. Eng.* **1** 1
[159] Antonov V A, Timoshkova E I and Holshevnikov K V 1988 *Introduction to the Theory of the Newton Potential* (Moscow: Nauka) (in Russian) ch 7
[160] Chebotarev G A 1967 *Analytical and Numerical Methods of Celestial Mechanics* (New York: Elsevier)
[161] Halpern F R 1957 *Phys. Rev.* **107** 1145
Halpern F R 1958 *Phys. Rev.* **109** 1836

Halpern F R, Sartori L, Nishimura K and Spitzer R 1959 *Ann. Phys.* **7** 154
[162] Chebyshev P L 1858 *J. Math. Pures Appl. Ser. II* **3** 289
Chebyshev P L 1859 *Mem. Acad. Imp. Sci. St Petersburg* **1** 1
Chebyshev P L 1874 *J. Math. Phys. Appl.* **19** 157
[163] Stieltjes T J 1858 *Ann. Fac. Sci. Toulouse* **3** 1
[164] Hamburger H 1920 *Mat. Ann.* **81** 235
Hamburger H 1921 *Mat. Ann.* **82** 120
Hamburger H 1921 *Mat. Ann.* **82** 168
[165] Geronimus J 1946 *Mat. Ann.* **47** 742
Geronimus J 1952 *C. R. (Dokl.) Acad. Sci., USSR* **83** 5
Geronimus J 1960 *Polynomials Orthogonal on a Circle and Interval (International Series of Monographs in Pure and Applied Mathematics)* ed I N Sneddon (Oxford: Pergamon)
Geronimus J 1961 *Orthogonal Polynomials* (New York: Consultants Bureau)
[166] Shohat J A and Tamarkin J D 1950 *The Problems of Moments (Mathematical Survey 1)* 2nd edn (Providence, RI: Rev. Ed. Am. Math. Soc.)
[167] Uspensky J V 1937 *Introduction to Mathematical Probability* (New York: McGraw-Hill)
[168] Szegö G 1959 Orthogonal polynomials *Am. Math. Soc. Coll. Publ.* **22** 285
[169] Rutishauser H 1957 *Der Quotienten-Differenzen-Algorithmus* (Basel & Stuttgart: Birkhäuser)
Henrici P 1963 *Proc. Symp. Appl. Math.* **15** 159
[170] Gordon R G 1968 *J. Math. Phys.* **2** 655
Wheeler J C and Gordon R G 1970 in *The Padé Approximant in Theoretical Physics* ed G A Baker and J L Gammel (New York: Academic) p 99
Reid C E 1967 *Int. J. Quantum Chem.* **1** 521
[171] Gragg W B 1972 *SIAM Rev.* **14** 1
Gragg W B 1974 *Rocky Mountain J. Math.* **4** 213
Wheeler J C and Blumstein C 1972 *Phys. Rev.* B **6** 4380
[172] Gautschi W 1970 *Math. Comput.* **24** 245
[173] Sack R A and Donovan A F 1969 *An Algorithm for Gaussian Quadrature for Given Generalized Moments* Department of Mathematics, University of Salford, Salford, England, unpublished
Sack R A and Donovan A F 1972 *Num. Math.* **18** 465
[174] Wheeler J C 1974 *Rocky Mountain J. Math.* **4** 287
[175] Gautschi W 1978 *Recent Advances in Numerical Analysis* ed C de Boor and G H Golub (New York: Academic) p 45
[176] Press W H, Teukolsky S A, Vetterling W T and Flannery B P 1992 *Numerical Recipes* 2nd edn (Cambridge: Cambridge University Press)
[177] Smith B T, Boyle J M, Dongarra J J, Garbow B S, Klema V C and Moler C B 1976 *Matrix Eigensystem Routines—EISPACK Guide (Lectures Notes in Computer Science 6)* 2nd edn (New York: Springer) p 277
[178] NAG Fortran Library *NAG: Numerical Algorithms Group*, 256 Banbury Road, Oxford OX2 7DE, UK
[179] IMSL Math. Library Users Manual, *International Mathematical Statistical Library*, IMSL Inc.: 2500 City West Boulevard, Houston TX 77042, USA
[180] Arfken G B and Weber H J 2000 *Mathematical Methods for Physicists* 5th edn (New York: Academic)

Korn G A and Korn T M 1961 *Mathematical Handbook for Scientists and Engineers* (New York: McGraw-Hill)
[181] Alacid M and Leforestier C 1998 *Int. J. Quantum Chem.* **68** 317
Kroes G-J and Neuhauser D 1996 *J. Chem. Phys.* **105** 9104
Löwdin P-O 1962 *Rev. Mod. Phys.* **34** 520
[182] Euler L 1753 *N. Comment. Acad. Sci. Petropol.* **3** 36
Euler L 1774 *Proc. Natl. Acad. Sci., St Petersburg* **14** 463 (Opera Omnia I, Series I) (in Latin) (Engl. transl. Wyman M F and Wyman B F 1958 *Math. System Theory* **18** 295)
[183] Gauss C F 1812 *Vide Abhandl. der Göttingen Gesellsch. der Wissensch.* **2**
Gauss C F 1866 *Werke* [Collected works of Gauss C F] vol 3 (Göttingen: König. Gesellsch. der Wissensch.) p 125
Muir T 1874 *Trans. R. Soc. Edinburgh* **8** 229
Muir T 1874 *Trans. R. Soc. Edinburgh* **8** 380
Muir T 1876 *Trans. R. Soc. Edinburgh* **27** 467
Muir T 1876 *Proc. London Math. Soc.* **7** 112
[184] Stieltjes T J 1889 *Ann. Fac. Sci. Toulouse* **3** 1
Stieltjes T J 1894 *Ann. Fac. Sci. Toulouse* **8** 1
Stieltjes T J 1895 *Ann. Fac. Sci. Toulouse* **9** 1
1918 *Oeuvres Complètes de Thomas Jan Stieltjes* vol 2 (Gröningen: Noordhoff) p 402
[185] Padé H 1899 *Ann. Fac. Sci. de l'École Norm. Suppl.* **16** 395
[186] Wall H S 1948 *Analytic Theory of Continued Fractions* (New York: Van Nostrand)
Wall H S 1929 *Trans. Am. Math. Soc.* **31** 91
[187] Olivier D 1961 *Comm. Assoc. Comput. Mach.* **4** 318
[188] Spielberg K 1960 *J. Assoc. Comput. Mach.* **7** 409
[189] Clendenin W W 1961 *Comm. Assoc. Comput. Mach.* **4** 354
[190] Abramowitz M and Stegun I A 1972 *Handbook of Mathematical Functions with Formulas, Graphs and Mathematical Tables* (Washington DC: US Government Printing Office)
[191] Gragg W B 1974 *Rocky Mountain J. Math.* **4** 213
Horáček J and Sasakawa T 1983 *Phys. Rev.* A **28** 2151
Makowski A J, Raczyńzky A and Staszewska G 1986 *Phys. Rev.* A **33** 733
Holler E W and Allen L J 1986 *Phys. Rev.* A **33** 4393
Staszewska G 1987 *Phys. Rev.* A **35** 3554
Lee M-T, Iga I, Fujimoto M M and Lara O 1995 *J. Phys. B: At. Mol. Opt. Phys.* **28** L299
Lee M-T, Iga I, Fujimoto M M and Lara O 1995 *J. Phys. B: At. Mol. Opt. Phys.* **28** 3325
[192] Thompson I J and Barnett A R 1985 *Comput. Phys. Commun.* **36** 363
Thompson I J 2004 *Comput. Phys. Commun.* **159** 55
Thompson I J 2004 *Comput. Phys. Commun.* **159** 241
[193] Löwdin P-O 1986 *I. J. Quantum Chem.* **29** 1651
[194] Levenberg K 1944 *Quart. Appl. Math.* **2** 164
Marquardt D W 1963 *SIAM J. Appl. Math.* **11** 431
Bevington P R 1969 *Data Reduction and Error Analysis for the Physical Sciences* (New York: McGraw-Hill) p 235

Householder A S 1970 *The Numerical Treatment of a Single Nonlinear Equation* (New York: McGraw-Hill)
[195] Voigt W 1912 *Münch. Ber.* p 603
[196] Reiche F 1913 *Verh. d. D. Phys. Ges.* **15** 3
[197] Lorentz H A 1914 *Proc. R. Acad. Amsterdam* **13** 134
[198] Zemansky M W 1930 *Phys. Rev.* **36** 219
Mitchell A C G and Zemansky M W 1932 *Resonance Radiation and Excited Atoms* (Cambridge: Cambridge University Press)
[199] Unsöld A 1968 *Physik der Sternatmosphären* (Berlin: Springer)
[200] Kielkopf J F 1973 *J. Opt. Soc. Am.* **63** 987
[201] Maltempo M M 1986 *J. Magn. Res.* **68** 102
[202] Freise A, Spencer E, Marshall I and Higinbotham J 1995 *Bull. Magn. Reson.* **17** 302
Marshall I, Huginbotham J, Bruce S and Freise A, 1997 *Magn. Reson. Med.* **37** 651
[203] Pauli W 1941 *Rev. Mod. Phys.* **13** 203
[204] Lill J V, Parker G A and Light J C 1982 *Chem. Phys. Lett.* **89** 483
Light J C, Hamilton I P and Lill J V 1985 *J. Chem. Phys.* **82** 1400
[205] Kosloff R 1988 *J. Phys. Chem.* **92** 2087
Kosloff R 1994 *Annu. Rev. Phys. Chem.* **45** 145
[206] Feit M D and Fleck Jr. J A 1989 *Opt. Lett.* **14** 662
[207] Zel'dovich Ya B 1961 *Sov. Phys.–JETP* **12** 542 (Engl. transl. 1960 *J. Exp. Theoret. Phys. (USSR)* **39** 776)
[208] Dykhne A M and Chaplik A V 1961 *Sov. Phys.–JETP* **13** 1002 (Engl. transl. 1961 *J. Exp. Theoret. Phys. (USSR)* **40** 1427)
[209] Lovelace C 1964 *Phys. Rev.* **135** B1225
[210] Baslev E and Combes J M 1971 *Commun. Math. Phys.* **22** 280
Resigno N and Reinhardt W P 1973 *Phys. Rev.* A **8** 2828
Junker B R and Huang C L 1978 *Phys. Rev.* A **18** 313
Simon B 1979 *Phys. Lett.* A **71** 211
[211] Kröger H 1992 *Phys. Rep.* **210** 45
[212] Slater J C 1960 *Quantum Theory of Atomic Structure* vol v (New York: McGraw-Hill) pp 113, 429
[213] Chisholm J S R 1973 *Math. Comput.* **27** 841
Cuyt A 1982 *Num. Math.* **40** 39
Cuyt A 1986 *J. Comput. Appl. Math.* **14** 289
Cuyt A and Verdonk B 1988 *J. Comput. Appl. Math.* **21** 145
Cuyt A and Verdonk B 1999 *J. Comput. Appl. Math.* **105** 25
[214] Levin D 1976 *J. Inst. Math. Applic.* **18** 1
Greif C and Levin D 1998 *Math. Comput.* **67** 695
[215] Siemaszenko W 1983 *J. Comput. Appl. Math.* **9** 137
[216] Feil T M and Homeier H H H 2004 *Comput. Phys. Commun.* **158** 124
[217] Faddeeva V N 1959 *Computational Methods of Linear Algebra* (New York: Dover)
Faddeev D K and Faddeeva V N 1963 *Computational Methods of Linear Algebra* (San Francisco, CA: Freeman) p 144
[218] Dahlquist F and Björk Å 1974 *Numerical Methods* (Englewood Cliffs, NJ: Prentice-Hall)
[219] Datta B N 1995 *Numerical Linear Algebra and Applications* (Pacific Grove, CA: Brooks and Cole)

[220] Gerald C F and Wheatley P O 1997 *Applied Numerical Analysis* 6th edn (Reading, MA: Addison-Wesley-Longman)
[221] Krylov A N 1931 *Ukraine Akademia Nauk.* **4** 491
van der Vorst H E 2000 *Comput. Sci. Eng.* **2** 32
[222] Hankel H 1861 *Leipziger Dissertation* (Göttingen, unpublished)
Kowalewski G 1909 *Determinantentheorie* (Lepzig: Veit und Companie) p 112
Muir T 1960 *The Theory of Determinants in the Historical Order of Developments* vols I–IV (New York: Dover)
[223] Householder A S 1964 *The Theory of Matrices in Numerical Analysis* (New York: Blaisdell)
[224] Geronimus J 1929 *Annais de Faculdade de Sciências do Porto* **16** 23
[225] Geronimus J 1930 *Proc. R. Soc. Edinburgh* **50** 304
[226] Clement P R 1963 *SIAM Rev.* **5** 131
[227] Kinoshita T 1957 *Phys. Rev.* **5** 1490
[228] Apostol T M 1969 *Multi-Variable Calculus and Linear Algebra, with Applications to Differential Equations and Probability* vol II, 2nd edn (New York: Wiley) p 31
[229] Löwdin P-O 1950 *J. Chem. Phys.* **18** 365
Löwdin P-O 1953 *Phys. Rev.* **90** 120
Löwdin P-O 1954 *Phys. Rev.* **94** 1600
Appel K and Löwdin P-O 1956 *Phys. Rev.* **103** 1746
Howarth D J 1955 *Phys. Rev.* **99** 469
[230] Levin D 1973 *Int. J. Comput. Math.* **3** 371
[231] W. Kohn 1947 *Phys. Rev.* **71** 635
Lippmann B A and Schwinger J 1950 *Phys. Rev.* **79** 469
[232] Tani S 1965 *Phys. Rev.* **139** B1011
Tani S 1966 *Ann. Phys.* **37** 411
Tani S 1966 *Ann. Phys.* **37** 451
Tani S 1979 *Phys. Rev.* D **20** 3131
Tani S 1963 *Bull. Am. Phys. Soc.* **8** 301 (Session CB)
[233] Bessis D and Talman J D 1974 *Rocky Mountain J. Math.* **4** 151
Bessis D 1973 *Padé Approximant* (Lectures Notes, Summer School, University of Kent, Canterbury), ed P R Graves-Morris (Bristol: Institute of Physics Publishing) p 19
Calamante F and Grinberg H 1995 *Int. J. Quantum Chem.* **54** 137
[234] Saraph H E and Seaton M J 1962 *Proc. Phys. Soc.* **80** 1057
[235] Coleman J P 1976 *J. Phys. B: At. Mol. Phys.* **9** 1079
[236] Weierstrass K T W 1885 *K. Akad. Wiss. Berlin* p 789
Weierstrass K T W 1903 *Werke (Berlin)* **3** 1
Courant R and Hilbert D 1953 *Methods of Mathematical Physics* (New York: Interscience)
[237] Main J, Dando P A, Belkić Dž and Taylor H S T 1999 *Europhys. Lett.* **48** 250
[238] Mayers D F 1965 *Methods of Numerical Approximations* (Oxford: Pergamon) p 106
Householder A S 1953 *Principles of Numerical Analysis* (New York: McGraw-Hill)
Householder A S 1964 *The Theory of Matrices in Numerical Analysis* (New York: Blaisdell)
[239] Nuttall J and Wherry C J 1978 *J. Inst. Math. Applic.* **21** 165
Prevost M 1983 *J. Comput. Appl. Math.* **9** 333
Stahl H 1986 *Constr. Approx.* **2** 225

Stahl H 1986 *Constr. Approx.* **2** 241
[240] Gradshteyn I S and Ryzhik I M 1980 *Tables of Integrals, Series and Products* (New York: Academic)
[241] Barrett W 1961 *The Computer Journal* **4** 272
[242] Geronimus J 1930 *Ann. Math.* **31** 681
Montegate G 1982 *SIAM Rev.* **24** 137
Brezinski C 1980 *Padé-Type Approximants and General Orthogonal Polynomials (Int. Series Numer. Math. 50)* (Basel: Bitrkhäuser)
Prevost M 1988 *J. Comput. Appl. Math.* **21** 133
[243] Box G E P and Jenkins G M 1970 *Time Series Analysis Forecasting and Control* (San Francisco, CA: Holden-Day)
[244] Sidi A 1975 *J. Comput. Appl. Math.* **1** 69
Fleischer J 1973 in *Padé Approximant* (Lectures Notes, Summer School, University of Kent, Canterbury), ed P R Graves-Morris (Bristol: Institute of Physics Publishing) p 126
[245] Aburdene M F, Zheng J and Kozick R J 1995 *IEEE Signal Process. Lett.* **2** 155
[246] Prevost M 1987 *J. Comput. Appl. Math.* **19** 89
[247] Ledsham F C 1961 *Math. Comput.* **1** 48
[248] Alaylioglu A, Evans G A and Hyslop J 1975 *Comput. J.* **18** 173
Alaylioglu A 1983 *J. Comput. Appl. Math.* **9** 305
[249] Belkić Dž 1999 *Nucl. Instrum. Methods* B **154** 220
[250] Bender C M and Orszag S A 1978 *Advanced Mathematical Methods for Scientists and Engineers* (New York: McGraw-Hill)
[251] Poncelet J V 1835 *J. Reine Angew. Math.* **13** 1
Lindeloöf E 1898 *Acta Soc. Sci. Fennicae* **24** 1
Ames L D 1901 *Ann. Math.* **3** 185
Jacobsthal E 1919 *Math. Z.* **6** 100
[252] Knopp K 1990 *Theory and Application of Infinite Series* (New York: Dover)
[253] Hurwitz Jr H and Zweifel P F 1956 *Math. Tables Aids. Comput.* **10** 140
[254] van Wijngaarden I 1961 *Modern Computing Methods* 2nd edn, ed E T Goodwin (New York: Philosophical Library)
[255] Schmittroth L A 1960 *Commun. Assoc. Comput. Mach.* **3** 171
[256] Wynn P 1971 *The Computer Journal* **14** 437
[257] Steck M, Beckert K, Eickloff H, Franzke B, Nolden F, Reich H, Schlitt B and Winkler T 1996 *Phys. Rev. Lett.* **77** 3803
Danared H, Källberg A, Rensfelt K-G and Simonsson A 2002 *Phys. Rev. Lett.* **88** 174801
[258] Okubo S and Feldman D 1960 *Phys. Rev.* **117** 279
[259] Gladwell G M L 1993 *Inverse Problems in Scattering* (Dordrecht: Kluwer Academic)
[260] Masson D R 1991 *Rocky Mountain J. Math.* **21** 489
[261] Schneider B I 1997 *Phys. Rev.* A **55** 3417
Schneider B I and Feder D L 1999 *Phys. Rev.* A **59** 2232
[262] Askey R and Wilson J 1985 *Mem. Am. Math. Soc.* **319** 1
[263] Forsythe G E 1957 *J. Soc. Indust. Appl. Math.* **5** 74
Ascher M and Forsythe G E 1958 *J. Soc. Indust. Appl. Math.* **5** 9
Peck J E L 1962 *SIAM Rev.* **4** 135
[264] Schoenberg I J 1946 *Quart. Appl. Math.* **4** 45

References 453

Schoenberg I J 1946 *Quart. Appl. Math.* **4** 112

Radamacher H A and Schoenberg I J 1946 *Quart. Appl. Math.* **4** 142

[265] Bakhalov N S 1977 *Numerical Methods: Analysis, Algebra, Ordinary Differential Equations* (Moscow: Mir) (in English)

[266] Clenshaw C W 1962 *Mathematical Tables* vol 5 (London: National Physical Laboratory, HMSO) p 7

[267] Gel'fand I M and Vilenkin N Ya 1964 *Generalized Functions* (New York: Academic)

[268] Vladimirov V S 1979 *Equations of Mathematical Physics* (New York: Dekker)

Vladimirov V S 1979 *Generalized Functions in Mathematical Physics* (Moscow: Mir) (in English)

[269] Ince E L 1925 *Proc. R. Soc. Edinburgh* **46** 20

Ince E L 1926 *Proc. R. Soc. Edinburgh* **46** 316

Ince E L 1927 *Proc. R. Soc. Edinburgh* **47** 294

Morse P M and Feshbach H 1953 *Methods in Theoretical Physics* (New York: McGraw-Hill)

Cadwell J H 1961 *The Computer Journal* **3** 266

Cadwell J H and Williams D E 1961 *The Computer Journal* **4** 260

Jansen P A 1972 *J. Opt. Soc. Am.* **62** 195

[270] Longman I M and Sharir M 1971 *Geophys. J. R. Astron. Soc.* **25** 299

[271] Campbell L L 1957 *J. Soc. Indust. Appl. Math.* **5** 244

[272] Wenrick R C and Houghton A V 1961 *J. Commun. ACM* **4** 314

[273] Lowe I J and Norberg R E 1957 *Phys. Rev.* **107** 46

[274] Hylleraas E A and Undheim B 1929 *Z. Phys.* **65** 759

[275] Mandelshtam V A and Taylor H S 1995 *J. Chem. Phys.* **102** 7390

Mandelshtam V A and Taylor H S 1995 *J. Chem. Phys.* **103** 2903 and references therein

[276] Chen R and Guo H 1999 *Comp. Phys. Commun.* **119** 19 and references therein

[277] Tanaka H, Kunishima W and Itoh M 2000 *RIKEN Rev.* **29** 20

Kunishima W, Itoh M and Tanaka H 2000 *Progr. Theor. Phys. (Suppl.)* **138** 149

Kunishima W, Itoh M and Tanaka H 2000 *AIP Conf. Proc.* **519** 350

Kunishima W, Tokihiro T and Tanaka H 2002 *Comput. Phys. Commun.* **148** 171

Index

Absorption, 18, 23, 30, 391, 393
Absorbing boundary conditions (ABC), 23
Acceleration, 75, 103, 115, 120, 142
Accuracy, 51, 147, 186, 295, 438
Acquisition time, 3, 30, 166, 352, 415
Adjoint matrix, 419
Aitken extrapolation, 78, 117, 130, 143, 153
Algebraic, 8, 123, 253, 353, 405
Aliasing, 17
All-poles model, 75
All-zeros model, 75
All-poles all-zeros model, 75, 268
Amplitude, 4, 21, 102, 159, 227, 310
Analysis, 10, 164, 237, 317, 350, 362
Analog (signal, representation, data, state vectors), 31, 33, 34, 50, 51, 60
Analytical continuation, 8, 75, 110, 153, 294, 388
Analytic function, 25, 26, 33, 82, 194, 423
Analytical formulae (expressions, results), 10, 48, 92, 237, 336, 436
Approximation in the mean, 39
Auto-regressive (AR) model, 37, 75, 268, 274

Auto-regressive moving average (ARMA), 3, 37, 73, 75, 76, 268
Asymptotic, 23, 85, 150, 166, 277, 438
Attenuated exponentials, 21, 53, 88, 293, 403, 408
Auto-correlation function, 10, 75, 146, 254, 300, 408
Auto-covariance, 91
Autonomous, 3, 215
Automated software, 413
Averaging, 66, 148, 149
Average separation (frequencies), 151, 399

Band limited signal, 68, 70, 71
Band limited decimated signal, 66, 70, 71, 72
Band limited spectrum, 32, 37, 66, 68, 69, 70
Band limited decimation, 67, 68, 70, 413
Bandwidth, 67, 68, 398
Basis functions, 58, 404, 423, 427
Bessel functions, 13
Bessel inequality, 40, 41
Beta-function, 437
Binomial expansion, 222, 244, 285, 404, 430
Block diagonal, 66, 244, 285, 404, 430
Boundary conditions, 23, 45, 55, 108, 421, 438

Bound states, 20, 24
Branch point singularities, 24, 76

Cayley-Hamilton theorem, 42, 44, 172, 208, 210, 319
Cauchy concept (of analytical continuation), 388, 413
Cauchy–Poincaré interlacing theorem, 427
Cauchy residue theorem, 26, 83, 195, 268, 275, 429
Cauchy sequence, 42
Cauchy expansion (of determinants), 42, 108, 144, 177, 220, 235
Cauchy contour integrals, 82, 195
Central nervous system, 1
Central processing unit (CPU), 33
Characteristic equation, 7, 44, 111, 200, 354, 404
Characteristic frequencies, 87
Characteristic numbers, 236, 420
Characteristic polynomial, 11, 102, 185, 288, 366, 421
Chebyshev algorithm, 12
Chebyshev–Gragg algorithm, 215, 303, 304, 307
Chebyshev polynomials, 65, 72, 89, 92, 99, 438
Chebyshev–Wheeler algorithm, 307
Chemical shift, 13, 393, 395, 397, 407, 413
Chemical shift imaging (CSI), 1,
Cholesky decomposition, 164
Christoffel coefficients, 83
Christoffel–Darboux, 198, 199, 429
Christoffel numbers, 351, 424, 429
Clenshaw sum rule, 191, 453
Classical polynomials, 38, 84, 307, 429, 433, 438
Closed form (formula), 10, 127, 197, 263, 366, 412
Closure relation, 26, 36, 38, 39, 200, 203

Completeness, 25, 38, 43, 200, 320, 435
Complex eigenvalues, 2, 413
Complex frequencies, 7, 29, 406, 412
Complex exponential, 21, 71, 124, 236, 315, 415
Complex spectrum, 14, 30, 147, 401, 407, 408
Complex frequency plane, 7
Concentrations (of metabolites), 6, 13, 14, 15
Concentration (of mass), 10
Constrained root reflection, 3, 80, 99, 101, 407, 415
Continuous (variables),
Continued fractions (CF), 10, 22, 221, 255, 347, 405
Contracted continued fractions (CCF), 264, 265, 348, 369, 411, 412
Contour integration, 39, 84, 194, 213, 275, 350
Convergence, 8, 29, 75, 125
Convergence acceleration, 120, 233
Convergence radius or region, 75, 100, 110, 121, 385, 413
Convergence pattern, 391
Convergence rate, 82, 149, 150, 151, 187, 406
Convergent series or sequences, 8, 75, 103, 154, 277, 383
Convergent series and sequences, 29
Convergent validation, 6
Cosine Fourier sum, 91
Coulomb potentials or functions, 13, 438
Coupling constants or parameters, 10, 52, 150, 203, 303, 436
Cramer's rule, 47, 104, 123, 224, 344, 418
Cross correlation plots, 10, 18
Cross validation, 7, 8, 9

Data matrix, 17, 59, 124, 353, 364, 415
Data processing, 67, 72, 439
Decay, 30, 88, 154, 310, 400, 421
Decaying off-diagonal elements, 66
Decaying transients, 120
Decimal places, 8, 10, 126, 147, 187
Decimation, 17, 68, 70, 71
Decimated linear predictor (DLP), 3, 72
Decimated Padé approximant (DPA), 2, 72, 414
Decimated Signal Diagonalization (DSD), 2, 72, 408, 414
Delayed continued fractions, 354, 358, 360, 363, 364, 366
Delayed Lanczos continued fractions, 368, 374, 377, 388
Delayed Green function, 356, 379
Delayed orthogonal polynomials, 174
Delayed PA (inside the unit circle), 384, 388
Delayed PA (outside the unit circle), 379, 383
Delayed Padé–Lanczos approximant, 375, 376, 377, 388
Delayed Lanczos polynomials, 375, 376
Delayed product difference algorithm, 366
Delayed time signals or series, 174, 350, 354, 364, 369
Delayed spectrum, 351, 357
Degenerate spectra or roots, 18, 110, 152, 277, 408, 415
Degree of polynomials, 8, 44, 112, 224, 380, 438
Density of states (DOS), 21, 165, 215, 345, 389, 434

Derivatives (of polynomials, functions), 79, 84, 174, 195, 317, 407
Diagnostics, 1, 7, 11, 15, 16, 152
Diagonalization, 2, 12, 72, 167, 210, 354
Diagonal matrix, 65, 162, 420
Diagonally dominated, 65
Difference equation, 45, 53, 107, 122, 307, 400
Differential equation, 44, 60, 172, 184, 402, 438
Diffusion in tissue, 5
Digital filter, 100
Digital representations, 33
Digital signals, 51
Digitized, 44, 50, 67
Dimensionality reductions, 17, 59, 60, 67, 166
Dirac bra-ket notation, 20
Dirac complex valued function, 195
Dirac delta function, 38, 195, 286
Dirac delta operator, 424
Dirac identity, 39
Dirac measure, 295
Direct problems, 45, 53, 54, 55, 316, 403
Dirichlet kernel, 62, 65, 71, 72
Dirichlet series, 133, 310
Discontinuous functions, 76, 287
Discrete auto-correlation function, 123
Discrete auto-covariance, 91
Discrete (variables), 21, 273
Discrete variant, 46
Discrete cosine transform, 92
Discretized (continuum, etc), 24, 31
Discrete data, 260
Discrete Fourier transform (DFT), 31, 33, 66, 75, 92, 146
Discrete negative energies, 24, 435
Discrete states, 24
Discrete variable representation (DVR), 273

Index 457

Discrete attenuated exponentials, 34
Discrete representation, 46
Discretization, 56, 134
Dispersion spectra, 30
Discrete time signal, 50, 67, 267, 402
Discrete wave function, 438
Distribution function, 18, 39
Divergent series or sequences, 8, 75, 103, 110, 154
Dynamics, 4, 20, 28, 45, 51, 60
Duality of representations, 203, 204, 293
Duration (of time signals), 401

Economization (of storage space), 165, 167, 298, 411, 435
Echo time (TE), 16
Eigenequation, 25
Eigenfunctions, 12, 25, 57
Eigenproblems, 2, 25, 61, 102
Eigenenergies, 2, 12, 24, 56
Eigenfrequencies, 12, 24, 25, 26, 58, 101
Eigenpolynomials, 11, 170
Eigenroots, 45
Eigensolution, 60
Eigenspace, 42
Eigenspectra, 4
Eigenstates, 25, 168, 171
Eigenvalues, 2, 22
Eigenvector, 57, 163, 171
Electric circuit theory, 3, 214
Energy, 20, 24, 39, 150, 167
Epoch, 30, 50, 150
Equidistant sampling, 67, 134
Equivalence, 4, 31, 51, 68, 73
Error analysis, 85, 432
Error bounds, 51
Error (probability) function (erf), 13, 437
Error estimates, 8, 82
Error-free algorithms, 296, 297, 364

Error of integrations, 81, 84, 85, 428, 432, 433
ESPRIT, 2
Estimation, 2, 235, 266, 352
Evolution matrix, 58, 65, 79, 161, 194, 219
Evolution operator, 2, 28, 33, 42, 56
Exact quantum-mechanical spectrum, 37, 222, 351
Expansion theorem, 43, 78, 108, 112
Exponential damping or decay, 124, 166, 412, 421
Extended harmonic inversion problem (EHIP), 125, 126, 127
Extraneous frequencies or roots, 2, 79, 87, 97, 100, 128
Extrapolation, 76, 78, 117, 129, 153, 223, 412
Euler linear transform, 148, 149, 151
Evanescent modes, 54, 110
Even parts (of CFs), 348, 368, 377, 388
Exact results, 18, 122, 154, 290
Explicit Lanczos algorithm, 208, 210

Fast Fourier transform (FFT), 2, 32, 67, 148, 166, 353
Fast Padé transform (FPT), 2, 22, 164, 265, 390, 411
First-order error, 153
Fitting, 4
Filter diagonalization (FD), 2, 3, 16, 22, 42, 60
Filtering out, 111, 123, 134
Finite differences, 153, 213
Finite rank expansions, 73
Forced convergence, 8, 75, 151, 388, 401, 413
Fourier grid, 31, 60, 66, 69, 92, 195
Fourier integral, 27, 29, 31

Fourier resolution or limit bound, 352, 353
Fourier spectrum, 67, 70
Fourier uncertainty principle, 352, 353
Fredholm determinant, 73
Frequency (domain or spectrum), 3, 14, 29, 50, 60, 159
Frequency response function, 8, 266, 405
Frequency analysis, 10
Frequency window, 59, 69
Free induction decay (FID), 2, 14, 18, 391, 410
Frobenius concept, 11, 74
Full-width-half-maximum (FWHM), 407
Functional, 97, 98, 302
Fundamental frequencies, 7, 17, 21, 29, 51, 147

Gamma function, 13, 438
Gaussians, 14, 17, 50
Gauss–Laguerre quadrature, 81
Gauss–Legendre quadrature, 81
Gauss–Hermite quadrature, 81
Gauss–Seidel method (relaxations), 103, 105, 106
Gegenbauer polynomials, 11, 65
Generalized eigenproblem, 61
Generalized harmonic inversion problem (GHIP), 133
Genuine resonances (or metabolites, frequencies, poles), 3, 15, 80, 86, 101, 164
Geometric sequence or progression, 34, 53, 71, 151, 297, 406
Geometrically decaying transients, 142, 148, 154, 158
Geronimus polynomials, 85
Ghost frequencies or roots, 87
Giant resonances (water peak), 16, 17

Gibbs oscillations or phenomena, 10, 17
Global spectral analysis, 17, 56
Gragg algorithm, 302, 303
Gram–Schmidt orthogonalization (GSO), 165, 323
Gravitational potential, 278
Green function, 278
Green operator, 7, 23, 73, 160, 214, 227

Hamiltonian, 13, 23, 24, 167, 273
Hankel determinant, 35, 115, 125, 206, 340, 405
Hankel matrix, 35, 49, 102, 256, 332, 402
Hankel–Lanczos singular value decomposition (HLSVD), 23, 127, 152, 403
Harmonic analysis, 133
Harmonic inversion problem (HIP), 34
Hausholder decomposition, 164
Healthy brain gray matter, 16, 393, 409
Heaviside partial fraction expansion, 9, 79, 215, 349
Heaviside step function, 63, 69, 192, 286
Heaviside symbolic calculus, 44
Height (of a resonance), 15, 16, 21, 195
Hermitean operators, 20, 24, 165
Hermite polynomials, 437
Hessenberg matrix, 12, 13, 170
Hilbert transform, 82, 292, 428
Homogeneous linear equations, 47, 49, 103
Hylleraas–Undheim theorem, 427
Hypergeometric Gauss function, 437

Identity matrix, 421
Ill-conditioning, 11, 28, 78, 170, 187, 210

Images, 1, 6, 7
Impulse train function, 195
Informational principle, 353
Incoming boundary conditions, 24
Incomplete beta-function, 438
Induced convergence, 121, 150
Inhomogeneous linear equations, 47, 102, 104, 120, 341, 348
Initial conditions, 44
Initial estimates, 173
Initial guesses, 14, 18, 105
Inner product, 20, 28, 42, 201
Input data (or parameters), 10, 35, 68, 76
Instability, 6, 79, 87
Instantaneous transition probability, 13
Integrated density of states (IDOS), 434, 435
Interlacing theorem (Poincaré–Cauchy), 427
Internal validation, 9
Interpolation, 412
Inverse discrete Fourier transform (IDFT), 31
Inverse matrices, 57, 222, 330, 336
Inverse problems, 51, 55
Ion cyclotron resonance (ICR), 1, 11, 31, 42
Iterative relaxations, 122, 123, 124

Jacobi matrix, 21, 163, 170, 273, 366, 411

Krylov basis functions, 33, 40, 56, 206
Krylov–Fourier basis functions, 60
Kronecker delta symbol, 26, 38, 62, 200

Lactate resonances, 6, 15
Lanczos algorithm, 2, 11, 22, 160
Lanczos continued fractions (LCF), 12, 161, 221, 232, 264, 346
Lanczos phenomenon, 163
Landau–Lifshitz convention, 58, 96
Laplace transform, 2, 3, 277
Least square minimizations, 14
Lebesgue measure, 292
Lebesgue–Stieltjes integral, 39
Lebesgue–Stieltjes space, 39
Legendre polynomials, 278, 279, 436, 437
Levenberg–Marquardt non-linear method, 14, 15
Linear algebra, 103
Linear combination of model spectra (LC Model), 14, 17
Linear independence, 12, 298
Linear predictor (LP), 2, 33, 37
Linear predictive coding (LPC), 75
Lipid resonances, 5, 6, 16
Local completeness, 26, 404
Local spectral analysis, 17
Localized orbitals or functions, 165
Local representations, 160, 166
Lorentzians, 6, 55, 105
Lorentz spectra, 4, 9, 14, 59, 125, 248
LU (left-upper) decomposition, 103

Machine accuracy, 12, 69
Maclaurin series, 9, 74, 76
Magnetic resonance (MR), 1, 9
Magnetic Resonance Spectroscopy (MRS), 1, 54, 127, 364, 391, 415
Magnetic resonance imaging (MRI), 1, 5, 6, 10, 11, 16
Magnetic Resonance Spectroscopic Imaging (MRSI), 1, 5, 6, 16, 152, 407
Magnetic field, 18, 391
Magnetic moments, 16, 19
Magnitude spectrum estimation, 2, 88, 99
Mass distribution, 10
Matrix notations, 418

460   Index

Metabolites, 7, 391
Minimum separation (frequencies), 69
Missing data, 352
Mixed moment problem, 305
Modified moment problem, 298
Modulation (of phase), 69
Monitoring convergence, 197
Moving average (MA), 37
Multiplicity (of roots), 412
Multiple signal classification (MUSIC), 2

Nearest neighbour approximation, 10
Newton–Raphson root finding, 126, 147
Noise, 14
Norm, 40
Natural frequencies, 17
Nodal frequencies, 16
Non-classical polynomials, 423, 435, 438
Non-degenerate spectra or roots, 79, 109, 152, 179, 282, 415
Nonlinear transformations, 29, 73, 116
Non-Lorentzians, 18, 79, 152
Non-uniqueness, 14, 51
Normalization, 46, 165
Nuclear magnetic resonance (NMR), 1
Numerical integrations (quadrature), 12, 290
Nyquist interval or range, 16, 17, 69

Occipital gray matter, 18, 393
Operator Padé approximant (OPA), 425
Ordinary differential or difference equations, 44, 45
Orthogonalization, 12, 38, 161
Orthogonal polynomials, 11, 13, 36

Orthogonal basis functions, 26, 40, 58
Orthonormalization, 160
Over-completeness, 163
Overlap determinant, 205, 206, 336
Over-fitting, 15
Overlap matrix, 35, 57
Overlapping resonances, 18, 79

Padé approximant (PA), 2, 73
Padé–Chebyshev approximant (PCA), 91, 100
Padé–Lanczos approximant (PLA), 12, 78, 213
Padé–Laplace transform, 2, 3
Padé–Schur approximant (PSA), 6, 18, 86
Parallel processing, 435
Parseval equality, 41
Partial fractions, 9, 79, 147, 215, 230, 349
Peak (of a resonance), 2, 15, 16, 21, 32
Peak parameters, 5, 15, 17
Periodic signal extension, 353
Perturbation methods, 150
Phase (or phase minimum), 80, 100
Phase (of a resonance), 15, 16, 21
Parts per million (ppm), 393
Position (of a resonance), 15, 16, 21
Principal branch (in the complex plane), 58
Principal components or resonances, 16
Principal Cauchy value (of integral), 82
Product-difference algorithm (PD), 412
Projection operator, 25, 26, 37
Power spectrum, 3, 17, 80, 88
Power moment problem, 10, 13, 278, 285
Pseudo Hamiltonian, 26

Quadratures, 10, 12, 38, 81

Quadratic convergence, 187
Quantification problem, 9, 13, 16, 18, 122
Quantum mechanics, 25, 73, 150, 236, 398, 433
Quantum-mechanical signal processing, 29, 60, 404
Quantum scattering theory, 10, 404
Quasi-degenerate spectrum, 18
Quotient difference algorithm (QD) 11, 260, 337, 358
QZ algorithm, 78

Radio frequency (RF) pulses, 5
Random noise, 9, 16, 86
Rational approximations, 73, 76
Rational Chebyshev polynomial, 91
Rational function, 412
Rational polynomials, 74, 86, 98, 100, 222, 352
Rational response (function), 3
Reconstructions, 5, 55, 133, 140, 266, 310
Rectangle function (rec), 69
Rectangular window (or filter), 66
Recursion, 11, 93, 127, 160, 174, 189
Recursive orthogonal polynomial expansion method (ROPEM), 438
Recursive residue generation method (RRGM), 221
Regularizations, 2, 6, 79, 80, 86, 100
Relaxation operator, 27
Relaxation matrix, 21, 33, 35, 102, 109
Relaxation method, 103, 105, 106, 112, 120, 122
Relaxation times, 5, 13, 16
Resampling, 6, 68
Residual absorption spectra, 391, 397

Residues, 25, 58, 169, 274, 402, 429
Resolution, 1, 67, 150, 352, 415, 435
Resolvent operator, 28, 160, 222, 233, 425
Resolving power, 6, 398
Resonances, 54, 164, 226, 310, 415, 435
Response function, 3
Resummation, 75, 151, 294
Retrospective validation, 266
Rayleigh limit or bounds, 37, 398, 399, 401
Riemann sum and/or integral, 412
Robust estimations, 13, 52, 103, 235, 309, 408
Root finding (searching), 12, 126, 186
Root means square value (rms), 9
Roots of characteristic equations or polynomials, 2, 413
Root reflections, 3, 80, 88, 97, 99, 100
Rotated coordinate method (RCM), 24
Round-off errors, 12, 28, 51, 82, 364, 438

Sampling rate or time or lag, 6, 21, 31, 44
Schrödinger basis, 4, 33, 40, 58
Schrödinger equation or eigenproblem, 20, 23, 27, 33, 56
Schrödinger states, 20, 23, 25, 27, 34
Schrödinger–Fourier basis, 60, 61, 65
Schwinger variational principle, 73, 415
Secular equation, 11, 170, 235, 354, 404, 422
Secular transients, 29

462   Index

Separability, 43
Separable potentials, 73
Sequential averaging, 148
Schmidt relaxations, 103, 105, 109, 112, 120, 123
Schur polynomials, 2, 3, 6, 407, 414
Shape spectra, 412
Signals, 21, 50, 126, 266, 350, 415
Signal length, 32, 50, 70, 146, 352, 415
Signal processing, 35, 78, 105, 215, 354, 413
Signal-to-noise ration, 9
Sinc function (sinc), 66
Single photon emission computerized tomography (SPECT), 1
Singularities, 73, 76
Singular value decomposition (SVD), 78
Sparse matrices, 12, 166, 168, 177
Special functions, 13
Spectral analysis or estimation, 17, 56, 102, 203, 317, 415
Spectral density, 2
Spectroscopy, 1, 6, 10, 20, 30, 150, 166, 391
Spectrum, 24, 88, 127, 210, 315, 435
Speech processing, 75
Spurious frequencies or poles, 2, 80, 86, 87, 97
Stable transients, 29
Staircase shaped functions, 232
State vectors, 10, 57, 160, 203, 362, 420
State space methods, 354, 356, 404
Stationary phenomena, 75
Stieltjes integrals, 194, 260, 292, 294, 351, 433
Stieltjes formula for continued fractions, 45, 260
Stieltjes measure, 350
Stieltjes series or sums, 294

Stieltjes polynomials, 85
Strong convergence (convergence in the norm), 42
Strikingly robust (or stable) convergence, 392
System response function, 8
Successive elimination approximations, 101
Survival probability amplitude, 21

Tensors, 161
Time domain, 14, 34, 67, 122, 226, 405
Time signals, 5, 50, 152, 277, 353, 412
Trace (of a matrix), 434
Transients, 54, 86, 122, 226, 310, 415
Transfer function, 75, 76
Transpose (of a matrix), 332, 418, 431
Tridiagonal matrix, 162, 163, 167, 168, 222, 341
Trigonometric functions, 13
Truncation errors (in series or sequences), 438
Two-dimensional MRS, 10, 17

Unattenuated, 32
Uncertainty principle (quantum mechanics), 398
Uncertainty principle (Fourier), 353
Under-fitting, 14
Uniqueness, 50, 51, 54, 74, 80, 88
Unitarity or unitary transformation, 28, 29, 349
Unit circle, 8, 88, 100, 186, 388, 413
Unity operator, 57, 98, 425
Unstable transients, 29
User-friendly software, 415

Vandermond determinant, 46, 48, 158, 337

Variational principle, methods, estimates, 73, 333, 334, 415
Voigt function (shape, profile), 18, 50

Weight function or factors, 40, 81, 151, 201, 317, 423
Wavefunctions, 10, 24, 25, 61, 167, 298
Wave packets, 20, 58, 165, 208, 277, 433

Weierstrass theorem, 73
Width (of a resonance), 15, 16, 24, 30, 407
Windowing, 16, 17, 60, 67, 72

Zero filling (padding), 14, 18, 353
Zero state vector, 28, 42, 170, 421
Z-transform, 7, 73